Nanotechnology in Sustainable Agriculture

Nanotechnology in Sustainable Agriculture

Edited by
M. Anwar Mallick, Manoj Kumar Solanki, Baby Kumari, and Suresh Kumar Verma

CRC Press
Taylor & Francis Group
Boca Raton London New York

CRC Press is an imprint of the
Taylor & Francis Group, an **informa** business

First edition published 2021
by CRC Press
6000 Broken Sound Parkway NW, Suite 300, Boca Raton, FL 33487-2742

and by CRC Press
2 Park Square, Milton Park, Abingdon, Oxon, OX14 4RN

CRC Press is an imprint of Taylor & Francis Group, LLC

ISBN: 9780367369408 (hbk)
ISBN: 9781032003771 (pbk)
ISBN: 9780429352003 (ebk)

Typeset in Nemilov
by Deanta Global Publishing Services, Chennai, India

Contents

List of Figures

List of Tables

Editor Biographies

Prof. M. A. Mallick is a Professor of Botany and Dean, Faculty of Science at Vinoba Bhave University, Hazaribag. He has had a first class academic career. He was a UGC teaching fellow in the Department of Botany, University of Delhi, during 1986–1990 and was awarded a Ph.D. degree in 1991. He was a UGC visiting associate at Bhabha Atomic Research Centre, Trombay, for short term postdoctoral work during 1994–1996. He has 37+ years of teaching and research experience to his credit. Dr Mallick is the man behind the establishment of the Undergraduate Biotechnology department at St. Columba's College, Hazaribag, the backward region of erstwhile Bihar and now Jharkhand State of India in 1994–95. He was also founder, course coordinator, and presently director of the Post Graduate Department of Biotechnology, VBU. He was elected as a member of the Jharkhand State Biotechnology Board during 2002–2004 and a member of Jharkhand State Higher Education Council (RUSA) during 2015–2017. Dr Mallick is a lifetime member of several academic societies and attended several seminars and conferences in India and abroad. He has authored one international book and two laboratory manuals in biotechnology and has published several research papers, including book chapters. He has supervised ten research scholars and is currently guiding seven students to a Ph.D. degree in Biotechnology/Botany. His current research interest lies in nanobiotechnology, and he has published papers in this field in scientific reports and nanomedicine.

Manoj Kumar Solanki (Ph.D.) is currently working as a scientist in the Institute of Biology, Biotechnology and Environment Protection of the Faculty of Natural Sciences, University of Silesia, Katowice, Poland, under the Police National Agency for Academic exchange program. He obtained his Master's degree in Microbiology from Barkatullah University in 2006 and a Ph.D. degree in Microbiology from Rani Durgawati University, India, in 2013. He has awarded a visiting scientist fellowship from the Guangxi Academy of Agriculture Sciences, China, in 2013–2015, and also awarded as visiting scientist from the Volcani Center, Agricultural Research Organization, Israel 2016-2020. During his Ph.D., he was awarded a Senior Research Fellowship from the Indian Council of Agriculture Research (ICAR), India. He also worked as a research associate in the International Crops Research Institute for the Semi-Arid Tropics (ICRISAT), India, in a Department of Biotechnology funded project. He has been engaged in various research activities on plants–microbes interaction, soil microbiology, plant disease management, enzymology, and published many book chapters and various research papers in the journals of international repute. His current research interest lies in post-harvest disease management, the study of bacterial and fungal microbiomes, nutrient management, plant disease management, bioformulation preparation and application, microbial genome analysis, microbial diversity, lytic enzymes such as chitinase and amylase, stress regulation by plant growth-promoting microbes, and soil microbial community analysis by modern biotechnological tools. He has vast experience in several plant diseases like root rot, wilt, charcoal rot, downy mildew, and their management. Dr Solanki has a great interest in agriculturally important microorganisms (such as bacteria, actinomycetes, fungi, and yeast), especially in the utilization of microorganisms in soil and crop health management and other allied areas.

Baby Kumari (Ph.D.) obtained her Master's degree in Biotechnology from the University of Rajasthan and a Ph.D. degree in Biotechnology from the Vinoba Bhave University (2017), Jharkhand, in collaboration with the Bhabha Atomic Research Centre, Mumbai, India. She has worked at the Bhabha Atomic Research Centre, Mumbai, Maharashtra, April 2013–April 2014, in Nuclear Agriculture and Plant Biotechnology Division. She also worked there in the Radiation and Health division. She worked at the different prestigious Teaching and Research Institutes of India (for 13 years). She worked in the Department of Biotechnology, St Columbas College, Jharkhand, December 2010–March 2011 as an Assistant Professor, Chemind Diagnosis and Biosolutions, Jaipur, Rajasthan, India, December 2009–November 2010, as a research scientist, Modi Institute of Management and Technology, Kota, Rajasthan, August 2007–December 2009, in the molecular biology specialization section. She has been engaged in various research activities on plant–microbes interaction, soil microbiology, plant disease management, nanotechnology, and published many research papers in journals of international repute. Dr Kumari has a great interest in plant–microbes interaction, especially in medicinal plants. Dr Kumari has a keen interest in nanoparticle biosynthesis and their application in plant disease management, biosensors, and bioremediation.

Suresh Kumar Verma (Ph.D.) is a research scientist in the group of Prof. M. Suar in the KIIT School of Biotechnology, KIIT University. He has served as a postdoctoral researcher in environmental nanotoxicology at the Department of Environmental Medicine, Karolinska Institutet, Stockholm, Sweden. He has been part of the South Asian Hub project as a research scientist. Currently, he is about to join Uppsala University. His main area of research interest includes nanotoxicology, environmental pollution, and nanomedicine. His research publications includes many publications in top reputed environmental journals like *Science of the Total Environment*, *Scientific Reports*, *Nanomedicine*, *Chemosphere*, and many others. He is also associated with many journals as a reviewer and editor.

List of Contributors

Isha Abhyankar
Biochemical Sciences Division
CSIR- National Chemical Laboratory
Pune, Maharashtra, India

Rohit Sam Ajee
Amity Institute of Biotechnology
Amity University Madhya Pradesh,
Gwalior, India

Shumaila Akram
Department of Physics
Jamia Millia Islamia University
New Delhi, India

Sharmistha Banerjee
Amity Institute of Biotechnology
Amity University Madhya Pradesh
Gwalior, India

Uma Chaurasia
Department of Botany
Jiwaji University, Gwalior
Madhya Pradesh, India

Seyma Dadi
Department of Analytical Chemistry
Faculty of Pharmacy
Erciyes University
Kayseri, Turkey

Priti Darne
Biochemical Sciences Division
CSIR-National Chemical Laboratory
Pune, India

Mainak Das
Department of Biological Sciences and Bioengineering
Design Program
Indian Institute of Technology Kanpur
Kanpur, Uttar Pradesh, India

Priyanka Debnath
Environmental Chemistry Laboratory
Department of Environmental Science
The University of Burdwan
Burdwan Rajbati, Raiganj, Bardhaman
West Bengal, India

Philem Pushparani Devi
Biochemical Sciences Division
CSIR-National Chemical Laboratory
Pune, Maharashtra, India

Rita S. Dhodapkar
Scientist, CSIR-National Environmental Engineering,
Research Institute (CSIR-NEERI)
Nagpur, India

Semiha Ekrikaya
Department of Analytical Chemistry
Faculty of Pharmacy
Erciyes University,
Kayseri, Turkey

Hemlata
Department of Biosciences
Jamia Millia Islamia (Central Univ)
New Delhi, India

Touseef Hussain
Department of Botany
Aligarh Muslim University,
Uttar Pradesh, India

Himanshi Jangir
Indian Institute of Technology Kanpur
Uttar Pradesh, India

Shaheen Husain
Cyanobacterial Biotechnology Lab
Department of Biosciences
Jamia Millia Islamia (Central Univ)
New Delhi, India

Zehra Karaagac
Department of Analytical Chemistry
Faculty of Pharmacy
Erciyes University
Kayseri, Turkey

Brijendra Kumar Kashyap
Department of Biotechnology Engineering
Institute of Engineering and Technology
Bundelkhand University, Jhansi
Uttar Pradesh, India

Shuchi Kaushik
State Forensic Science Laboratory
Madhya Pradesh, India

Anand Kumar
University Department of Biotechnology
Vinoba Bhave University, Hazaribag,
Jharkhand, India

Ashutosh Kumar
ICAR-Indian Institute of Seed Science
Kushmaur, Mau,
Uttar Pradesh, India

M. Pavan Kumar
DOS in Biochemistry
Jnanakaveri PG Centre
Mangalore University
Mangalore, India

Baby Kumari
University Department of Biotechnology
Vinoba Bhave University
Hazaribag, Jharkhand, India

Puja Kumari
Biotechnology & Advanced Science &Technology Research Centre
Vinoba Bhave University, Hazaribag
Jharkhand, India

Sandeep Kumar
ICAR-National Dairy Research Institute
Karnal, Haryana, India

M.A. Mallick
Biotechnology & Advanced Science &Technology Research Centre
Vinoba Bhave University, Hazaribag
Jharkhand, India

Deepak Kumar Maurya
Aghrakar Research Institute
Pune, Maharashtra, India

Vineet Kumar Maurya
Department of Botany and Microbiology
H. N. B. Garhwal University
Srinagar Garwal, India

Arghadip Mondal
Environmental Chemistry Laboratory
Department of Environmental Science
The University of Burdwan
BurdwanRajbati, Raiganj, Bardhaman
West Bengal, India

Naba Kumar Mondal
Environmental Chemistry Laboratory
Department of Environmental Science
The University of Burdwan
BurdwanRajbati, Raiganj, Bardhaman
West Bengal, India

Aditya Nandi
School of Biotechnology,
KIIT University
Bhubaneswar, India

Ismail Ocsoy
Department of Analytical Chemistry
Faculty of Pharmacy
Erciyes University
Kayseri, Turkey

Sukdeb Pal
CSIR-National Environmental Engineering
Research Institute (CSIR-NEERI)
Nagpur, India

Deepak Panchal
AcSIR- CSIR-National Environmental Engineering
Research Institute (CSIR-NEERI)
Nagpur, India

Anand Kumar Pandey
Department of Biotechnology Engineering
Institute of Engineering and Technology
Bundelkhand University, Jhansi
Uttar Pradesh, India

Bijayananda Panigrahi
School of Biotechnology, Kalinga Institute of Industrial Technology
Deemed to be University, Bhubaneswar
Odisha, India

Paritosh Patel
School of Biotechnology
KIIT University
Bhubaneswar, India

Asmita Prabhune
Green Pyramid Biotech Pvt Ltd Pune
Maharashtra, India

Narendhran Sadasivam
Department of Biotechnology
Sri Krishan Arts and Science College
Kuniamuthur, Coimbatore
Tamil Nadu, India

Shruti Shandilya
University Hospital Carl Gustav Carus
Technische Universität Dresden
Dresden, Germany

Anjney Sharma
ICAR-National Bureau of Agriculturally Important Microorganisms
Kusmaur, Mau
Uttar Pradesh, India

Neha Sharma
Amity Institute of Biotechnology
Amity University Madhya Pradesh
Gwalior, India

Pradeep Kumar Singh
Nano-Science Center, Department of Chemistry
University of Copenhagen Denmark
Copenhagen, Denmark

A.H. Sneharani
Department of Biochemistry
Acharya Institute of Technology
Jnana Kaveri PG Centre
Mangalore University
Bangalore, India

Manoj Kumar Solanki
Plant Cytogenetics and Molecular Biology Group
Institute of Biology
Biotechnology and Environmental Protection
Faculty of Natural Sciences University of Silesia in Katowice
Katowice, Poland

Rajesh Singh Tomar
Amity Institute of Biotechnology
Amity University Madhya Pradesh
Gwalior, India

Vinoth Velusamy
Department of Biotechnology
Sri Krishan Arts and Science College
Kuniamuthur, Coimbatore
Tamil Nadu, India

Shalja Verma
School of Biochemical Engineering
Indian Institute of Technology
Banaras Hindu University,
Varanasi, Uttar Pradesh, India

Suresh Kumar Verma
KIIT University
Bhubaneshwar
Orissa, India

Sudheer Kumar Yadav
Department of Botany
Banaras Hindu University
Varanasi, Uttar Pradesh, India,

Durdana Yasin
Department of Biosciences
Jamia Millia Islamia (Central Univ)
New Delhi, India

Part 1

Introduction

1 The Emerging Role of the Fungus Mediated Synthesis of Nanoparticles in Agriculture

Shruti Shandilya, Sandeep Kumar, and Anjney Sharma

CONTENTS

1.1 INTRODUCTION

Nanoparticles hold a special place in the area of nanoscience and nanotechnology not only because of their nanosize (1–100 nm) but also because of their promising role in building complex nanostructures. Conventional techniques for synthesizing nanoparticles are capital intensive and inefficient in terms of material and energy use. An environmentally friendly and non-toxic approach harnessing fungi has been developed to overcome this problem, mainly emerging as myconanotechnology. This green chemistry-based approach utilizes the enzyme secreted by fungus for the biosynthesis of nanoparticles of a predefined chemical composition (Zhao et al, 2018).

The broad spectrum and diversity of fungi potentiate the role of myconanotechnology in interdisciplinary science (Gautam and Avasthi, 2020). Fungi are observed as excellent secretors of proteins in comparison to bacteria and actinomycetes, which not only results in a higher yield of nanoparticles (Rai et al., 2009) but also adds an advantage to the downstream processing and handling of biomass when compared to bacterial fermentation (Gade et al., 2008; Wani et al., 2020). Properties such as slower kinetics (Vaidyanathan et al., 2009), the ability to withstand flow pressure and agitation in bioreactors (Saha et al., 2010), and the capability of reducing metal ions quickly through a non-hazardous process makes fungi advantageous for the extracellular and intracellular synthesis of nanoparticles (Moaveni et al., 2011; Moudgil and Chaudhari, 2019). Moreover, the large-scale secretion of reductive proteins by fungi minimizes the use of unnecessary cellular components, which are utilized for the outside precipitation of nanoparticles. Therefore, fungi act as natural bio-factories for the biosynthesis of nanoparticles (Kashyap et al., 2012).

As an eco-friendly approach, the fungi mediated synthesis of nanoparticles has gained importance in recent years in the biomedical field (Rajeshkumar and Sivapriya, 2020) and culture (Jabeen and Anum, 2020). Wide varieties of fungal species have been utilized for the synthesis of nanoparticles, which includes *Aspergillus*, *Penicillium*, *Fusarium*, *Verticillium*, *Trichoderma*, *Pleurotus*, etc. (Khan et al., 2018). Salts of metals such as gold, silver, selenium, platinum, silica, magnetite, palladium, zirconia, cadmium, and tellurium are commonly used for the biosynthesis of myconanoparticles; from of these silver myconanparticles have been observed to be widely used (Salem and Fouda, 2020). Myconanoparticles show a tremendous antioxidant and antimicrobial potential (Hulikere et al., 2019) along with its use in various fields such as the treatment of plant diseases (Jabeen and Anum, 2020), in cancer therapy (Jeyaraj et al., 2013), drug delivery, DNA analysis, and also as biosensors (Satyanarayana et al., 2019).

Myconantechnology provides a wide range of applications, from the nanomaterial synthesis used for pathogen suppression and targeted delivery in the agricultural sector to synthesizing nanocomposite materials for food packaging (Dhillon et al., 2012). Thus, it is believed that the mycogenic synthesis of nanoparticles might bring an exciting transformation in the agriculture and food industry. Despite advantages and high applicability, the approaches and mechanisms to synthesize nanoparticles using fungi are still in the development stage, which potentiates the need for research efforts for the large-scale use of myconanotechnology in agriculture and allied sectors. This chapter emphasizes the current strategies, research challenges, and opportunities used for the mycogenic synthesis of nanoparticles with a focus on mechanism and potential application in the agriculture and allied sector.

1.2 WHICH STRATEGY IS FOLLOWED FOR THE MYCOGENIC SYNTHESIS OF NANOPARTICLES?

Bottom-up approaches are usually followed for the mycogenic synthesis of nanoparticles (Singh et al., 2011). This strategy involves constructing a structure atom by atom, molecule by a molecule, or by self-organization. In contrast, the use of physical or chemical means for building nanostructures is a top-down approach (Marchiol, 2012). The primary reaction involved in the bottom-up approach of mycosynthesis is the reduction/oxidation of substrates by fungal enzymes or antioxidants, reducing them to their respective nanoparticles, which are in a state of thermodynamic equilibrium due to a decrease in Gibbs free energy (Mohanpuria et al., 2008). The nanostructures obtained with the bottom-up approach have a homogenous chemical composition and minor defects, which makes this approach more advantageous in comparison to the top-down method (Zhang et al., 2011).

1.3 MECHANISM OF NANOPARTICLE FABRICATION BY FUNGI

The two main mechanisms involved in the fungi mediated synthesis of nanoparticles include:

1.3.1 *In vivo*/Intracellular Mechanism

The *in vivo*/intracellular mechanism consists of the integration of metal nanoparticles inside the living mycelia. The procedural steps include the electrostatic interaction of metal ions with enzymes or proteins on a fungal cell wall leading to its trapping in the fungal cell surface. Subsequently, the trapped metal ions are reduced by fungal cell surface enzymes causing their aggregation and therefore the formation of nanoparticles (Mukherjee et al., 2001; Rajput et al., 2016). It is also observed that metal ions may diffuse through the fungal cell wall and then be reduced by reductase present in the cytoplasmic membrane and cytoplasm (Sastry et al., 2003; Molnár et

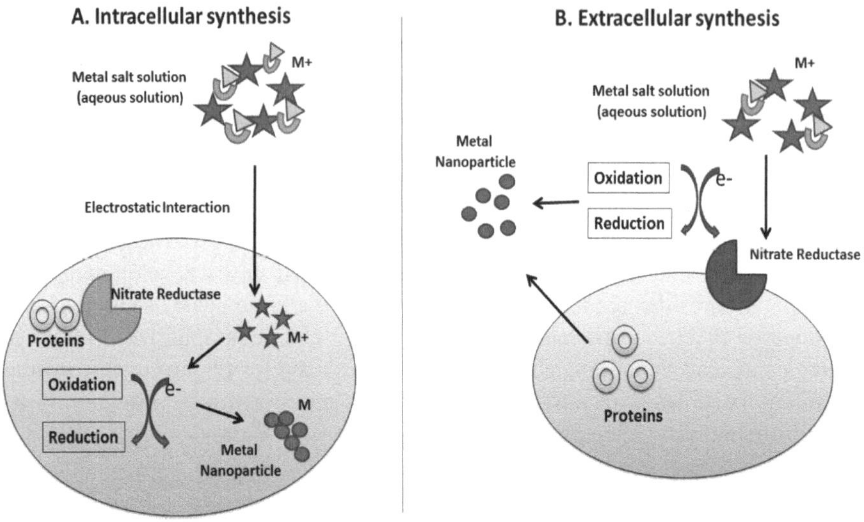

FIGURE 1.1 The mechanism of nanoparticle fabrication by fungi.

al., 2018). A representative picture depicting the *in vivo*/intracellular mechanism is shown in Figure 1.1A. This phenomenon is a type of survival strategy that is followed by fungal mycelium to convert toxic transition metal ions into a less toxic form. The *in vivo* method is the most basic approach that is used for the mycosynthesis of Au NPs (Molnár et al., 2018), which further involves the extraction of nanoparticles from mycelia prior to any downstream applications (Mukherjee et al., 2001; Rajput et al., 2016; Molnár et al., 2018).

1.3.2 *In vitro*/Extracellular Mechanism

The *in vitro*/extracellular mechanism involves the synthesis of metal nanoparticles using the fungal cell-free extract. In this process, extracellular enzymes and metabolites secreted by fungus convert toxic metal ions into non-toxic metallic nanoparticles (Ahmad et al., 2002). There are three possible *in vitro* methods:

i) Extracellular enzymes and other compounds present in the supernatant of fermented fungi can be used to produce metal nanoparticles such as Au NPs. But it is observed that the remaining components of the fermented media may cause hindrance during the reaction
ii) Sometimes, autolysates, i.e. intracellular components obtained after the lysis of fungal cells, can be used to produce functional bio-Au NPs
iii) Third, aqueous extract in which mycelia is suspended can be used for the conversion of toxic metal ions into non-toxic metal nanoparticles.

This method is limited by the occurrence of artefacts due to bacterial contamination because it is challenging to wash and resuspend the fungal mycelia in sterile conditions (Molnár et al., 2018). A representative picture depicting the extracellular mechanism is shown in Figure 1.1B.

Following the nitrate reductase assays performed in previous studies (Kumar et al., 2007), it can be concluded that along with the electrostatic interaction of metal ions with the fungal cell wall, enzymes specific to fungi, such as NADPH-dependent reductase and hydroxyquinoline, and phytochelatin plays an essential role in the fungus mediated synthesis of metal nanoparticles (Ingle et al., 2008; Gautam and Avasthi, 2020). At present, the simple mechanism of the mycogenic synthesis of nanoparticles is not precise, which shows that myconanotechnology is still in its infancy.

1.4 INFLUENCE OF PHYSICAL, CHEMICAL, BIOLOGICAL, AND TECHNICAL FACTORS ON THE MYCOGENIC SYNTHESIS OF NANOPARTICLES

1.4.1 Physical Factors Which Affect the Synthesis of Green Nanoparticles

1.4.1.1 Temperature

Temperature is a very promising parameter as it affects all three methods (physical, chemical, and biological) utilized for the synthesis of green nanoparticles. The mycogenic synthesis of nanoparticles usually requires an ambient temperature or temperatures of less than 100°C (Rai et al., 2006). Different temperature conditions are required for different methods used; for example, a temperature higher than 350°C is essential for the physical approach to mycogenic synthesis. On the other hand, a temperature of less than 350°C is required for the chemical-based method to synthesis (Patra and Baek, 2014). Therefore, temperature regulates the mycogenic synthesis by regulating the fungal growth but also the uptake of metal ions from the surrounding environment (Elamawi et al., 2018). It thus influences the rate of formation of the nanoparticles. Temperature also affects the size of the nanoparticles (Shahzad et al., 2019). Gericke and Pinches (2006) showed that when fungi were incubated for 1 h in a gold solution at a lower temperature, the majority of nanoparticles were found to be smaller in size (< 10 nm) and spherical in shape, though increasing the incubation period up to 24 h decreased the number of smaller nanoparticles; on the other hand, it increased the number of larger nanoparticles with a well-defined shape. Thus nanoparticle size can be controlled by incubation at a lower temperature, but it may lead to a slower synthesis rate.

1.4.1.2 pH

PH is one of the predominant factors affecting the size and shape of the green nanoparticles. Changes in the shape of the nanoparticles have been observed by varying the pH of the incubation medium (Gericke and Pinches, 2006). At a pH of 3 nanoparticles produced by *V. luetoalbum* were found to be smaller and spherical in shape; increasing the pH to 5 resulted in bigger sized nanoparticles of well-defined shapes, such as hexagons, rods, triangles and spheres. Whereas a further increase in pH to 7–9 results in the synthesis of bigger green nanoparticles of an irregular and undefined morphology. The reduction of metallic ions by fungi is also affected by pH (Sanghi and Verma, 2010). It has been observed that *Coriolus versicolor* reduces the metal silver to its individual nanoparticles in 1 h at an alkaline pH of 10 in comparison to its normal reduction time of 72 h (Prakash and Soni, 2011).

1.4.1.3 Pressure

Pressure applied during the synthesis of nanoparticles by fungi affects its size and shape. A faster reduction of metal ions by fungi has been observed at ambient pressure conditions (Tran and Le, 2013).

1.4.1.4 Time

Incubation time of the reaction medium also influences the type and quality of green nanoparticles. A long incubation time affects the characteristics of the nanoparticles formed as it may lead to the aggregation of the nanoparticles or may result in their shrinkage and growth, thus affecting their shelf life and potential (Kuchibhatla et al., 2012).

1.4.2 Chemical Factors

1.4.2.1 Metal Ion Concentration

A high concentration of metal ions usually influences the size and monodispersity of the nanoparticles from the normal range. Previous reports showed that a high silver ion concentration retards the synthesis of nanoparticles mediated by *Penicillium fellutanum* (Kathiresan et al., 2009). The optimum concentration of the reactants decides the extent of the reaction along with the nanoparticle size and its monodispersity. Studies showed that $Aucl^{4-}$ at a concentration of 500 mg l^{-1} in the presence of *Verticillium luteoalbum* results in the formation of gold nanoparticles of a uniform size of ~20 nm. However, the increase in particle size along with the polydispersity and aggregation of the gold nanoparticles was observed at a concentration > 500 mg l^{-1} (Gericke and Pinches, 2006).

1.4.3 Biological Factors

1.4.3.1 Fungi Specific Enzymes

Fungi specific enzymes are one of the predominant factors in the formation of predefined nanoparticles of varying chemical composition, size, and shape. One example is the synthesis of silver nanoparticles by *F. oxysporum*, where silver ion solution is converted into silver nanoparticles using the specific NADH-dependent reductase enzyme produced by *F. oxysporum*, but no reduction of the silver ion solution was observed using the reductase of *F. moniliforme* (Ahmad et al., 2003). Similarly, it was observed that *F. oxysporum* specifically causes protein-mediated extracellular hydrolysis of aqueous K_2ZrF_6 resulting in the formation of silicon nanoparticles

(Bansal et al., 2006). However, further research is still needed into the different types of enzymes used for the mycogenic synthesis of nanoparticles.

1.4.4 Technical Factors

1.4.4.1 Technique

The techniques utilized for the mycogenic synthesis of nanoparticles include physical methods using mechanical procedures or chemical or biological approaches using organic or inorganic chemicals and organisms (Kharissova et al., 2013). Despite the traditional methods used, the biological approach is usually preferred as it is non-toxic and eco-friendly in conjunction with green technology (Vadlapudi and Behara, 2014).

1.4.4.2 Expenditure

The total cost required for nanoparticle synthesis using the different methods is an essential consideration for the large-scale production of green nanoparticles. Chemical and physical synthesis results in a high yield over a short time, but they are limited by their high-cost requirement. On the other hand, the biological method is commonly used because it has fewer set up costs when used on a large scale. Therefore, it is essential to regulate the cost associated with green nanoparticle synthesis to facilitate their potential applications (Patra and Baek, 2014).

1.4.4.3 Environment

The surrounding environment greatly influences the physical and chemical nature of the synthesized nanoparticles. Various oxidizing and corrosive reactants present in the environment converts a myconanoparticle into large sized and thicker core shell nanoparticle (Sarathy et al., 2008). Another example showed a change in the crystalline nature of zinc sulfide nanoparticles when its environment was changed from wet to dry conditions. Similarly, the presence of peroxide in the suspension solution varies the chemical nature of the cerium nitrate nanoparticles (Kuchibhatla et al., 2012).

1.5 IMPORTANCE OF MYCONANOTECHNOLOGY IN AGRICULTURE

Green nanotechnology research is growing at a rapid rate which is having a profound impact on the agricultural sector. Different types of myconanoparticles with wide applications in agriculture are described in the following section. The fungi that are used for nanoparticle synthesis with a particular applicability in agriculture are listed in Table 1.1.

1.5.1 Antifungal and Antibacterial Green Nanoparticles

Fungi are common phytopathogens when compared to viruses and bacteria. For the potential solution of this problem, antifungal green nanoparticles are the essential players. Previous reports showed that silver nanoparticles are used as pesticides to control the phytopathogenic fungi that causes sclerotia in plants (Min et al., 2009).

TABLE 1.1
The Application of Fungi Derived Nanoparticle in Agriculture

Fungi	Nanoparticle	Mode of synthesis	Size (nm)	Application	References
F. accuminatum	Ag	Extracellular	4–50	Antibacterial	Ingle et al. (2008)
L. lecanii	Ag	Extracellular	45–100	Textile fabrics	Namasivayam and Avimanyu et al. (2011)
R. oryzae	Au	Extracellular	10	Water hygiene management	Das et al. (2009)
T. viride	Ag	Extracellular	5–40	Fruit preservation	Mohammed Fayaz et al. (2009); de la Cruz Quiroz et al. (2019)
A. alternata	Ag	Extracellular	20–60	Antifungal	Gajbhiye et al. (2009)
F. oxysporum	CdS	Extracellular	9–15	Live cell imaging	Kumar et al. (2007)
F. oxysporum	Ag	Extracellular	20–50	Textile fabrics	Duran et al. (2007)
P. infestans	Ag	Extracellular	Not determined	antibacterial	Thirumurugan et al. (2009)
T. crassum	Ag	Extracellular	5–50	antimicrobial	Ray et al. (2011)
A. niger	Ag	Extracellular	15–20	antibacterial	Gade et al. (2008)
A. niger	Ag	Extracellular	820	Wound healing	Sundaramoorthi et al. (2009), Kalaiselvan and Rajasekaran, (2019)
F. solani	Ag	Extracellular	3–8	Textile fabrics	El-Rafie et al. (2010)

Green nanoparticles can be synthesized from species of fungi which are common plant pathogens, such as *Aspergillus*, *Fusarium*, and *Phytophthora* (Yadav et al., 2015). It was observed that when antifungal double capsulized silver nanoparticles (1.5 nm) were sprayed over plants suffering from rose powdery mildew, caused by *Sphaerotheca pannosa var rosae*, 95% of the mildew faded with no recurrence (Kim et al., 2008). Similarly, silver nanoparticles (5–24 nm) were found to be effective against anthracnose caused by *Colletotrichum gloesporioides* in fruits (Aguilar-Mendez et al., 2010). Kim et al. (2012) also reported that silver nanoparticles inhibited the growth of 18 plant pathogenic fungi at different levels *in vitro*. Recent report also highlighted the potential antibacterial and antioxidant activity of silver nanoparticles synthesized using marine endophytic fungus Cladosporium cladosporioides (Hulikere and Joshi, 2019). Apart from the silver nanoparticles, other formulations were also found to be antifungal. It was reported that silica-silver nanoparticles possess potent antifungal activity against many fungal diseases, and most importantly it was a cost-effective, non-toxic, and safe nanoformulation (Bhattacharyya et al., 2016). These studies show that silver nanoparticles can act as an alternative to costly fungicides against many phytofungal diseases

Myconantechnology also serves as a new paradigm in the development of antibacterial nanoparticles. Duran et al. (2007) reported that green silver nanoparticles (1.6 nm) synthesized using *F. oxysporum*, when incorporated into cloth, prevents the infection caused by pathogenic bacteria *Staphylococcus aureus*. Also, antibacterial fabrics effective against *S. aureus* and *E. coli* are produced by coating silver nanoparticles (45–100 nm) synthesized using *Lecanicillium lecanii* on bleached cotton fibres (Namasivayam and Avimanyu, 2011). The mechanism of action of these antibacterial fabrics involves the attachment of silver nanoparticles to the cell surface of the pathogenic bacteria, thus inhibiting the permeability of the membrane and the normal respiration process.

Similarly, nanofibrous mats can inhibit the growth of phytopathogens due to their high surface area and the porous structure of their electrospin fibres. These nanofibrous mats were prepared using chitosan and *T.viridae*,and was reported that the incorporated *T.viridae* spores in nanofibrous mats remain viable and directlyattack the cell wall of phytopathogens by producing extracellular hydrolytic enzymes (Spasova et al., 2011; Pirzada et al., 2020).

1.5.2 Detection of Adulterants and Pathogenicity in Agriculture

Mycogenically synthesized nanowires are used to produce nanosize electronic devices that are very useful for detecting adulterants and pathogenicity in the agriculture, food, and textile industries (Chau et al., 2007; Shawon et al., 2020). Synthesis of gold nanowires using fungi was reported by Sugunan et al. (2007), where fungi of the genus *Aspergillus* and *Neurospora* were incubated with gold nanoparticles of 15 nm for a week, resulting in the coating of the fungi by gold nanoparticles and the formation of gold microwires. Using a similar pattern, silver and platinum nanowires can be synthesized. Previously, it was reported that pathogenic fungi could be detected with a higher sensitivity using a combination of gold nanowire and film surface-enhanced resonance Raman scattering with the target DNA recycling reaction mediated by exonuclease III (Yoo et al., 2011). Similarly, nanowire-based SERRS sensors are used to detect toxic metal ions like Hg2+, Ag+ and Pb2+ in the solution (Kang et al., 2012). Furthermore, an improved water quality for irrigation purposes was observed using nanowire-based papers which possessed the ability to filter organic pollutants, oil, and soil sediments from water (Reguera et al., 2005; Griesche and Baeumner, 2020). Also, ZnCdSe quantum dots synthesized using fungi can be used to filter pesticides in water (Bakar et al., 2011; Singh et al., 2019).

1.5.3 Green Nanopesticides

As a replacement for chemical-based pesticides and to reduce the risk of environmental hazards caused by pesticides, fungus derived nanoparticles formulated with biopesticides are known to be used in agriculture these days (Kumar et al., 2019), but they are not used extensively. Due to their small size and large surface area to volume ratio, they cause the spatial distribution of biopesticides onto leaf surfaces, which also increases their ability to make contact with microbes and their ability to permeate (Liu et al., 2008; Kumar et al., 2019). It has been observed in *in vitro* studies that Ag nanoparticles are effective against phytopathogenic fungi *Raffaelea* sp., which causes oak wilt (Kim et al., 2009). In another study, it was shown that Ag nanoparticles inhibited the hyphal growth of *R. solani*, *S. sclerotiorum*, and *S. minor in vitro*, when treated in a dose-dependent manner (Min et al., 2007). These studies reveal the potential of silver nanoparticles for use as nanopesticides. Insecticide RPA 107382, when encapsulated in polymeric nanospheres, resulted in better insecticidal action against aphids in cotton plants when compared to the control (Boehm et al., 2003). Similarly, pesticides such as aluminosilicates encapsulated in nanotubes were found to be more effective against insect pests because they were able to stick to the insect's hairs and enter through the skin, resulting in the impairment of the normal physiological functions (Patil, 2009). Other metal nanoparticles made up of aluminium oxide, titanium dioxide, and zinc oxide were found to reduce rice weevil infestations caused by *Sitophilus oryzae* and grassierie disease caused by baculovirus BmNPV in silkworms (Goswami et al., 2010). Likewise, both silver ions and silver nanoparticles inhibit the colony formation of plant pathogenic fungi, as observed in the case of fungi *B. sorokiniana* and *M. grisea* (Jo et al., 2009).

1.5.4 Green Nanofertilizers

Fertilizers are known to increase nutrient efficiency and improve plant growth. Chemical fertilizers, which are traditionally used in agriculture, decrease the fertility of the soil and subsequent crop production.

In order to improve nutrient efficiency in the field of agriculture nanofertilizers are meant to play a role. Nano fertilizers are any product that uses nanotechnology or is made up of nanoparticles to improve plant growth and reduce soil toxicity (Prasad et al., 2014). Nanofertilizers can also be a nanoscale additive (traditional fertilizer with a nanoscale additive) or they can be a nanoscale coating type (traditional fertilizer coated or loaded with nanoparticles) (Mikkelsen, 2018). Zinc oxide nanoparticles were found to supply a sufficient amount of zinc to the plants, which is very important for their growth (Pandey et al., 2010). Green nanofertilizer formation was reported by Tarafdar et al. (2014), where zinc oxide solution was incubated with fungus *R. bataticola* for 62 h, resulting in the fungus mediated extracellular synthesis of zinc nanoparticles (18.5 nm). Mycogenically synthesized zinc nanoparticles (18.5 nm) were found to improve the production of pearl millet after four weeks of its spraying in comparison to the spraying of normal size zinc oxide. Significant improvement in the shoot and root length was observed.

Recently, one report showed that ZnO nanotubes, in association with the fungus *Piriformospora indica* DSM 11827, resulted in the formation of a nano-embedded fungus, which improved the growth of broccoli (Singhal et al., 2017). These green nanofertilizers are new for agronomists because these fertilizers are still in the early stages of development.

1.5.5 Nanodelivery Vehicles

Nanotechnology is used in agriculture for the time-controlled, self-regulated, and targeted delivery of insecticides, pesticides, and fungicides. This has improved the efficiency of chemicals (fungicides, insecticides) for the treatment of plant diseases (Bouwmeester et al., 2009). The benefit of nanotechnology to agriculture occurs with the successful penetration and transport of nanoparticles into the plants. It has been observed that nanofungicides, nanopesticides, and nanoherbicides have been commercialized and are now extensively used in agricultural practices (Owolade et al., 2008; Singh et al., 2019; Gautam and Avasthi, 2020). Christou et al. (1988), created a transgenic plant through gene transfer using gold nanoparticles. Similarly, silica nanoparticles were also utilized for the efficient delivery of DNA and chemicals to plant cells (Torney et al., 2007; Wang et al., 2019). A nanofiber array was found to have a high potential in the site-directed and slow delivery of agrochemicals as well as the other macromolecules needed for improved plant disease resistance (Moaveni et al., 2011; Guha et al., 2020). The cotransfection of nucleic acid with semiconductor quantum dots leads to the formation of fluorescent photostable nanoparticles, which are utilized in tracking the delivery of nucleic acids (Elumalai et al., 2011), but this technique still needs to be improved for agricultural applications. Thus, nanodelivery vehicles can assist agricultural growers in the future to minimize the use of fungicides and pesticides (Wani et al., 2020).

1.6 LIMITATIONS OF MYCONANOTECHNOLOGY IN AGRICULTURE

Metallic nanoparticle production using fungi is a relatively simple process and also a non-toxic and safe approach to synthesis, which overcome the limitations of toxic chemical-based and labour intensive physical methods. The green synthesis of nanoparticles has attracted great interest with significant applications in diagnostics, precision farming, and targeted delivery systems in agriculture. However, there are some limitations and challenges that need to be addressed in the future to access its full potential in agriculture (Moudgil and Chaudhari, 2019). The limitations which are commonly observed are:

1.6.1 Size and Monodispersity of Green Nanoparticles

Mycogenic nanoparticles are usually assessed by their size and monodispersity. Therefore, it is essential that the size and monodispersity of the fungus derived nanoparticles are investigated using changing factors like pH, temperature, metal ion concentration, and enzyme concentration before its application in agricultural practices (Hnamte et al., 2017).

1.6.2 Limited Study on the Mechanism Involved

The exact mechanism of the fungus mediated synthesis of nanoparticles is still unknown (Gautam and Avasthi, 2020). Therefore, a deep and elucidated study will be required to discover more of the fungal proteins and enzymes involved in the synthesis of nanoparticles, which will further widen their scope in agriculture. After the synthesis of nanoparticles by fungi, its safe and effective recovery is essential for downstream applications, but the recovery techniques used are very costly (Molnár et al., 2018). Thus, it is also required to develop a cost-effective method for the safe recovery of nanoparticles to make it commercially feasible.

1.6.3 Mycogenic Synthesis Limited to Metals

Reports until now have only demonstrated the mycogenic synthesis of nanoparticles using metals, metal oxides, and metal sulphides, which shows that the mycogenic synthesis of nanoparticles is only confined to metals. Therefore, the procedure of the mycogenic synthesis of nanoparticles using other nitrides, carbides, sulphides, and oxides, requires further elucidation and standardization to allow the use of this approach for more extensive agricultural applications (Jabeen and Anum, 2020).

1.6.4 Limited Study on the Mode of Action

Myconanotechnology has enabled the use of new nanotechnological tools for the detection and molecular treatment of plant diseases (Gautam Avasthi, 2020). But still, there is a lack of understanding of how these myconanoparticles are absorbed and accumulated in plants, and their targeted mode of action. Therefore, further focused research is needed to understand the possible uptake, translocation, and toxicity of myconanparticles in plant cells (Shawon et al., 2020).

1.7 CONCLUSION AND FUTURE OUTLOOK

The microbial synthesis of nanoparticles has attracted many scientists over the years. As a more eco-friendly approach than the toxic chemical and costly physical methods of metal nanoparticle synthesis, fungi are most suitable. Fungi can be used for the production of novel nanomaterials of varying compositions using a straightforward and biocompatible approach utilizing its enzymes and proteins either extracellularly or intracellularly. In the form of a sustainable natural nanofactory, myconantechnology also shows tremendous applicability in the medicine, textile, and food industries, but its use in agriculture is growing fast. Despite various advantages, the feasibility of myconanotechnology on a commercial scale is limited. Myconantechnology is still in its infancy and requires more focused and defined research goals for its safe and secure use in agricultural practices. With the recent advancements and development for improving the strategies involved in the mycogenic synthesis of nanoparticles and exploring their role in agriculture, it is believed that myconanotechnology will be used on a larger scale for improving agriculture in the near future.

REFERENCES

Abd-Elsalam, K., & Prasad, R. (2019) *Nanobiotechnology Applications in Plant Protection*. Volume 2. Springer International Publishing.

Aguilar-Méndez, M. A., San Martín-Martínez, E., Ortega-Arroyo, L., Cobián-Portillo, G., & Sánchez-Espíndola, E. (2011). Synthesis and characterization of silver nanoparticles: Effect on phytopathogen *Colletotrichum gloesporioides*. *Journal of Nanoparticle Research*, 13(6), 2525–2532.

Ahmad, A., Mukherjee, P., Mandal, D., Senapati, S., Khan, M. I., Kumar, R., & Sastry, M. (2002). Enzyme mediated extracellular synthesis of CdS nanoparticles by the fungus, *Fusarium oxysporum*. *Journal of the American Chemical Society*, 124(41), 12108–12109.

Ahmad, A., Mukherjee, P., Senapati, S., Mandal, D., Khan, M. I., Kumar, R., & Sastry, M. (2003). Extracellular biosynthesis of silver nanoparticles using the fungus *Fusarium oxysporum*. *Colloids and Surfaces, Part B: Biointerfaces*, 28(4), 313–318.

Bakar, N. A., Salleh, M. M., Umar, A. A., & Yahaya, M. (2011). The detection of pesticides in water using ZnCdSe quantum dot films. *Advances in Natural Sciences: Nanoscience and Nanotechnology*, 2(2), 025011.

Bansal, V., Poddar, P., Ahmad, A., & Sastry, M. (2006). Room-temperature biosynthesis of ferroelectric barium titanate nanoparticles. *Journal of the American Chemical Society*, 128(36), 11958–11963.

Bhattacharyya, A., Duraisamy, P., Govindarajan, M., Buhroo, A. A., & Prasad, R. (2016). Nano-biofungicides: Emerging trend in insect pest control. In: *Advances and Applications Through Fungal Nanobiotechnology* (pp. 307–319). Springer, Cham.

Boehm, A. L., Martinon, I., Zerrouk, R., Rump, E., & Fessi, H. (2003). Nanoprecipitation technique for the encapsulation of agrochemical active ingredients. *Journal of Microencapsulation*, 20(4), 433–441.

Bouwmeester, H., Dekkers, S., Noordam, M. Y., Hagens, W. I., Bulder, A. S., De Heer, C., ten Voorde, S. E., Wijnhoven, S. W., Marvin, H. J., & Sips, A. J. (2009). Review of health safety aspects of nanotechnologies in food production. *Regulatory Toxicology and Pharmacology: RTP*, 53(1), 52–62.

Chau, C. F., Wu, S. H., & Yen, G. C. (2007). The development of regulations for food nanotechnology. *Trends in Food Science and Technology*, 18(5), 269–280.

Christou, P., McCabe, D. E., & Swain, W. F. (1988). Stable transformation of soybean callus by DNA-coated gold particles. *Plant Physiology*, 87(3), 671–674.

Das, S. K., Das, A. R., & Guha, A. K. (2009). Gold nanoparticles: Microbial synthesis and application in water hygiene management. *Langmuir : The ACS Journal of Surfaces and Colloids*, 25(14), 8192–8199.

Dhillon, G. S., Brar, S. K., Kaur, S., & Verma, M. (2012). Green approach for nanoparticle biosynthesis by fungi: Current trends and applications. *Critical Reviews in Biotechnology*, 32(1), 49–73.

Durán, N., Marcato, P. D., De Souza, G. I., Alves, O. L., & Esposito, E. (2007). Antibacterial effect of silver nanoparticles produced by fungal process on textile fabrics and their effluent treatment. *Journal of Biomedical Nanotechnology*, 3(2), 203–208.

e la Cruz Quiroz, R., Maldonado, J. J. C., Alanis, M. D. J. R., Torres, J. A., & Saldívar, R. P. (2019). Fungi-based biopesticides: Shelf-life preservation technologies used in commercial products. *Journal of Pest Science*, 92(3), 1003–1015.

Elamawi, R. M., Al-Harbi, R. E., & Hendi, A. A. (2018). Biosynthesis and characterization of silver nanoparticles using *Trichoderma longibrachiatum* and their effect on phytopathogenic fungi. *Egyptian Journal of Biological Pest Control*, 28(1), 28.

El-Rafie, M. H., Mohamed, A. A., Shaheen, T. I., & Hebeish, A. (2010). Antimicrobial effect of silver nanoparticles produced by fungal process on cotton fabrics. *Carbohydrate Polymers*, 80(3), 779–782.

Elumalai, E. K., Prasad, T. N. V. K. V., Nagajyothi, P. C., & David, E. (2011). A bird's eye view on biogenic silver nanoparticles and their applications. *Der Chemica Sinica*, 2(2), 88–97.

Gade, A. K., Bonde, P., Ingle, A. P., Marcato, P. D., Duran, N., & Rai, M. K. (2008). Exploitation of *Aspergillus niger* for synthesis of silver nanoparticles. *Journal of Biobased Materials and Bioenergy*, 2(3), 243–247.

Gajbhiye, M., Kesharwani, J., Ingle, A., Gade, A., & Rai, M. (2009). Fungus-mediated synthesis of silver nanoparticles and their activity against pathogenic fungi in combination with fluconazole. *Nanomedicine: Nanotechnology, Biology and Medicine*, 5(4), 382–386.

Gautam, A. K., & Avasthi, S. (2020). Myconanoparticles as potential pest control agents. In: *Nanotechnology for Food, Agriculture, and Environment* (pp. 189–226). Springer, Cham.

Gericke, M., & Pinches, A. (2006). Biological synthesis of metal nanoparticles. *Hydrometallurgy*, 83(1–4), 132–140.

Goswami, A., Roy, I., Sengupta, S., & Debnath, N. (2010). Novel applications of solid and liquid formulations of nanoparticles against insect pests and pathogens. *Thin Solid Films*, 519(3), 1252–1257.

Griesche, C., & Baeumner, A. J. (2020). Biosensors to support sustainable agriculture and food safety. *TrAC Trends in Analytical Chemistry*, http://www.ncbi.nlm.nih.gov/pubmed/115906.

Guha, T., Gopal, G., Kundu, R., & Mukherjee, A. (2020). Nanocomposites for delivering agrochemicals: A comprehensive review. *Journal of Agricultural and Food Chemistry*, 68(12), 3691–3702.

Hnamte, S., Siddhardha, B., & Sarma, V.V. (2017). Myconanotechnology in agriculture. In: *Fungal Nanotechnology* (pp. 77–88). Springer, Cham.

Hulikere, M. M., & Joshi, C. G. (2019). Characterization, antioxidant and antimicrobial activity of silver nanoparticles synthesized using marine endophytic fungus – *Cladosporium cladosporioides*. *Process Biochemistry*, 82, 199–204.

Ingle, A., Gade, A., Pierrat, S., Sonnichsen, C., & Rai, M. (2008). Mycosynthesis of silver nanoparticles using the fungus *Fusarium acuminatum* and its activity against some human pathogenic bacteria. *Current Nanoscience*, 4(2), 141–144.

Jabeen, K., & Anum, F. (2020). Myco-nanotechnology in agriculture. In S. Javad (Ed.) *Nanoagronomy* (pp. 53–68). Springer, Cham.

Jeyaraj, M., Sathishkumar, G., Sivanandhan, G., MubarakAli, D., Rajesh, M., Arun, R., Kapildev, G., Manickavasagam, M., Thajuddin, N., Premkumar, K., & Ganapathi, A. (2013). Biogenic silver nanoparticles for cancer treatment: An experimental report. *Colloids and Surfaces B: Biointerfaces*, 106, 86–92.

Jo, Y. K., Kim, B. H., & Jung, G. (2009). Antifungal activity of silver ions and nanoparticles on phytopathogenic fungi. *Plant Disease*, 93(10), 1037–1043.

Kalaiselvan, V., & Rajasekaran, A. (2009). Biosynthesis of silver nanoparticles from Aspergillus niger and evaluation of its wound healing activity in experimental rat model. *International Journal of PharmTech Research*, 4, 1523–1529.

Kang, T., Yoo, S. M., Kang, M., Lee, H., Kim, H., Lee, S. Y., & Kim, B. (2012). Single-step multiplex detection of toxic metal ions by Au nanowires-on-chip sensor using reporter elimination. *Lab on a Chip*, 12(17), 3077–3081.

Kashyap, P. L., Kumar, S., Srivastava, A. K., & Sharma, A. K. (2012). Myconanotechnology in agriculture: A perspective. *World Journal of Microbiology and Biotechnology*, 29(2), 191–207.

Kathiresan, K., Manivannan, S., Nabeel, M. A., & Dhivya, B. (2009). Studies on silver nanoparticles synthesized by a marine fungus, *Penicillium fellutanum* isolated from coastal mangrove sediment. *Colloids and Surfaces, Part B: Biointerfaces*, 71(1), 133–137.

Khan, A. U., Malik, N., Khan, M., Cho, M. H., & Khan, M. M. (2018). Fungi-assisted silver nanoparticle synthesis and their applications. *Bioprocess and Biosystems Engineering*, 41(1), 1–20.

Kharissova, O. V., Dias, H. R., Kharisov, B. I., Pérez, B. O., & Pérez, V. M. J. (2013). The greener synthesis of nanoparticles. *Trends in Biotechnology*, 31(4), 240–248.

Kim, H. S., Kang, H. S., Chu, G. J., & Byun, H. S. (2008). Antifungal effectiveness of nanosilver colloid against rose powdery mildew in greenhouses. In: *Solid State Phenomena* (pp. 15–18). Trans Tech Publications, 135.

Kim, S. W., Jung, J. H., Lamsal, K., Kim, Y. S., Min, J. S., & Lee, Y. S. (2012). Antifungal effects of silver nanoparticles (AgNPs) against various plant pathogenic fungi. *Mycobiology*, 40(1), 53–58.

Kim, S. W., Kim, K. S., Lamsal, K., Kim, Y. J., Kim, S. B., Jung, M. S., Kim, H. S., Chang, S. J., Kim, J. K., & Lee, Y. S. (2009). An in vitro study of the antifungal effect of silver nanoparticles on oak wilt pathogen Raffaelea sp. *Journal of Microbiology and Biotechnology*, 19(8), 760–764.

Kuchibhatla, S. V., Karakoti, A. S., Baer, D. R., Samudrala, S., Engelhard, M. H., Amonette, J. E., Thevuthasan, S., & Seal, S. (2012). Influence of aging and environment on nanoparticle chemistry: Implication to confinement effects in nanoceria. *The Journal of Physical Chemistry. C, Nanomaterials and Interfaces*, 116(26), 14108–14114.

Kumar, S. A., Abyaneh, M. K., Gosavi, S. W., Kulkarni, S. K., Pasricha, R., Ahmad, A., & Khan, M. I. (2007). Nitrate reductase-mediated synthesis of silver nanoparticles from $AgNO_3$. *Biotechnology Letters*, 29(3), 439–445.

Kumar, S., Nehra, M., Dilbaghi, N., Marrazza, G., Hassan, A. A., & Kim, K. H. (2019). Nano-based smart pesticide formulations: Emerging opportunities for agriculture. *Journal of Controlled Release: Official Journal of the Controlled Release Society*, 294, 131–153.

Liu, Y., Tong, Z., & Prud'homme, R. K. (2008). Stabilized polymeric nanoparticles for controlled and efficient release of bifenthrin. *Pest Management Science: Formerly Pesticide* Science, 64(8), 808–812.

Marchiol, L. (2012). Synthesis of metal nanoparticles in living plants. *Italian Journal of Agronomy*, e37–e37.

Mikkelsen, R. (2018). Nanofertilizer and nanotechnology: A quick look. *Better Crops with Plant Food*, 102(3), 18–19.

Min, J. S., Kim, K. S., Kim, S. W., Jung, J. H., Lamsal, K., Kim, S. B., Jung, M., & Lee, Y. (2009). Effects of colloidal silver nanoparticles on Sclerotium-forming phytopathogenic fungi. *The Plant Pathology Journal*, 25(4), 376–380.

Moaveni, P., Karimi, K., & Valojerdi, M. Z. (2011). The nanoparticles in plants: Review paper. *Journal of Nanostructure in Chemistry*, 2, 59–78.

Mohammed Fayaz, A., Balaji, K., Girilal, M., Kalaichelvan, P. T., & Venkatesan, R. (2009). Mycobased synthesis of silver nanoparticles and their incorporation into sodium alginate films for vegetable and fruit preservation. *Journal of Agricultural and Food Chemistry*, 57(14), 6246–6252.

Mohanpuria, P., Rana, N. K., & Yadav, S. K. (2008). Biosynthesis of nanoparticles: Technological concepts and future applications. *Journal of Nanoparticle Research*, 10(3), 507–517.

Molnár, Z., Bódai, V., Szakacs, G., Erdélyi, B., Fogarassy, Z., Sáfrán, Varga, T., Kónya, Z., Tóth-Szeles, E., Szűcs, R., & Lagzi, I. (2018). Green synthesis of gold nanoparticles by thermophilic filamentous fungi. *Scientific Reports*, 8(1), 3943.

Moudgil, A., & Chaudhari, B. P. (2019). Fungi the crucial contributors for nanotechnology: A green chemistry perspective. In: *Advancing Frontiers in Mycology & Mycotechnology* (pp. 279–298). Springer, Singapore.

Mukherjee, P., Ahmad, A., Mandal, D., Senapati, S., Sainkar, S. R., & Khan (2001). Bioreduction of AuCl4– ions by the fungus, Verticillium sp. and surface trapping of the gold nanoparticles formed. *Angewandte Chemie International Edition*, 40(19), 3585–3588.

Namasivayam, S.K.R., Avimanyu (2011). Silver nanoparticle synthesis from *Lecanicilliumlecanii* and evalutionary treatment on cotton fabrics by measuring their improved antibacterial activity with antibiotics against.*Staphylococcus aureus* (ATCC 29213) and *E. coli* (ATCC 25922) strains. *International Journal of Pharmacy and Pharmaceutical Sciences* 3, 190–195.

Owolade, O. F., Ogunleti, D. O., & Adenekan, M. O. (2008). Titanium dioxide affects disease development and yield of edible cowpea. *Agricultural and Food Chemistry*, 7(50), 2942–2947.

Pandey, A. C., Sanjay, S., Yadav, R. (2010). Application of ZnO nanoparticles in influencing the growth rate of *Cicer arietinum*. *Journal of Experimental Nanoscience*, 5(6), 488–497.

Patil, S. A. (2009). *Economics of agri poverty: Nano-bio solutions*. Indian Agricultural Research Institute, New Delhi, Indian. http://www.assocham.org/nanobio2007/.

Patra, J. K., & Baek, K. H. (2014). Green nanobiotechnology: Factors affecting synthesis and characterization techniques. *Journal of Nanomaterials*, 219.

Pirzada, T., de Farias, B. V., Mathew, R., Guenther, R. H., Byrd, M. V., Sit, T. L., Pal, L., Opperman, C. H., & Khan, S. A. (2020). Recent advances in biodegradable matrices for active ingredient release in crop protection: Towards attaining sustainability in agriculture. *Current Opinion in Colloid and Interface Science*.

Prakash, S., & Soni, N. (2011). Factors affecting the geometry of silver nanoparticles synthesis in *Chrysosporium tropicum* and *Fusarium oxysporum*. *American Journal of Nanotechnology*, 2(1), 112–121.

Prasad, R., Kumar, V., & Prasad, K. S. (2014). Nanotechnology in sustainable agriculture: Present concerns and future aspects. *African Journal of Biotechnology*, 13(6), 705–713.

Rai, M., Yadav, A., Bridge, P., & Gade, A. (2009). Myconanotechnology: A new and emerging science. In: M.K. Rai and P.D. Bridge (Eds.) *Applied Mycology* (pp. 258–267). CAB International, New York.

Rai, A., Singh, A., Ahmad, A., & Sastry, M. (2006). Role of halide ions and temperature on the morphology of biologically synthesized gold nanotriangles. *Langmuir : The ACS Journal of Surfaces and Colloids*, 22(2), 736–741.

Rajeshkumar, S., & Sivapriya, D. (2020). Fungus-mediated nanoparticles: Characterization and biomedical advances. In: *Nanoparticles in Medicine* (pp. 185–199). Springer, Singapore.

Ray, S., Sarkar, S., & Kundu, S. (2011). Extracellular biosynthesis of silver nanoparticles using the mycorrhizal mushroom *Tricholoma crassum* (Berk.) Sacc: Its antimicrobial activity against pathogenic bacteria and fungus, including multidrug resistant plant and human bacteria. *Digest Journal of Nanomaterials and Biostructures*, 6, 1289–1299.

Reguera, G., McCarthy, K. D., Mehta, T., Nicoll, J. S., Tuominen, M. T., & Lovley, D. R. (2005). Extracellular electron transfer via microbial nanowires. *Nature*, 435(7045), 1098.

Saha, S., Sarkar, J., Chattopadhyay, D., Patra, S., Chakraborty, A., & Acharya, K. (2010). Production of silver nanoparticles by a phytopathogenic fungus Bipolaris nodulosa and its antimicrobial activity. *Digest Journal of Nanomaterials and Biostructures*, 5(4), 887–895.
Salem, S. S., & Fouda, A. (2020). Green synthesis of metallic nanoparticles and their prospective biotechnological applications: An overview. *Biological Trace Element Research*, 1–27.
Sanghi, R., & Verma, P. (2009). A facile green extracellular biosynthesis of CdS nanoparticles by immobilized fungus. *Chemical Engineering Journal*, 155(3), 886–891.
Sanghi, R., & Verma, P. (2010). PH dependant fungal proteins in the 'green' synthesis of gold nanoparticles. *Advanced Materials Letters*, 1(3), 193–199.
Sarathy, V., Tratnyek, P. G., Nurmi, J. T., Baer, D. R., Amonette, J. E., Chun, C. L., Penn, R. L., & Reardon, E. J. (2008). Aging of iron nanoparticles in aqueous solution: Effects on structure and reactivity. *The Journal of Physical Chemistry C*, 112(7), 2286–2293.
Sastry, M., Ahmad, A., Khan, M. I., & Kumar, R. (2003). Biosynthesis of metal nanoparticles using fungi and actinomycete. *Current Science*, 85(2), 162–170.
Satyanarayana, T., Deshmukh, S. K., & Deshpande, M. V. (Eds.) (2019). *Advancing Frontiers in Mycology & Mycotechnology: Basic and Applied Aspects of Fungi*. Springer Nature.
Shahzad, A., Saeed, H., Iqtedar, M., Hussain, S. Z., Kaleem, A., Abdullah, R., Sharif, S., Naz, S., Saleem, F., Aihetasham, A., & Chaudhary, A. (2019). Size-controlled production of silver nanoparticles by Aspergillus fumigatus BTCB10: Likely antibacterial and cytotoxic effects. *Journal of Nanomaterials*, 2019.
Shawon, Z. B. Z., Hoque, M. E., & Chowdhury, S. R. (2020). Nanosensors and nanobiosensors: Agricultural and food technology aspects. In: *Nanofabrication for Smart Nanosensor Applications* (pp. 135–161). Elsevier.
Singh, M., Manikandan, S., & Kumaraguru, A. K. (2011). Nanoparticles: A new technology with wide applications. *Research Journal of Nanoscience and Nanotechnology*, 1(1), 1–11.
Singh, T., Singh, A., Wang, W., Yadav, D., Kumar, A., & Singh, P. K. (2019). Biosynthesized nanoparticles and its implications in agriculture. In: *Biological Synthesis of Nanoparticles and Their Applications* (pp. 257–274). CRC Press.
Singhal, U., Khanuja, M., Prasad, R., & Varma, A. (2017). Impact of synergistic association of ZnO-nanorods and symbiotic fungus *Piriformospora indica* DSM 11827 on *Brassica oleracea* var. botrytis (Broccoli). *Frontiers in Microbiology*, 8, 1909.
Spasova, M., Manolova, N., Naydenov, M., Kuzmanova, J., & Rashkov, I. (2011). Electrospun biohybrid materials for plant biocontrol containing chitosan and *Trichoderma viride* spores. *Journal of Bioactive and Compatible Polymers*, 26(1), 48–55.
Suchitra, D., Nageswara Rao, A. B. N., Ravindranath, A., Sakunthala Madhavendra, S., & Jayathirtha Rao, V. (2011). Silver nanoparticle synthesis from *Lecanicillium Lecanii* and evalutionary treatment on cotton fabrics by measuring their improved antibacterial activity with antibiotics against *Staphylococcus aureus* (ATCC 29213) and *E. Coli* (ATCC 25922) strains. *International Journal of Pharmacy and Pharmaceutical Sciences*, 3(4), 190–195.
Sugunan, A., Melin, P., Schnürer, J., Hilborn, J. G., & Dutta, J. (2007). Nutrition-driven assembly of colloidal nanoparticles: Growing fungi assemble gold nanoparticles as microwires. *Advanced Materials*, 19(1), 77–81.
Sundaramoorthi, C., Kalaivani, M., Mathews, D.M., Palanisamy, S., Kalaiselvan, V., & Rajasekaran, A. (2009). Biosynthesis of silver nanoparticles from *Aspergillus niger* and evaluation of its wound healing activity in experimental rat model. *International Journal of PharmTech Research* 1, 1523–1529.

Tarafdar, J. C., Raliya, R., Mahawar, H., & Rathore, I. (2014). Development of zinc nanofertilizer to enhance crop production in pearl millet (*Pennisetum americanum*). *Agricultural Research*, 3(3), 257–262.

Thirumurugan, G., Shaheedha, S. M., & Dhanaraju, M. D. (2009). In vitro evaluation of antibacterial activity of silver nanoparticles synthesised by using Phytophthora infestans. *International Journal of ChemTech Research*, 1(3), 714–716.

Torney, F., Trewyn, B. G., Lin, V. S. Y., & Wang, K. (2007). Mesoporous silica nanoparticles deliver DNA and chemicals into plants. *Nature Nanotechnology*, 2(5), 295.

Tran, Q. H., & Le, A. T. (2013). Silver nanoparticles: Synthesis, properties, toxicology, applications and perspectives. *Advances in Natural Sciences: Nanoscience and Nanotechnology*, 4(3), 033001. http://www.ncbi.nlm.nih.gov/pubmed/033001.

Vadlapudi, V., & Behara, M. (2014). Green synthesis and biocompatibility of titanium nanoparticles. *Nano Science and Nano Technology: An Indian Journal*, 8, 367–372.

Vaidyanathan, R., Kalishwaralal, K., Gopalram, S., & Gurunathan, S. (2009). RETRACTED: Nanosilver – The burgeoning therapeutic molecule and its green synthesis. *Biotechnology Advances*, 27(6), 924–937.

Wang, P., Zhao, F. J., & Kopittke, P. M. (2019). Engineering crops without genome integration using nanotechnology. *Trends in Plant Science*, 24(7), 574–577.

Wani, T. A., Rather, G. A., Ahmad, M., & Kaloo, Z. A. (2020). Embodiment of nanobiotechnology in agriculture: An overview. In: *Nanobiotechnology in Agriculture* (pp. 113–128). Springer, Cham.

Yadav, A., Kon, K., Kratosova, G., Duran, N., Ingle, A. P., & Rai, M. (2015). Fungi as an efficient mycosystem for the synthesis of metal nanoparticles: Progress and key aspects of research. *Biotechnology Letters*, 37(11), 2099–2120.

Yoo, S. M., Kang, T., Kang, H., Lee, H., Kang, M., Lee, S. Y., & Kim, B. (2011). Combining a nanowire SERRS sensor and a target recycling reaction for ultrasensitive and multiplex identification of pathogenic fungi. *Small*, 7(23), 3371–3376.

Zhang, X., Yan, S., Tyagi, R. D., & Surampalli, R. Y. (2011). Synthesis of nanoparticles by microorganisms and their application in enhancing microbiological reaction rates. *Chemosphere*, 82(4), 489–494.

Zhao, X., Zhou, L., Riaz Rajoka, M. S., Yan, L., Jiang, C., Shao, D., Zhu, J., Shi, J., Huang, Q., Yang, H., & Jin, M. (2018). Fungal silver nanoparticles: Synthesis, application and challenges. *Critical Reviews in Biotechnology*, 38(6), 817–835.

2 Endophytes as Nanofactories

Philem Pushparani Devi, Isha Abhyankar, and Pradeep Kumar Singh

CONTENTS

2.1 INTRODUCTION

In recent times, as nanotechnology has found its way in almost all areas of life, including health care, agriculture, non-food industries, and environmental applications, the need for a clean route for the synthesis of nanoparticles is growing. An ideal synthesis approach should provide a safe, efficient, eco-friendly, and cost-effective nanoparticle synthesis method. There are different nanoparticle (NP) synthesis approaches that can be considered under three heads – chemical, physical, and biosynthetic methods. The biosynthetic approach is driven by different biological species, such as plant, plant products, and microbes, including fungi, algae, yeast, bacteria, and viruses, to synthesize nanoparticles (Malik et al., 2017). The microbial synthesis of nanoparticles has received unequivocal attention among all the biological systems for the synthesis of nanoparticles, including plants, due to the various advantages it offers, which includes being eco-friendly as it does not use hazardous chemicals, high pressure, high temperature, or high amounts of energy (Hulkoti and Taranath, 2014). It is also cost-effective, as microorganisms are abundant in nature, and easy to grow and maintain. Moreover, the method can be used for large-scale production with easy downstream processing of the final product with a minimal use of toxic chemicals (Monowar et al., 2018; Malik et al., 2017).

Several bacteria, actinomycetes, fungi, yeasts, and viruses have been explored for the biosynthesis of gold, silver, cobalt, zinc, selenium, tellurium, platinum, palladium, silica, titania, zirconia, quantum dots (QDs), magnetite, and uraninite NPs (Hulkoti and Taranath, 2014). Among the microbial community, 'endophytes' are

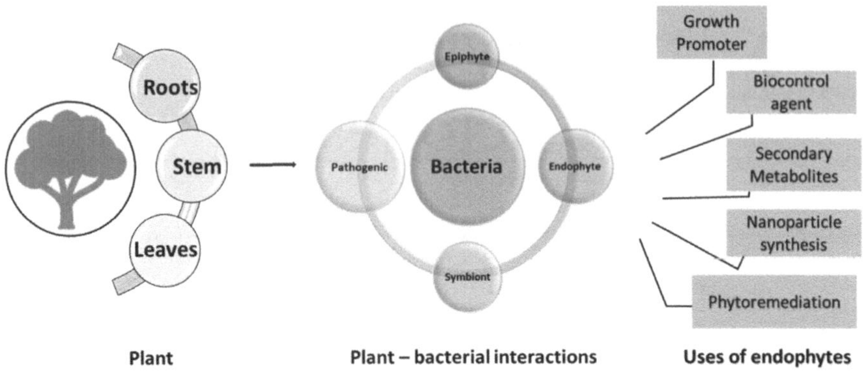

FIGURE 2.1 Endophyte and their applications.

appreciated as being a rich source of diverse secondary metabolites, and they are currently a relatively less explored yet rapidly rising approach for the synthesis of NPs (Baker and Satish, 2018).

De Bary (1866) coined the term 'endophyte' as a microorganism that grows within plant tissue. However, the definition of endophyte has evolved over the years and is expected to continue evolving. Endophytes are endosymbiont, basically, either a fungus or a bacterium, that spend a part of their life cycle in plant tissue without harming the host (Petrini, 1991, Hardoim et al., 2015, Gouda et al., 2016). They can be either 'obligate' or 'facultative' in nature (Gouda et al., 2016). Endophytes provide its host with a wide range of benefits by promoting growth, increasing yield, and as biocontrol agents for the plant (Figure 2.1). Endophytes produce numerous natural products that can be harnessed for applications in agriculture, medicine, and other industries; thus, the interest in endophytes has grown multi-fold over the last two decades and is still growing. This chapter aims to provide a detailed overview of how far we have come regarding the application of endophytes as nanofactories.

2.2 THE GREEN ROUTE TO NANOPARTICLE SYNTHESIS

Microbes employ two strategies for the synthesis of NPs: 1) an *in vitro* biosynthesis approach that uses biological extracts or specific molecules and 2) an *in vivo* biosynthesis using living cells. Moreover, an *in vivo* biosynthesis can be classified into a intracellular and extracellular synthesis, according to the synthesis of nanoparticles (Li et al., 2011). Using the intracellular route, ions from the environment are transported into the microbial cell to form nanoparticles in the presence of enzymes and other reducing agents. Whereas the extracellular route for NP synthesis involves trapping the ions on the surface of the cells and reducing the ions in the presence of enzymes and reducing agents (Zhang et al., 2011). The extracellular synthesis of nanoparticles has an advantage over the intracellular process of being able to bypass the need for cell lysis and purification of the final product, and hence it is the most commonly preferred mode of synthesis when using endophytes. Particle size and mono disparity are two crucial parameters in the evaluation of nanoparticle quality. Using a microbial biosynthetic approach, particle size and monodispersity control are achieved to no small extent by changing the microorganism used, and parameters like the growth phase of the microbes, growth medium, and synthesis conditions, such as pH, temperature, substrate concentrations, source compound, reaction time (Li et al., 2011).

Microorganisms can synthesize both organic and inorganic nanoparticles. Organic nanoparticles include the NPs of carbohydrates, proteins, lipids, nucleic acids, and a mixture of biomolecules, while inorganic NP can be metals and oxides (Malik et al., 2017). Organic NPs often use biomolecules such as DNA, proteins as biological templates to synthesize various complex nanostructures (Zhou et al., 2017; Qi et al., 2019). The synthesis of nanoparticles in solution generally requires a metal precursor, reducing agents, and a stabilizing/capping agent. However, a detailed molecular mechanism is still indistinct. Microbial NP synthesis seems to take place with the help of biomolecules, including enzymes such as nitrate reductase, peptides, proteins, exopolysaccharides or the triggering of the SOS system for reducing cell toxicity, or it may involve electron shuttle quinones or an enzymatic metal reduction process (Gahlawat and Choudhury, 2019; Malik et al., 2017).

Numerous biomolecules, such as phenolics, flavonoids, terpenoids, polysaccharides, alkaloids, proteins, enzymes, amino acids, carbohydrates, vitamins, alcoholic compounds, phytosterols, quinol and chlorophyll pigments, linalool, methyl chavicol, eugenol, caffeine, theophylline, ascorbic acid, carboxylic acids, flavonols, flavonones, saponins, tannins, triterpenes, sugars, sulfhydryls, quinones, etc., affect the reduction and the capping of the nanoparticles (Monowar et al., 2018). These biomolecules have unique chemical properties for reducing and effectively wrapping nanoparticles, and thus preventing their agglomeration. Hydroxyl and carboxyl groups of phenolic compounds bind to metals, making them useful for reducing and capping metal nanoparticles agents.

Although not essential for an organism's growth, secondary metabolites play an adaptive role in functioning as the defence compound or the signalling molecule during ecological interactions and environmental stresses. When it comes to applying plants and endophyte, secondary metabolites can be considered the most important agents for their various activities. Endophytic bacteria are known to produce low-molecular-weight secondary metabolites that include antimicrobial compounds, phytohormones, or their precursors, vitamins like B12 (Ivanova et al., 2006), and bioprotectants (Hardoim et al., 2015). Several secondary metabolites, such as alkaloids, steroids, terpenoids, peptides, polyketones, flavonoids, quinols, and phenols, are produced by endophytic bacteria (Gahlawat and Choudhury, 2019).

Terpenoids, also referred to as isoprenoids, belong to a class of organic compounds that are synthesized naturally by plants. As reported by Shankar et al. (2003), the plant extract of geranium leaves contains terpenoids which play a crucial role in the transformation of silver ions into nanoparticles. There are different classes of terpenoid which aid in the biotransformation process. Eugenol, the prime terpenoid of *Cinnamomum zeylanisum* (cinnamon) extracts, causes the bioreduction of $HAuC_{14}$ and $AgNO_3$ into nanoparticles. This is due to the dissociation of a proton from the eugenol OH group, resulting in the formation of resonance structures that catalyze a further oxidation process during nanoparticle synthesis (Singh et al., 2017).

Another important class of secondary metabolites is flavonoids consisting of polyphenolic compounds that include various subclasses like anthocyanins, isoflavonoids, flavonols, chalcones, flavones, and flavanones, which can actively chelate and reduce metal ions into nanoparticles. The tautomeric transformations of flavonoids from the enol-form to the keto-form cause the release of the reactive hydrogen atom that reduces metal ions into nanoparticles. For example, extracts of *Ocimum basilicum* (sweet basil) contain luteolin and rosmarinic acid, which transform the enol-group into the keto-group, thereby leading to the formation of silver nanoparticles (Ahmad et al., 2010).

Interestingly, flavonoids are also able to chelate metal ions with their carbonyl groups or π-electrons. Quercetin is a flavonoid possessing a strong chelating activity at three positions, also involving the carbonyl and hydroxyls groups at the C3 and C5 positions and the catechol group at the C3' and C4' sites. These groups are known to chelate various metal ions like Fe2+, Fe3+, Cu2+, Zn2+, Al3+, Cr3+, Pb2+, and Co2+. The presence of such mechanisms indeed explains the ability of flavonoids to be adsorbed onto the surface of a nascent nanoparticle. Apart from secondary metabolites, nanoparticles' biotransformation also happens due to certain enzymes present within the bacterial system. Microbes have the inherent ability to synthesize inorganic material using intercellular or extracellular mechanisms. In the intracellular system, the bacterial cell possesses an overall negative charge, thereby attracting positive ions towards it. Due to this electrostatic interaction, the metal ions are incorporated inside the cell through the ion transport system. Once within the cell, the entrapped metal ions are reduced into nanoparticles by different enzymes. Generally, the nanoparticles synthesized via this method are accumulated into the cytoplasm, periplasmic space, and cell wall. Whereas in the case of extracellular synthesis, the bacterial cell secretes reductase enzymes into the culture supernatant, which causes the biotransformation into nanoparticles. Extracellular synthesis is a more preferred method for commercial use due to its minimal downstream processing when compared to intracellular methods (Reverberi et al., 2017). Either intracellular or extracellular, both the methods have enzymes associated with them.

The majority of the enzymes associated with nanoparticle synthesis are dependent on cofactors nicotinamide adenine dinucleotide and its reduced form nicotinamide adenine dinucleotide phosphate. Vaidyananthan et al. (2010) reported the process of silver nanoparticle formation, starting with the reduction of silver complexes to silver atoms followed by their agglomeration into oligomeric clusters, leading to the formation of colloidal silver nanoparticles. The reduction is mediated by NADPH-dependent nitrate reductase. This was confirmed by comparing the activity with an inhibitor like sodium azide, a known inhibitor of nitrate reductase enzyme. In the inhibitor's presence, the biomass failed to synthesize the nanoparticles, indicating the role of nitrate reductase in silver nanoparticle synthesis (Vaidyanathan, 2009). Similarly, gold nanoparticle synthesis from *Stenotrophomonas maltophilia* using a nitrate reductase enzyme was studied by Nangia et al. (2009). A possible mechanism for the biosynthesis of gold nanoparticles from gold chloride involves the use of an NADPH-dependent reductase enzyme that promotes reduction through an electron shuttle enzymatic metal reduction (Nangia et al., 2009). Enzymes also play an essential role in transporting electrons from certain donors into the positive metal ion. These enzymes are considered to be responsible for the stabilization of the resultant NPs. For example, the enzyme nitrate reductase present in the supernatant facilitates electron transfer to Ag^+ ions, resulting in the formation of Ag NPs (Mishra et al., 2017). Mishra et al. (2017) reported on the role of amylase enzyme for synthesizing silver nanoparticles. The probable mechanism lies in the amino acid structure. The free thiol group of cysteine binds to the silver metal forming Ag–S bonds, thereby facilitating the reduction of the silver cation into the silver atom (Mishra and Sardar, 2012).

Amino acids differ in their ability to bind to metal ions and to cause a reduction. After amino acids are linked together in the peptide chain, their ability to bind and reduce metal ions drastically changes in terms of the functionality of the R-carbon of the carboxylic acids and amines of some of the amino acid residues. However, the free side chains of amino acids can still participate in the binding and reducing of metal ions. The suitability of side chains for reduction depends on the amino acid sequence, which affect the accessibility of individual groups. The work by Tan et al. (2010) explains in detail the effects of the amino acid sequence on the protein's ability to chelate and/or reduce metal ions. Amino acids like leucine, phenylalanine, and proline bind weakly to metal ions, thereby being ineffective in reducing tetrachloroauric acid anions (Tan et al., 2010).

The structural characterization of nanoparticles is routinely done using SEM, FTIR or TEM. The elucidation of the functional groups responsible for the transition is assessed using FTIR. A shift in peaks is definitive proof of nanoparticles. Various functional groups are associated with nanoparticle synthesis. It is reported that some functional groups, such as $-NH_2$, –OH, –SH, –COOH, and C-O-C, secreted by bacteria, play important roles in the reduction and stabilization of NPs. The functional groups' mechanism provides binding sites for the fixing of metal ions, followed by a reduction of the metal ions outside the cells, on the cell wall, or in the periplasmic space (Mishra et al., 2017). Wang et al. (2006) studied the effect of different functional groups on silver nanoparticle synthesis. The results conclude that amide, ether, carboxyl, and sulphate groups generally result in the formation of small-sized nanoparticles. Whereas hydroxyl and sulfonic groups are beneficial for synthesizing larger particles. Small-sized particles are formed through a burst nucleation reaction followed by Ostwald ripening. Large-sized particles are synthesized through the adhesion process, followed by subsequent coalescence of small particles (Wang et al., 2006). Cytochrome protein is another integral approach for reducing metal ions to NPs. The cytoplasmic components, such as NADH/NADPH, vitamins, and organic acids, act as electron donors to reduce the metal ions. Lv et al. (2018) reported on copper nanoparticle synthesis using a bioprocess involving cytoplasmic cytochrome C as the primary reductive factor. A thorough understanding of the myriad of mechanisms and cellular pathways involved in the biosynthesis of nanoparticles would facilitate optimization of production parameters for large-scale production and manipulate the desired characteristics of the synthesized nanoparticles.

2.3 ENDOPHYTEAS NANOPARTICLE FACTORY

The most frequently encountered endophytes are fungi and bacteria that coexist with each other (Sunkar and Nachiyar, 2012), and they are also the most common choice for nanoparticle synthesis, as is evident from the literature on the endophytic synthesis of NPs. We have presented below the available literature on the endophytic synthesis of NPs under both bacterial and fungal categories.

2.3.1 Endophytic Bacteria

Bacteria localize into the plant using different four means: as pathogens, symbionts, epiphytes, and endophytes. Endophytic bacteria generally colonize the intracellular spaces of the plants. They can be found in all types of different plants: monocotyledonous, dicotyledonous, conifers, and herbaceous plants (Rahman et al., 2019). Endophytic bacteria are rich in hydrolytic enzymes, like pectinases, cellulases, and endoglucanases, which help them colonize the endorhizosphere (Ferreira et al., 2008). Endophytic bacteria have a wide range of potential applications in phytoremediation, plant growth promotion, nanoparticle synthesis, and biocontrol agents. Of these, nanoparticle synthesis from endophytes has gained interest over time, owing to endophytes' ability to synthesize an array of secondary metabolites which catalyze the bioreduction process. The major advantage associated with the use of endophytic bacteria for NP synthesis is its large-scale production and minimal use of toxic chemicals, along with the improved stability and oxidation resistance of synthesized nanoparticles (Monowar et al., 2018). Table 2.1 describes the different kinds of bacteria and their nanoparticles.

Generally, metal ions can be lethal to the bacterial cells, and their ability to synthesize NPs is an outcome of their mechanism that evolved to resist this metal stress. The biosynthesis of NPs involve the use of metal ions along with the liquid media for microbial growth. As is the case with all microbial synthesis of NPs, bacteria can intracellularly or extracellularly perform NP synthesis. In either system, the reduction of ions is facilitated by proteins, enzymes, and cell wall components. For intracellular processes, the process is carried out by functional groups like the carboxyl groups present on the cell wall, which attract the metal ions due to electrostatic interactions (Reverberi et al., 2017). The ions then enter the cell and interact with intracellular proteins, thereby leading to reduction. Similarly, the carbonyl groups of some proteins and enzymes stabilize the NPs by binding to the NP surfaces. It has been reported that some functional groups, such as –NH2, –OH, –SH, and –COOH, of the proteins secreted by bacteria, play essential roles in the reduction and stabilization of NPs. These functional groups provide binding sites for the fixing of metal ions, followed by the reduction of the metal ions outside the cells, on the cell wall, or in the periplasmic space (Mishra et al., 2017).

The methodology used for nanoparticle synthesis using endophytes is either by utilizing the biomass or by cell-free extract (Figure 2.2). Briefly, the endophytic bacteria is grown in nutrient media for 24 hours, after which the culture is centrifuged. The cell-free supernatant or biomass obtained after centrifugation is used for the

TABLE 2.1
List of Endophytic Bacteria Reported for Nanoparticle Synthesis

Plant name	Endophytic bacteria	Nanoparticle	Size	Application	Reference paper
Garcinia xanthocymus	*Bacillus cereus*	Silver	20–40 nm	Antibacterial	Sunkar and Nachiyar (2012)
Adhatodabeddomei	*Bacillus cereus*	Silver	11–16 nm	Antibacterial	Sunkar and Nachiyar (2012)
Piper nigrum	*Bordetella* sp.	Silver	63–90 nm	Antibacterial	Thomas et al. (2012)
Euphorbia hirta L.	*Endophyticbacterium* EH 419	Silver	10–60 nm	Bactericidal	Baker et al. (2016)
B.aralocaspica	*Isoptericoladokdonensis*	Silver	11–40 nm	Antibacterial	Dong et al. (2017)
NA	*Pantoeaananatis*	Silver	30–40 nm	Antibacterial	Thamina Monowar (2018)
Mimosa pudica L.	*Aneurinibacillusmigulanus*	Silver	20–25 nm	Bactericidal	Baker et al. (2016a)
Coraindrumsativum	*Bacillus siamensis*	Silver	25–50 nm	Plant growth promotion, antibacterial	Ibrahim et al. (2019)
Allium sativum	*Pseudomonas poae*	Silver	20–45 nm	Antifungal	Ibrahim et al. (2020)
Coffea arabica	*Pseudomonas fluorescence*	Gold	5–50 nm	Bactericidal	Baker et al. (2016b)
Annona squamosa L.	*Pseudomonas veronii* AS41G	Gold	5–50 nm	Antibacterial	Baker et al. (2015)
Azadirachta indica	*Saccharomonaspora sp*	Gold	40–60 nm	NA	Verma et al. (2013)
Withaniasomnifera	*Sphingobacteriumthalpophilum*	Zinc oxide	35–40 nm	Antibacterial	Rajabairavi et al. (2017)
Oxalis corniculata L	*Streptomyces zaomyceticus, Streptomyces pseudogriseolus*	Zinc oxide	75–80 nm	Antibacterial, antioxidant	Hassan et al. (2019)
Convolvulus arvensis	*Streptomyces capillispiralis*	Copper	5–50 nm	Antibacterial, Biocontrol agent	Hassan et al. (2018)
Justiciabeddomei	*Bacillus cereus*	Titanium oxide	70–140 nm	Phytotoxicity	Sunkar et al. (2014)
Annona squamosa L.	*Pseudomonas veronii* AS41G	Silver – gold	5–50 nm	Antibacterial	Baker (2017)

NA: Not available

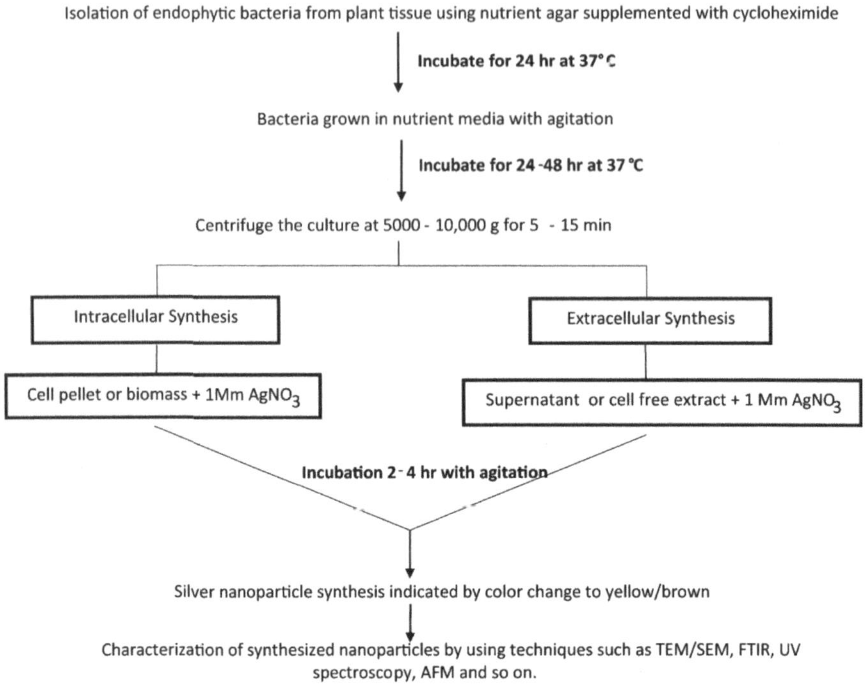

FIGURE 2.2 Schematic representation of nanoparticle biosynthesis by endophytic bacteria, modified from Baker et al. (2016).

synthesis of nanoparticles, which is then cultured with nutrient media supplemented with the salt solution. The media is then incubated at specific temperatures until there is a visible colour change in the media. Samples are removed at the periodic interval and assessed for the synthesis of the desired nanoparticles.

Endophytic bacteria are predominantly used to synthesize silver nanoparticles (Ag NPs) mainly because of their varied applicability. For example, Sunkar and Nachiyar (2012) reported on the synthesis of spherical Ag NPs of 20–40 nm in size using *Bacillus cereus* isolated from the plant *Garcinia xanthochymus*. The synthesized nanoparticles exhibited antibacterial activity against *Escherichia coli*, *Pseudomonas aeruginosa*, *Staphylococcus aureus*, *Salmonella typhi*, and *Klebsiella pneumonia* (Sunkar and Nachiyar, 2012). Dong et al. (2017) isolated *Isoptericola dokdonensis* from the roots of *B. aralocaspica* and used it for synthesizing spherical Ag NPs of sizes ranging from 11–40 nm. The nanoparticles have potent antimicrobial activity against *Staphylococcus warneri* (Dong et al., 2017). Similarly, Ag NPs were synthesized using endophytic bacteria *B. aralocaspica* with an average size of 35 nm (Monowar et al., 2018). These bacteria showed antibacterial activity against multidrug-resistant bacteria *S. pneumonia*, *E. faecium*, *S. aureus*, and *E. coli* with a concentration of 500 μg/ml and antifungal activity against *C. albicans*. Baker et al. (2016) synthesized silver nanoparticles of 20–60 nm in size from endophytic bacteria *Aneurnibacillus migulanus* inhabiting *Mimosa pudica*. The nanoparticles exhibited antibacterial activity against Gram-positive *S. aureus* and *B. subtilis* and Gram-negative *E. coli* and *P. aeruginosa* with MIC value of 12.5 µg/ml (Baker et al., 2016a).

Apart from Ag NPs, endophytic bacteria are also used for synthesizing various other metal nanoparticles. For example, Baker et al. (2016b) reported the synthesis of gold nanoparticles with sizes ranging from 5–25 nm using endophytic bacteria *Pseudomonas fluorescens*, inhabiting plant *Coffea arabica* L. The synthesized nanoparticles showed promising activity against Gram-positive *S. aureus* and Gram-negative *E. coli* bacteria (Baker et al., 2016b). Similarly, gold nanotriangles 40–60 nm in size synthesized from endophytic actinomycetes *Saccharomonospora* inhabiting *Azadirachta indica* were reported by Verma et al. (2013). Rajabairavi et al. (2017) reported the synthesis of zinc oxide nanoparticles with an average size of 37 nm from endophytic bacteria *Sphingobacterium thalpophilum* isolated from *Withania somnifera*, and the synthesized nanoparticles were antibacterial against *P. aeruginosa* and *E. aerogens.*

Hassan et al. (2018) reported on synthesis of copper nanoparticles using actinomycetes, *Streptomyces capillispiralis*, isolated from *Convolvulus arvensis.* Synthesized nanoparticles showed antibacterial activity against Gram-positive bacteria, *Bacillus dimenuta*, *Bacillus subtilis* and *Staphylococcus aureus*, Gram-negative bacteria *Pseudomonas aeruginosa* and *E. coli* and unicellular fungi *Candida albicans*, and multicellular fungi, *A. brasiliensis.* Nanoparticles also possessed larvicidal activity against house fly (*M. domestica*) and mosquito (*C. pipiens*) (Hassan et al., 2018). Similarly, actinomycetes, *Streptomyces zaomyceticus*, and *Streptomyces pseudogriseolus* were used for synthesizing copper nanoparticles isolated from *Oxalis corniculata* L. These synthesized nanoparticles were sized 78–80 nm and exhibited promising antibacterial activity against *Bacillus subtilis*, *Bacillus diminuta*, *Staphylococcus aureus*, and *Pseudomonas aeruginosa*, antifungal activity against *Fusarium oxysporum*, *Pythium ultimum*, *Aspergillus niger*, and *Alternaria alternata* and anti-larvicidal activity against *Musca domestica* and *Culex pipiens* (Hassan et al., 2019).

Titanium oxide nanoparticles with an average size of 89 nm were synthesized using endophytic bacteria *Bacillus cereus* inhabiting *Justicia beddomei*, and toxicity assessment during seed germination further reveal a reduction of root length, without hampering germination rate (Sunkar et al., 2014). Metalloid nanoparticles, such as spherical selenium NPs, have also been reported by Xu et al. (2020) using endophytic bacteria employing *Herbaspirillum* isolated from *Camellia sinensis* L. Interestingly, bimetallic nanoparticles have also been synthesized using *Pseudomonas veronii*, isolated from *Annona squamosa* L. (Baker et al., 2017). The bimetallic nanoparticles were synthesized by mixing the culture supernatant and salt solution ($AgNO_3$ and $HAuC_{14}$) together and measured 5–50 nm in size. The NPs showed potent antibacterial activity against various antibiotic-resistant species (Baker et al., 2017).

2.3.2 Endophytic Fungus

Endophytic fungi are a source of various bioactive compounds. So far, few endophytic fungi have been engaged in the biosynthesis of nanoparticles. Some authors refer to the process of synthesis of nanoparticles using fungi as myconanotechnology (Hulikere and Joshi, 2017). Fungi as a nanoparticle factory provide several advantages over other microbes, including bacteria. Most fungi have modest nutrient requirements, simple handling, metal uptake capabilities and high surface-binding capacities (Dias et al., 2002; Sanghi and Verma, 2009). Fungi allow for the large-scale production of nanoparticles when compared to other biosynthetic systems, as they secrete large amounts of enzymes, exhibit easier and cost-effective downstream processing, and, unlike bacterial broth, fungal broth can be easily filtered by a filter press (Dias et al., 2002; Sanghi and Verma, 2009).

The mechanism of NP synthesis by fungus exploits the natural survival response of fungus mycelium when exposed to stress, including osmotic stress (Mehra and Winge, 1991). The metal salt solution creates an osmotic stress environment for the fungus and prompts it to produce extracellular enzymes and metabolites which catalyze the reduction of toxic metal ions into non-toxic metallic solid nanoparticles (Gericke and Pinches, 2006). Among the metabolites, several naphthoquinones and anthraquinones with excellent redox potential have been reported in some fungi that could act as an electron shuttle in metal reductions (Gericke and Pinches, 2006).

So far, an endophytic fungus of the genus *Aspergillus* (Devi and Joshi, 2014), *Pestaloptiopsis pauciseta* (Vardhana and Kathiravan, 2015), *Penicillium* (Rahi and Parmar, 2014), and *Phomosis* sp. (Bharathidasan and Panneerselvam, 2012, Table 2.2) have been reported for NP synthesis. The earliest report on the endophytic fungus synthesis of NP by Shankar and co-workers (2003) on the extracellular synthesis of gold nanoparticles using geranium leaves (*Pelargonium graveolens*) and their endophytic fungus (*Colletotrichum* sp.) indicated that polypeptides/enzymes were the reducing and capping agents. Different fungal endophytes have been utilized maximally for silver nanoparticle (Ag NPs) synthesis. This might be due to the demand for silver NPs, which exhibit a broad range of antibiotic-resistant microorganisms attributed to a higher surface to volume ratio compared to standard metallic silver (Sharma et al., 2009).

The genus *Aspergillus* has been explored extensively for the synthesis of Ag NPs of different sizes and shapes. Verma et al. (2010) performed the extracellular synthesis of spherical to hexagonal Ag NPs of 10–20 nm, using *Aspergillus clavatus* (AzS-275) isolated from *Azadirachta indica* A. Juss, which showed promising antifungal and antibacterial activity against *Candida albicans*, *P. fluorescens*, and *E. coli* at a low concentration of 5.8 μg ml^{-1}. Smaller sized nanoparticles ranging from 3.5 to 8.7 nm were synthesized using another two endophytic *Aspergillus* sp., namely *A. tamarii* PFL2, *A. niger* PFR6, along with *Peniclliumo chrochloron* PFR8 (Devi and Joshi, 2015). The nanoparticles synthesized using the fungus *A. tamarii* PFL2 were found to have the smallest average particle size (3.5 ± 3 nm) as compared to the nanoparticles biosynthesized using two other fungi *A. niger* PFR6 and *P. ochrochloron* PFR8, which produced average particle sizes of 8.7 ± 6 nm and 7.7 ± 4.3 nm, respectively.

Another endophytic *Aspergillus*, *A. terreus*, isolated from *Calotropis procera* was reported to synthesize 16.54 nm Ag NPs, which significantly inhibited the growth of *Salmonella typhi*, *Staphylococcus aureus*, and *Escherichia coli* (Rani et al., 2017).

TABLE 2.2
List of Endophytic Fungus Reported for Nanoparticle Synthesis

Plant name	Endophytic fungi	Nanoparticle	Size	Application	Reference paper
Withania somnifera	*Fusariumsemitectum*	Silver	10–20 nm	Antibacterial	Halkai et al. (2017)
Azadirachta indica	*Aspergillus clavatus*	Silver	10–25 nm	Antimicrobial	Verma et al. (2010)
Syzygium cumini (L)	*Pestalotia* sp.	Silver	10–40 nm	Antibacterial	Raheman et al. (2011)
Phellodendronamurense	*Epicoccumnigrum*	Silver	1–22 nm	Antifungal	Abdel Hafez et al. (2013)
Curcuma longa	*Penicillium* sp.	Silver	25 nm	Antibacterial	Singh et al. (2013)
Aloe vera	*Penicillium* sp.	Silver	16–39 nm	Antibacterial	Rahi and Parmar (2014)
Potentillafulgens L	*Cryptosporiopsisericae*	Silver	2–15 nm	Antimicrobial, Growthkinetics	Devi and Joshi (2014)
Sidaacuta	*Geotrichum syncephalis*	Silver	6.40–16.9 nm	Antimicrobial	Rahi et al. (2014)
Potentillafulgens L.	*Aspergillus tamarii* PFL2	Silver	3.5 nm	NA	Devi and Joshi (2015)
Catharanthusroeus	*Curvularialunata*	Silver	10–50 nm	Antimicrobial	Ramalingmam et al. (2015)
Calophyllumapetalum	*Penicillium* sp.	Silver	100 nm	NA	Chandrappa et al. (2016)
Madhucalongifola	*Pestalotia* sp.	Silver	Below 40 nm	Antibacterial activity	Verma et al. (2016)
Tomato	*Cladosporiumsphaerospermum*	Silver	15–17 nm	NA	Abdel-Hafez et al. (2016)
Gymnema sylvestre	*Pestalotiopsis microspore*	Silver	2–10 nm	Antioxidant, Cytotoxicity study, Celllines and cell culture	Netala et al. (2016)
Centella asiatica	*Aspergillus versicolor* ENT7	Silver	3–40 nm	Antibacterial, Antifungal	Netala et al. (2016)
Calotropisprocera	*Aspergillus terreus*	Silver	16.5 nm	Antibacterial, Cell leakage analysis	Rani et al. (2017)
Raphanussativus	*Alternaria*	Silver	4–30 nm	Antibacterial	Singh et al. (2017)
Phlogacanthusthyrsiflorus	*Penicilliumoxalicum*	Silver		Antibacterial	Bhattacharjee et al. (2017)
Sargassumwightii	*Cladosporiumcladosporioides*	Silver	60 nm	Antimicrobial, Antioxidant	Joshi (2017)

(*Continued*)

TABLE 2.2 (CONTINUED)
List of Endophytic Fungus Reported for Nanoparticle Synthesis

Plant name	Endophytic fungi	Nanoparticle	Size	Application	Reference paper
Simaroubaglauca	*Aspergillus niger*	Silver	41 nm	Antibacterial, Antioxidant	Hemashekhar et al. (2017)
Casuarina junhuhniana	*Aspergillus tamaraii*	Silver	31–40 nm	Antimicrobial, Germination study, Photocatalyticdegradation	Prasad et al. (2018)
Chetomorphaantennina	*Penicilliumpolonicum*	Silver	10–15 nm	Antibacterial	Neethu et al. (2018)
Solanumnigrum	*Setosphaeria* sp.	Silver	20–50 nm	Antibacterial	Akther et al. (2018)
Chonemorphafragrans	*Fusariumsolani*	Gold	40–45 nm	Anticancer	Clarance et al. (2020)
Pelargonium graveolens	*Colletotrichum* sp.	Gold	20–40 nm	NA	Shankar et al. (2003)
Azadirachta indica A. Juss	*Aspergillus clavatus*	Gold	20–35 nm	Antibacterial	Verma (2011)
Nothapodytesfoetida	*Aspergillus nidulans*	Cobalt	20 nm	NA	Vijayanandan and Balakrishna (2018)
NA	*Alternariatenuissima*	Zinc oxide	46 nm	Antibacterial, anticancer, antioxidant	Abdelhakim et al. (2020)
Nothapodytesfoetida	*Cochliobolusgeniculatus*	Zinc oxide	2–6 nm	Semiconductor	Kadam (2019)

NA: Not available

An increase in protein leakage level and nucleic acid degradation was detected in cell leakage analysis as an outcome of the exposure to Ag NPs (Rani et al., 2017). Hemashekhar et al. (2017) also isolated *Aspergillus niger* from *Simarouba glauca* leaf for the synthesis of 41.9 nm sized Ag NPs which showed antimicrobial activity against *Escherichia coli*, *Pseudomonas aeruginosa*, *Staphylococcus aureus*, and *Klebsiella pneumonia*, as well as good antioxidant activity. Another report on *Aspergillus* is *A. tamaraii* isolated from the root of *Casuarina junghuhniana* Miq which synthesized Ag NPs of 1–40 nm, and it has also found to be effective in inhibiting several bacterial and fungal species, including *Bacillus subtilis*, *Pseudomonas aeruginosa*, *Fusarium solani*, *Fusarium oxysporum*, *Alternaria solani*, *Rhizoctonia solani*, *Curvularia lunata*, and *Alternaria alternate* (Bose and Gowrie, 2017). Netala et al. (2016) also isolated *Aspergillus versicolor* ENT7 from the ethnomedicinal plant *Centella asiatica* that was used to synthesize Ag NPs with a size range of 3–40 nm. The Ag NPs exhibited effective antimicrobial and antifungal activity against *Staphylococcus aureus*, *Streptococcus pneumonia* (Gram-negative), *Pseudomonas aeruginosa*, and Gram-negative *Klebsiella pneumonia*; *Candida albicans* and *Candida nonalbicans*. Moreover, the *Aspergillus versicolor* ENT7 synthesized Ag NPs also possessed a significant free radical scavenging activity with an IC50 value of 60.64 lg/ml. The antimicrobial mechanism used by nanoparticles is still elusive, and there are several mechanisms proposed which include changes in the membrane permeability affecting transport, genetic regulation, production of reactive oxygen species, cation release, biomolecule damages, ATP depletion, and inhibition of biofilm formation (Slavin et al., 2017; Wang et al., 2017).

Polydispersed Ag NPs of 10–50 nm were synthesized extracellularly from endophytic fungus *Curvularialunata* (Ramalingmam et al., 2015), and it enhanced the antibacterial activities of ampicillin, rifampicin, chloramphenicol, erythromycin, and kanamycin against test strains such as *E. coli*, *S. paratyphi*, and *B. subtillis*. Such antibiotic enhanced efficacy of Ag NPs has been previously reported in different non-endophytic biosynthesis studies against different antibiotics (Birla et al., 2009; Gajbhiye et al., 2009; Masri et al., 2018; Kaur and Kumar, 2019). Ag NPs of 4–30 nm in size were synthesized from *Alternaria* sp. isolated from the leaves of *Raphanus sativus* (Singh et al., 2017). Verma et al. (2016) isolated endophytic fungus *Pestalotia* sp. from *Madhucalongi folia* to synthesize Ag NPs of less than 40 nm, which significantly inhibited the growth of pathogenic bacteria such as *Aeromonas hydrophila*, *Staphylococcus aureus*, *Escherichia coli*, *Morganella morganii*, *Salmonella typhi*, *Enterococcus faecalis*, and *Klebsiella pneumonia*. The Ag NPs also exhibited a strong scavenging activity on 2,2′-diphenyl-1-picrylhydrazyl (DPPH) with an inhibitory concentration (IC50) at 47 μg /ml. Ag NPs of less than 100 nm in size were reported from *Cladosporium cladosporioides* isolated from seaweed *S. wighttii*, which showed inhibitory action against *C. albicans* (Hulikere et al., 2017). An NADPH-dependant reductase dependent mechanism of Ag NPs synthesis was proposed in this study. *Setosphaeria* sp. BAB-4662 isolated from the stem of *Solanum nigrum* synthesized Ag NPs with an average size of about 35 nm and they were potent against *K. pneumonia*, *A. Baumannii*, *P. mirabilis*, *S. Typhimurium*, *P. aeruginosa*, and *E. Coli* (Akther et al., 2018).

Ag NPs were synthesized from *Penicillium* sp., from leaves of the turmeric plant (*Curcumin longa*), with a size of 25 nm and they showed antibacterial activity against *P. aeruginosa* and *K. pneumonia* (Singh et al., 2013). Neethu et al. (2018) isolated *Penicillium polonicum* ARA 10 from *Chetomorpha antennina*, for the synthesis of 10–15 nm Ag NPs with an inhibitory action against major foodborne bacterial pathogens such as *Salmonella enterica serovar* and *Typhimurium* MTCC 1251. *Pestalotiopsismicrospora* VJ1/VS1 isolated from the leaves of *Gymnema sylvestre* was used for the synthesis of 2–10 nm sized Ag NPs with effective radical scavenging activity against DPPH and H_2O_2 radicals, and they showed IC50 values of 76.95 ± 2.96 and 94.95 ± 2.18 μg/mL, respectively (Netala et al., 2016). Similar size Ag NPs were synthesized by Halkai et al. (2017) using *Fusarium semitectum* isolated from the leaf of *Withania somnifera* (Ashwagandha) and they demonstrated antibacterial activity against *Porphyromonas gingivalis*, an important periodontal pathogen. Ag NPs of below 10 nm in size were synthesized using two fungal root endophytes of the genus, *Syncephalis* and *Geotrichum*, isolated from *Sidaacuta syncephalis* sp. (Rahi et al., 2014). The size of the *Syncephalis* synthesized Ag NPs ranged from 6.40 nm to 16.9 nm, whereas that of *Geotrichum* sp. was in the range of 3.63 nm to 14.2 nm. These nanoparticles augmented the antibacterial activity of standard antibiotics such as Penicillin, Methicillin, Ampicillin, and Amoxicillin against the *E. coli*, *MRSA*, *Pseudomonas aeruginosa*, and *Staphylococcus aureus*. Ag NPs of an average size of 5.5 ± 3.1 nm were synthesized using *Cryptosporiopsisericae* PS4, isolated from *Potentilla fulgens* L. by Devi and Joshi (2015) showed anti-bactericidal and anti-fungicidal activity with enhanced antimicrobial activity when used in combination with chloramphenicol and fluconazole, against five pathogens – *Staphylococcus aureus* MTCC96, MTCC735, *E. coli* MTCC730, *Enterococcus faecalis* MTCC2729, and *Candida albicans* MTCC 183, supported by SEM analysis depicting pits and ruptures in the bacterial cells, which indicated a fragmented cell membrane and severe cell damage (Devi and Joshi, 2014). The authors reported a delay of the exponential and stationary phases by exposure to Ag NPs (Devi and Joshi, 2015). Bhattacharjee et al. (2017) reported Ag NP synthesis from *Penicillium oxalicum* isolated from the medicinal plant *Phlogacanthus thyrsiflorus* with anti-streptomycin activity; there was no information on the size of the Ag NPs.

While Ag NPs are the most extensively studied NP using endophytic fungi, there are also a few reports on gold nanoparticles and very even fewer on cobalt and zinc nanoparticles. In a very recent study, Clarance et al. (2020) synthesized gold nanoparticles of 40–45 nm using *Fusarium solani* ATLOY – eight isolated from the plant *Chonemorpha fragrans*. These gold NPs were cytotoxic against cervical He La and human breast cancer cells by inducing apoptosis in these cell lines (Clarance et al., 2020). Another report on gold NP synthesis used *Colletotrichum* sp. isolated from the leaf of *Pelargonium graveolens* (Shankar et al., 2003). The gold NPs were decahedral and icosahedral in shape with sizes ranging from 20–40 nm. Vijayanandan and Balakrishna (2018) synthesized cobalt NPs extracellularly using *Aspergillus nidulans* from *Nothapodytes foetida*, which were predominantly spherical in shape with an average size of 20.29 nm. FTIR analysis of the cobalt NPs revealed the presence of protein capping stabilizing the NPs.

Spherical zinc oxide NPs were synthesized from *Alternaria tenuissima,* with an average size of 46.58 nm (Abdelhakim et al., 2020). The ZnO NPs were inhibitory against HepG-2 cancer cell lines. Kadam et al. (2019) synthesized quasi-spherical 2–6 nm sized ZnO using *Cochliobolus geniculatus* obtained from the leaf of *Nothapodytes foetida.* The UV-vis spectroscopy, fluorescence spectroscopy, FTIR and SDSPAGE analysis revealed the possibility of proteins with a size range of 97–36 kDa as the capping material in ZnO nanoparticle synthesis (Kadam et al., 2019). *Aspergillus flavus* was used by Uddandarao et al. (2019) for the synthesis of rare-earth Gadolinium (Gd)-doped ZnS NPs, hexagonal in shape and 10–18 nm in size. ZnS doping with Gd was shown to enhance the biosensing capacity of ZnS nanoparticles (Uddandarao et al., 2019). Figure 2.3 represents the nanoparticle synthesis by fungi.

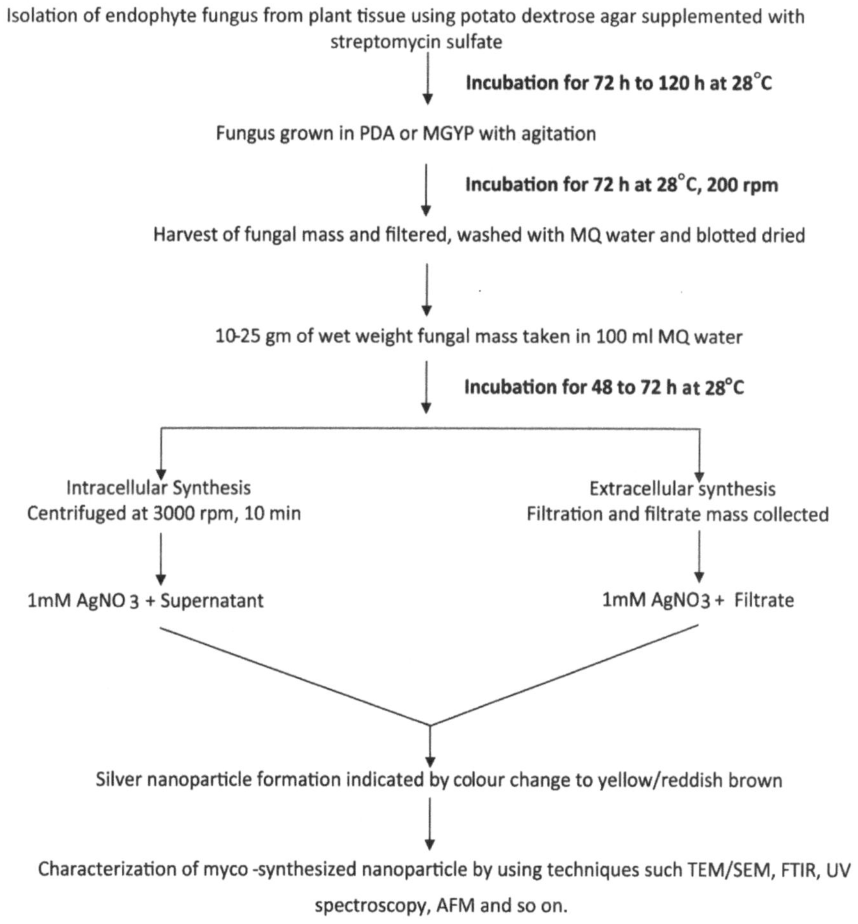

FIGURE 2.3 Schematic representation of nanoparticle biosynthesis by the endophytic fungus, modified from Shukla and Sandhu, 2017.

2.4 PROS AND CONS

The biosynthesis of NPs, including the endophytic approach, offers several advantages for economic and environmental concerns over chemical-based methods. One of the most desirable features of biosynthesis methods is that they sometimes impart a lipid layer coating on the NP that confers physiological solubility and stability, which is desirable for biomedical applications (Razavi et al., 2019). When it comes to controlling particle size, chemical and physical synthesis approaches still have challenges to meet and such this is an ongoing area of research. Biological synthesis offers several parameters that can be manipulated easily to strictly control particle morphology and would therefore offer considerable advantage over other non-biosynthetic methods.

However, many studies have shown the poor stability of the nanoparticles synthesized by microorganisms as it has been observed that they decompose after a certain period of time when tested for different applications, such as drug delivery systems and diagnostic agents (Hergt et al., 2006). Moreover, the microbial biosynthesis of NPs is a slow process, lasting for several hours and even a few days when compared to chemical and physical methods. Thus, the enhancement of the stability of nanoparticles produced by biological methods and strategies to reduce synthesis time would make this biosynthesis route more lucrative. A better understanding of the synthesis mechanism on a cellular and molecular level would facilitate the identification of the compounds responsible for the reduction of the nanoparticles, improving the output of isolation and reaction time, and therefore giving a higher synthesis efficiency. Advances in multi-Omics research are currently assisting researchers to design faster and more efficient biosynthetic strategies.

2.5 CONCLUDING REMARKS

The microbial synthesis of nanoparticles has been in demand over the last two decades for its green approach. Microorganisms, including bacteria, yeast, fungi, and actinomycetes, either free-living or endophytic, continue to be explored for the biosynthesis of nanoparticles. There are still challenges and requirements for improving stability and synthesis efficiency, and a need to explore biomedical applications. The current research on the endophytic synthesis of NPs focus more on exploring candidate endophytes with the capacity for nanoparticle synthesis. As a result, the mechanistic understanding of synthesis and biomedical applications is still largely lacking. So there is a significant scope left for exploring the applicability and mechanisms of synthesis associated with the green synthesis of NPs. Some of the open questions on the topic include: are there any mechanistic differences to NP synthesis between free-living and endophytic microbes? Does the shape and size of the NPs impact their application in both the industrial and medical industries? Current advances in the multi-Omics approach provides useful tools that can aid in a deeper understanding of the synthesis pathways, the impact of the nanoparticles on the body, and also optimize the desired applications of these NPs by manipulating microorganisms at the genomic and proteomic levels. Hopefully, the implementation of these endophytic systems as nanofactories on a large scale and their commercial applications in medicine, health care, and industries will take place in the coming years.

REFERENCES

Abdel-Hafez, S.; Nafady, N.A.; Abdel-Rahim, I.R.; Shaltout, A.M.; Mohamed, M.A. Biogenesis and optimisation of silver nanoparticles by the endophytic fungus cladosporium sphaerospermum. *International Journal of Nanomaterials and Chemistry* 2016, 2(1), 11–19. https://doi.org/10.18576/ijnc/020103.

Abdelhakim, H.K.; El-Sayed, E.R.; Rashidi, F.B. Biosynthesis of zinc oxide nanoparticles with antimicrobial, anticancer, antioxidant and photocatalytic activities by the endophytic *Alternaria tenuissima*. *Journal of Applied Microbiology*, 2020, 128, 1634–1646. https://doi.org/10.1111/jam.14581.

Ahmad, N.; Sharma, S.; Alam, M.K.; Singh, V.N.; Shamsi, S.F.; Mehta, B.R.; Fatma, A. Rapid synthesis of silver nanoparticles using dried medicinal plant of basil. *Colloids and Surfaces, Part B: Biointerfaces*, 2010, 81(1), 81–86.

Akther, T.; Khan, M.S.; Srinivasan, H.A. Facile and rapid method for green synthesis of silver myco nanoparticles using endophytic fungi. *International Journal of Nano Dimension*, 2018, 9, 435–441.

Baker, S.; Pasha, A.; Satish, S. Biogenic nanoparticles bearing antibacterial activity and their synergistic effect with broad spectrum antibiotics: Emerging strategy to combat drug resistant pathogens. *Saudi Pharmaceutical Journal: SPJ: The Official Publication of the Saudi Pharmaceutical Society*, 2017, 25(1), 44–51.

Baker, S.; Prasad, N.N.M.; Satish, S. Synthesis and characterization of silver nanobactericides produced by Aneurini bacillus migulanus 141, a novel endophyte inhabiting *Mimosa pudica* L. *Arabian Journal of Chemistry*, 2016a, 12(8), 3743–3752.

Baker, S.; Prasad, N.N.; Satisha, S. Endogenic mediated synthesis of gold nanoparticles bearing bactericidal activity. *Journal of Microscopy and Ultrastructure*, 2016b, 4(3), 162. https://doi.org/10.1016/j.jmau.2016.01.004.

Baker, S.; Rao, H.C.Y.; Nagendra-Prasad, M.N.; Prasad, A.; Harini, B.P.; Azmath, P.; Rakshith, D.; Satish, S. Biomimetic synthesis of silver nanoparticles using endosymbiotic bacterium inhabiting *Euphorbia hirta* L. and their bactericidal potential. *Scientifica*, 2016c. https://doi.org/10.1155/2016/9020239.

Baker, S.; Satish, S. Endophytes : Toward a vision in synthesis of nanoparticle for future international journal of bio-inorganic hybrid nanomaterial endophytes: Toward a vision in synthesis of nanoparticle for future therapeutic agents, 2018, 1(2), 67–77. https://doi.org/10.1002/9781118369920.ch1.

Bharathidasan, R.; Panneerselvam, A. Biosynthesis and characteriza- tion of silver nanoparticles using endophytic fungi *Aspergillus concius*, *Penicillium janthinellum* and Phomopsis sp. *International Journal of Pharmaceutical Sciences Review and Research*, 2012, 3(9), 3163–3169.

Bhattacharjee, S.; Debnath, G.; Das, A.R.; Saha, A.K.; Das, P. Characterization of silver nanoparticles synthesized using an endophytic fungus, *Penicillium oxalicum* having potential antimicrobial activity. *Advances in Natural Sciences: Nanoscience and Nanotechnology*, 2017, 8(4). https://doi.org/10.1088/2043-6254/aa84ec.

Birla, S.S.; Tiwari, V.V.; Gade, A.K.; Ingle, A.P.; Yadav, A.P.; Rai, M.K. Fabrication of silver nanoparticles by *Phoma glomerata* and its combined effect against *Escherichia coli*, *Pseudomonas aeruginosa* and *Staphylococcus aureus*. *Letters in Applied Microbiology*, 2009, 27, 76–83.

Bose, P.; Nadu, T., Research article mycosynthesis, optimisation and characterization of silver nanoparticles by endophytic fungus isolated from the root of *Casuarina junghuhniana* Miq. *International Journal of Pharmaceutical Sciences*, 2017, 43, 107–115.

Chandrappa, C.P.; Govindappa, M.; Chandrasekar, N.; Sarkar, S.; Ooha, S.; Channabasava, R. Endophytic synthesis of silver chloride nanoparticles from *Penicillium* sp. of Calophyllum apetalum. *Advances in Natural Sciences: Nanoscience and Nanotechnology* 2016, 7(2). https://doi.org/10.1088/2043-6262/7/2/025016.

Clarance, P.; Luvankar, B.; Sales, J.; Khusro, A.; Agastian, P.; Tack, J.C.; Al Khulaifi, M.M.; AL-Shwaiman, H.A.; Elgorban, A.M.; Syed, A.; Kim, H.J. Green synthesis and characterization of gold nanoparticles using endophytic fungi *Fusarium solani* and its in-vitro anticancer and biomedical applications. *Saudi Journal of Biological Sciences*, 2020, 27(2), 706–712. https://doi.org/10.1016/j.sjbs.2019.12.026.

De Bary, A. *Morphologie und Physiologie der Pilze, Flechten und Myxomyceten Hofmeister's Handbook of Physiological Botany*. Leipzig: Wilhelm Engelmann, 1866. https://doi.org/10.5962/bhl.title.120970.

Devi, L.S.; Joshi, S.R. Evaluation of antimicrobial potency of silver nanoparticles biosynthesized by using endophytic fungus *Cryptosporiopsis ericae* PS4. *Journal of Microbiology*, 2014, 52(8), 667–674.

Devi, L.; Joshi, S. Ultrastructures of silver nanoparticles biosynthesized using endophytic fungi. *Journal of Microscopy and Ultrastructure*, 2015, 3(1), 29. https://doi.org/10.1016/j.jmau.2014.10.004.

Dias, M.A.; Lacerda, I.C.A.; Pimentel, P.F.; Castro, D.; Rosa, H.F. Removal of heavy metals by an *Aspergillus terreus* strain immobilized in a polyurethane matrix. *Letters in Applied Microbiology*, 2002, 34(1), 46–50.

Dong, Z.Y.; Rao, M.P.N.; Xiao, M.; Wang, H.F.; Hozzein, W.N.; Chen, W.; Li, W.J. Antibacterial activity of silver nanoparticles against Staphylococcus warneri synthesized using endophytic bacteria by photo-irradiation. *Frontiers in Microbiology*, 2017, 8, 1090. https://doi.org/10.3389/fmicb.2017.01090.

Ferreira, A.; Quecine, M.C.; Lacava, P.T.; Oda, S.; Azevedo, J.L.; Araújo, W.L. Diversity of endophytic bacteria from Eucalyptus species seeds and colonization of seedlings by *Pantoea agglomerans*. *FEMS Microbiology Letters*, 2008, 287(1), 8–14.

Gahlawat, G.; Choudhury, A.R. A review on the biosynthesis of metal and metal salt nanoparticles by microbes. *RSC Advances*, 2019, 9, 12944–12967. https://doi.org/10.1039/c8ra10483b.

Gajbhiye, M.; Kesharwani, J.; Ingle, A.; Gade, A.; Rai, M. Fungus-mediated synthesis of silver nanoparticles and their activity against pathogenic fungi in combination with fluconazole. *Nanomedicine: Nanotechnology, Biology and Medicine*, 2009, 5(4), 382–386.

Gericke, M.; Pinches, A. Biological synthesis of metal nanoparticles. *Hydrometallurgy*, 2006, 83(1–4), 132–140.

Gouda, S.; Das, G.; Sen, S.K.; Shin, H.S.; Patra, J.K. Endophytes: A treasure house of bioactive compounds of medicinal importance. *Frontiers in Microbiology*, 2016, 7(SEP), 1–8. https://doi.org/10.3389/fmicb.2016.01538.

Halkai, K.R.; Mudda, J.A.; Shivanna, V.; Rathod, V.; Halkai, R.S. Biosynthesis characterization and antibacterial efficacy of silver nanoparticles derived from endophytic fungi against *P. Gingivalis*. *Journal of Clinical and Diagnostic Research: JCDR*, 2017, 11, 92–96.

Hardoim, P.R.; Van Overbeek, L.S., Berg, G.; Pirttilä, M.; Compant, S.; Campisano, A.; Döring, M. The hidden world within plants: evolutionary considerations for defining functioning of microbial endophytes. *Microbiology and Molecular Biology Reviews*, 2015, 79(3), 293–320. https://doi.org/10.1128/MMBR.00050-14.

Hassan, S.E.D.; Fouda, A.; Radwan, A.A.; Salem, S.S.; Barghoth, M.G.; Awad, M.A.; Abdo, A.M.; El-Gamal, M.S. Endophyticactinomycetes Streptomyces spp mediated biosynthesis of copper oxide nanoparticles as a promising tool for biotechnological applications. *Journal of Biological Inorganic Chemistry: JBIC: A Publication of the Society of Biological Inorganic Chemistry*, 2019, 24, 377–393. https://doi.org/10.1007/s00775-019-01654-5.

Hassan, S.E.-D.; Salem, S.S.; Fouda, A.; Awad, M.A.; El-Gamal, M.S.; Abdo, A.M. New approach for antimicrobial activity and bio-control of various pathogens by biosynthesized copper nanoparticles using endophyticactinomycetes. *Journal of Radiation Research and Applied Sciences*, 2018, 11(3), 262–270.

Hemashekhar, B., Chandrappa, C.P.; Govindappa, M.; Chandrasekhar, N. Green synthe sis of silver nanoparticles from endophytic fungus *Aspergillus niger* isolated from *Simarouba glauca* leaf and its antibacterial and antioxidant activity Green synthesis of silver nanoparticles from endophytic fungus *Aspergillus niger* isolated from *Simarouba glauca* leaf and its antibacterial and antioxidant activity. *International Journal of Engineering Research and Application*, 2017, 7(8), 17–24. https://doi.org/10.9790/9622-0708011724.

Hergt, R.; Dutz, S.; Müller, R.; Zeisberger, M. Magnetic particle hyperthermia: Nanoparticle magnetism and materials development for cancer therapy. *Journal of Physics: Condensed Matter*, 2006, 18(38), 2019.

Hulikere, M.M.; Joshi, C.G. WITHDRAWN: Characterization, antioxidant and antimicrobial activity of silver nanoparticles synthesized using marine endophytic fungus – *Cladosporium cladosporioides. Biochemistry and Biophysics Reports*, 2017. https://doi.org/10.1016/j.bbrep.2017.08.011.

Hulkoti, N.I.; Taranath, T.C. Colloids and surfaces B : Biointerfaces biosynthesis of nanoparticles using microbes – A review. *Colloids and Surfaces, Part B: Biointerfaces*, 2014, 121, 474–483.

Ibrahim, E.; Fouad, H.; Zhang, M.; Zhang, Y.; Qiu, W.; Yan, C.; Li, B.; Mo, J.; Chen, J. Biosynthesis of silver nanoparticles using endophytic bacteria and their role in inhibition of rice pathogenic bacteria and plant growth promotion. *RSC Advances*, 2019, 9(50), 29293–29299. https://doi.org/10.1039/c9ra04246f.

Ibrahim, E.; Zhang, M.; Zhang, Y.; Hossain, A.; Qiu, W.; Chen, Y.; Wang, Y.; Wu, W.; Sun, G.; Li, B. Green-synthesization of silver nanoparticles using endophytic bacteria isolated from garlic and its antifungal activity against wheat fusarium head blight pathogen fusariumgraminearum. *Nanomaterials*, 2020, 10(2). https://doi.org/10.3390/nano10020219.

Ivanova, E.G.; Fedorov, D.N.; Doronina, N.V.; Trotsenko, Y.A. Production of vitamin B12 in aerobic methylotrophic bacteria. *Microbiologiya*, 2006, 75, 494–496.

Kadam, V.V.; Ettiyappan, J.P.; Balakrishnan, M.R. Mechanistic insight into the endophytic fungus mediated synthesis of protein capped ZnO nanoparticles. *Materials Science and Engineering B: Solid-State Materials for Advanced Technology*, 2019, 243, 214–221. https://doi.org/10.1016/j.mseb.2019.04.017.

Kaur, A.; Kumar, R. Enhanced bactericidal effcacy of polymer stabilized silver nanoparticles in conjugation with different classes of antibiotics, 2019, 9(2), 1095–1105. https://doi.org/10.1039/c8ra07980c.

Li, X.; Xu, H.; Chen, Z.S.; Chen, G. Biosynthesis of nanoparticles by microorganisms and their applications. *Journal of Nanomaterials*, 2011. https://doi.org/10.1155/2011/270974.

Lv, Q.; Zhang, B.; Xing, X.; Zhao, Y.; Cai, R.; Wang, W.; Gu, Q. Biosynthesis of copper nanoparticles using *Shewanella loihica* PV-4 with antibacterial activity: Novel approach and mechanisms investigation. *Journal of Hazardous Materials*, 2018, 347, 141–149.

Malik, T.; Chauhan, G.; Rath, G.; Murthy, R.; Goya, A. Fusion and binding inhibition key target for HIV-1 treatment and pre-exposure prophylaxis: Targets, drug delivery and nanotechnology approaches. *Drug Delivery*, 2017, 24(1), 608–621.

Masri, A.; Anwar, A.; Ahmed, D.; Siddiqui, R.B.; Shah, M.R.; Khan, N.A. Silver nanoparticle conjugation-enhanced antibacterial efficacy of clinically approved drugs cephradine and vildagliptin. *Antibiotics*, 2018, 7, 100. https://doi.org/10.3390/antibiotics7040100.

Mehra, R.K.; Winge, D.R. Metal ion resistance in fungi: Molecular mechanisms and their regulated expression. *Journal of Cellular Biochemistry*, 1991, 45(1), 30–40.

Mishra, A.; Sardar, M. Alpha-amylase mediated synthesis of silver nanoparticles. *Science of Advanced Materials*, 2012, 4(1), 143–146. https://doi.org/10.1166/sam.2012.1263.

Mishra, S.; Singh, B.R.; Naqvi, A.H.; Singh, H.B. Potential of biosynthesized silver nanoparticles using Stenotrophomonas sp. BHU-S7 (MTCC 5978) for management of soil-borne and foliar phytopathogens. *Scientific Reports*, 2017, 7, 1–15. https://doi.org/10.1038/srep45154.

Monowar, T.; Rahman, S.; Bhore, S.J.; Raju, G.; Sathasivam, K.V. Silver nanoparticles synthesized by using the endophytic bacterium *Pantoea ananatis* are promising antimicrobial agents against multidrug resistant bacteria. *Molecules*, 2018, 23(12), 1–17.

Nangia, Y.; Wangoo, N.; Goyal, N.; Shekhawat, G.; Suri, C.R. A novel bacterial isolate *Stenotrophomonas maltophilia* as living factory for synthesis of gold nanoparticles. *Microbial Cell Factories*, 2009, 8(1), 1–7.

Neethu, S.; Midhun, S.J.; Sunil, M.A.; Soumya, S.; Radhakrishnan, E.K.; Jyothis, M. Efficient visible light induced synthesis of silver nanoparticles by *Penicillium polonicum* ARA 10 isolated from *Chetomorpha antennina* and its antibacterial efficacy against *Salmonella enterica* serovar Typhimurium. *Journal of Photochemistry and Photobiology, Part B: Biology*, 2018, 180, 175–185. https://doi.org/10.1016/j.jphotobiol.2018.02.005.

Netala, V.R.; Kotakadi, V.S.; Bobbu, P.; Gaddam, S.A.; Tartte, V. Endophytic fungal isolate mediated biosynthesis of silver nanoparticles and their free radical scavenging activity and antimicrobial studies. *3 Biotech*, 2016, 6(2), 1–9. https://doi.org/10.1007/s13205-016-0433-7.

Petrini, O. Fungal endophytes of tree leaves, In: Andrews J.H., Hirano S.S. (eds) *Microbial Ecology of Leaves*. Brock/Springer Series in Contemporary Bioscience. Springer, New York, NY. https://doi.org/10.1007/978-1-4612-3168-4_9, 1991, 179–180.

Prasad, R.; Kumar, V.; Kumar, M.; Wang, S. Fungal nanobionics: Principles and applications. 2018, 1–316. https://doi.org/10.1007/978-981-10-8666-3.

Qi, C.; Musetti, S.; Fu, L.; Zhu, Y.; Huang, L. Biomolecule-assisted green synthesis of nanostructured calcium phosphates and their biomedical applications. *Chemical Society Reviews*, 2019, 48(10), 2698.

Rahi, D.K.; Parmar, A.S. Mycosynthesis of silver nanoparticles by an endophytic Penicillium species of Aloe vera root, evaluation of their antibacterial and antibiotic enhancing activity. *International Journal of Nanomaterials and Biostructures*, 2014, 4, 46–51.

Rahi, D.K.; Parmar, A.S.; Tiwari, V. Biosynthesis of silver nanoparticles from fungal root endophytes of Sidaacuta plant and evaluation of their antibacterial and antibiotic enhancing activity. *International Journal of Pharmacy and Pharmaceutical Sciences*, 2014, 6, 160–166.

Raheman, F.; Deshmukh, S.; Ingle, A.; Gade, A.; Rai, M. Silver nanoparticles: Novel antimicrobial agent synthesized from an endophytic fungus *Pestalotia* sp. isolated from leaves of Syzygium cumini (L). *Nano Biomedicine and Engineering* 2011, 3(3), 174–178. https://doi.org/10.5101/nbe.v3i3.p174-178.

Rahman, S.; Rahman, L.; Khalil, A.T.; Ali, N.; Zia, D.; Ali, M.; Shinwari, Z.K. Endophyte-mediated synthesis of silver nanoparticles and their biological applications. *Applied Microbiology and Biotechnology*, 2019, 103(6), 2551–2569.

Rajabairavi, N.; Raju, C.S.; Karthikeyan, C.; Varutharaju, K.; Nethaji, S.; Hameed, A.S.H.; Shajahan, A. Biosynthesis of novel zinc oxide nanoparticles (ZnO NPs) using endophytic bacteria *Sphingobacterium thalpophilum*. *Springer Proceedings in Physics*, 2017, 189, 245–254. https://doi.org/10.1007/978-3-319-44890-9_23.

Ramalingmam, P.; Muthukrishnan, S.; Thangaraj, P. Biosynthesis of silver nanoparticles using an endophytic fungus, *Curvularia lunata* and its antimicrobial potential. *Journal of Nanoscience and Nanoengineering*, 2015, 1, 241–247. https://doi.org/10.1002/anie.200460990.

Rani, R.; Sharma, D.; Chaturvedi, M.; Yadav, J.P. Green synthesis, characterization and antibacterial activity of silver nanoparticles of endophytic fungi *Aspergillus terreus*. *Journal of Nanomedicine and Nanotechnology*, 2017, 08(4). https://doi.org/10.4172/2157-7439.1000457.

Razavi, S.; Lin, B.; Lee, K.Y.; Tu, R. S.; Kretzschmar, I. Impact of surface amphiphilicity on the interfacial behavior of Janus particle layers under compression. *Langmuir: The ACS Journal of Surfaces and Colloids*, 2019, 35(48), 15813–15824. https://doi.org/10.1021/acs.langmuir.9b01664.

Reverberi, A.P.; Vocciante, M.; Lunghi, E.; Pietrelli, L.; Fabiano, B. New trends in the synthesis of nanoparticles by green methods. *Chemical Engineering Transactions*, 2017, 61, 667–672. https://doi.org/10.3303/CET1761109.

Sanghi, R.; Verma, P. Biomimetic synthesis and characterization of protein capped silver nanoparticles. *Bioresource Technology*, 2009, 100(1), 501–504.

Shankar, S.S.; Ahmad, A.; Pasricha, R.; Sastry, M. Bioreduction of chloroaurate ions by geranium leaves and its endophytic fungus yields gold nanoparticles of different shapes. *Journal of Materials Chemistry*, 2003, 13(7), 1822–1826. https://doi.org/10.1039/b303808b.

Sharma, V.K.; Yngard, R.A.; Lin, Y. Silver nanoparticles: Green synthesis and their antimicrobial activities. *Advances in Colloid and Interface Science*, 2009, 145(1–2), 83–96.

Shukla, H.; Sandhu, S.S. Mycofabrication and characterization of silver nanoparticles by using some endophytic fungi with special reference to their antimicrobial potential. *International Journal of Nanotechnology and Application*, 2017, 7, 7–22.

Singh, T.; Jyoti, K.; Patnaik, A.; Singh, A.; Chauhan, R.; Chandel, S. Biosynthesis, characterization and antibacterial activity of silver nanoparticles using an endophytic fungal supernatant of *Raphanus sativus*. *Journal of Genetic Engineering and Biotechnology*, 2017, 15(1), 31–39. https://doi.org/10.1016/j.jgeb.2017.04.005.

Singh, D.; Rathod, V.; Ninganagouda, S.; Herimath, J.; Kulkarni, P. Biosynthesis of silver nanoparticle by endophytic fungi *Pencillium* sp. isolated from *Curcuma longa* (turmeric) and its antibacterial activity against pathogenic gram negative bacteria. *Journal of Pharmacy Research* 2013, 7(5), 448–453. https://doi.org/10.1016/j.jopr.2013.06.003.

Slavin, Y.N.; Asnis, J.; Häfeli, U.O.; Bach, H. Metal nanoparticles : Understanding the mechanisms behind antibacterial activity. *Journal of Nanobiotechnology*, 2017, 15, 1–20. https://doi.org/10.1186/s12951-017-0308-z.

Sunkar, S.; Nachiyar, C.V. Microbial synthesis and characterization of silver nanoparticles by using endophytic bacterium *Bacillus cereus*: A novel source in benign synthesis. *Global Journal of Medical Research*, 2012, 2(12), 953–959.

Sunkar, S.; Nachiyar, C.V.; Lerensha, R.; Renugadevi, K. Biogenesis of TiO2 nanoparticles using endophytic *Bacillus cereus*. *Journal of Nanoparticle Research*, 2014, 16, 2681–2691.

Tan, Y.N.; Lee, J.Y.; Wang, D.I.C. Uncovering the design rules for peptide synthesis of metal nanoparticles. *Journal of the American Chemical Society*, 2010, 132(16), 5677–5686.

Thomas, R.; Jasim, B.; Mathew, J.; Radhakrishnan, E.K. Extracellular synthesis of silver nanoparticles by endophytic Bordetella sp. isolated from *Piper nigrum* and its antibacterial activity analysis. *Nano Biomedicine and Engineering*, 2012, 4(4), 183–187. https://doi.org/10.5101/nbe.v4i14.p183-187.

Uddandarao, P.; Balakrishnan, R.M.; Ashok, A.; Swarup, S.; Sinha, P. Bioinspired ZnS : Gd nanoparticles synthesized from an endophytic fungi *Aspergillus flavus* for fluorescence-based metal detection. *Biomimetics*, 2019, 4(1), 11. https://doi.org/10.3390/biomimetics4010011.

Vaidyanathan, R.; Gopalram, S.; Kalishwaralal, K.; Deepak, V.; Pandian, S.R.K.; Gurunathan, S. Enhanced silver nanoparticle synthesis by optimization of nitrate reductase activity. *Colloids Surfaces B Biointerfaces*, 2010, 75(1), 335–341. https://doi.org/10.1016/j.colsurfb.2009.09.006.

Vardhana, J.; Kathiravan, G. Biosynthesis of silver nanoparticles by endophytic fungi *Pestaloptiopsis pauciseta* isolated from the leaves of *Psidium guajava* Linn. *International Journal of Pharmaceutical Sciences Review and Research*, 2015, 31, 29–31.

Verma, S.K.; Gond, S.; Mishra, A. Biofabrication of antibacterial and antioxidant silver nanoparticles (AgNPs) by an endophytic fungus Pestalotia sp. isolated from *Madhuca longifolia*. *Journal of Nanomaterials and Molecular Nanotechnology*, 2016, 5(4). https://doi.org/10.4172/2324-8777.1000189.

Verma, V.C.; Kharwar, R.N.; Gange, A.C. Biosynthesis of antimicrobial silver nanoparticles by the endophytic fungus aspergillus clavatus. *Nanomedicine* 2010, 5(1), 33–40. https://doi.org/10.2217/nnm.09.77.

Verma, V.C.; Singh, S.K.; Solanki, R.; Prakash, S. Biofabrication of anisotropic gold nanotriangles using extract of endophytic aspergillus clavatus as a dual functional reductant and stabilizer. *Nanoscale Research Letters* 2011, 6(1), 1–7. https://doi.org/10.1007/s11671-010-9743-6.

Verma, V.C.; Anand, S.; Ulrichs, C.; Singh, S.K. Biogenic gold nanotriangles from Saccharomonospora sp., an endophyticactinomycetes of *Azadirachta indica* A. Juss. *International Nano Letters*, 2013, 3, 12.

Vijayanandan, A.S.; Balakrishnan, R.M. Biosynthesis of cobalt oxide nanoparticles using endophytic fungus *Aspergillus nidulans*. *Journal of Environmental Management*, 2018, 218, 442–450. https://doi.org/10.1016/j.jenvman.2018.04.032.

Wang, A.L.; Yin, H.B.; Ren, M.; Cheng, X.N.; Zhod, Q.F.; Zhang, X.F. Effects of different functional group-containing organics on morphology-controlled synthesis of silver nanoparticles at room temperature. *Acta Metallurgica Sinica (English Letters)*, 2006, 19(5), 362–370.

Wang, L.; Hu, C.; Shao, L. The antimicrobial activity of nanoparticles: Present situation and prospects for the future. *International Journal of Nanomedicine*, 2017, 12, 1227.

Xu, X.; Cheng, W.; Liu, X.; You, H.; Wu, G.; Ding, K.; Tu, X.; Yang, L.; Wang, Y.; Li, Y.; Gu, H.; Wang, X. Selenate reduction and selenium enrichment of tea by the endophytic Herbaspirillum sp. Strain WT00C. *Current Microbiology*, 2020, 77(4), 588–601.

Zhang, Q.; Li, N.; Goebl, J.; Lu, Z.; Yin, Y. A systematic study of the synthesis of silver nanoplates: Is citrate a 'magic' reagent? *Journal of the American Chemical Society*, 2011, 133(46), 18931–18939.

Zhou, D.; Qi, C.; Chen, Y.X.; Zhu, Y.J.; Sun, T.W.; Chen, F..; Zhang, C.Q. Comparative study of porous hydroxyapatite/chitosan and whitlockite/chitosan scaffolds for bone regeneration in calvarial defects. *International Journal of Nanomedicine*, 2017, 12 (12), 2673–2687.

3 Nano-Biotechnological Products and Their Beneficial Linkages towards Sustainable Wheat Production

Baby Kumari, Manoj Kumar Solanki, Suresh Kumar Verma, and M. A. Mallick

CONTENTS

3.1 INTRODUCTION

The rapidly growing world population was expected to grow to 9.1 billion by the year 2020 (Jasrotia et al., 2018). A changing climate and overpopulation have led to a crisis in nutrient availability and food security for humans, especially in developing countries (Gouda et al., 2018; Thakur, Thakur, and Kumar, 2018; Prasad, Bhattacharyya, and Nguyen, 2017). In this scenario, the increased demand for high yield of crop production with fewer adverse effects on the soil system is a big challenge for the modern agricultural system. Most of the developed and developing countries have intensified their agricultural methods based on new sustainable and eco-friendly technologies (Usman et al., 2020).

Wheat (*Triticumaestivum* L.) is the most popular and widely cultivated crop in agricultural practice. Wheat production is a major source of economy in various countries, such as the USA, Russia, India, and Australia (Abdelsalam et al., 2019; Ubeyitogullari et al., 2019). Wheat is considered a staple food (a food which is regularly/routinely consumed by a large population to meet its nutrient and energy demands) (Kashyap et al., 2020). Wheat products are especially widely used in food processing industries. Biochemical analysis of wheat has proven it to be one of the most important nutritional choices among all cereals (Jan, Pirzadah, and Malik, 2020; Kashyap et al., 2020). It is composed of well-balanced nutrients like-starch (67%), protein (13%), β-glucan (0.8%), arabinoxylan (8%), and 2% cellulose. Precision agriculture is a farming management concept which involves measuring and responding to inter- and intra-field variations in crops to form a decision support system for whole-farm management and to reap the maximum output from the available resources (Servin and White, 2016; Duhan et al., 2017).

Currently, the major issues include agricultural scarcity and the nutritional process of getting quality food. Thus, innovative technology has to be designed that particularly emphasizes the improvement of wheat production. Nowadays, most agricultural researchers want to utilize different novel methods to manage nutritional security. Modern agricultural practices need sufficient improvisation to reduce human and environmental health risks caused by the overuse of agrochemicals, such as fertilizers and pesticides. A nanotechnology-based strategy could be an alternative solution to develop a high tech and smart agricultural system that will help to reduce the impact of modern agricultural practices on the environment.

3.2 MAJOR PROBLEMS OF WHEAT CULTIVATION

The major wheat-producing countries (USA, France, Russia, China, and India) suffer production losses due to various abiotic stresses, mainly ozone depletion and global warming (Mills et al., 2018). Nowadays, studies are focused on the assessment of climatic changes in wheat production per hectare of agricultural land (Iavicoli et al., 2017). Recently it has been reported that elevated CO_2 levels have a high impact on the yield losses of wheat crops (Wang et al., 2018; Mills et al., 2018). So, despite increased overall yield, the world is facing a major crisis with this valuable cereal crop (Foreign Agricultural Service, USDA, 2018). According to a recent article, the worldwide production of wheat is 855 million tons (765 million metric tons) (Zou et al., 2019), which is the third-largest cereal produced worldwide. As per a recent Scopus data search, the European Union (137,600 metric tonnes/year) is the leading state among the major wheat-producing countries, like China, India Russia, United States, Canada, Pakistan Ukraine, Argentina, and Turkey (Figure 3.1). Despite the sizeable production of wheat globally, wheat agriculture is falling short of fulfilling the demand: maybe this is due to its dependency on various conditions like biotic and abiotic factors.

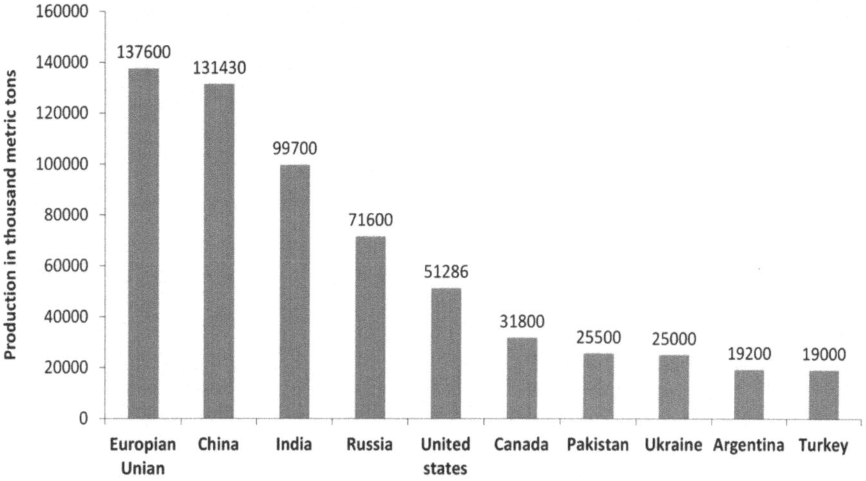

FIGURE 3.1 Leading wheat-producing countries worldwide (https://www.statista.com).

Because of current climatic conditions, wheat agriculture is exposed to high temperature, heat stress, salt, and drought stress.

3.3 A BRIEF HISTORY OF NANOTECHNOLOGY APPLICATION IN WHEAT

Nanotechnology occupies an important place in transforming agriculture and food production (Liu and Lal, 2015; Mukhopadhyay, 2014; Thakur, Thakur, and Kumar, 2018; Prasad, Bhattacharyya, and Nguyen, 2017). The development of nanofertilizers, nano-pesticides, and nano-sensors could open up novel applications in plant biotechnology and agriculture (Thakur et al., 2018; Timmusk et al., 2018). Nanotechnology promises enhancements in food production/capita. Scientific innovations in agricultural practices have led to revolutionary devolvement in the production of major crops, especially wheat. Nanotechnology is a new technology to design and develop nano-bioformulations, such as nanoparticles (NPs), nano-gel, or nano-colloids, for applications in different areas of science and agriculture (Baby Kumari, Hora, and Mallick, 2017a; Thakur, Thakur, and Kumar, 2018). Nanotechnology-based applications form the most innovative aspect of the recent progress in farming practices (Gilbertson et al., 2020).

According to a review of the available literature, Richard Feyman (USA) is credited with the first use of the term *nanotechnology* in the late 1950s (Toumey, 2009). The word 'nanotechnology' was coined by N. Feynman (Japan) in 1974 (Fanfair and Kelty, 2005). The emergence of nano-science took place in late 1980s (China), though the application of this revolutionary science started in the 1990s. The 1990s–2010s witnessed the foundation of many national and international programmes in Japan, China, and the USA (e.g. the NNI-National Nano-technological Initiative, USA).

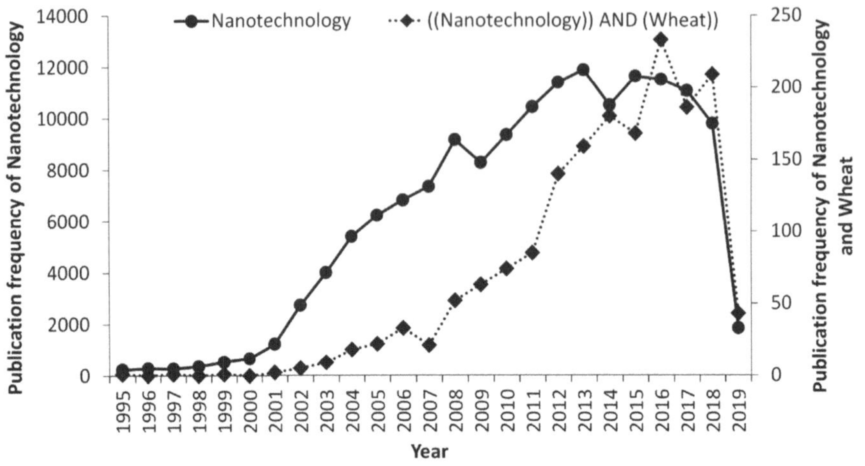

FIGURE 3.2 Showing the current status of 'Nanotechnology' research (year wise) in wheat agriculture, of publication frequency searching by (Nanotechnology and wheat), (source data – Scopus, access date 6 October 2020).

The fabrication of different nanoscale materials like NPs, nanotubes, nanofibres, and some early nanodevices had begun in the year 2000 onwards (Figure 3.2).

Nanotechnology has been well applied in the field of wheat agriculture to justify methods of enhancing global wheat productions. Nano-biosensors are flourishing and have many profitable effects on agricultural productivity. Recently a variety of nano-sensors (synthesized through physical, chemical, and biosynthetic pathways) were implemented in wheat agriculture practices to attain effective plant production, plant protection, and plant disease management. These nano-sensors were also implemented in toxicity assessment studies. Precision farming, a potential technique that involves the use of nano-sensors to enhance the quality of wheat grains (by monitoring the growth yield, disease management, abiotic stress tolerance, and food packaging), is the leading area of wheat agriculture research (Majumdar and Keller, 2020; Fu et al., 2020; Svendsen et al., 2020).

Given future climatic changes, elevated carbon dioxide level, soil quality, and other abiotic stress in wheat cultivation, nanotechnology is considered as one of the most promising sciences to deal with it. The potential uses and benefits of nanotechnology are enormous (Prasad, Bhattacharyya, and Nguyen, 2017). These include the improvement of wheat growth via formulations of nanomaterial-based fertilizers and a manifold increase in agricultural productivity using NP-encapsulated fertilizers for a slow and sustained release of nutrients and water (Wani and Kothari, 2018; Gouda et al., 2018; Thakur, Thakur, and Kumar, 2018; Mukhopadhyay, 2014).

This review will contain detailed information that illustrates the recent progress in the applied aspects of nanobiotechnology for eco-friendly agriculture through sustainable approaches in wheat crop technology (Table 3.1). Nanotechnological applications for the detection and treatment of plant diseases, the enhancement of

TABLE 3.1
Nano-Nutrients in Wheat Crop Production

Nanoparticles (NPs)	Country	Mechanism of action	Significance	Reference
Multiwall Carbon nanotubes,	India, Spain, China	Increases the intake of Fe and Ca, water retention, enhance seed germination, induces shoot and root growth	Increased growth and yield of wheat, seedling vigour enhanced	(Jasrotia et al., 2018) (Chang et al., 2020)
magnetic carbon-coated nanofluids	Spain	Absorption through the xylem and translocated in aerial parts	Increased growth and yield of wheat	(Cifuentes et al., 2010)
Zinc and Iron particles (ZnO, Fe)	Pakistan	Absorption through the seed treatment	Induced plant growth by increasing Chlorophyll content and reduced the oxidative stress	(Rizwan et al., 2019)
ZnO particles prepared by green method (using plant leaf), ZnO + Zn salt	India	Seed samples were treated	Induced improved seedling germination, uptake of Zn, and growth	(Dimkpa et al., 2018; J. Singh et al., 2019; Rizwan et al., 2019)
Nickel oxide NP (NiO NPs)	Egypt	Seed samples were treated	Induces increased photosynthesis and reduced the level of oxidative stress in future climate CO_2 level	(Saleh et al., 2019)
Hyper feed Amino NPs	Egypt	Foliar application	Showed increased growth and yield of wheat along with some genotoxicity	(Abdelsalam et al., 2019)
Supercritical CO_2 Nano aerosols	USA	Applied on nonporous starch aerosols of wheat	Enhanced the bioaccessibility of phytosterol	(Ubeyitogullari et al., 2019)
Metallic Silver(Ag)NPs, sulfidized (Ag_2S) silver NPs	China, Pakistan, France	Applied on seedling in liquid form, Ag thiols	Induced improved seedling germination and growth under heat stress, phytotoxicity were evaluated	(Iqbal et al., 2017; Saira, 2018; Yang et al., 2018; Pradas del Real et al., 2017)

(Continued)

TABLE 3.1 (CONTINUED)
Nano-Nutrients in Wheat Crop Production

Nanoparticles (NPs)	Country	Mechanism of action	Significance	Reference
Cerium oxide (CeO_2) NPs	USA	Seeds were exposed to different conc. of CeO_2	Improved the nutrient uptake profile and physiology of plants in the second generation	(Rico et al., 2017)
Cerium oxide (CeO_2) NPs	USA	Seeds were exposed to low, moderate, and high conc. of cerium oxide	There was no change in biomass yield overall, but vacuole formation, chlorophyll content, and structure of nuclei were greatly influenced	(Du et al., 2015)
Selenium (Se NPs)	China	Seedlings were treated with Se NPs in a hydroponic system in the presence of an aquaporin inhibitor	Mechanism of Se absorption and biotransformation were evaluated and also showed increased root-shoot growth	(Hu et al., 2018)
Magnesium oxide (MgO) NPs	China	Applied through the leaf surface in a different concentration	Evidence of increased root-shoot growth, biomass, and overall yield	(Cai et al., 2018)
ZnO - CuO NPs	USA	Exposed at different concentrations in wheat seeds	Induced improved seedling germination and growth under metal stress, improved the overall yield	(Anderson et al., 2018)
FeO NPs	Germany, Pakistan, China	Exposed at different concentrations in wheat seeds, under Cd contamination in soil	Induced improved seedling germination and growth under cadmium metal stress	(Rizwan et al., 2019)
Titanium (Ti NPs)	China, USA	Applied in liquids and solid form in root, leaves, and on seedlings	Increased growth, quality, and yield through enhancing chlorophyll content, photosynthesis, nutrient uptake, tolerance	(Lyu et al., 2017)
Cu-Zn NPs	Ukraine	Seed treated with the colloidal solution of NPs	Increased growth and diseases resistance in drought conditions	(Taran et al., 2017)

nutrients absorption by plants, the delivery of active ingredients to specific sites, and water treatment processes are elaborated in this review.

3.4 NANO-BIOPRODUCTS

In the present decade 2010-onwards, wheat production is greatly increased through the successful application of nano-nutrients. A variety of nanostructured materials like NPs, nanofluids, nanogels, nanotubes, nanofibres, and many more were synthesized for applications in wheat crops (Bhattacharyya et al., 2010; Roco, 2003; Rathinamoorthy and Senthilkumar, 2009; Min et al., 2009; Mishra and Kumar, 2009; Vineet Kumar and Yadav, 2009; Alejandro and Rubiales, 2009; Nair et al., 2010). Nano-bioformulation has emerged as a potential tool to clean up the metal-contaminated/polluted environment. Reducing the bioavailability of metal contaminants in the rhizosphere (phytostabilization) as well as improving plant establishment, growth, and health could significantly speed up plant growth and indeed productivity (Tables 3.1 and 3.2) (Kashyap et al., 2013; Vejan et al., 2016; Duhan et al., 2017; Wani and Kothari, 2018; Prasad, Bhattacharyya, and Nguyen, 2017; Iavicoli et al., 2017).

3.4.1 Nanoparticle Fabrication

The NPs and other nanomaterials which are employed in wheat agriculture are synthesized using three methodologies: physical processes, chemical processes, and biological synthesis. Above all green technology is gaining popular due to its eco-friendly and cost-effective methods (Iravani, 2014; Yadav et al., 2013; Linsinger, Roebben, and Gilliland, 2012; Nair et al., 2010; Rathinamoorthy and Senthilkumar, 2009).

The synthesis of NPs employs various methods depending on the size of the NPs, the source (Inorganic and organic), and requirements (Linsinger, Roebben, and Gilliland, 2012). The method involves physical, chemical, and biogenesis processes. The nanotechnology-based synthesis of biomolecules is widely applicable in almost all areas of the agricultural system and in the environmental cleanup programmes (Mocan et al., 2017; Oliveira et al., 2018; Duhan et al., 2017; Prasad, Bhattacharyya, and Nguyen, 2017). The replacement of biogenetic NPs with inorganic NPs is advantageous as they do not require complex hazardous chemicals, sol-gels, or laser applications. These techniques lead to toxicity in the environment (Siddiqi, Husen, and Rao, 2018; Bansal, Duhan, and Gahlawat, 2014; Iavicoli et al., 2017) and they are also very costly (Bansal, Duhan, and Gahlawat, 2014). Physical conditions like pH, temperature, and substrate concentration play a deciding role in the orientation, size, and shape of the synthesized NPs (Oliveira et al., 2018). Despite the beneficial novel properties, there is also the possibility of the emergence of adverse novel effects. Some NPs have been reported to have toxicity at higher concentrations, and a few reports are available on the presence of these NPs in the food chain (Singh et al., 2014; Gui et al., 2017; Servin and White, 2016). So, because of these human and plant health concerns, these NPs require proper authentication and validation approval. Over the past few years, the use of NPs in wheat agriculture has increased. Among all the nonmaterials, silver, iron, and Zn are the most incorporated elements in their

TABLE 3.2
Table Showing the Uses of Nonmaterial to Improve Wheat Production

Products Used as Nano-Nutrients in Wheat Agriculture

Nanoproducts	Example	Source + Method of fabrication	Mechanism of action	Reference
Nanotubes (Carbon)	Graphene, GO, graphene ribbon	Graphite, Chemical vapour deposition	Enhancement of nutrients absorption by plants and the delivery of nutrients to specific sites,	(Zaytseva and Neumann, 2016)
Chitosan				
NPs (Physical)	Ag, Au, Zn, Cu, Fe, Ti, Se, Si		Improved delivery of agrochemicals in the field, better efficacy, and better control of application/dose, Growth enhancement	(Ogawa, 2011; Agrawal and Rathore, 2014; Parisi, Vigani, and Rodríguez-Cerezo, 2015; Asadi et al., 2016; Iqbal et al., 2017; Dimkpa et al., 2018; Hu et al., 2018; Ubeyitogullari et al., 2019)
NPs (Chemical)	Ag, Au, Zn, Cu, Fe, Ti, Se, Si	Metallic weathering (a top-down and bottom-up process)		(Jasrotia et al., 2018; J. Singh et al., 2019; Cifuentes et al., 2010; Rizwan et al., 2019; Rico et al., 2017)
NPs (Biological)				
Plant origin	Ag, Au, Zn, Cu, Fe,	*Artocarpusgomezianus* fruit, *Citrus medica* fruit, orange and pineapple fruits, *Lawsoniainermis* Leaves *Gardenia jasminoides*		(Huang et al., 2015; Ubeyitogullari et al., 2019; Hu et al., 2018)
Microbial origin NPs	Ag, Au, Zn, Cu, Fe Si	Products used for the breeding programme in wheat agriculture		(Barazani and Friedman, 1999; Rathinamoorthy and Senthilkumar, 2009; Bhattacharyya et al., 2010; Liu and Lal, 2015)

(Continued)

TABLE 3.2 (CONTINUED)
Table Showing the Uses of Nonmaterial to Improve Wheat Production

Products Used as Nano Pesticide in Wheat Agriculture

Nanoproducts	Example		Mechanism of action	Reference
NPs	Ag (10–20nm)	Biofabrication	Controls the infection of wheat caused by *B. sorokiniana* (spot blotch disease)	(Jasrotia et al., 2018)
	Zn, Fe, Mn, Strontium phosphate, Starch, Calcium, Mn, Silica, chitosan Gold, etc.	Chemical methods	Controls *Fusarium* infection in wheat	(Seyed Sharifi, Khalilzadeh, and Soltanmoradi, 2017)
Microbial NPs	Ag, Au, Zn, Cu, Fe Si	Biogenesis method, source: *Bacillus megaterium* NCIM 2326, *Geobacillus stearothermophilus*, *Stenotrophomonas strain* GSG-2, *Pseudomonas aeruginosa*, Bacillus sp. GP-23	Controls the infection of wheat caused by many fungi, like *Fusarium*	(Siddiqi, Husen, and Rao, 2018; Iravani, 2014; Min et al., 2009)
Plant origin NPs	Ag, Au, Zn, Cu, Fe,	*Artocarpusgomezianus* fruit, *Citrus medica* fruit, orange and pineapple fruits, *Lawsoniainermis* leaves, *Gardenia jasminoides*	It controls the pests during packaging of wheat	(D. Kumar and Kalita, 2017; Zaytseva and Neumann, 2016; Yang et al., 2018; White, Frisbie, and Dorfman, 2018)

(Continued)

TABLE 3.2 (CONTINUED)

Table Showing the Uses of Nonmaterial to Improve Wheat Production

Products Used for the Breeding Programme in Wheat Agriculture

Nanoproducts	Example	Source/Method of fabrication	Mechanism of action	Reference
Nano-formulations, NPs	Zn, Fe, Mn, Strontium phosphate, Starch, Calcium, Mn, Silica Gold, etc.	Metals and their phosphates, chemical methods	Improves the gene delivery system, creating new varieties in less time as in conventional methods of breeding	(Sokolova and Epple, 2008)
Nanofibre	Carbon nanofibre		Micro-injection with carbon nanofibres (CNFs) containing foreign DNA has been used to genetically modify golden rice enriched with extra vitamin A	(Abd-Elsalam and Alghuthaymi, 2014)

Products used as Nano-sensors in wheat agriculture

Nanoproducts	Example	Source Method of fabrication	Mechanism of action	Reference
Nanoparticles	Nanogold	Gold (physical)	Surface Plasmo-resonance based detection of the pathogen in wheat	(Sadhna Singh et al., 2010)
EN (electronic nose)	Metal oxide-based	Metal Oxide Semiconductor	Metal oxide Semiconductor sensor detects the presence deoxynivalenol in durum wheat	(Campagnoli et al., 2011)

(Continued)

TABLE 3.2 (CONTINUED)
Table Showing the Uses of Nonmaterial to Improve Wheat Production

Nanoproducts	Example	Source Method of fabrication	Mechanism of action	Reference
Products used as Nano-sensors in wheat agriculture				
E-nose	Metal-based	Metalloporphyrin-based	Detects the *Fusarium* infections in wheat grain	(Eifler et al., 2011)
Nose	Alpha MOX, FOX 300	Metal-based nano-sensors	Used for detection of Red Floor Beetle (RFB)	(Lin and Guo, 2016)
Nano-sensors (commercial product)	Liposome-based	A liposome-based nano-biosensor	Pesticide detection	(Justino, Duarte, and Rocha-Santos, 2017; Bucur et al., 2018; Liu and Lal, 2015; Gui et al., 2017; Mocan et al., 2017)
Nanodevices nanowire, nano-formulated particle, nano-encapsulated beds	Nanoparticles-Ag, Au Cu, Zn, etc. in the microorganism	Bacteria, virus, and Fungus	characterization of diseases in plants and plays an important role in cleanup strategies	(Galati et al., 2018; Hora et al., n.d.; Jafarizadeh-Malmiri et al., 2019; Kumar and Kalita, 2017; Kumari et al., 2016; Parisi et al., 2015; White et al., 2018)

oxide form (Mufamadi and Sekhejane, 2017; Iqbal et al., 2017; Saleh et al., 2019; Rizwan et al., 2019; Saira, 2018; Anderson et al., 2018; Rico et al., 2017). These NPs in different forms have resulted in an increased growth in wheat production, quality, and yield through enhancing chlorophyll content, photosynthesis, nutrient uptake, tolerance and germination and growth under different metal stresses (Galati et al., 2018; White, Frisbie, and Dorfman, 2018; Hussain et al., 2019a; Mills et al., 2018; Saira, 2018; Rizwan et al., 2019; Abdelsalam et al., 2019).

3.4.2 Nanofertilizer

Nanotechnology has led to enormous benefits to the field of sustainable agriculture. This includes the improvement of plant growth via formulations of nanomaterial-based fertilizers and a manifold increase in agricultural productivity using nanoparticle encapsulated fertilizers for the slow and sustained release of nutrients and water (Tilman et al., 2002; Ma et al., 2011; Alejandro and Rubiales, 2009; Huang et al., 2015). Currently, nano-scaled biofertilizers have gained popularity in the field of wheat agriculture. Nano-nutrients have also emerged as potential cleanup methods, especially in metal-contaminated soil systems. These NP-based systems reduce the bioavailability of the metal contaminants in the rhizosphere (phytostabilization), and impacts growth parameters such as plant establishment, growth, and health. Some nanoparticles like Au, Al, Ag, Cu, Ti, and Zn are well reported in wheat agriculture to promote wheat plant growth, and they also inhibit the growth of harmful fungal parasites within the rhizosphere, thus acting as potential nano-biofertilizers (Table 3.1).

These nano-biofertilizers are potent in wheat agriculture production as they are fabricated so they control the release of the fertilizer into the target cell without any unintended losses (De et al., 2014; Wilson et al., 2008; Oliveira et al., 2018; Liu and Lal, 2015). These nanoproducts (Table 3.1 and 3.2) improve plant growth with formulations of nanomaterial-based fertilizers and produce a manifold increase in the agricultural productivity of wheat. These nanofertilizers (Ag, Au, Cu, Cd, and Zn) release their effect in controlled conditions. This controlled release of nano-nutrients results in the increased growth of wheat seedlings irrespective of abiotic stresses like salinity, acidity, and low water holding capacity (Saleh et al., 2019; Zou et al., 2019; Yang et al., 2018). Nano-bioformulations have emerged as potential tools to clean up the metal-contaminated/polluted environment (Timmusk, Seisenbaeva, and Behers, 2018; Hussain et al., 2019a; Thakur, Thakur, and Kumar, 2018). These nano products create phytostabilization as well as improving plant establishment, growth, and health, and they could significantly speed up plant growth and indeed increase productivity (Ma et al., 2011). The rate of seed germination in wheat could be improved by a pretreatment of ZnO, Ag, and Cu NPs (Mishra and Kumar, 2009). In the present scenario, the context of nano-biofertilizers offers a great opportunity to develop eco-friendly compounds (Pantidos, 2014; Contado, 2015; Servin and White, 2016; Singh et al., 2014; Hussain et al., 2019a; Huang et al., 2015; Timmusk, Seisenbaeva, and Behers, 2018; Yadav et al., 2013).

NP-based nutrients are made available to the wheat plant through different mechanisms like absorption through the seedlings via soaking, agitation, etc. (Hu et al.,

2018; Dimkpa et al., 2018; Du et al., 2015; Ogawa, 2011; Cifuentes et al., 2010). These nano-formulations are absorbed through the roots, and the leaves are greatly helped by the aquaporins present in the cell membrane. This absorbed nano-bioformulation then reaches the various aerial parts of the plant through the transpiration stream and vascular system. The efficacy of the uptake of these particles is also greatly influenced by the method of nano-nutrient synthesis as well as their size (Razzaq et al., 2016; Abhinandan et al., 2018; Mushtaq et al., 2017; Mufamadi and Sekhejane, 2017; Iqbal et al., 2017; Yang et al., 2018; Asadi et al., 2016; Huang et al., 2015; Seyed Sharifi, Khalilzadeh, and Soltanmoradi, 2017; Hu et al., 2018; Saira, 2018) (Table 3.2).

The rate of seed germination in different wheat varieties have also be documented to be improved by a pretreatment with ZnO, Se, Fe, and Ti nanoparticles (Hu et al., 2018; Seyed Sharifi, Khalilzadeh, and Soltanmoradi, 2017; Cifuentes et al., 2010; Zou et al., 2019; Saira, 2018). In the present scenario, the context of nano-biofertilizers offers a great opportunity to develop eco-friendly compounds that can easily replace chemical pesticides. Nanoencapsulation of these nanoscale biofertilizers may enhance their adaptability.

An improved understanding of the interactions between NPs and plant responses, including their uptake, localization, and activity, could revolutionize crop production through increased disease resistance, nutrient utilization, and crop yield (Huang et al., 2015). A variety of elements (Ag, Cu, Fe, Ti, Se, and Zn) have recently been documented as being used as a nano-biofertilizer in wheat agriculture (Table 3.2). These elements proved to have a potential for enhancing the growth of the wheat plant by stimulating photosynthetic enzymes, chlorophyll content, and nutrient uptake mechanisms. They have also been reported to boost the plant's disease resistance and abiotic stress tolerance efficiency (Hussain et al., 2019b; Hu et al., 2018; Seyed Sharifi, Khalilzadeh, and Soltanmoradi, 2017; Cifuentes et al., 2010; Zou et al., 2019; Saira, 2018; Rizwan et al., 2019; Abhinandan et al., 2018; Asadi et al., 2016). Iqbal et al. (2017) studied the effect of Ag, Cu, and Fe NPs on wheat to examine its effect on the germination and vigour index. Fe NPs improved the germination percentage while root-shoot growths were also greatly enhanced by the above treatment, but contradictory Fe and Cu NPs reduced the growth of the root and shoot. So, Fe NPs were more growth stimulatory when compared to Ag and Cu NPs in wheat (Iqbal et al., 2017). According to a study by Singh et al. (2019) in India, Zn NPs synthesized via green technology methods (the biosynthesis of ZnO NPs from a plant leaf extract, *Aloe barbadensis* Mill) showed an increase in the overall growth of the wheat plant. This study has also showed the effect of NPs for enhancing seed germination and seedling emergence in wheat plants (Razzaq et al., 2016; Mufamadi and Sekhejane, 2017; Iqbal et al., 2017; Abhinandan et al., 2018; Yang et al., 2018; Asadi et al., 2016; Huang et al., 2015; Seyed Sharifi, Khalilzadeh, and Soltanmoradi, 2017). Recently, a large number of studies on wheat production have been focused on the synthesis and application of reducing the bioavailability of metal contaminants in the rhizosphere (phytostabilization) as well as improving plant establishment, growth, and health, as this could significantly speed up the plant growth and indeed its productivity (Ma et al., 2011). In the present scenario, nano-biofertilizers offer a great opportunity to develop eco-friendly compounds that can easily replace chemical fertilizers and improve wheat production (Cheng et al., 2016; Subramanian and Tarafdar, 2011; Mocan et al., 2017; Vejan et al., 2016; Servin and White, 2016; Hussain et al., 2019a).

3.4.3 Plant Biosensors

Biosensors have also been classified using their most sensitive carrier/recognition elements: immune-sensors, aptasensors, genosensors, and enzymatic biosensors (Pretty and Pervez Bharucha, 2015; Mocan et al., 2017; De et al., 2014; Oliveira et al., 2018; Huang et al., 2015). These sensors are very budget-friendly, rapid, specific, and selective when used for the continuous monitoring of various toxic substances in agricultural wheat systems as well as for the detection of plants diseases (Lin and Guo, 2016; Galati et al., 2018; Wang et al., 2018; Mills et al., 2018). Various NPs have demonstrated their adaptability for gluten level detection in wheat crops (White, Frisbie, and Dorfman, 2018). Biologically originated (especially bacterial) biosensors are nowadays developed by incorporating different NPs, such as Ag, Au Cu, Zn, etc., in the microbes (bacteria, virus, and fungi) (Liu and Lal, 2015; Mishra and Kumar, 2009; Min et al., 2009; Siddiqi, Husen, and Rao, 2018) (Table 3.2).

Biosensors have revolutionized the agricultural systems for pathogen detection, high throughput analysis, and the high-quality monitoring of crops (Justino, Duarte, and Rocha-Santos, 2017; Gui et al., 2017). Nano-biosensors are special sensors, which are fabricated using an assembly of different transducers. These transducers may be physical, chemical, biological, electrochemical, etc. These transducers are based on principles such as optical, piezoelectric, or electrochemical (Justino, Duarte, and Rocha-Santos, 2017; Mocan et al., 2017; Duhan et al., 2017). Biosensors are also classified based on their carrier sensitivity or recognition elements: immune-sensor, aptasensors, genosensors, and enzymatic biosensors (Mocan et al., 2017; De et al., 2014; Oliveira et al., 2018; Pretty and Pervez Bharucha, 2015). These sensors are very rapid, specific, and selective when used for the detection of various toxic substances in agricultural systems and they also detect plant diseases. Biologically originated (especially bacterial) biosensors are developed by incorporating different NPs, such as Ag, Au Cu, Zn, etc., in the microorganism (bacteria, virus, and fungi). Nanodevices (nanowires, nano-formulated particles, nano-encapsulated beds, and gels) consist of the different forms of biosensors that are used in wheat agricultural systems (Zou et al., 2019; Hussain et al., 2019b; White, Frisbie, and Dorfman, 2018) (Table 3.2). These sensors are designed such that they can characterize the diseases in plants and play an important role in the cleanup strategies related to pesticide and insecticide accumulation in the agricultural system (Baby Kumari, Mallick, and Hora, 2016; Baby Kumari, Hora, and Mallick, 2017b; Anand et al., 2016). As it is accurate and easy to analyze, it is in great demand in the wheat agriculture and food sectors, as it accelerates the quality of food and crops.

The role of NPs in the diagnostics of many pathogenic bacteria has been demonstrated (Contado, 2015; Mocan et al., 2017; Gui et al., 2017; Justino, Duarte, and Rocha-Santos, 2017) Recently, Bucur reviewed the role of biosensors based on enzymatic inhibition (microbial origin) (Bucur et al., 2018). These enzymatic-based

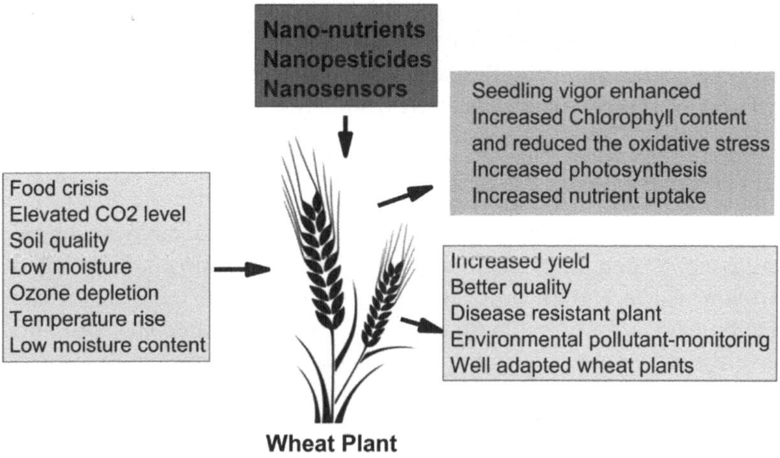

FIGURE 3.3 Showing the potential cause, mechanism, and advantages of Nanotechnological applications on wheat crops.

biosensors could be very helpful in the quantitative toxicity detection of numerous insecticides presently being used in the agro-system, such as organophosphorus compounds, carbamate compounds, etc., as these insecticides are banned or reported as having the highest risk by the European Food Safety Authority (EFSA) for foods (Peiyan et al., 2018; Linsinger, Roebben, and Gilliland, 2012; Singh et al., 2014; Subramanian and Tarafdar, 2011; Roco, 2003) (Figure 3.3).

A variety of bacterial species converted into biogenetic nano-biosensors were field-tested for a disease diagnosis and toxicity assessment analysis (Table 3.2). Nanodevices for identity preservation and tracking are highly sought after, as one of the major constraints in Indian agriculture is the quality maintenance of agricultural produce. Proper monitoring of the production system through nanotechnology would be very suitable to promote quality and make a clear distinction between organic products. Identity preservation (IP) is a system that creates increased value by providing customers with information about the practices and activities used to produce a particular crop or other agricultural product (Lin and Guo, 2016).

3.4.4 Abiotic Stress

Abiotic stresses are a major concern around the globe for future agricultural practices. According to recent climate change models, wheat production (Mills et al., 2018; Wang et al., 2018) would also be negatively affected in terms of growth and productivity worldwide. Wheat is the largest growing cereal (area wise) among the top three kinds of cereal globally, i.e. rice, maize, and wheat (Appels et al., 2018; Zou et al., 2019). Abiotic stresses like heat stress, salinity, the rise in temperature drought, and rise in CO_2 levels in climatic conditions are major factors that may greatly affect wheat-growing conditions (Fu et al., 2020; Hofmann et al., 2020). Thus, understanding the effects of these stresses becomes indispensable for wheat improvement programmes. Nanotechnology has been used in many fields of science, like physics, chemistry, pharmaceutical science, material science, medicine, and agriculture (Iavicoli et al., 2017; Wani and Kothari, 2018). Promising results in other fields have opened up a lot of scope in the wheat agriculture field (Figure 3.3).

3.4.5 Elevated Carbon Dioxide

Recently the rise in carbon dioxide (CO_2) level is considered a major environmental issue related to global warming. Global warming and elevated CO_2 (eCO_2) would affect grain production negatively; in this case, nano-science is considered a boon for wheat cultivation under super-elevated CO_2 conditions. In a recent report, the effect of future climatic carbon dioxide conditions was studied in wheat plants (Laboratory level) (Jasrotia et al., 2018). These studies were conducted in future climatic conditions, super-elevated CO_2 (5000 mg/L CO_2), and under normal CO_2 conditions (400 mg/L CO_2); wheat seeding's treated with TiO_2 NPs showed an increase in root mass and lateral root numbers. These studies could give us a physiological understanding of the interaction between grain crops and NPs in future CO_2 conditions. Interestingly other studies have also supported that NP absorption, accumulation, and translocation is positively induced in higher CO_2 conditions as compared to normal level conditions, and this may be due to the physiological stress, the increase in abscisic acid (ABA), and decrease in jasmonic acid (JA) contents as compared to wheat grown in normal conditions. NiO NPs have also been reported to have negative effects on wheat plant growth when applied in normal CO_2 conditions, but a different pattern was observed when wheat seedings were treated in future CO_2 conditions (Figure 3.3). NiO NPs reduced the level of phototoxic hazards by reducing the oxidative damage level (Saleh et al., 2019).

3.4.6 Saline Resistance

The excessive use of poor quality water used for irrigation and soil contamination lead to salinity stress in agricultural wheatlands. According to a recent report more than 6% of the world's total land and approximately 20% of irrigated land are affected by salinity, and therefore this is a serious concern for agriculture (Shukla et al., 2018). This abiotic stress affects the growth and productivity of wheat plants in the soil. Plants do adapt to these stresses through changes in their physiological, biochemical, and molecular pathways. Notably, recent biotechnological breakthroughs in the nanobiotechnology field have opened up new possibilities for crop improvement in wheat production. Various reports have evidenced the positive effect of NPs in mitigating the harmful effects of salt stress. ZnO, Ag, TiO, and silicon NPs have proved their potential to improve the growth of wheat by ultimately improving plant growth and developmental processes through seed germination, improving plant resistance against abiotic and biotic stresses, and augmenting nutrient utilization efficiency, with reduced environmental impact compared to traditional approaches to salinity stress.

3.4.7 Metal Contamination

In a study, wheat seed was primed with ZnO (0–100 mg L^{-1}) and Fe (0–20 mg L^{-1}) NPs and cultivated in Cd-contaminated soil. This contamination was due to the accumulation of long-term sewage disposal in the particular soil studied. With a 60–70% moisture content, the results showed a positive effect on the treated wheat plants as there were increased plant heights, spike lengths, and dry weights of shoots, roots, spikes, and grains. The reduction of Cd levels demonstrated the effect of NPs for reducing the soil toxicity and oxidative stress level in the wheat (Rizwan et al., 2019). A similar study on wheat was also conducted by Hussain et al. (2019b) which showed that Fe and ZnO NPs reduced the Cd contamination level in the soil and boosted wheat production in terms of overall morphological parameters. These studies suggest possibilities for future wheat production methods to overcome contaminated soil problems through the application of NPs (Hussain et al., 2019a) (Table 3.2).

3.5 LIMITATIONS

Nanotechnology-based wheat production could lead to a lack of employment around the world, especially in developing countries as nanodevices/machines would replace human fieldwork. The biggest challenge here is the bioavailability of nanomaterials in the food chain, which may lead to toxic effects on the customer's health. Another big challenge is to overcome the lack of knowledge concerning smart delivery systems, i.e. using nanotechnology for increased wheat production in a smarter way. The biggest limitation of nano-biotechnologically based agricultural advancement is its ethical acceptance.

3.6 FUTURE PROSPECT AND CONCLUSION

The role of nanotechnology in wheat farming has immense potential to meet the global crisis of calorie-provider foods like wheat. However, despite the emerging picture of the potential applications of nanotechnology in wheat improvement so far, their relevance has not reached field conditions. Farmers need more knowledge about this product, such as why it is better than chemical fertilizers because beneficial effects attract farmers' interest. Nanoproducts must be cost-effective as they are required in large quantities for improved results in wheat production. Shortly, it is expected that fabrication methods will involve the detailed study of wheat's genetic, physiological, and morphological variations, and their mode of action towards nanomaterials. These studies will undoubtedly help researchers to understand the requirements of NPs as per plants and climatic conditions. Another major issue is the toxicity related to the nano-based agricultural products, and more efficient biosensors in the future could play a role in inducing the adaptability of these products by the field growers (farmers). Further metaproteomics and rhizoengineering will develop significantly and will allow progress in the understanding of the activity and ecological behaviour of nonmaterial within the rhizosphere. We can conclude that the transit of the nano sector from a marginal level to the mainstream will require eco-friendly methods of fabrication, cost-effective processes, and increasing safety database availability. This would help us to achieve our goal and change farmers' perceptions of nano-based products. Including all these points, we can say that the role of nanotechnology will play an important role in sustainability and climate change.

REFERENCES

Abd-Elsalam, K A, and M A Alghuthaymi. 2014. 'Nanodiagnostic Tools in Plant Breeding'. *Plant Pathology and Microbiology* 5: 5. https://doi.org/10.4172/2157-7471.1000e107.

Abdelsalam, Nader R, Essam E Kandil, Mohammed A F Al-Msari, Malik A M Al-Jaddadi, Hayssam M Ali, Mohamed Z M Salem, and Mohamed S Elshikh. 2019. 'Effect of Foliar Application of NPK Nanoparticle Fertilization on Yield and Genotoxicity in Wheat (*Triticum aestivum* L.)'. *Science of the Total Environment* 653: 1128–39. https://doi.org/10.1016/j.scitotenv.2018.11.023.

Abhinandan, Kumar, Logan Skori, Matija Stanic, Neil M N Hickerson, Muhammad Jamshed, and Marcus A Samuel. 2018. 'Abiotic Stress Signaling in Wheat – An Inclusive Overview of Hormonal Interactions During Abiotic Stress Responses in Wheat'. *Frontiers in Plant Science*. https://doi.org/10.3389/fpls.2018.00734.

Agrawal, Shweta, and Pragya Rathore. 2014. 'Nanotechnology Pros and Cons to Agriculture : A Review'. *International Journal of Current Microbiology and Applied Sciences* 3(3): 43–55.

Alejandro, Pérez De Luque, and Diego Rubiales. 2009. 'Nanotechnology for Parasitic Plant Control'. *Pest Management Science* 65(5): 540–45. https://doi.org/10.1002/ps.1732.

Anand, K, B Kumari, and M A Mallick. 2016. 'Phosphate Solubilizing Microbes: An Effective and Alternative Approach as Biofertilizers'. *International Journal of Pharmacy and Pharmaceutical Sciences* 8(2): 37–40.

Anderson, Anne J, Joan E McLean, Astrid R Jacobson, and David W Britt. 2018. 'CuO and ZnO Nanoparticles Modify Interkingdom Cell Signaling Processes Relevant to Crop Production'. *Journal of Agricultural and Food Chemistry* 66(26): 6513–24. https://doi.org/10.1021/acs.jafc.7b01302.

Appels, Rudi, Kellye Eversole, Catherine Feuillet, Beat Keller, Jane Rogers, Nils Stein, Curtis J Pozniak et al. 2018. 'Shifting the Limits in Wheat Research and Breeding Using a Fully Annotated Reference Genome'. *Science* 361(6403): eaar7191. https://doi.org/10.1126/science.aar7191.

Asadi, M, M Zahedi, M H Ehtemam, and A H Khoshgoftarmanesh. 2016. 'Effects of Nano Zinc Oxide on the Growth and Ion Content of Four Wheat Cultivars under Salinity Stress'. *Journal of Science and Technology of Greenhouse Culture* 7(25): Pe, 25–33. *En34*.

Bansal, P, J S Duhan, and S K Gahlawat. 2014. 'Biogenesis of Nanoparticles: A Review'. *African Journal of Biotechnology* 13(28): 2778–85. https://doi.org/10.5897/AJB2013.13458.

Barazani, Oz, and Jacob Friedman. 1999. 'Is IAA the Major Root Growth Factor Secreted from Plant-Growth-Mediating Bacteria?' *Journal of Chemical Ecology* 25(10): 2397–406. https://doi.org/10.1023/A:1020890311499.

Bhattacharyya, Atanu, Asim Bhaumik, Pathipati Usha Rani, Suvra Mandal, and Timothy T Epidi. 2010. 'Nanoparticles – A Recent Approach to Insect Pest Control'. *African Journal of Biotechnology* 9(24): 3489–93. https://doi.org/10.5897/AJBx09.021.

Bucur, Bogdan, Florentina Daniela Munteanu, Jean Louis Marty, and Alina Vasilescu. 2018. 'Advances in Enzyme-Based Biosensors for Pesticide Detection'. *Biosensors* 8(2): 1–28. https://doi.org/10.3390/bios8020027.

Cai, Lin, Juanni Chen, Zhongwei Liu, Hancheng Wang, Huikuan Yang, and Wei Ding. 2018. 'Magnesium Oxide Nanoparticles: Effective Agricultural Antibacterial Agent Against Ralstonia solanacearum'. *Frontiers in Microbiology* 9 (April): 790. https://doi.org/10.3389/fmicb.2018.00790.

Campagnoli, Anna, Federica Cheli, Carlo Polidori, Mauro Zaninelli, Oreste Zecca, Giovanni Savoini, Luciano Pinotti, and Vittorio Dell'Orto. 2011. 'Use of the Electronic Nose as a Screening Tool for the Recognition of Durum Wheat Naturally Contaminated by Deoxynivalenol: A Preliminary Approach'. *Sensors* 11(5): 4899–916. https://doi.org/10.3390/s110504899.

Chang, Xipeng, Zhengguo Song, Yalei Xu, and Minling Gao. 2020. 'Effects of Carbon Nanotubes on Growth of Wheat Seedlings and Cd Uptake'. *Chemosphere* 240: 124931. https://doi.org/10.1016/j.chemosphere.2019.124931.

Cheng, H N, K T Klasson, Tetsuo Asakura, and Qinglin Wu. 2016. 'Nanotechnology in Agriculture'. In *ACS Symposium Series* 1224(7): 233–42. https://doi.org/10.1021/bk-2016-1224.ch012.

Cifuentes, Zuny, Laura Custardoy, Jesús M de la Fuente, Clara Marquina, M Ricardo Ibarra, Diego Rubiales, and Alejandro Pérez-de-Luque. 2010. 'Absorption and Translocation to the Aerial Part of Magnetic Carbon-Coated Nanoparticles through the Root of Different Crop Plants'. *Journal of Nanobiotechnology* 8(1): 26. https://doi.org/10.1186/1477-3155-8-26.

Contado, Catia. 2015. 'Nanomaterials in Consumer Products: A Challenging Analytical Problem'. *Frontiers in Chemistry* 3 (August): 1–20. https://doi.org/10.3389/fchem.2015.00048.

De, Arnab, Rituparna Bose, Ajeet Kumar, and Subho Mozumdar. 2014. *Targeted Delivery of Pesticides Using Biodegradable Polymeric Nanoparticles.* https://doi.org/10.1007/978-81-322-1689-6.

Dimkpa, Christian O, Upendra Singh, Prem S Bindraban, Wade H Elmer, Jorge L Gardea-Torresdey, and Jason C White. 2018. 'Exposure to Weathered and Fresh Nanoparticle and Ionic Zn in Soil Promotes Grain Yield and Modulates Nutrient Acquisition in Wheat (*Triticum aestivum* L.)'. *Journal of Agricultural and Food Chemistry* 66(37): 9645–56. https://doi.org/10.1021/acs.jafc.8b03840.

Du, Wenchao, Jorge L Gardea-Torresdey, Rong Ji, Ying Yin, Jianguo Zhu, Jose R Peralta-Videa, and Hongyan Guo. 2015. 'Physiological and Biochemical Changes Imposed by CeO_2 Nanoparticles on Wheat: A Life Cycle Field Study'. *Environmental Science and Technology* 49(19): 11884–93. https://doi.org/10.1021/acs.est.5b03055.

Duhan, Joginder Singh, Ravinder Kumar, Naresh Kumar, Pawan Kaur, Kiran Nehra, and Surekha Duhan. 2017. 'Nanotechnology: The New Perspective in Precision Agriculture'. *Biotechnology Reports* 15 (December 2016): 11–23. https://doi.org/10.1016/j.btre.2017.03.002.

Eifler, Jakob, Eugenio Martinelli, Marco Santonico, Rosamaria Capuano, Detlev Schild, and Corrado Di Natale. 2011. 'Differential Detection of Potentially Hazardous Fusarium species in Wheat Grains by an Electronic Nose'. *PLoS ONE* 6(6): e21026. https://doi.org/10.1371/journal.pone.0021026.

Fanfair, Devon, and Christopher Kelty. 2005. *The Early History of Nanotechnology*, 1–15.

Fu, Lin, Zhenyu Wang, Om Parkash Dhankher, and Baoshan Xing. 2020. 'Nanotechnology as a New Sustainable Approach for Controlling Crop Diseases and Increasing Agricultural Production'. *Journal of Experimental Botany* 71(2): 507–19. https://doi.org/10.1093/jxb/erz314.

Galati, C, M G Manera, A Colombelli, M De Pascali, P Rampino, C Perrotta, and R Rella. 2018. 'Opto-Plasmonic Biosensors for Monitoring Wheat End-Products Quality BT'. In *Sensors*, edited by Bruno Andò, Francesco Baldini, Corrado Di Natale,

Giovanna Marrazza, and Pietro Siciliano, 194–99. Cham: Springer International Publishing.

Gilbertson, Leanne M, Leila Pourzahedi, Stephanie Laughton, Xiaoyu Gao, Julie B Zimmerman, Thomas L Theis, Paul Westerhoff, and Gregory V Lowry. 2020. 'Guiding the Design Space for Nanotechnology to Advance Sustainable Crop Production'. *Nature Nanotechnology* 15(9): 801–10. https://doi.org/10.1038/s41565-020-0706-5.

Gouda, Sushanto, Rout George Kerry, Gitishree Das, Spiros Paramithiotis, Han Seung Shin, and Jayanta Kumar Patra. 2018. 'Revitalization of Plant Growth Promoting Rhizobacteria for Sustainable Development in Agriculture'. *Microbiological Research* 206 (July 2017): 131–40. https://doi.org/10.1016/j.micres.2017.08.016.

Gui, Qingyuan, Tom Lawson, Suyan Shan, Lu Yan, and Yong Liu. 2017. 'The Application of Whole Cell-Based Biosensors for Use in Environmental Analysis and in Medical Diagnostics'. *Sensors* 17(7): 1623. https://doi.org/10.3390/s17071623.

Hofmann, Thilo, Gregory Victor Lowry, Subhasis Ghoshal, Nathalie Tufenkji, Davide Brambilla, John Robert Dutcher, Leanne M Gilbertson, et al. 2020. 'Technology Readiness and Overcoming Barriers to Sustainably Implement Nanotechnology-Enabled Plant Agriculture'. *Nature Food* 1(7): 416–25. https://doi.org/10.1038/s43016-020-0110-1.

Hora, Amandeep, C P Malik, Baby Kumari, and Guru Nanak College. n.d. *Potential of Chloroplast Simple Sequence Repeats (CpSSRs) in Genetic Analysis and Crop Improvement.*

Hu, Ting, Huafen Li, Jixiang Li, Guishen Zhao, Wenliang Wu, Liping Liu, Qi Wang, and Yanbin Guo. 2018. 'Absorption and Bio-Transformation of Selenium Nanoparticles by Wheat Seedlings (*Triticum aestivum L.*)'. *Frontiers in Plant Science.*

Huang, Shiwen, Ling Wang, Lianmeng Liu, Yuxuan Hou, and Lu Li. 2015. 'Nanotechnology in Agriculture, Livestock, and Aquaculture in China. A Review'. *Agronomy for Sustainable Development* 35(2): 369–400. https://doi.org/10.1007/s13593-014-0274-x.

Hussain, Afzal, Shafaqat Ali, Muhammad Rizwan, Muhammad Zia ur Rehman, Muhammad Farooq Qayyum, Hailong Wang, and Jörg Rinklebe. 2019a. 'Responses of Wheat (*Triticum aestivum*) Plants Grown in a Cd Contaminated Soil to the Application of Iron Oxide Nanoparticles'. *Ecotoxicology and Environmental Safety* 173: 156–64. https://doi.org/10.1016/j.ecoenv.2019.01.118.

Hussain, Afzal, Muhammad Rizwan, Qasim Ali, and Shafaqat Ali. 2019b. 'Seed Priming with Silicon Nanoparticles Improved the Biomass and Yield While Reduced the Oxidative Stress and Cadmium Concentration in Wheat Grains'. *Environmental Science and Pollution Research International.* https://doi.org/10.1007/s11356-019-04210-5.

Iavicoli, Ivo, Veruscka Leso, Donald H Beezhold, and Anna A Shvedova. 2017. 'Nanotechnology in Agriculture: Opportunities, Toxicological Implications, and Occupational Risks'. *Toxicology and Applied Pharmacology* 329: 96–111. https://doi.org/10.1016/j.taap.2017.05.025.

Iqbal, Muhammad, Naveed Iqbal Raja, Zia-Ur-Rehman Mashwani, Mubashir Hussain, Muhammad Ejaz, and Farhat Yasmeen. 2017. 'Effect of Silver Nanoparticles on Growth of Wheat Under Heat Stress'. *Iranian Journal of Science and Technology, Transactions A: Science.* https://doi.org/10.1007/s40995-017-0417-4.

Iravani, Siavash. 2014. 'Bacteria in Nanoparticle Synthesis: Current Status and Future Prospects'. *International Scholarly Research Notices* 2014: 1–18. https://doi.org/10.1155/2014/359316.

Jafarizadeh-Malmiri, Hoda, Zahra Sayyar, Navideh Anarjan, and Aydin Berenjian. 2019. 'Nanobiotechnology at a Glance BT'. In *Nanobiotechnology in Food: Concepts, Applications and Perspectives*, edited by Hoda Jafarizadeh-Malmiri, Zahra Sayyar, Navideh Anarjan, and Aydin Berenjian, 1–17. Cham: Springer International Publishing. https://doi.org/10.1007/978-3-030-05846-3_1.

Jan, Aarifa, Tanveer Bilal Pirzadah, and Bisma Malik. 2020. 'Nanotechnology: An Innovative Tool to Enhance Crop Production BT'. In *Nanobiotechnology in Agriculture: An Approach Towards Sustainability*, edited by Khalid Rehman Hakeem, and Tanveer Bilal Pirzadah, 163–70. Cham: Springer International Publishing. https://doi.org/10.1007/978-3-030-39978-8_9.

Jasrotia, Poonam, Prem Lal Kashyap, Ajay Kumar Bhardwaj, Sudheer Kumar, and G P Singh. 2018. 'Scope and Applications for Wheat Production and Quality Enhancement : A Review of Recent Advances'. *Wheat and Barley Research* 10(1): 1–13. https://doi.org/10.25174/2249-4065/2018/76672.

Justino, Celine, Armando Duarte, and Teresa Rocha-Santos. 2017. 'Recent Progress in Biosensors for Environmental Monitoring: A Review'. *Sensors* 17(12): 2918. https://doi.org/10.3390/s17122918.

Kashyap, Prem, S Kumar, P Jasrotia, D P Singh, and G P Singh. 2020. 'Nanotechnology in Wheat Production and Protection BT'. In *Environmental Nanotechnology Volume 4*, edited by Nandita Dasgupta, Shivendu Ranjan, and Eric Lichtfouse. Cham: Springer International Publishing. 165–94. https://doi.org/10.1007/978-3-030-26668-4_5.

Kashyap, Prem, S Kumar, A K Srivastava, and A K Sharma Lal. 2013. 'Myconanotechnology in Agriculture: A Perspective'. *World Journal of Microbiology and Biotechnology* 29(2): 191–207. https://doi.org/10.1007/s11274-012-1171-6.

Kumar, Deepak, and Prasanta Kalita. 2017. 'Reducing Postharvest Losses during Storage of Grain Crops to Strengthen Food Security in Developing Countries'. *Foods Basel, Switzerland* 6(1): 8. https://doi.org/10.3390/foods6010008.

Kumar, Vineet, and Sudesh Kumar Yadav. 2009. 'Plant-Mediated Synthesis of Silver and Gold Nanoparticles and Their Applications'. *Journal of Chemical Technology and Biotechnology* 84(2): 151–57. https://doi.org/10.1002/jctb.2023.

Kumari, B, M A Mallick, and A Hora. 2016. 'Plant Growth-Promoting Rhizobacteria (PGPR): Their Potential for Development of Sustainable Agriculture'. In: *Bio-Exploitation for Sustainable Agriculture*, edited by P. C. Trivedi, 1st edition, 1–19. Jaipur, India: Avinskar Publishing House.

Kumari, Baby, Amandeep Hora, and Muhammad Mallick. 2017a. Nanomedicines in Cancer Research: An Overview. *An International Journal of Life Sciences* 6(1): 7. https://doi.org/10.5958/2319-1198.2017.00002.1.

Kumari, Baby, Amandeep Hora, and Muhammad Mallick 2017b. 'Stimulatory Effect of PGPR (Plant Growth Promoting Rhizospheric Bacteria) on Medicinal and Growth Properties of a Potential Medicinal Herb *Chlorophytum borivilianum*: A Review'. *Journal of Plant Science Research* 33.

Lin, Xialu, and Xiong Guo. 2016. 'Advances in Biosensors, Chemosensors and Assays for the Determination of Fusarium Mycotoxins'. *Toxins* 8(6): 161. https://doi.org/10.3390/toxins8060161.

Linsinger, Thomas, Gert Roebben, and Douglas Gilliland. 2012. *Requirements on Measurements for the Implementation of the European Commission Definition of the Term 'Nanomaterial'*. *EUR* Report. Vol. EUR 25404. https://doi.org/10.2787/63490.

Liu, Ruiqiang, and Rattan Lal. 2015. 'Potentials of Engineered Nanoparticles as Fertilizers for Increasing Agronomic Productions'. *Science of the Total Environment* 514: 131–39. https://doi.org/10.1016/j.scitotenv.2015.01.104.

Lyu, Shiheng, Xiangying Wei, Jianjun Chen, Cun Wang, Xiaoming Wang, and Dongming Pan. 2017. 'Titanium as a Beneficial Element for Crop Production'. *Frontiers in Plant Science* 8: 597. https://doi.org/10.3389/fpls.2017.00597.

Ma, Y, M N V Prasad, M Rajkumar, and H Freitas. 2011. 'Plant Growth Promoting Rhizobacteria and Endophytes Accelerate Phytoremediation of Metalliferous Soils'. *Biotechnology Advances* 29(2): 248–58. https://doi.org/10.1016/j.biotechadv.2010.12.001.

Majumdar, Sanghamitra, and Arturo A Keller. 2020. 'Omics to Address the Opportunities and Challenges of Nanotechnology in Agriculture'. *Critical Reviews in Environmental Science and Technology, June*: 1–42. https://doi.org/10.1080/10643389.2020.1785264.

Mills, Gina, Katrina Sharps, David Simpson, Håkan Pleijel, Malin Broberg, Johan Uddling, Fernando Jaramillo, et al. 2018. 'Ozone Pollution Will Compromise Efforts to Increase Global Wheat Production'. *Global Change Biology* 24(8): 3560–74. https://doi.org/10.1111/gcb.14157.

Min, Ji Seon, Kyoung Su Kim, Sang Woo Kim, Jin Hee Jung, Kabir Lamsal, Seung Bin Kim, Mooyoung Jung, and Youn Su Lee. 2009. 'Effects of Colloidal Silver Nanoparticles on Sclerotium-Forming Phytopathogenic Fungi'. *Plant Pathology Journal* 25(4): 376–80. https://doi.org/10.5423/PPJ.2009.25.4.376.

Mishra, Vijendra Kumar, and Ashok Kumar. 2009. 'Impact of Metal Nanoparticles on the Plant Growth Promoting Rhizobacteria'. *Digest Journal of Nanomaterials and Biostructures* 4(3): 587–92.

Mocan, Teodora, Cristian T Matea, Teodora Pop, Ofelia Mosteanu, Anca Dana Buzoianu, Cosmin Puia, Cornel Iancu, and Lucian Mocan. 2017. 'Development of Nanoparticle-Based Optical Sensors for Pathogenic Bacterial Detection'. *Journal of Nanobiotechnology* 15(1): 1–14. https://doi.org/10.1186/s12951-017-0260-y.

Mufamadi, M S, and P R Sekhejane. 2017. 'Nanomaterial-Based Biosensors in Agriculture Application and Accessibility in Rural Smallholding Farms: Food Security BT'. In *Nanotechnology: An Agricultural Paradigm*, edited by Ram Prasad, Manoj Kumar, and Vivek Kumar, 263–78. Singapore: Springer Singapore. https://doi.org/10.1007/978-981-10-4573-8_12.

Mukhopadhyay, Siddhartha S. 2014. 'Nanotechnology in Agriculture: Prospects and Constraints'. *Nanotechnology, Science and Applications* 7(2): 63–71. https://doi.org/10.2147/NSA.S39409.

Mushtaq, Ayesha, Nelofer Jamil, Musarrat Riaz, G L Hornyak, and Nazeer Ahmed. 2017. 'Synthesis of Silica Nanoparticles and Their Effect on Priming of Wheat (*Triticum aestivum* L.) under Salinity Stress'. *Biological Forum Anais an International Journal* 9(1), 150–57.

Nair, Remya, Saino Hanna Varghese, Baiju G Nair, T Maekawa, Y Yoshida, and D Sakthi Kumar. 2010. 'Nanoparticulate Material Delivery to Plants'. *Plant Science* 179(3): 154–63. https://doi.org/10.1016/j.plantsci.2010.04.012.

Ogawa, Atsushi. 2011. 'Rational Design of Artificial Riboswitches Based on Ligand-Dependent Modulation of Internal Ribosome Entry in Wheat Germ Extract and Their Applications as Label-Free Biosensors'. *RNA* 17: 478–88. https://dx.doi.org/10.1261%2Frna.2433111.

Oliveira, Jhones L De, Estefânia V R Campos, Anderson E S Pereira, Tatiane Pasquoto, Renata Lima, Renato Grillo, Daniel Junior De Andrade, Fabiano Aparecido Dos Santos, and Leonardo Fernandes Fraceto. 2018. 'Zein Nanoparticles as Eco-Friendly Carrier Systems for Botanical Repellents Aiming Sustainable Agriculture'. *Journal of Agricultural and Food Chemistry* 66(6): 1330–40. https://doi.org/10.1021/acs.jafc.7b05552.

Pantidos, Nikolaos. 2014. 'Biological Synthesis of Metallic Nanoparticles by Bacteria, Fungi and Plants'. *Journal of Nanomedicine and Nanotechnology* 05(05). https://doi.org/10.4172/2157-7439.1000233.

Parisi, Claudia, Mauro Vigani, and Emilio Rodríguez-Cerezo. 2015. 'Agricultural Nanotechnologies: What Are the Current Possibilities?'. *Nano Today* 10(2): 124–27. https://doi.org/10.1016/j.nantod.2014.09.009.

Peiyan, Yuan, Ding Xin, Yang Yi Yan, and Xu Qing-Hua. 2018. 'Metal Nanoparticles for Diagnosis and Therapy of Bacterial Infection'. *Advanced Healthcare Materials* 7(13): 1701392. https://doi.org/10.1002/adhm.201701392.

Pradas del Real, Ana E, Vladimir Vidal, Marie Carrière, Hiram Castillo-Michel, Clément Levard, Perrine Chaurand, and Géraldine Sarret. 2017. 'Silver Nanoparticles and Wheat Roots: A Complex Interplay'. *Environmental Science and Technology* 51(10): 5774–82. https://doi.org/10.1021/acs.est.7b00422.

Prasad, Ram, Atanu Bhattacharyya, and Quang D Nguyen. 2017. 'Nanotechnology in Sustainable Agriculture: Recent Developments, Challenges, and Perspectives'. *Frontiers in Microbiology* 8 (June): 1–13. https://doi.org/10.3389/fmicb.2017.01014.

Pretty, Jules, and Zareen Pervez Bharucha. 2015. 'Integrated Pest Management for Sustainable Intensification of Agriculture in Asia and Africa'. *Insects* 6(1): 152–82. https://doi.org/10.3390/insects6010152.

Rathinamoorthy, R, and P Senthilkumar. 2009. 'Synthesis of Nanoparticles and Their Application in Textiles'. *Melliand International* 15(3): 10–3.

Razzaq, A, R Ammara, H M Jhanzab, T Mahmood, A Hafeez, and S Hussain. 2016. 'A Novel Nanomaterial to Enhance Growth and Yield of Wheat'. 2(1): 55–8.

Rico, Cyren M, Mark G Johnson, Matthew A Marcus, and Christian P Andersen. 2017. 'Intergenerational Responses of Wheat (*Triticum aestivum* L.) to Cerium Oxide Nanoparticles Exposure'. *Environmental Science: Nano* 4(3): 700–11. https://doi.org/10.1039/C7EN00057J.

Rizwan, Muhammad, Shafaqat Ali, Basharat Ali, Muhammad Adrees, Muhammad Arshad, Afzal Hussain, Muhammad Zia ur Rehman, and Aisha Abdul Waris. 2019. 'Zinc and Iron Oxide Nanoparticles Improved the Plant Growth and Reduced the Oxidative Stress and Cadmium Concentration in Wheat'. *Chemosphere* 214: 269–77. https://doi.org/10.1016/j.chemosphere.2018.09.120.

Roco, Mihail C. 2003. 'Nanotechnology: Convergence with Modern Biology and Medicine'. *Current Opinion in Biotechnology* 14(3): 337–46. https://doi.org/10.1016/S0958-1669(03)00068-5.

Saira, Hafiza. 2018. 'Assessment of AgNPs Exposure on Physiological and Biochemical Changes and Antioxidative Defence System in Wheat (*Triticum aestivum* L) under Heat Stress'. *IET Nanobiotechnology* 13: 230–6. https://doi.org/10.1049/iet-nbt.2018.5041.

Saleh, Ahmed M, Yasser M Hassan, Samy Selim, and Hamada AbdElgawad. 2019. 'NiO-Nanoparticles Induce Reduced Phytotoxic Hazards in Wheat (*Triticum aestivum* L.) Grown under Future Climate CO_2'. *Chemosphere* 220: 1047–57. https://doi.org/10.1016/j.chemosphere.2019.01.023.

Servin, Alia D, and Jason C White. 2016. 'Nanotechnology in Agriculture: Next Steps for Understanding Engineered Nanoparticle Exposure and Risk'. *NanoImpact* 1: 9–12. https://doi.org/10.1016/j.impact.2015.12.002.

Seyed Sharifi, Raouf, Razieh Khalilzadeh, and Soraya Soltanmoradi. 2017. 'Evaluation Effects of Mycorrhizal Fungi (AM) and Nano Zinc Oxide on Seed Yield and Dry Matter Remobilization of Wheat (*Triticum aestivum* L.) under Salinity Stress'. *Journal of Crop Nutrition Science* 3(3): 61–9. http://jcns.iauahvaz.ac.ir/article_540857.html.

Shukla, Pradeep Kumar, Saumya Shukla, Preeti Rajoriya, and Pragati Misra. 2018. 'Enhancing Crop Productivity in Saline Environment Using Nanobiotechnology BT'. In *Salinity Responses and Tolerance in Plants, Volume 2: Exploring RNAi, Genome Editing and Systems Biology*, edited by Vinay Kumar, Shabir Hussain Wani, Penna Suprasanna, and Lam-Son Phan Tran, 289–305. Cham: Springer International Publishing. https://doi.org/10.1007/978-3-319-90318-7_12.

Siddiqi, Khwaja Salahuddin, Azamal Husen, and Rifaqat A K Rao. 2018. 'A Review on Biosynthesis of Silver Nanoparticles and Their Biocidal Properties'. *Journal of Nanobiotechnology* 16(1). https://doi.org/10.1186/s12951-018-0334-5.

Singh, Jagpreet, Sumit Kumar, Anshu Alok, Santosh Kumar Upadhyay, Mohit Rawat, Daniel C W Tsang, Nanthi Bolan, and Ki-Hyun Kim. 2019. 'The Present Study Demonstrated the Possible Impacts of Iron Oxide Nanoparticles (Fe NPs) on the Alleviation of Toxic Effects of Cadmium (Cd) in Wheat and Enhance Its Growth, Yield, and Fe Biofortification. A Pot Experiment Was Conducted in Historically Cd'. *Journal of Cleaner Production* 214: 1061–70. https://doi.org/10.1016/j.jclepro.2019.01.018.

Singh, Sachin, Govind Gupta, Ekta Khare, K K Behal, and Naveen K Arora. 2014. 'Effect of Enrichment Material on the Shelf Life and Field Efficiency of Bioformulation of Rhizobium sp. and P-Solubilizing Pseudomonas fluorescens'. *Science Research Reporter* 4(1): 44–50.

Singh, Sadhna, Manoj Singh, Ved Varun Agrawal, and Anil Kumar. 2010. 'An Attempt to Develop Surface Plasmon Resonance Based Immunosensor for Karnal Bunt (*Tilletia indica*) Diagnosis Based on the Experience of Nano-Gold Based Lateral Flow Immuno-Dipstick Test'. *Thin Solid Films* 519(3): 1156–59. https://doi.org/10.1016/j.tsf.2010.08.06 1.

Sokolova, Viktoriya, and Matthias Epple. 2008. 'Inorganic Nanoparticles as Carriers of Nucleic Acids into Cells'. *Angewandte Chemie International Edition* 47(8): 1382–95. https://doi.org/10.1002/anie.200703039.

Subramanian, K S, and J C Tarafdar. 2011. 'Prospects of Nanotechnology in Indian Farming'. *Indian Journal of Agricultural Sciences* 81(10): 887–93.

Svendsen, Claus, Lee A Walker, Marianne Matzke, Elma Lahive, Samuel Harrison, Alison Crossley, Barry Park et al. 2020. 'Key Principles and Operational Practices for Improved Nanotechnology Environmental Exposure Assessment'. *Nature Nanotechnology* 15(9): 731–42. https://doi.org/10.1038/s41565-020-0742-1.

Taran, Nataliya, Volodymyr Storozhenko, Nataliia Svietlova, Ludmila Batsmanova, Viktor Shvartau, and Mariia Kovalenko. 2017. 'Effect of Zinc and Copper Nanoparticles on Drought Resistance of Wheat Seedlings'. *Nanoscale Research Letters* 12(1): 60. https:/ /doi.org/10.1186/s11671-017-1839-9.

Thakur, S, S K Thakur, and R R Kumar. 2018. 'Bio-Nanotechnology and Its Role in Agriculture and Food Industry'. *Journal of Molecular and Genetic Medicine* 12(1): 1–5. https://doi.org/10.4172/1747-0862.1000324.

Tilman, David, Kenneth G Cassman, Pamela A Matson, Rosamond Naylor, and Stephen Polasky. 2002. 'Agricultural Sustainability and Intensive Production Practices'. *Nature* 418 (August): 671. http://doi.org/10.1038/nature01014.

Timmusk, Salme, Gulaim Seisenbaeva, and Lawrence Behers. 2018. 'Titania (TiO2) Nanoparticles Enhance the Performance of Growth-Promoting Rhizobacteria'. *Scientific Reports* 8(1): 1–13. https://doi.org/10.1038/s41598-017-18939-x.

Toumey, Chris. 2009. 'Plenty of Room, Plenty of History'. *Nature Nanotechnology* 4 (December): 783. https://doi.org/10.1038/nnano.2009.357.

Ubeyitogullari, Ali, Régis Moreau, Devin J Rose, Jian Zhang, and Ozan N Ciftci. 2019. 'Enhancing the Bioaccessibility of Phytosterols Using Nanoporous Corn and Wheat Starch Bioaerogels'. *European Journal of Lipid Science and Technology* 121(1): 1700229. https://doi.org/10.1002/ejlt.201700229. http://www.ncbi.nlm.nih.gov/pubmed/ 1700229.

Usman, Muhammad, Muhammad Farooq, Abdul Wakeel, Ahmad Nawaz, Sardar Alam Cheema, Hafeez ur Rehman, Imran Ashraf, and Muhammad Sanaullah. 2020. 'Nanotechnology in Agriculture: Current Status, Challenges and Future Opportunities'. *Science of the Total Environment* 721: 137778. https://doi.org/10.1016/j.scitotenv.2020 .137778.

Vejan, Pravin, Rosazlin Abdullah, Tumirah Khadiran, Salmah Ismail, and Amru Nasrulhaq Boyce. 2016. 'Role of Plant Growth Promoting Rhizobacteria in Agricultural Sustainability-A Review'. *Molecules* 21(5): 1–17. https://doi.org/10.3390/molecules21 050573.

Wang, Bin, De L, Liu, Garry J O'Leary, Senthold Asseng, Ian Macadam, Rebecca Lines-Kelly, Xihua Yang, et al. 2018. 'Australian Wheat Production Expected to Decrease by the Late 21st Century'. *Global Change Biology* 24(6): 2403–15. https://doi.org/10.1111/gcb.14034.

Wani, Khursheed Ahmad, and Richa Kothari. 2018. 'Agricultural Nanotechnology : Applications and Challenges'. *Annals of Plant Sciences*, 7(3): 2146–48.

White, Scott P, C Daniel Frisbie, and Kevin D Dorfman. 2018. 'Detection and Sourcing of Gluten in Grain with Multiple Floating-Gate Transistor Biosensors'. *ACS Sensors* 3(2): 395–402. https://doi.org/10.1021/acssensors.7b00810.

Wilson, Michael A, Nguyen H Tran, Adrian S Milev, G SKamali Kannangara, Herbert Volk, and G Q Max Lu. 2008. 'Nanomaterials in Soils'. *Geoderma* 146(1–2): 291–302. https://doi.org/10.1016/j.geoderma.2008.06.004.

Yadav, Archana, T Theivasanthi, P K Paul, and K C Upadhyay. 2013. *Extracellular Biosynthesis of Silver Nanoparticles from Plant Growth Promoting Rhizobacteria Pseudomonas sp.*

Yang, Jie, Fuping Jiang, Chuanxin Ma, Yukui Rui, Mengmeng Rui, Muhammad Adeel, Weidong Cao, and Baoshan Xing. 2018. 'Alteration of Crop Yield and Quality of Wheat upon Exposure to Silver Nanoparticles in a Life Cycle Study'. *Journal of Agricultural and Food Chemistry* 66(11): 2589–97. https://doi.org/10.1021/acs.jafc.7b04904.

Zaytseva, Olga, and Günter Neumann. 2016. 'Carbon Nanomaterials: Production, Impact on Plant Development, Agricultural and Environmental Applications'. *Chemical and Biological Technologies in Agriculture* 3(1): 17. https://doi.org/10.1186/s40538-016-0070-8.

Zou, Liqiang, Anqi Xie, Yuqing Zhu, and David Julian McClements. 2019. 'Cereal Proteins in Nanotechnology: Formulation of Encapsulation and Delivery Systems'. *Current Opinion in Food Science* 25: 28–34. https://doi.org/10.1016/j.cofs.2019.02.004.

United State Department of Agriculture Forigen Agriculture Service. Available at: https://www.fas.usda.gov/data/grain-world-markets-and-trade.

4 Role of Nano-Biotechnology in Agriculture and Allied Sciences

Uma Chaurasia, Ashutosh Kumar, Deepak Kumar Maurya, Sudheer Kumar Yadav, Touseef Hussain, and Vineet Kumar Maurya

CONTENTS

4.1 INTRODUCTION

The current global population of 7.2 billion is anticipated to be approximately 9.6 billion by the end of 2050. This would require more land for homes and therefore a decrease of the net land area available for agricultural practices. Food security for this growing population, via ensuring the production of good quality food in an adequate amount, is a prime concern for every nation. Gaining a higher yield from less arable land, an ever-changing environment, and an increasing pollution are significant challenges for scientists (Yadav et al., 2017). The uncontrolled use of chemical fertilizers and chemical pesticides to enhance agricultural production has severely affected the topsoil, groundwater, and ultimately the whole environment. Hence, there is a need for new approaches to significantly enhance agriculture production that are also environment friendly. Nanotechnology in agriculture is one of the top emerging approaches to overcome all the problems highlighted above (Danish and Hussain, 2019).

Nanotechnology can enhance food production in many ways, with a simultaneous reduction in production costs and the amount of waste generated (Kah, 2015). One of the approaches is the use of nanotechnologies for fertilizers, pesticides, and insecticides, in which either nanoparticles (NPs) themselves or their encapsulation in nanomaterials are used. This enables the slow and sustained release of these chemicals, thus reducing the amount required for desired effects and decreasing environmental pollution (Ghormade et al., 2011; Hussain, 2017a, 2017b). The second approach using nanosensors for detecting soil conditions (pH, temperature, moisture, and nutrient availability) and microbial pathogens. Timely detection of microbial pathogens would help with their treatment and minimum loss of crop productivity; besides, useful information about soil health enables for a proper treatment of the soil and to choose the right crop for that soil. After crop production, the preservation and processing of food are the next steps to ensure food security, and nanotechnology is also being applied in this field too. Therefore, nanotechnology in agriculture and food processing is a better way for producing quality food with a high yield for the global population with less available arable lands (Singh et al., 2018). However, a major concern about nanotechnology research and its application in the agricultural field is the minimal agricultural field conditions, which are out of human control beyond a specific limit (Yadav et al., 2018).

Toxic by-products, generated during the synthesis of NPs, are another concern regarding the application of nanotechnology because these toxic by-products result in severe problems to human health (Servin and White, 2016). The biological methods for NP synthesis are better as they are cost-effective and eco-friendly because they do not produce toxic by-products like the chemical method of NPs synthesis (Vigneshwaran et al., 2007). The remediation of chemical and metal toxicity by adopting a green synthesis of NPs is an efficient way to overcome environmental hazards. This chapter will enhance comprehension of the equilibrium between the significance and application of nanotechnology in agriculture, in genetic engineering-based crop improvement, in food preservation and processing technologies, in sensor technologies, and for nanoremediation. In addition, various forms of nanomaterials and their usage are also briefly discussed.

4.2 NANOTECHNOLOGY FOR AGRICULTURE

Better agricultural management practices should aim towards sustainable agricultural production, higher productivities, and profitability. Nanotechnology has given value to agriculture practices and leads to the enhancement of agriculture productivity. Different nanotechnologies and their various applications that strengthen agriculture production are summarized in Figure 4.1. Precision agriculture is an area where nanotechnology can be used to enhance agriculture production. The use of nanofertilizers, nanopesticides/insecticides and nanosensors/nano-biosensors, microbial nanoformulations for heavy metal remediation, and use of nanotechnology in biotechnology and genetic engineering are the areas of precision agriculture where nanotechnology has a wide scope of application.

4.2.1 Nanopesticides/Insecticides and Nanofertilizers

4.2.1.1 Nanopesticides

Pesticides are used for pest management in plant defence, including fungicides, herbicides and insecticides, and conventional plant protection strategies. The major drawback of pesticide use is its loss into the environment; according to a study, > 90% of pesticides are lost into the environment before reaching their target site; thus, pesticides fail to achieve the desired effect (Nuruzzaman et al., 2016). Hence, there is an urgent need to develop a formulation or methodology that could provide better protection to pest invasion plants without being lost into the environment. Recently, industries are moving towards developing and applying nanoformulations and nanoencapsulations of pesticides (Figure 4.2), which would release pesticides/insecticides in a more slow and sustained manner, thus minimizing their loss into the environment. This technology has a better output and the potential to revolutionize the plant protection sector. Nanoformulations are mixtures of a small number of active ingredients of pesticides along with the useful pesticidal properties of modified nanostructures/particles (Haq and Ijaz, 2019; Danish and Hussain, 2019).

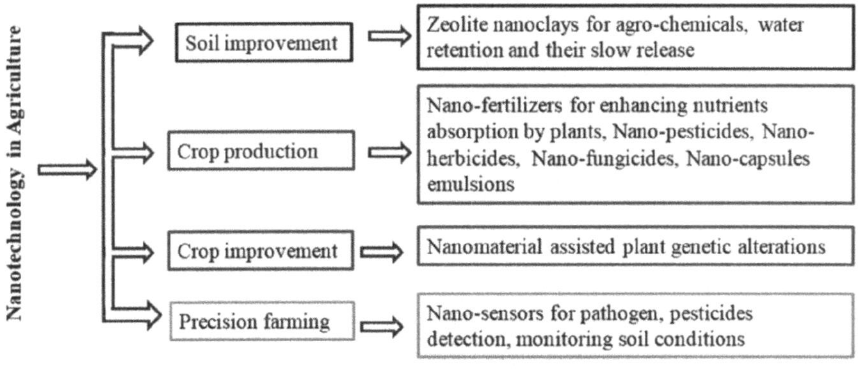

FIGURE 4.1 Different nanotechnological aspects of agriculture.

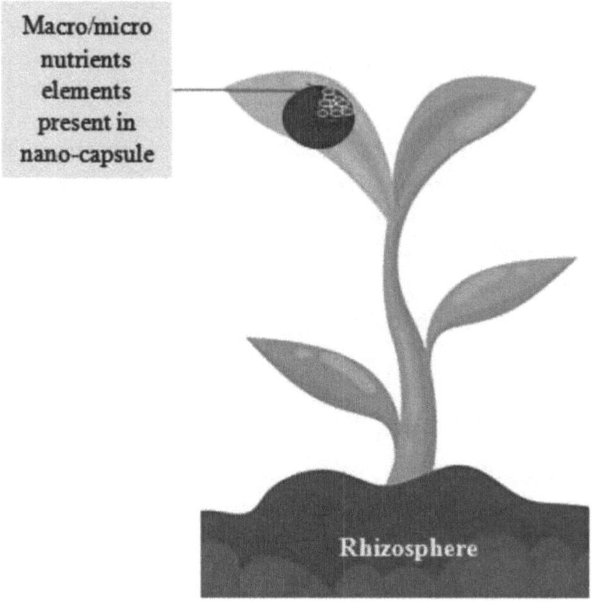

FIGURE 4.2 Prototype of nanocapsule containing macro/microelements.

Encapsulation is the coating of the active ingredient of the pesticide in various sizes of nanomaterial.

Conventional formulations of pesticides or herbicides have certain inherent limitations like solubility issues in water, adverse effects on other organisms and on the plant, and the development of resistance in target organisms. All these limitations are mitigated by the nanoformulations and nanoencapsulations of pesticides, which can be included under category of integrated pest management (IPM). Nanoformulations or nanoencapsulations release the active ingredient of a pesticide in a controlled manner near to the root zone, and thus not causing harm to other organisms or the plant itself (Patil and Solanki, 2016; Dwivedi et al., 2016). Due to its controlled and targeted release, there is very little chance of the development of resistance against them. Other advantages of nanoformulated and encapsulated pesticides are accuracy, controlled solubility, permeability, stiffness, thermal stability, crystallinity, and that they are sustainable with the agro-environmental system (Khan and Rizvi, 2014; Haq and Ijaz, 2019). The accuracy of nanopesticides, their timely and targeted release at the desired site, reduces the total amount of pesticide required for pest and disease control (Figure 4.3).

Engineered nanomaterials, such as NPs of ZnO, Cu, MgO, CaO, Cu, TiO2, MnO, SiO2, and Ag, have potential applications in agricultural disease and weed management. (Servin et al., 2015; Elmer and White, 2016). Nanoparticles like copper (Cu), silver (Ag), titanium (Ti), and chitosan have shown their potential as novel antimicrobials for the management of pathogenic microorganisms that affect agricultural crops (Rai and Kratosova, 2015). Details about the roles of these NPs, their effects, and their targeted pathogens are mentioned in Table 4.1. Thus, the roles and

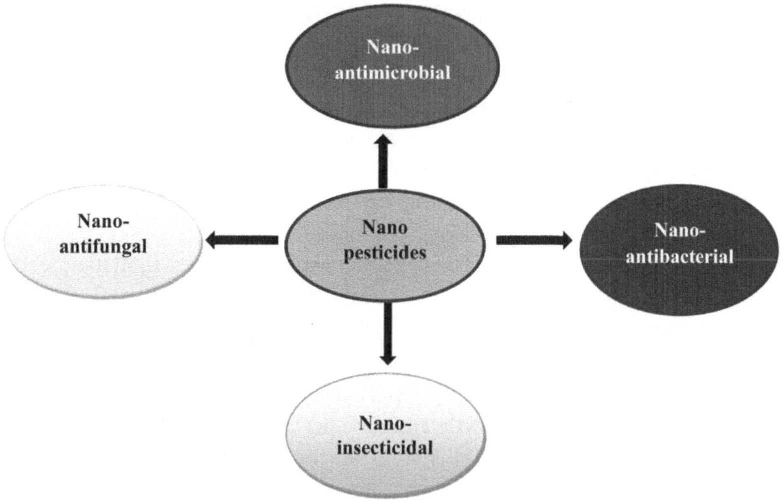

FIGURE 4.3 Uses of nanoparticles in pest control (against microbial pathogens, fungi, bacteria, and insects) (Source: Rai and Ingle, 2012; Worrall et al., 2018).

TABLE 4.1
Role of Nanoparticle in Disease Management

NPs	Role	Effects	References
ZnO	Effective growth control	*Fusariumgraminearum, F. oxysporum, Penicillium expansum, Alternaria alternate, Rhizopus stolonifer, Mucor plumbeus* and *Aspergillus flavus*	Dwivedi et al., 2016; Servin et al., 2015; Vanathi et al., 2016
Gal NPs	Pathogenic bacteria	*Pseudomonas aeruginosa*	Flockton et al., 2019
NanoCu	Effective growth control	*Phytophthora infestans*	Giannousi et al., 2013
Si andTiO_2	Suppress crop diseases	Antimicrobial activity	Servin et al., 2015
Ag/Cu/ ZnO	Inhibit the development	Fungal conidia and conidiophores	Panpatte and Jhala, 2019
TiO_2 NPs	Plant protection	Pathogen disinfection efficiency	Duhan et al., 2017

applications of nanomaterials in pesticides and pathogen controlling agents have tremendous scope for food production and sustainable agricultural development.

4.2.1.2 Nanofertilizers

The use of fertilizer and its management is very important for sustainable agricultural production (Davarpanah et al., 2016; Huang et al., 2017). Farmers use fertilizers to increase their agricultural yields, but most of the fertilizers used nowadays are chemical fertilizers, and their extensive use is resulting in decreased soil fertility. This is one of the major reasons for soil fertility loss in approximately 40% of the world's agricultural land. Due to the negative effect of the prolonged use of chemical fertilizers, even when using a higher dose of fertilizer, the agricultural yield decreases or becomes constant (Bharti et al., 2017; Kale and Gawade, 2016). A large portion of the chemical fertilizer remains in the soil or is transformed into another form and enters the environment causing environmental pollution, which affects the normal growth of flora and fauna (Tomer et al., 2016; Mahanty et al., 2017).

In modern agriculture and agri-food production, biofertilizers and nanofertilizers are the most important tools for meeting the demand for macro- and micro-nutrients by crops in an eco-friendly manner. They play an important role in supplementing essential nutrients, improving crop productivity, reducing environmental pollution, and improving soil fertility (Li et al., 2018; Danish and Hussain, 2019). Biofertilizers enhance plant growth and increase the bioavailability of nutrients using various mechanisms like atmospheric nitrogen fixation, siderophore production, phosphorus solubilization, and phytohormones synthesis (Devi et al., 2016; Raghavendra et al., 2017).

Implementing nanotechnology with fertilizers and biofertilizers could also be useful for sustainable agriculture and food production (Godfray et al., 2010). Nanofertilizers are nanomaterial-based fertilizers that have the potential to change the global agriculture scenario. Nanofertilizers help with the slow sustained and targeted release of nutrients, which decreases the overall demand for fertilizers and the cost incurred for them. Besides, less use of fertilizers also protects the environment. The role of nano and biofertilizers in enhancing plant growth and crop yield have been tabulated in Table 4.2. Nanospheres, nanocapsules, nanogels, and micelles are the various forms used for nanomaterials as nanopesticides and nanofertilizers (Ragaei and Sabry, 2014), which are discussed in the last section of this chapter.

TABLE 4.2
Role of Nano and Biofertilizers for Plant Growth Promotion

Role	References
Protective effects of nanofertilizers against biotic and abiotic stresses and minimization the negative effects of stresses	Shang et al., 2019
Increase availability and solubility of nutrients like phosphate, nitrogen and other micro/macronutrients	Suyal et al., 2016; Raghavendra et al., 2017; Devi et al., 2016
Production of phytohormones (i.e. auxins, gibberellins, cytokinins, abscisic acid) in soils enhancing plant growth	Mehnaz, 2015;
Plant growth promotion by nitrogen fixation	Syiem et al., 2017
management of soil fertility	Bharti et al., 2017
Sustaining the environment	Tomer et al., 2016; Mahanty et al., 2017

4.2.2 Nanosensors and Nano-Biosensors

Biotic and abiotic stresses, including diseases caused by pathogens, genetics, or nutrient deficiencies, are a major problem for agriculture, negatively affecting the quantity and quality of food production. Nanosensor-based identification of diseases and stress conditions at an early stage of crop growth would be much helpful for enhancing the yield and quality of food production (Li et al., 2005). A nanosensor is a sensor in which one or more essential parts of a sensor, such as a transducer, recognition element, or amplifier, are made up of nanomaterials for their enhanced output (Srivastava et al., 2018). A nano-biosensor is a nanosensor with a biological entity as its bio-recognition element. It can detect and respond to the desired analyte's physicochemical and biological signals in a more efficient manner. These nanosensors or nano-biosensors can sense a very small amount of analyte and can transduce it into a detectable output or signal that can be used for better agriculture output. Nanomaterial engineering is used to develop nanosensors and nano-biosensors, and the sensors designed for agricultural uses are useful for agricultural research.

The various types of nanosensors are: 1) nanoparticle-based sensors, 2) nanostructured materials, e.g. porous silicon, 3) nanoprobes, 4) nanowire nanosensors, 5) nanosystems: cantilevers and nanoelectromechanical systems (NEMS) (Fraceto et al., 2016). Nano-biosensors are used for more sustainable approaches to the management of agricultural practices (Panpatte et al., 2016). Nanosensor-based monitoring of crop and soil health is easy, cost-effective, eco-friendly, and ensures sustainable development. Nano-biosensors have an excellent potential in the area of automated irrigation systems to maximize the efficient use of water (Fraceto et al., 2016), as well as for the detection of insect attacks (Afsharinejad et al., 2015) or pathogens so that pesticides, insecticides or fertilizers can be appropriately applied to ensure the proper yield of a crop. The various applications of nanosensors and nano-biosensors are listed in Table 4.3. A massive number of research reports in this area indicate the importance of this field for future research.

Nanosensors reduce processing times and keep track of sustainable agricultural developments. Nanosensors have multiple agriculture applications, including testing soil quality, soil fertility, soil health, microbial population, and the microenvironment of the soil. Important soil quality parameters include soil moisture, soil temperature, soil pH, and the residual amount of pesticides, insecticides, and fertilizers. The study of root growth using nanotechnology can detect compatibility issues with marginal soils (Ghormade et al., 2011). The use of carbon nanotubes as a vector for the dissemination of chemicals desired for seed germination is another aspect of nanotechnology in agriculture (McLamore et al., 2010). The availability of different types of nanosensor or nano-biosensor provides the opportunity for their application in the agriculture system.

Nanosensors for soil quality and soil fertility provide information about the crop suitable for cultivation in that soil. Accurate information about nutrient deficiency provided by nanosensors could be useful for enhancing total crop production, and the integration of nanosensors with any remote sensing device can also enable the monitoring of crops from distant locations. Nanosensors can play a crucial role in crop protection by providing timely information about plant pathogens. Surface Plasmon-coupled emission biosensors modified with gold nanoparticles are also used

TABLE 4.3
Role of Nanosensor in Disease Detection

Nanomaterial exploited	Sensing strategy	References
Nanogold based immunosensor	To detect karnal bunt disease in wheat	Singh et al., 2010a
Potassium niobate (KNO_3) nanofibers	Sensing the humidity	Ganeshkumar et al., 2016
Single-walled carbon nanotubes (SWNTs)	Development of bionic plants	Ghorbanpour and Fahimirad, 2017; Kwak et al., 2017: Giraldo et al., 2014
Gold electrode nanosensor with copper nanoparticles	Detect the pathogenic fungus infestation in oilseed	Wang et al., 2010
Multiwalled carbon nanotubes (MWCNTs)	Plant growth study by regulation of hormones	McLamore et al., 2010
UC NPs/Au NPs (label-free glutathione regulated dual-functional)	Fluorescence detection of acetylcholinesterase (AChE) activity and toxic Cd^{2+} in real water	Fang et al., 2017

to estimate pathogenic organisms (Malic et al., 2011). These days, nano-biosensors are gaining thrust in food science to analyze and monitor bacteria and viruses in soil and crop products. They can also be used for assaying seed viability and shelf life of the seeds and other agricultural products. Nanosensors are eco-friendly in nature and have reduced the cost and effort required in agriculture. Their use in agriculture would help to assure soil and plant health, product quality, and overall safety for sustainable agriculture and environmental systems (Cheng et al., 2016).

4.2.3 Nanotechnology in the Food Sciences

According to a FAO report (FAO, 2011), approximately 6.0 Gtonnes of global agricultural production (for food and non-food uses) is produced every year. About one third (1.6 G tonnes) of food produced globally for human consumption is wasted every year, out of which 1.3 Gtonnes are edible food materials only. This wastage of food products causes not only financial losses but also exerts negative effects on the environment. According to a 2011 report by the FAO, the carbon footprint of this annual food wastage is equivalent to 3.3 Gtonnes of CO_2, which can be ranked as the third top emitter of greenhouse gases after the USA and China. In terms of blue water footprint (i.e. the consumption of surface and groundwater resources) food wastage accounts for 250 km^3 globally. Moreover, this wasted food occupies almost 1.4 billion hectares of land, which is 30% of the world's arable land. This food wastage also negatively affects global biodiversity (FAO, 2011). If this wastage could be avoided by following proper food preservation and food processing technologies, it would ensure global food security. Besides, wastage accounted for by food products; agricultural wastage also affects the environment. Post-harvest waste contains about

80% of the biomass in agricultural areas and this is often burned in fields, resulting in large amounts of green gas emissions and smog problems. These immense emissions have severe health effects and researchers have already seen an impact in New Delhi over the last few years, as the New Delhi-NCR region has suffered air pollution problems. Although these agricultural residues are being used for the production of manure and some biofuels on a smaller scale, these can also be used for the synthesis of valuable NPs.

Food procession is the comprehension of food preservation practices and the processes and technologies used for transforming and enhancing the shelf life of food. Food processing technology protects the flavour and quality of food and prevents it from infestation by harmful microorganisms which cause food spoilage. The production and processing of fresh food is not the only target of the food industry, as it also wants to produce healthier food (fortified food) is also a part of the current food industries objectives. Irradiation, autoclaving, canning, smoking, heating, and elevated hydrostatic pressure are methods commonly used for food preservation (Duncan, 2011; Neethirajan and Jayas, 2011; Athinarayanan et al., 2014). Methods and techniques for food processing that use nanomaterials cover the incorporation of nutraceuticals, gel preparation and enhancement of viscosity through various agents, fortification of vitamin, mineral, and flavours through their encapsulations in nanomaterials (Kumar et al., 2019; Bratovčić et al., 2015; Couch et al., 2016). Some of the nanomaterials and techniques used in food processing are listed in Table 4.4.

4.2.4 Nanoscale Biosensors

Nanoscale biosensors can detect and diagnose foodborne pathogens. Nanoscale bioactive ingredients in foodstuffs enhance the food quality as well as the health of the consumer (Martirosyan and Schneider, 2014; Pradhan et al., 2015). Simultaneously, nanoscale filtration systems also manipulate and improve food texture and hence preserve the glaziers and colours of processed food. Nanotechniques are widely used in the modern food industries; areas where nanotechniques are used include the use of nano-additives for processing and packaging, the authentication and identification purposes of food products (Khond and Kriplani, 2016; Savina et al., 2016), the controlled release of flavours in foodstuffs by mixing flavours in the form of nano-additives (Ghaani et al., 2016), the use of nanoprint for smart and intelligent packaging, the use of magnetic nanocomposites for tagging (Khond and Kriplani, 2016), and the nanocoding of plastics and paper materials (Bhushani and Anandharamakrishnan, 2014).

Nanosensors have higher sensitivity and selectivity than traditional sensors, and there are reports on the use of nanosensors to detect food spoilage and foodborne pathogens. Nanosensors are used for detecting changes, like the production of gases formed in the food or colour alterations due to spoilage in the food industry. Commonly used nanosensors employed in the food packaging industries include time-temperature integrators, gas detectors, array biosensors, nanoparticles in solution, nanoparticle-based sensors, electronic noses, nanotest strips, and nanocantilevers (Tang et al., 2009; Srivastava et al., 2018), etc. Generally, the sensors are sensitive to gas production, such as SO_2, NH_3, CO_2, H_2S, H_2, and NO_2 (Yotova et

TABLE 4.4
Nanomaterials and Techniques Used in Food Processing

Nanotechniques	Types	Functions	References
Nanoencapsulation (Food processing)	Nanocapsule, Nanoliposomes, Colloiodosomes, Nanochelates, Nanoceuticals	Increase stability, protection from oxidation, Taste making, pH control release, delivery of enzymes, pesticides, fertilizers, antioxidants, vitamins	Ezhilarasi et al., 2013; Fathi et al., 2014; Jafari and McClements, 2017
Nanoemulsions (Food processing)	Nanoemulsions	Salad dressing, Flavouring agent, Sweeteners, Quality enhancement	O'Sullivan et al., 2017; McClements, 2018
Nanosensors (Food packaging)	Metal-based nanosensor, nanosmart -dust, nanobarcodes, Biometric sensors, Nano-biocomposites Enzyme immobilization	Antimicrobial, use as a gas barrier, O_2 scavengers, Use to laminate meat, cheese, fruits, vegetables, pest control	Srivastava et al., 2018; Jain et al., 2018; Yaseen et al., 2018
Nanoparticles (Food preservation)	(Silicon, Silver, Zinc, Titanium) Nanoparticles	Anticaking, drying, Reducing moisture, Food Whiting, Antimicrobial	Khan et al., 2019; Sharma et al., 2020

al., 2015). Nanosensors installed at the food packaging plant can detect microorganisms on the spot and save costs incurred in quality control. These gas sensors are composed of metals like gold (Au), palladium (Pd), and platinum (Pt) (Deng et al., 2015). Nanosensors composed of gold nanoparticles are used to sense toxins like aflatoxin B1 in processed and unprocessed milk (Mao et al., 2006; Srivastava et al., 2018). A nanosensor is used to detect pesticides on the surface and inside fruits and vegetables. Nanosensors are confined to the detection of residual pesticides (Kumar et al., 2018) and used to detect carcinogens in food materials (Xiang et al., 2018). Nanobarcodes are another competent mechanism for monitoring the quality of agricultural products (He et al., 2019).

4.3 NANOTECHNOLOGY FOR WASTE AND POLLUTION MANAGEMENT

4.3.1 Nanotechnology in Agricultural Wastes/Residues Recycling

Waste management represents a significant challenge in the agro-food industries and requires an integrated approach in terms of recycling, reuse, and recovery

(Krishnaswamy et al., 2014). Agricultural waste globally accounts for 0.3 Gtonnes of biomass every year (FAO, 2011). Besides traditional uses as fuel, fodder, or biofertilizer, agricultural waste may have multiple other uses like in separation, adsorption, catalysis, detection, semi-conductivity, and thermal insulation. These alternative uses of agro-wastes need further research and development for their broader application. However, recent developments in the plant-based eco-friendly synthesis of NPs has opened a new window to the large-scale synthesis of metal or organo-functional nanostructures. Plants, in general, are known as bioreactors because of their ability to synthesize nanoparticles. The synthesis of nanoparticles through parts of a plant is a one-step and straightforward process. The bio-based economy and renewable biomass as a raw material have been considered sustainable alternatives to address the problems associated with local and global pollution rises. The various types of nanoparticle which have been synthesized using agro-waste are mentioned in Table 4.5.

Biological methods for NP synthesis are a new field of chemistry called 'green chemistry', which is cheaper and more eco-friendly than chemical methods. In this context, crop plants also play an important role in the synthesis of NPs using biological methods and can help phytoremediation (Ahamd et al., 2019). Crop plants are a cheap and renewable resource for NP synthesis because the extract of the plant biomass, such as flowers, fruits, leaves, stems, and seeds, is used to synthesize NPs. This biological method for NP synthesis using plant extracts can be the best alternate to the physical and chemical methods for NP synthesis as it overcomes the toxicity and safety issues that arise due to toxic by-products (Shankar et al., 2004; Shekhawat et al., 2012; Hussain and Hussain, 2015). Plant-based NP synthesis supports applications in agriculture and biotechnology and has attracted scientists to explore its use as a way to provide quality food for ever-growing populations. Hence, many countries in the world are looking at nanotechnology as a reliable technology to implement in agriculture (Yadav et al., 2018; Hussain et al., 2017c). Some of the plant extracts used for NP synthesis are summarized in Table 4.6.

4.3.2 Microorganism Assisted Nanoparticle-Based Metal Remediation

Microorganisms produce NPs from the metals present in their surroundings as a by-product of their metal resistance mechanism. This helps them to survive in environments with a high metal concentration. As a useful by-product of their survival mechanism, microorganisms reduce the metal toxicity from the environment by bioremediation of the metals. There are two mechanisms of NP-based metal bioremediation adopted by microorganisms. In the intracellular process, microorganisms grab the toxic ionic forms of metal from their environment and reduce them to nontoxic elemental forms with the help of their cellular enzymatic activities. In the alternative extracellular process, the metal ions are trapped on the microbial surface and reduce to an elemental form for NP formation through the enzymes present in the microbes' cell envelope (Parsad et al., 2015). The biosynthesis of gold, silver, gold-silver alloy, selenium, tellurium, platinum, palladium, silica,

TABLE 4.5
Synthesis of Nanoparticles From Different Agrowaste

Type of Agrowaste	Nanoparticle Produced	Size of Nanoparticles	Reference
Citrus Aurantifolia peel extract	Ag NPs	17.3–61.2 nm	Dauthal and Mukhopadhyay, 2015
Moringaoleifera	Ag NPs	5–80 nm	Sathyavathi et al., 2011
Sugarcane Bagasse	Ag NPs	50–150 nm	Mishra and Sardar, 2013
Banana peel Extract	Au NPs	20 nm	Bankar et al., 2010
Grape Waste	Au NPs	20–25 nm	Krishnaswamy et al., 2014
Cauliflower Broth	Ag NPs	42–83 nm	Sridhara et al., 2012
Mango peel	Au NPs	6–18 nm	Yang et al., 2014
Annona squamosal peel	Ag NPs	35 nm	Kumar et al., 2013
Timber industry Waste	Ag NPs	28 nm	Maria et al., 2015
Rice husk and straw	SiO_2 NPs Ag NPs	70 nm, 25 nm	Wang et al., 2017Andas and Adam, 2016
Bamboo leaf	SiO_2 NPs	13.8 nm	Rangaraj and Venkatachalam, 2017
Egg Shell waste	Ca_2SiO_4 NPs	75 nm	Pandit and Fulekar, 2017
Walnut shell waste	Cu/Cu_2O NPs CeO_2 NPs	11 nm 9–21 nm	Saadati et al., 2017Zamani et al., 2018
Wheat straw	ChCl-LA-DES	70–90 nm	Lou et al., 2019
Coconut shell waste	Ag NPs	14–22 nm	Sinsinwar et al., 2018
Tea waste	Ag NPs	2–34 nm	Kaur and Jayral, 2018
Olive oil waste	Ti NPs	30 nm	Mihankhah et al., 2018
Groundnut shell waste	SiO_2 NPs	20–200 nm	Yaro et al., 2016

titanium, zirconia, quantum dots (QDs), magnetite, and uraninite NPs by bacteria, actinomycetes, fungi, yeasts, and viruses have been reported. Several fungal strains, like *Fusarium*, *Aspergillus*, *Verticillium*, and *Penicillium*, have been used as promising nanoparticle fabrication resources. Alghuthaymi et al. (2015) reported that *Fusarium oxysporum* synthesized the smallest size of Ag nanoparticles. They found that *F. oxysporum* can extracellularly reduce aqueous Ag ions in water to generate Ag nanoparticles. Some of the microorganisms used for the synthesis of metal NPs are summarized in Table 4.6.

TABLE 4.6
Characteristics and Function of Metal Nanoparticles From Microbes and Plant Extract

Species	Metal	Size	Shape	Uses	References
Lactobacillus casei	Silver	20–50	Spherical	Drug delivery, bio-labelling	Verma et al., 2019
E. coli	Cadmium	2–5	Wurtzite structures	Fluorescent labels	Patel et al., 2019
Saccharomyces cerevisiae	Gold, Silver	4–15	Spherical	Catalysis	Verma et al., 2020
Penicillium brevicompactum	Silver	23–105	Spherical	Antimicrobial activity	Mishra et al., 2011
Aspergillus niger	Silver	3–30	Spherical	Antimicrobial activity	Alani et al., 2012
Trichoderma viride	Silver	5–40	Spherical	Antibacterial activity	Mohammed Fayaz et al., 2009
Helminthosporum solani	Gold	2–70	Irregular	Anti-cancer drug	Kumar et al., 2008
Fusarium oxyporum	Silver	20–50	Spherical	Antibacteria	Durán et al., 2005
Syzygium aromaticum	Gold	5–100	Irregular	Detection/ destruction of cancer cells	Verma et al., 2020
Azadirachta indica	Gold, Silver	5–35	Spherical, hexagonal	Remediation of toxic metal	Kumari et al., 2019
Mentha piperita	Silver	5–30	Spherical	Antimicrobial	Verma et al., 2020

4.4 NANOTECHNOLOGY FOR GENETIC ENGINEERING AND CROP IMPROVEMENT

Nano-biotechnology, a part of nanotechnology, provides gene manipulation techniques using nanoparticles, nanofibers, and nanocapsules. Nanomaterials have the capability to carry many genes for their controlled release and specific gene expression in plants (McLamore et al., 2010; Hussain et al., 2017c). Nano-biotechnology plays an essential role in improving agricultural productivity through the (i) transfer of target genes and drug molecules to target sites at cellular levels; (ii) genetic improvement of seed germination, water management, and the enhanced production of plant hormones and metabolites; (iii) nanoarray based gene-technologies for gene expression profiling in plants; and (iv) development of disease and pest-resistant varieties through the breeding of nanoparticle-based gene transfer (Worrall et al., 2018).

4.5 NANOTOXICITY: CAUSE AND REMEDIATION

While nanotechnology holds excellent future potential, its novelty and speed of development have lead to uncertainty about the long-term effects of NPs on the environment and human health. In the short term, no risks are identified, but in the long term they can affect humans through the bio-accumulation of toxins in plants and animals (Kah et al., 2013; Rana and Kalaichelvan, 2013; Kah and Hofmann, 2015). NPs and nono-devices are being manufactured frequently and being released directly or indirectly into the environment without any safety testing. NPs induce cytotoxicity by stimulating inflammatory and stress responses at molecular levels. Nanomaterials can travel more freely and easily, hence they may be more toxic compared to their larger counterparts. They can enter the human body through water, air, and via contact, and can cause damage to the internal tissue and organs of the body (Wani et al., 2011). The smaller the NPs, the deeper their response and the slower their elimination from the body. Some NPs are known to exhibit high levels of vascular thrombosis too. This has been observed in a study in which lactic-acid-based NPs of 121–343 nm were used to encapsulate a drug (meso-tetra carboxy-phenyl porphyrin). In a short period of time, research suggested that the larger NPs were more quickly removed from the bloodstream, and therefore they were found to be more effective and less toxic (Pegaz et al., 2006). The toxicity of NPs depend on their concentration, properties, and size, which affect the performance and mobility of NPs within the environment (Kah et al., 2013). After entering into the human body, the metal (Ag, Au, and Cu) based NPs disturb the body's antioxidant mechanism by interacting with the antioxidant enzymes and proteins of the mammalian cells. This generates reactive oxygen species (ROS) in the cell and results in the onset of inflammatory reactions and, sometimes, even destruction of mitochondria, which, in turn, cause apoptosis and cellular necrosis (Schrand et al., 2010). NPs present in UV-protective cosmetic creams can accumulate on the skin's surface and in the 'stratum corneum' between the keratinized cells, over a period of time. Generally, it may not cause toxicity, but if the skin is compromised due to burns and injuries, there is a possibility that these NPs may enter the body and cause adverse effects (Barnard, 2010).

The impacts and effects of nanomaterial generated toxicity on human health should be evaluated because it is necessary to know how toxic the nanomaterial is before its use for various applications. The field of nanotoxicology has grown rapidly, and a review of the literature suggests that results are diverse and fascinating. The International Nanotechnology Council (ICON) has created a database of all publications of several nanomaterials, along with their impact on environmental health and safety (Haynes, 2010). Besides the toxicity of NPs themselves, toxic chemicals generated during the synthesis of NPs are also harmful to both the environment and humans. In order to mitigate the deleterious effects of some of the NPs on the environment, the toxicity of the NPs can be removed or reduced by redesigning them (Dasgupta et al., 2016; Neal, 2008; Jain et al., 2018). The use of agricultural waste for the synthesis of NPs can be another approach to mitigate NPs generated toxicity (Smith and Granqvist, 2011; Oomen et al., 2015).

4.6 QUANTUM EFFECT OF NANOMATERIALS AND THEIR DIFFERENT FORMS

When a material is moved to the nanoscale, the surface area to mass ratio of the material increases; therefore, it can interact with a larger amount of the surrounding material, and its reactivity is increased. This effect is known as a quantum effect, due to which lower activation energy is required for chemical reactions to occur, compared to crude and non-nano sized materials. Surface Plasmon resonance is a phenomenon which is accountable for the increased sensitivity of the NPs. Many metallic and non-metallic nanomaterials of different sizes and shapes can be used for environmental clean-up. NPs, nanowires, nanosheets, nanocomposites, etc. are examples of different nanomaterials. These materials are converted into various forms like nanogels, nanoencapsulations, nanoemulsions, etc., for the ease of their application.

4.6.1 Nanoencapsulation

Nanoencapsulation is a nano range vesicular system in which a typical core and shell-like structure is present. Basically, the encapsulated material contains two phases, i.e. the internal phase of the core material (drug/pesticide/nutrients/any desired products) and the external phase of the capsulation materials (coated nanomaterials) (Nuruzzaman et al., 2016). The core material is kept within a cavity to restrict it in the core, and this core is surrounded by a coating or nanopolymer membrane (Sekhon, 2010; Biswal et al., 2012; Pradhan et al., 2015). The cavity created within the encapsulation holds the active substance either in a liquid/solid or molecular dispersion form. The prime aim of a nanocapsule is to transfer the desired product in a slow and sustained manner, such as the prolonged release of pleasant odour by deodorizing gels compared to sprays. Within the biological system, nanocapsules entrap the desired product (such as drug or nutrient or food supplement) and move via the gastrointestinal tract, and thus result in the availability of that substance for a prolonged duration. Basically, there are six construction methods for nanocapsules: layer-by-layer, polymer coating emulsion-diffusion, double emulsification, and nanoprecipitation emulsion-coacervation (Maynard et al., 2006; Huang et al., 2010; Pradhan et al., 2015; Singh et al., 2017).

Nanoencapsulation has several advantages, like protection against lipid peroxidation, conservation of volatile ingredients, enhancement of taste, controlled release of moisture, enhanced stability, ease of handling, pH-triggered controlled release, multiple active ingredients delivery, alteration of flavour, long-lasting organoleptic perception, and enhanced bioavailability and efficacy (Chaudhry et al., 2008; Dasgupta et al., 2015; Gupta et al., 2016a). Nanoencapsulation is also applicable for transferring lipophilic health supplements, such as minerals and vitamins, fatty acids, and growth hormones, into the food, resulting in a value addition to the foodstuffs (Huang et al., 2010; Jain et al., 2018). The prime advantage of encapsulation is to defend the hidden component for its transfer to the targeted site, even in adverse conditions. Zein fibres (the fibres made up of a principal protein of maize, amended with gallic acid) defend the lipids from getting degraded early within the food system before reaching the target (Moomand and Lim, 2014); it is the most widely used nanotechnique for encapsulation in the food packaging industry (Thomas and Sayre, 2005; Pinto et al., 2013).

Liposome, a nano-based carrier considered for nanoencapsulation as nanoliposomes, assists in the controlled and specific transfer of different foodstuffs components. Nanoliposomes are used to transfer additives, vitamins, enzymes, nutraceuticals, and antimicrobial compounds (Godwin et al., 2009; Pradhan et al., 2015). Lipid-based encapsulation techniques are more efficient than other encapsulation techniques due to the encapsulated components' high solubility and specificity. Nanoliposome encapsulation restricts the component from being mixed with the food components; thus, the food's original quality remains intact, and the at same time the components to be delivered within the biological system are also unaltered (Klaine et al., 2008; Inbaraj and Chen, 2015). Different components like food supplements and drugs can be placed and transported successfully inside the nanoliposomal, making nanoliposomes a good carrier for the delivery of desired components within the biological system. Similarly, nanocochleates enhance the quality of processed food. Nanocochleates are nanocoils composed of soy-based phospholipids, which enclose micro-nutrients and provide stability in encapsulation (Abbas et al., 2009; Gupta et al., 2016a).

4.6.2 Nanoemulsions

Nanoemulsions are generated by dissipating liquid hydrophobic (lipophilic) material in the continuous aqueous phase. Lipophilic components are employed for the construction of nanoemulsion, and the lipophilic component is blended within the oil phase (McClements, 2012; Gupta et al., 2016b). Further, the lipophilic component's stability and delivery within the nanoemulsion rely on various factors like the chemical, molecular, and physical properties of the lipophilic component. The physical and chemical properties include surface activity, hydrophobicity, solubility, oil-water partition coefficient, and melting point (SinhaRay and Okamoto, 2003; Mungure et al., 2018). Some standard lipophilic components that are used in encapsulation and nanoemulsion formation are capsaicin, citral, flaxseed oil, β-carotene, tri-butyrin, and different oil-soluble vitamins (Vo-Dinh et al., 2001; Pathak, 2017). Nanoemulsions are more stable than conventional emulsions for gravitational separation and droplet assemblage. In addition, nanoemulsions are thermally more stable than conventional emulsions and are in higher demand than conventional emulsions due to their larger surface area, rapid digestion, and ease of absorption (Asghar et al., 2018).

Proteins (milk, vegetable, soy protein, and egg) and carbohydrates (starch, pectin, alginate, and xanthan) enhance the appearance and texture of the emulsion (Hogan et al., 2001; Bouchaala et al., 2017/2018). Nanoemulsions also have antimicrobial activity and are found more effective against Gram-positive bacteria than Gram-negative bacteria (Pinto et al., 2013); hence, nanoemulsions are applied for the sterilization of food packaging articles. Microbial growth control can be achieved by applying nanoemulsions made up of tributyl phosphate (a non-ionic surfactant) and soybean oil (Sanguansri and Augustin, 2006; Sanzari et al., 2019). Nanoemulsions form a shield via self-assemblage toward functional compounds (vitamins A, vitamin D, vitamin E, β-carotene, iso-flavones, lutein, and lycopene) of the foodstuff and restrict the action of degrading enzymes responsible for food spoilage. Nanoemulsions inhibit the effect of food degrading enzymes and inhibit the process of hydrolysis, oxidation, temperature, and the pH alteration of foods (Bhattacharya et al., 2007; Pinto et al., 2013; Mungure et al., 2018). High energy-based and low energy-based approaches are both used for nanoemulsion formation. High energy-based approaches include the ultrasound technique, homogenization under high pressure, and liquid coaxial jets at high speed, while low energy-based approaches include phase inversion point, spontaneous emulsification, solvent displacement emulsification, and emulsion inversion point (Yu et al., 2012; Mungure et al., 2018).

4.6.3 Nanocomposites

Nanocomposite materials are composed of polymers mixed with NPs and are used to improve the polymer's properties through association with the NPs (Yotova et al., 2013; Mousavi et al., 2018). Nanocomposites help in maintaining the sterility and freshness of food products for a reasonable time. In the bottles of carbonated beverages, nanocomposites behave as barriers to minimize the leakage of carbon dioxide (Shi et al., 2013; Mousavi et al., 2018) and thus enhance shelf life. Nanoclay is an excellent example of a nanocomposite that has an application in the creation of gas barriers. NPs combined with cellulose results in a low cost, light-weight nanocomposite (Guo et al., 2018). Polymers that are reinforced with metals in the nano range act as antimicrobials, such as nanozinc oxide and nanomagnesium oxide mixed polymers. Pediocin incorporated ZnO_2 nanoparticles in nanocomposite films have been reported to show antimicrobial activity (GharoyAhangar et al., 2015). Similarly, silver-coated nanocomposites also work as antimicrobial agents. Silver oxide attaches to the cell surface, dissolves the lipopolysaccharide, and results in an enhanced permeability of the bacteria as well as the potential to damage the bacterial DNA. Polyethene glycol (PEG) coated with garlic oil nanocomposites has been found useful in pest control at food stores (Sung et al., 2013; Mishra et al., 2019). Bio-nanocomposites, like poly-butylene succinate, poly-caprolactone, poly-lactic acid, and poly-hydroxy-butyrate, are composed of starch and cellulose derivatives. These bio-nanocomposites show antibacterial properties and have been used as laminating materials for the food packaging (Xie et al., 2013; Karmee, 2018).

4.6.4 Nanospheres

The spherical, homogenous mixture of an active compound and polymeric nanomaterials (nanospheres), hydrophilic, generally cross-linked polymers, which absorb a high volume of water and take the form of a gel-like consistency (nanogels), and spherical droplets of molecules containing hydrophilic and hydrophobic moieties (micelles) are few other forms in which nanomaterials are used for their various applications.

4.7 CONCLUSION

Significant numbers of nanomaterials behave as growth regulators and hence show remarkable enhancement in plant biomass (leaves, fruit, and flower, etc.) yields, as well as value additions to nutritional quality (fortification). Nanomaterial toxicity may happen at physiological, biochemical, and molecular levels. Further, the nanotoxicology researcher should emphasize several aspects like how NPs go through plant cells and are then dispersed across different parts, affecting growth, yield, and deciphering the mechanisms involved. Nanophytoremediation technology is a comparatively newer area for the environmental remediation process; however, currently, a lot of research is still limited to the laboratory. The various nanophytoremediation processes help with environmental remediation. However, the remediation process rate in different regions varies due to differences in soil type, pH, temperature, humidity, moisture, and available nutrients in the soil. Hence, studies are required for the optimization of the efficiencies and compatibilities of nanomaterials with suitable plant species. Modern food packaging holds the scope for nanosensors that convey information on food conditions inside the packing, and nanopackaging should prevent the release of enzymes, flavours, antioxidants, nutraceuticals, and the entry of microorganisms to enhance the shelf life of packed food. It is also essential to conduct studies on the migration of NPs in the human body and their human toxicological aspects to develop safe nanotechnologies for food packaging.

ACKNOWLEDGEMENT

Authors are thankful to their respective institutes for extending the departmental facilities for completion of this work. All the authors express their gratitude towards Director (ICAR-IISR, Kushmaur, Mau), Director (Agharkar Research Institute, Pune), Department of Botany and Microbiology, HNB Garhwal University; Department of Botany, Jiwaji University; Gwalior and Department of Botany, Aligarh Muslim University for their support.

REFERENCES

Abbas, K. A., Saleh, A. M., Mohamed, A., MohdAzhan, N. 2009. The recent advances in the nanotechnology and its applications in food processing: A review. *Journal of Food, Agriculture and Environment* 7(3–4):14–17.

Afsharinejad, A., Davy, A., Jennings, B., Brennan, C. 2015. Performance analysis of plant monitoring nanosensor networks at THz frequencies. *IEEE Internet of Things Journal* 3(1):59–69.

Ahamd, G., Nishat, Y., Haris, M., Danish, M., Hussain, T. 2019. Efficiency of soil, plant and microbes for the healthy plant immunity and sustainable agricultural system. In *Plant-Microbe Interface*, edited by A. Varma, S. Tripathi, R. Prasad. Springer, Cham.

Alani, F., Moo-Young, M., Anderson, W. 2012. Biosynthesis of silver nanoparticles by a new strain of *Streptomyces* sp. compared with *Aspergillus fumigatus*. *World Journal of Microbiology and Biotechnology* 28(3):1081–1086. https://doi.org/10.1007/s11274-011-0906-0.

Alghuthaymi, M. A., Almoammar, H., Rai, M., Said-Galiev, E., Abd-Elsalam, K. A. 2015. Myconanoparticles: Synthesis and their role in phytopathogens management. *Biotechnology and Biotechnological Equipment* 29(2):221–236.

Andas, J., Adam, F. 2016. One-pot synthesis of nanoscale silver supported biomass-derived silica. *Materials Today: Proceedings* 3(6):1345–1350.

Asghar, A., Randhawa, M. A., Masood, M. M., Abdullah, M., Irshad, M. A. 2018. Nutraceutical formulation strategies to enhance the bioavailability and efficiency: An overview. In *Role of Materials Science in Food Bioengineering* (pp. 329–352). Academic Press.

Athinarayanan, J., Periasamy, V. S., Alsaif, M. A., Al-Warthan, A. A., Alshatwi A. A. 2014. Presence of nanosilica (E551) in commercial food products: TNF-mediated oxidative stress and altered cell cycle progression in human lung fibroblast cells. *Cell Biology and Toxicology* 30(2):89–100. https://doi.org/10.1007/s10565-014-9271-8.

Bankar, A., Joshi, B., RaviKumar, A., Zinjarde, S. 2010. Banana peel extract mediated synthesis of gold nanoparticles. *Colloids and Surfaces, Part B: Biointerfaces* 80(1):45–50.

Barnard, S. A. 2010. Barnard, A. S. (2010). One-to-one comparison of sunscreen efficacy, aesthetics and potential nanotoxicity. *Nature Nanotechnology* 5(4):271–274.

Bharti, V. S., Dotaniya, M. L., Shukla, S. P., Yadav, V. K. 2017. Managing soil fertility Through microbes: Prospects, challenges and future strategies. In *Agro- Environmental Sustainability*, edited by J. S. Singh, G. Seneviratne (pp. 81–111). Springer International Publishing AG. https://doi.org/10.1007/978-3-319-49724-2_5.

Bhattacharya, S., Jang, J., Yang, L., Akin, D., Bashir, R. 2007. BioMEMS and nanotechnology-based approaches for rapid detection of biological entities. *Journal of Rapid Methods and Automation in Microbiology* 15(1):1–32.

Bhushani, J. A., Anandharamakrishnan, C. 2014. Electrospinning and electrospraying techniques: Potential food based applications. *Trends in Food Science and Technology* 38(1):21–33.

Biswal, S. K., Nayak, A. K., Parida, U. K., Nayak, P. L. 2012. Applications of nanotechnology in agriculture and food sciences. *International Journal of Science Innovations and Discoveries* 2(1):21–36.

Bouchaala, R., Anton, N., Anton, H., Vandamme, T., Vermot, J., Smail, D. et al. 2017. Light-triggered release from dye-loaded fluorescent lipid nanocarriers *in vitro* and *in vivo*. *Colloids and Surfaces, Part B: Biointerfaces* 156:414–421.

Bratovčić, A., Odobašić, A., Ćatić, S., Šestan, I. 2015. Application of polymer nanocomposite materials in food packaging. *Croatian Journal of Food Science and Technology* 7:86–94. https://doi.org/10.17508/CJFST.2015.7.2.06.

Chaudhry, Q., Scotter, M., Blackburn, J., Ross, B., Boxall, A., Castle, L. et al. 2008. Applications and implications of nanotechnologies for the food sector. *Food Additives and Contaminants. Part A, Chemistry, Analysis, Control, Exposure and Risk Assessment* 25(3):241–258.

Cheng, H. N., Klasson, K. T., Asakura, T., Wu, Q. 2016. *In Nanotechnology: Delivering on the Promise Vol. 2*. In ACS Symposium Series, *Am. Chem. Soc* (pp. 233–242).

Couch, L. M., Wien, M., Brown, J. L., Davidson, P. 2016. Food nanotechnology: Proposed uses, safety concerns and regulations. *Nutraceuticals, Functional Foods and Healthy Ingredients* 27:36–39.

Danish, M., Hussain, T. 2019. Nanobiofertilizers in Crop production. In *Nanoparticles and microbes-Next generation Technology for Agricultural* (pp. 107–118), edited by D. G. Panpatte et al. Springer nature.. https://doi.org/10.1007/978-981-32-9374-8_6.

Dasgupta, N., Ranjan, S., Mundekkad, D., Ramalingam, C., Shanker, R., Kumar, A. 2015. Nanotechnology in agro-food: From field to plate. *Food Research International* 69:381–400. https://doi.org/10.1016/j.foodres.2015.01.005.

Dasgupta, N., Ranjan, S., Rajendran, B., Manickam, V., Ramalingam, C., Avadhani, G. S., Kumar, A. 2016. Thermal co-reduction approach to vary size of silver nanoparticle: Its microbial and cellular toxicology. *Environmental Science and Pollution Research International* 23(5):4149–4163.

Dauthal, P., Mukhopadhyay, M. 2015. Agro-industrial waste-mediated synthesis and characterization of gold and silver nanoparticles and their catalytic activity for 4-nitroaniline hydrogenation. *Korean Journal of Chemical Engineering* 32(5):837–844.

Davarpanah, S., Tehranifar, A., Davarynejad, G., Abadía, J., Khorasani, R. 2016. Effects of foliar applications of zinc and boron nanofertilizers on pomegranate (*Punicagranatum cv. Ardestani*) fruit yield and quality. *Scientia Horticulturae* 210:57–64.

Deng, J., Lu, Q., Hou, Y., Liu, M., Li, H., Zhang, Y., Yao, S. 2015. Nanosensor composed of nitrogen-doped carbon dots and gold nanoparticles for highly selective detection of cysteine with multiple signals. *Analytical Chemistry* 87(4):2195–2203.

Devi, S. N. P., Kumara, K. S., Vasandha, S. 2016. Assessment of competence of the *Pseudomonas aeruginosa* to solubilize insoluble form of zinc under various cultural parameters. *Arabian Journal for Science and Engineering* 41(6):2117–2121. https://doi.org/10.1007/s13369-015-1907-3.

Duhan, J. S., Kumar, R., Kumar, N., Kaur, P., Nehra, K., Duhan, S. 2017. Nanotechnology: The new perspective in precision agriculture. *Biotechnology Reports* 15:11–23.

Duncan, T. V. 2011. Applications of nanotechnology in food packaging and food safety: Barrier materials, antimicrobials and sensors. *Journal of Colloid and Interface Science* 363(1):1–24.

Durán, N., Marcato, P. D., Alves, O. L., De Souza, G. I., Esposito, E. 2005. Mechanistic aspects of biosynthesis of silver nanoparticles by several *Fusarium oxysporum* strains. *Journal of Nanobiotechnology* 3(1):8. https://doi.org/10.1186/1477-3155-3-8.

Dwivedi, S., Saquib, Q., Al-Khedhairy, A. A., Musarrat, J. 2016. Understanding the role of nanomaterials in agriculture. In *Microbial Inoculants in Sustainable Agricultural Productivity*, edited by D. P. Singh, H. B. Singh, R. Prabha (pp. 271–288). Springer, New Delhi, India.

Elmer, W. H., White, J. C. 2016. The use of metallic oxide nanoparticles to enhance growth of tomatoes and eggplants in disease infested soil or soilless medium. *Environmental Science: Nano* 3(5):1072–1079.

Ezhilarasi, P. N., Karthik, P., Chhanwal, N., Anandharamakrishnan, C. 2013. Nanoencapsulation techniques for food bioactive components: A review. *Food and Bioprocess Technology* 6(3):628–647. https://doi.org/10.1007/s11947-012-0944-0.

Fang, A., Chen, H., Li, H., Liu, M., Zhang, Y., Yao, S. 2017. Glutathione regulation-based dual-functional upconversion sensing-platform for acetylcholinesterase activity and cadmium ions. *Biosensors and Bioelectronics* 87:545–551.

FAO 2011. Food wastage footprint: Impact on Natural Resources, Summary report, accessed on 22.02.2020. http://www.fao.org/3/i3347e/i3347e.pdf.

Fathi, M., Martin, A., McClements, D. J. 2014. Nanoencapsulation of food ingredients using carbohydrate based delivery systems. *Trends in Food Science and Technology* 39(1):18–39. https://doi.org/10.1016/j.tifs.2014.06.007.

Flockton, T. R., Schnorbus, L., Araujo, A., Adams, J., Hammel, M., Perez, L. J. 2019. Inhibition of *Pseudomonas aeruginosa* biofilm formation with surface modified polymeric nanoparticles. *Pathogens* 8(2):55.

Fraceto, L. F., Grillo, R., de Medeiros, G. A., Scognamiglio, V., Rea, G., Bartolucci, C. 2016. Nanotechnology in agriculture: Which innovation potential does it have? *Frontiers in Environmental Science* 4:20.

Ganeshkumar, R., Sopiha, K. V., Wu, P., Cheah, C. W., Zhao, R. 2016. Ferroelectric $KNbO_3$ nanofibers: Synthesis, characterization and their application as a humidity nanosensor. *Nanotechnology* 27(39):395607. https://doi.org/10.1088/0957-4484/27/39/395607.

Ghaani, M., Cozzolino, C. A., Castelli, G., Farris, S. 2016. An overview of the intelligent packaging technologies in the food sector. *Trends in Food Science and Technology* 51:1–11.

GharoyAhangar, E., Abbaspour-Fard, M. H., Shahtahmassebi, N., Khojastehpour, M., Maddahi, P. 2015. Preparation and characterization of PVA/ZnO nanocomposite. *Journal of Food Processing and Preservation* 39(6):1442–1451.

Ghorbanpour, M., Fahimirad, S. 2017. Plant nanobionics a novel approach to overcome the environmental challenges. In *Medicinal Plants and Environmental Challenges* (pp. 247–257). Springer, Cham.

Ghormade, V., Deshpande, M. V., Paknikar, K. M. 2011. Perspectives for nanobiotechnology enabled Protection and nutrition of plants. *Biotechnology Advances* 29(6):792–803.

Giannousi, K., Avramidis, I., Dendrinou-Samara, C. 2013. Synthesis, characterization and evaluation of copper-based nanoparticles as agrochemicals against *Phytophthora infestans*. *RSC Advances* 3(44):21743–21752.

Giraldo, J. P., Landry, M. P., Faltermeier, S. M., McNicholas, T. P., Iverson, N. M., Boghossian, A. A., et al. 2014. Plant nanobionics approach to augment photosynthesis and biochemical sensing. *Nature Materials* 13(4):400.

Godfray, H. C. J., Beddington, J. R., Crute, I. R., Haddad, L., Lawrence, D., Muir, J. F. et al. 2010. Food security: The challenge of feeding 9 billion people. *Science* 327(5967): 812–818.

Godwin, H. A., Chopra, K., Bradley, K. A., Cohen, Y., Harthorn, B. H., Hoek, E. M. et al. 2009. The University of California Center for the environmental implications of nanotechnology. *Environmental Science and Technology* 43(17):6453–6457.

Guo, F., Aryana, S., Han, Y., Jiao, Y. 2018. A review of the synthesis and applications of polymer–nanoclay composites. *Applied Sciences* 8(9):1696.

Gupta, A., Eral, H. B., Hatton, T. A., Doyle, P. S. 2016. Nanoemulsions: Formation, properties and applications. *Soft Matter* 12(11):2826–2841. https://doi.org/10.1039/c5sm02958a.

Haq, I. U., Ijaz, S. 2019. Use of metallic nanoparticles and nanoformulations as nanofungicides for sustainable disease management in plants. In *Nanobiotechnology in Bioformulations*, edited by R. Prasad, V. Kumar, M. Kumar, D. Choudhary (pp. 289–316). Springer, Cham.

Haynes, C. L. 2010. The emerging field of nanotoxicology. *Analytical and Bioanalytical Chemistry* 398(2):587–588. https://doi.org/10.1007/s00216-010-3972-5.

He, X., Deng, H., Hwang, H. M. 2019. The current application of nanotechnology in food and agriculture. *Journal of Food and Drug Analysis* 27(1):1–21.

Hogan, S. A., McNamee, B. F., Dolores O'Riordan, E., O'Sullivan, M. 2001. Microencapsulating properties of sodium caseinate. *Journal of Agricultural and Food Chemistry* 49(4):1934–1938.

Huang, M., Wang, Z., Luo, L., Wang, S., Hui, X., He, G., et al. 2017. Soil testing at harvest to enhance productivity and reduce nitrate residues in dryland wheat production. *Field Crops Research* 212:153–164.

Huang, Q., Yu, H., Ru, Q. 2010. Bioavailability and delivery of nutraceuticals using nanotechnology. *Journal of Food Science* 75(1):50–56.

Hussain, T. 2017. Nanotechnology: Diagnosis of plant diseases. *Agricultural Research and Technology: Open Access Journal* 10(1):555777. http://www.ncbi.nlm.nih.gov/pubmed/555777.

Hussain, K., Hussain, T. 2015. Gold Nanoparticles: A boon to Drug delivery system. *South Indian Journal of Biological Sciences* 1(3):127–133.

Hussain, K., Hussain, T., Khan, A. A., Anwar, F., Khan, M. A. 2017. Nanobiotechnology for the Management of Pest and Diseases Incidence in Agricultural as well as Horticultural crops. *Trends in Biosciences* 10(39):8095–8101.

Inbaraj, B. S., Chen, B. H. 2015. Nanomaterial-based sensors for detection of foodborne bacterial pathogens and toxins as well as pork adulteration in meat products. *Journal of Food and Drug Analysis* 24:15–28. https://doi.org/10.1016/j.jfda.2015.05.001.

Jafari, S. M., McClements, D. J. 2017. Nanotechnology approaches for increasing nutrient bioavailability. In *Advances in Food and Nutrition Research* (Vol. 81, pp. 1–30). Academic Press.

Jain, A., Ranjan, S., Dasgupta, N., Ramalingam, C. 2018. Nanomaterials in food and agriculture: An overview on their safety concerns and regulatory issues. *Critical Reviews in Food Science and Nutrition* 58(2):297–317.

Kah, M., Beulke, S., Tiede, K., Hofmann, T. 2013. Nanopesticides: State of knowledge, environmental fate, and exposure modeling. *Critical Reviews in Environmental Science and Technology* 43(16):1823–1867.

Kah, M. 2015. Nanopesticides and nanofertilizers: Emerging contaminants or opportunities for risk mitigation? *Frontiers in Chemistry* 3:1–6.

Kah, M., Hofmann, T. 2015. The Challenge: Carbon nanomaterials in the environment: New threats or wonder materials? *Environmental Toxicology and Chemistry* 34(5):954–954.

Kale, A. P., Gawade, S. N. 2016. Studies on nanoparticle induced nutrient use efficiency of fertilizer and crop productivity. *Green Chemistry & Technology Letters* 2(2):88–92.

Karmee, S. K. 2018. A spent coffee grounds based biorefinery for the production of biofuels, biopolymers, antioxidants and biocomposites. *Waste Management* 72:240–254.

Kaur H., Jaryal, N. 2018. Utilization of biogenic tea waste silver nanoparticles for the reduction of organic dyes. *Materials Research Express* 5(5):055402. https://doi.org/10.1088/2053-1591/aabf4b

Khan, I., Saeed, K., Khan, I. 2019. Nanoparticles: Properties, applications and toxicities. *Arabian Journal of Chemistry* 12(7):908–931. https://doi.org/10.1016/j.arabjc.2017.05.011

Khan, M. R., Rizvi, T. F. 2014. Nanotechnology: Scope and application in plant disease management. *Journal of Plant Pathology* 13(3):214–231.

Khond, V. W., Kriplani, V. M. 2016. Effect of nanofluid additives on performances and emissions of emulsified diesel and biodiesel fueled stationary CI engine: A comprehensive review. *Renewable and Sustainable Energy Reviews* 59:1338–1348.

Klaine, S. J., Alvarez, P. J., Batley, G. E., Fernandes, T. F., Handy, R. D., Lyon, D. Y., et al. 2008. Nanomaterials in the environment: Behavior, fate, bioavailability, and effects. *Environmental Toxicology and Chemistry* 27(9):1825–1851.

Krishnaswamy, K., Vali, H., Orsat, V. 2014. Value-adding to grape waste: Green synthesis of gold nanoparticles. *Journal of Food Engineering* 142:210–220.

Kumar, A., Chisti, Y., Chand, U. 2013. Synthesis of metallic nanoparticles using plant extracts. *Biotechnology Advances* 31:346–356.

Kumar, A., Agarwal, D. K., Kumar, S., Reddy, Y. M., Chintagunta, A. D., Saritha, K. V. et al. 2019. Nutraceuticals derived from seed storage proteins: Implications for health wellness. *Biocatalysis and Agricultural Biotechnology* 17:710–719.

Kumar, A., Imran, M., Kumar, S. P. J., Kumar, M., Singh, A. N., Srivastava, A. K. 2018. Assessment of soil urease activity and its kinetic parameters in chilli rhizosphere under the influence of pesticides. *Biopesticides International* 14(1):47–57.

Kumar, S. A., Peter, Y. A., Nadeau, J. L. 2008. Facile biosynthesis, separation and conjugation of gold nanoparticles to doxorubicin. *Nanotechnology* 19(49):495101. https://doi.org/10.1088/0957-4484/19/49/495101.

Kumari, S., Panda, P. K., Pramanik, N., Verma, S. K., Mallick, M. A., Mallick, M. A. 2019. Molecular aspect of phytofabrication of gold nanoparticle from *Andrographis peniculata* photosystem II and their in vivo biological effect on embryonic zebrafish (*Danio rerio*). *Environmental Nanotechnology, Monitoring and Management* 11:100201. https://doi.org/10.1016/j.enmm.2018.100201.

Kwak, S. Y., Giraldo, J. P., Wong, M. H., Koman, V. B., Lew, T. T. S., Ell, J., et al. 2017. A nanobionic light-emitting plant. *Nano Letters* 17(12):7951–7961.

Li, C. Li, Y., Li, Y., Fu, G. 2018. Cultivation techniques and nutrient management strategies to improve productivity of rain-fed maize in semi-arid regions. *Agricultural Water Management* 210:149–157.

Li, Y., Cu, Y. T. H., Luo, D. 2005. Multiplexed detection of pathogen DNA with DNA-based fluorescence nanobarcodes. *Nature Biotechnology* 23(7):885–889.

Lou, R., Ma, R., Kuan-ting, Lin, Ahamed, A., Xiao, Z. 2019. Facile extraction of wheat straw by deep eutectic solvent (DES) to produce lignin nanoparticles. *ACS Sustainable Chemistry and Engineering* 7(12):10248–10256.

Mahanty, T., Bhattacharjee, S., Goswami, M., Bhattacharyya, P., Das, B., Ghosh, A., Tribedi, P. 2017. Biofertilizers: A potential approach for sustainable agriculture development. *Environmental Science and Pollution Research International* 24(4):3315–3335. https://doi.org/10.1007/s11356-016-8104-0.

Malic, L., Sandros, M. G., Tabrizian, M. 2011. Designed biointerface using near-infrared quantum dots for ultrasensitive surface plasmon resonance imaging biosensors. *Analytical Chemistry* 83(13):5222–5229.

Mao, X., Huang, J., Leung, M. F., Du, Z., Ma, L., Huang, Z. et al. 2006. Novel core-shell nanoparticles and their application in high-capacity immobilization of enzymes. *Applied Biochemistry and Biotechnology* 135(3):229–246.

Maria, B. S., Devadiga, A., Kodialbail, V. S., Saidutta, M. B. 2015. Synthesis of silver nanoparticles using medicinal *Zizyphus xylopyrus* bark extract. *Applied Nanoscience* 5(6):755–762.

Martirosyan, A., Schneider, Y. J. 2014. Engineered nanomaterials in food: Implications for food safety and consumer health. *International Journal of Environmental Research and Public Health* 11(6):5720–5750.

Maynard, A. D., Aitken, R. J., Butz, T., Colvin, V., Donaldson, K., Oberdörster, G., et al. 2006. Safe handling of nanotechnology. *Nature* 444(7117):267.

McClements, D. J. 2012. Nanoemulsions versus microemulsions: Terminology, differences, and similarities. *Soft Matter* 8(6):1719–1729.

McClements, D. J. 2018. Encapsulation, protection, and delivery of bioactive proteins and peptides using nanoparticle and microparticle systems: A review. *Advances in Colloid and Interface Science* 253:1–22. https://doi.org/10.1016/j.cis.2018.02.002.

McLamore, E. S., Diggs, A., Calvo Marzal, P., Shi, J., Blakeslee, J. J., Peer, et al. 2010. Non-invasive quantification of endogenous root auxin transport using an integrated flux microsensor technique. *The Plant Journal: For Cell and Molecular Biology* 63(6):1004–1016.

Mehnaz, S. 2015. *Azospirillum*: A biofertilizer for every crop. In *Plant Microbes Symbiosis: Applied Facets* (pp. 297–314). Springer, New Delhi.

Mihankhah, T., Delnavaz, M, GhaffariKhaligh, N. 2018. Application of TiO_2 nanoparticles for eco-friendly biodiesel production from waste olive oil. *International Journal of Green Energy* 15(2):69–75.

Mishra, A., Sardar, M. 2013. Rapid biosynthesis of silver nanoparticles using sugarcane bagasse-an industrial waste. *Journal of Nanoengineering and Nanomanufacturing* 3(3):217–219.

Mishra, A., Tripathy, S. K., Wahab, R., Jeong, S. H., Hwang, I., Yang, Y. B., et al. 2011. Microbial synthesis of gold nanoparticles using the fungus *Penicillium brevicompactum* and their cytotoxic effects against mouse mayo blast cancer C2C12 cells. *Applied Microbiology and Biotechnology* 92(3):617–630. https://doi.org/10.1007/s00253-011-3556-0.

Mishra, P., Seenivasan, R., Mukherjee, A., Chandrasekaran, N. 2019. Polymer/layered silicate nanocomposites as matrix for bioinsecticide formulations. In *NanoBiopesticides Today and Future Perspectives* (pp. 161–178). Academic Press.

Mohammed Fayaz, A., Balaji, K., Girilal, M., Kalaichelvan, P. T., Venkatesan, R. 2009. Mycobased synthesis of silver nanoparticles and their incorporation into sodium alginate films for vegetable and fruit preservation. *Journal of Agricultural and Food Chemistry* 57(14):6246–6252. https://doi.org/10.1021/jf900337h.

Moomand, K., Lim, L. T. 2014. Oxidative stability of encapsulated fish oil in electrospunzein fibres. *Food Research International* 62:523–532.

Mousavi, M., Habibi-Yangjeh, A., Pouran, S. R. 2018. Review on magnetically separable graphitic carbon nitride-based nanocomposites as promising visible-light-driven photocatalysts. *Journal of Materials Science: Materials in Electronics* 29(3):1719–1747.

Mungure, T. E., Roohinejad, S., Bekhit, A. E. D., Greiner, R., Mallikarjunan, K. 2018. Potential application of pectin for the stabilization of nanoemulsions. *Current Opinion in Food Science* 19:72–76.

Neal, A. L. 2008. What can be inferred from bacterium–nanoparticle interactions about the potential consequences of environmental exposure to nanoparticles? *Ecotoxicology* 17(5):362–371.

Neethirajan, S., Jayas, D. S. 2011. Nanotechnology for the food and bioprocessing industries. *Food and Bioprocess Technology* 4(1):39–47.

Nuruzzaman, M. Rahman, M. M., Liu, Y. J., Naidu, R. 2016. Nanoencapsulation, nanoguard for pesticides: A new window for safe application. *Journal of Agricultural and Food Chemistry* 64(7):1447–1483.

O'Sullivan, J. J., Park, M., Beevers, J., Greenwood, R. W., Norton, I. T. 2017. Applications of ultrasound for the functional modification of proteins and nanoemulsion formation: A review. *Food Hydrocolloids* 71:299–310. https://doi.org/10.1016/j.foodhyd.2016.12.037.

Oomen, A. G., Bleeker, E. A., Bos, P. M., Van Broekhuizen, F., Gottardo, S., Groenewold, M., et al. 2015. Grouping and read-across approaches for risk assessment of nanomaterials. *International Journal of Environmental Research and Public Health* 12(10):13415–13434.

Pandit, P. R., Fulekar, M. H. 2017. Egg shell waste as heterogeneous nanocatalyst for biodiesel production: Optimized by response surface methodology. *Journal of Environmental Management* 198(1):319–329.

Panpatte, D. G., Jhala, Y. K. 2019. *Nanotechnology for Agriculture: Crop Production and Protection*. Springer Nature Singapore. https://doi.org/10.1007/978-981-32-9374-8.

Panpatte, D. G., Jhala, Y. K., Shelat, H. N., Vyas, R. V. 2016. Nanoparticles: The next generation technology for sustainable agriculture. In *Microbial Inoculants in Sustainable Agricultural Productivity* (pp. 289–300). Springer, New Delhi.

Patel, P., Kumari, P., Verma, S. K., Mallick, M. A. 2019. Cellular and molecular impact of green synthesized silver nanoparticles. In *Silver Nanoparticles-Health and Safety*. IntechOpen. https://doi.org/10.5772/intechopen.90717.

Pathak, M. 2017. Nanoemulsions and their stability for enhancing functional properties of food ingredients. In *Nanotechnology Applications in Food* (pp. 87–106). Academic Press.

Patil, H. J., Solanki, M. K. 2016. Microbial inoculant: Modern era of fertilizers and pesticides. In *Microbial Inoculants in Sustainable Agricultural Productivity* (pp. 319–343). Springer, New Delhi.

Pegaz, B., Debefve, E., Ballini, J. P., Konan-Kouakou, Y. N., van den Bergh, H. 2006. Effect of nanoparticle size on the extravasation and the photothrombic activity of meso (p-tetracarboxyphenyl) porphyrin. *Journal of Photochemistry and Photobiology, Part B: Biology* 85(3):216–222.

Pinto, R. J. B., Daina, S., Sadocco, P., Neto, C. P., Trindade, T. 2013. Antibacterial activity of nanocomposites of copper and cellulose. *BioMed Research International* 6:280512. https://doi.org/10.1155/2013/280512.

Pradhan, N., Singh, S., Ojha, N., Shrivastava, A., Barla, A., Rai, V., et al. 2015. Facets of nanotechnology as seen in food processing, packaging, and preservation industry. *BioMed Research International* 36:5672. http://dx.doi.org/10.1155/2015/365672.

Prasad, R., Pandey, R., Barman, I. 2015. Engineering tailored nanoparticles with microbes: Quo vadis? *Wiley Interdisciplinary Reviews. Nanomedicine and Nanobiotechnology* 8(2):316–330.

Ragaei, M., Sabry, A. K. H. 2014. Nanotechnology for insect pest control. International Journal of Science, *Environmental and Technology* 3(2):528–545.

Raghavendra, M. P., Nayaka, S. Chandra, Nuthan, B. R. 2017. Role of rhizosphere microflora in potassium solubilization. In *Potassium Solubilizing Microorganisms for Sustainable Agriculture*, edited by V. S. Meena et al. Springer, India. http://dx.doi.org/10.1007/978-81-322-2776-2_4.

Rai, M., Ingle, A. 2012. Role of nanotechnology in agriculture with special reference to management of insect pests. *Applied Microbiology and Biotechnology* 94(2):287–293.

Rai, M., Kratosova, G. 2015. Management of phytopathogens by application of green nanobiotechnology: Emerging trends and challenges. *Acta AgrariaDebreceniensis* 66(66):15–22.

Rana, S., Kalaichelvan, P. T. 2013. Ecotoxicity of nanoparticles. *ISRN Toxicology* 2:1–11.

Rangaraj, S., Venkatachalam, R. 2017. A lucrative chemical processing of bamboo leaf biomass to synthesize biocompatible amorphous silica nanoparticles of biomedical importance. *Applied Nanoscience* 7(5):145–153.

Saadati, F., Leghaei, V., Zamani, A. 2017. Environmentally benign copper nanoparticles supported on walnut shell as a highly durable nanocatalyst for the synthesis of propargylamines. *Journal of the Serbian Chemical Society* 82:1–12.

Sanguansri, P., Augustin, M. A. 2006. Nanoscale materials development – A food industry perspective. *Trends in Food Science and Technology* 17(10):547–556.

Sanzari, I., Leone, A., Ambrosone, A. 2019. Nanotechnology in plant science: To make a long story short. *Frontiers in Bioengineering and Biotechnology* 7:120.

Sathyavathi, R., Krishna, M. B., Rao, D. N. 2011. Biosynthesis of silver nanoparticles using *Moringaoleifera* leaf extract and its application to optical limiting. *Nanoscience and Nanotechnology* 11(3):2031–2035.

Savina, I. N., Ingavle, G. C., Cundy, A. B., Mikhalovsky, S. V. 2016. A simple method for the production of large volume 3D macroporous hydrogels for advanced biotechnological, medical and environmental applications. *Scientific Reports* 6:21154.

Schrand, M. A., Rahman, M. F., Hussain, S. M., John, J. S., Schlager, D. A., Syed, A. F. 2010. *WIREs*, Vol. 2. John Wiley & Sons, pp. 544–568.

Sekhon, B. S. 2010. Food nanotechnology – An overview. *Nanotechnology,Science and Applications* 3(1):1–15.

Servin, A., Elmer, W., Mukherjee, A., De la Torre-Roche, R., Hamdi, H., White, J. C. et al. 2015. A review of the use of engineered nanomaterials to suppress plant disease and enhance crop yield. *Journal of Nanoparticle Research* 17(2):92.

Servin, A. D., White, J. C. 2016. Nanotechnology in agriculture: Next steps for understanding engineered nanoparticle exposure and risk. *NanoImpact* 1:9–12.

Shang, Y., Hasan, M., Ahammed, G. J., Li, M., Yin, H., Zhou, J. 2019. Applications of nanotechnology in plant growth and crop protection: A review. *Molecules* 24(14):2558.

Shankar, S. S., Rai, A., Ankamwar, B., Singh, A., Ahmad, A., Sastry, M. 2004. Biological synthesis of triangular gold nanoprisms. *Nature Materials* 3(7):482–488.

Sharma, R., Jafari, S. M., Sharma, S. 2020. Antimicrobial bio-nanocomposites and their potential applications in food packaging. *Food Control* 112:107086. https://doi.org/10.1016/j.foodcont.2020.107086

Shekhawat, M. S., Kannan, N., Manokari, M. 2012. Biogenesis of silver nanoparticles using leaf extract of *Turnera ulmifolia* L. and screening of their antimicrobial activity. *Journal of Ecobiotechnology* 4(1):54–57.

Shi, S., Wang, W., Liu, L., Wu, S., Wei, Y., Li, W. 2013. Effect of chitosan/nanosilica coating on the physicochemical characteristics of longan fruit under ambient temperature. *Journal of Food Engineering* 118(1):125–131.

Singh, P., Hussain, T., Patel, S., Akhtar, N. 2018. Impact of climate change on root–pathogen interactions. In *Root Biology. Soil Biology* (Vol. 52), edited by B. Giri, R. Prasad, A. Varma. Springer, Cham. https://doi.org/10.1007/978-3-319-75910-4_16.

Singh, R., Singh, D., Singh Mani, J. K., Karwasra, S. S., Beniwal, M. S. 2010. Effect of weather parameters on Karnal bunt disease in wheat in Karnal region of Haryana. *Journal of Agrometeorology* 12(1):99–101.

Singh, T., Shukla, S., Kumar, P., Wahla, V., Bajpai, V. K., Rather, I. A. 2017. Application of nanotechnology in food science: Perception and overview. *Frontiers in Microbiology* 8:1501. https://doi.org/10.3389/fmicb.2017.01501.

Sinha Ray, S., Okamoto, M. 2003. Polymer/layered silicate nanocomposites: A review from preparation to processing. *Progress in Polymer Science* 28(11):1539–1641.

Sinsinwar, S., Sarkar, M. K., Suriya, K. R., Nithyanand, P., Vadivel, V. 2018. Use of agricultural waste (coconut shell) for the synthesis of silver nanoparticles and evaluation of their antibacterial activity against selected human pathogens. *Microbial Pathogenesis* 124:30–37.

Smith, G. B., Granqvist, C. G. 2011. *Green Nanotechnology: Solutions for Sustainability and Energy in the Built Environment.* CRC Press, Boca Raton, FL.

Sridhara, V., Ali, B., Shaziya, K., Satapathy, L. N., Khandelwal, P. 2012. Biosynthesis and antibacterial activity of silver nanoparticles. *Research Journal of Biotechnology* 8(1):11–17.

Srivastava, A. K., Dev, A., Karmakar, S. 2018.Nanosensors and nanobiosensors in food and agriculture. *Environmental Chemistry Letters* 16(1):161–182.

Sung, S. Y., Sin, L. T., Tee, T. T., Bee, S. T., Rahmat, A. R., Rahman, W. A. W. A., et al. 2013. Antimicrobial agents for food packaging applications. *Trends in Food Science and Technology* 33(2):110–123.

Suyal, D. C., Soni, R., Sai, S.., Goel, R. 2016. Microbial inoculants as biofertilizer. In *Microbial Inoculants in Sustainable Agricultural Productivity*, edited by D. P. Singh et al. Springer, India. https://doi.org/10.1007/978-81-322-2647-5_18.

Syiem, M. B., Singh, A. K.., Rai, A. N. 2017. N2–fixing cyanobacterial systems as biofertilizer. In *Agro-Environmental Sustainability* (pp. 43–61), edited by J. S. Singh, G. Seneviratne. Springer International Publishing AG. https://doi.org/10.1007/978-3-319-49724-2_3

Tang, D., Sauceda, J. C., Lin, Z., Ott, S., Basova, E., Goryacheva, I. et al. 2009. Magnetic nanogold microspheres-based lateral-flow immunodipstick for rapid detection of aflatoxin b2 in food. *Biosensors and Bioelectronics* 25(2):514–518.

Thomas, K., Sayre, P. 2005. Research strategies for safety evaluation of nanomaterials, Part I: Evaluating the human health implications of exposure to nanoscale materials. *Toxicological Sciences: An Official Journal of the Society of Toxicology* 87(2):316–321.

Tomer, S., Suyal, D. C., Goel, R. 2016. Biofertilizers: A timely approach for sustainable agriculture. In *Plant-Microbe Interaction: An Approach to Sustainable Agriculture*, edited by D. K. Choudhary et al. (eds.). Springer Nature Singapore Pvt. Ltd. https://doi.org/10.1007/978-981-10-2854-0_17.

Vanathi, P., Rajiv, P., Sivaraj, R. 2016. Synthesis and characterization of *Eichhornia*-mediated copper oxide nanoparticles and assessing their antifungal activity against plant pathogens. *Bulletin of Materials Science* 39(5):1165–1170.

Verma, S. K., Jha, E., Panda, P. K., Thirumurugan, A., Suar, M. 2019. Biological effects of green-synthesized metal nanoparticles: A mechanistic view of antibacterial activity and cytotoxicity. In *Advanced Nanostructured Materials for Environmental Remediation* (pp. 145–171). Springer, Cham. https://doi.org/10.1007/978-3-030-04477-0_6.

Verma, S. K., Nisha, K., Panda, P. K., Patel, P., Kumari, P., Mallick, M. A., et al. 2020. Green synthesized MgO nanoparticles infer biocompatibility by reducing in vivo molecular nanotoxicity in embryonic zebrafish through arginine interaction elicited apoptosis. *Science of the Total Environment* 713:136521. https://doi.org/10.1016/j.scitotenv.2020.136521.

Vigneshwaran, N., Ashtaputre, N. M., Varadarajan, P. V., Nachane, R. P., Paralikar, K. M., Balasubramanya, R. H. 2007. Biological synthesis of silver nanoparticles using the fun gus *Aspergillus flavus*. *Materials Letters* 61(6):1413–1418.

Vo-Dinh, T., Cullum, B. M., Stokes, D. L. 2001. Nanosensors and biochips: Frontiers in biomolecular diagnostics. *Sensors and Actuators. Part B: Chemical* 74(1–3):2–11.

Wang, Z., Liu, J., Wang, W., Wei, Z., Wang, F., Gong, P. et al. 2017. Photoluminescent carbon quantum dot grafted silica nanoparticles directly synthesized from rice husk biomass. *Journal of Materials Chemistry B* 5(24):4679–4689.

Wang, Z., Wei, F., Liu, S. Y., Xu, Q., Huang, J. Y., Dong, X. Y. et al. 2010. Electrocatalytic oxidation of phytohormone salicylic acid at copper nanoparticles-modified gold electrode and its detection in oilseed rape infected with fungal pathogen *Sclerotinia sclerotiorum*. *Talanta* 80(3):1277–1281.

Wani, Y. M., Hashim, M. A., Nabi, F., Malik, M. A. 2011. Nanotoxicity: Hindawi Publishing Corporation. *Advances in Physical Chemistry* Article ID 450912. https://doi.org/10.1155/2011/450912.

Worrall, E. A., Hamid, A., Mody, K. T., Mitter, N., Pappu, H. R. 2018. Nanotechnology for plant disease management. *Agronomy* 8(12):285.

Xiang, Y., Camarada, M. B., Wen, Y., Wu, H., Chen, J., Li, M., Liao, X. 2018. Simple voltammetric analyses of ochratoxin A in food samples using highly-stable and anti-fouling black phosphorene nanosensor. *Electrochimica Acta* 282:490–498.

Xie, F., Pollet, E., Halley, P. J., Averous, L. 2013. Starch-based nanobiocomposites. *Progress in Polymer Science* 38(10–11):1590–1628.

Yadav, S. K., Patel, J. S., Kumar, G., Mukherjee, A., Maharshi, A., Sarma, B. K., et al. 2018. Factors affecting the fate, transport, bioavailability and toxicity of nanoparticles in the agroecosystem. In *Emerging Trends in Agrinanotechnology Fundamental and Applied Aspects* (pp. 118–134), edited by H. B. Singh, S. Mishra, L. F. Fraceto, R. de Lima. CAB International.

Yadav, S. K., Singh, S., Singh, H. B., Sarma, B. K. 2017. Compatible rhizosphere competent microbial consortium adds value to the nutritional quality in edible parts of chickpea. *Journal of Agricultural and Food Chemistry* 65(30):6122–6130.

Yang, N., WeiHong, L., Hao, L. 2014. Biosynthesis of Au nanoparticles using agricultural waste mango peel extract and its in vitro cytotoxic effect on two normal cells. *Materials Letters* 134:67–70.

Yaro, S. A., Olajide, O. S., Asuke, F., Popoola, A. P. I. 2016. Synthesis of groundnut shell nanoparticles: Characterization and particle size determination. *The International Journal of Advanced Manufacturing Technology* 91(1–4):1111–1116.

Yaseen, T., Pu, H., Sun, D. W. 2018. Functionalization techniques for improving SERS substrates and their applications in food safety evaluation: A review of recent research trends. *Trends in Food Science and Technology* 72:162–174. https://doi.org/10.1016/j.tifs.2017.12.012.

Yotova, L., Hassaan, A., Yaneva, S. 2015. Covalent immobilization of peroxidase onto hybrid membranes for the construction of optical biosensor. *International Journal Bioautomation* 19(2):177–186.

Yotova, L., Yaneva, S., Marinkova, D. 2013. Biomimetic nanosensors for determination of toxic compounds in food and agricultural products. *Journal of Chemical Technology and Metallurgy* 48(3):215–227.

Yu, L., Li, C., Xu, J., Hao, J., Sun, D. 2012. Highly stable concentrated nanoemulsions by the phase inversion composition method at elevated temperature. *Langmuir: The ACS Journal of Surfaces and Colloids* 28(41):14547–14552.

Zamani, A., PoursattarMarjani, A., Alimoradlu, K. 2018. Walnut shell-templated ceria nanoparticles: Green synthesis, characterization and catalytic application. *International Journal of Nanoscience* 17(6):1850008. http://www.ncbi.nlm.nih.gov/pubmed/1850008.

5 Phytochemicals and Microbial Metabolites as Capping Agents in Nanoparticle Formulation and Its Implications in Agriculture

Neha Sharma, Sharmistha Banerjee, Rajesh Singh Tomar, and Shuchi Kaushik

CONTENTS

5.1 INTRODUCTION

Nanoparticles (NPs) are nanosized particles which have dimensions of 100 nm or less. They have gained a lot of attention and are used in various applications due to their fascinating morphological properties. They have many advantages over their bulk counterparts (Daniel and Astruc, 2004; Kato, 2011). NPs have unique physical properties like high surface area, shape, and size as well as chemical properties such as catalytic activity. Study of these potential properties help to open new paradigms and introduce new strategies for agriculture. New paradigms also introduce new terms such as 'agri-nanotechniques', although no proper definitions have yet been provided. It has been indicated that the use of NPs and nanosystems for the delivery of various nutrients in crops (Mani et al., 2016).

There are various methods that are used to synthesize NPs with specific morphological properties. These methods are biological, chemical, and physical as well as hybrid of the above methods (Tiwari et al., 2008; Mohanpuria et al., 2008; Luechinger et al., 2010; Liu et al., 2011). Chemical and physical methods utilize several types of toxic solvent and therefore they are prone to the release of hazardous by-products. Therefore, there is the need for an alternative and eco-friendly approach to synthesize nanoparticles. Different natural resources like plants, algae, fungi, and bacteria can be used to achieve this goal. Many researchers are attracted to the green synthesis of NPs because of its feasibility and less toxic nature. Biological methods are primarily preferred as they are environmentally friendly and cost-effective, and they can also act as capping agents due to their reducing properties. Medicinal plants have a variety of phytochemical constituents such as alcohols, terpenes, phenols, alkaloids, and saponins as secondary metabolites, whereas microbes can also act as stabilizing and reducing agents, due to their specific enzymes, present in intracellular and extracellular forms, for the synthesis of nanoparticles (Ahmed et al., 2016; Parveen et al., 2016; Muhammad et al., 2018).

The primary and secondary metabolites in plants are frequently involved in a reduction and oxidation reaction when synthesizing nanosized particles, which is also eco-friendly. Several previously published research papers have indicated that plant- and microbe-based synthesized NPs can effectively control changes related to apoptosis, genotoxicity, and oxidative stress (Kim et al., 2007). Novel secondary metabolites or compounds present in plant crude extracts are predominantly involved in the reduction of ionic compounds into zero-valent corresponding metallic NPs (Aromal and Philip, 2012), and they are also involved in the synthesis of green NPs (Dubey et al., 2010). Primary and secondary biomolecules primarily participate in the reduction reaction. Plant primary biomolecules consist of carbohydrates, proteins, organic compounds, and pigments. Secondary metabolites have various compounds like resins, phenols, saponins, antioxidant enzymes, terpenoids, and alkaloids. These secondary metabolites are recognized as decisive resources for controlling many acute diseases such as cancer and diabetes. They also have many other applications, including drug carriers for targeted drug delivery, gene therapy, and as antimicrobial agents (Huang and Yang, 2004; Wei and Qian, 2008; Sanghi and Verma, 2009; Dubey et al., 2010; Moulton et al., 2010).

Plant secondary metabolites are the main factors in the synthesis of metallic NPs due to their reducing capacity. Reducing potential of a plant extract depends on various functional groups, such as amide, alkenyl, phenolic, alcoholic, amine, and carboxylic groups (Jha et al., 2009). The functional and chemical groups participate in the plant-based synthesis of NPs. The functional group present in plant secondary metabolites acts as a reducing and stabilizing agent for NPs synthesis (Awwad et al., 2013).

NPs have many applications in various industries, such as pharmacy, agriculture and the plant sciences. Only a few studies are found with usages of the phytochemical and microbial agents as capping material to maintain the size and shape of NPs; hence, the mechanism of nanoparticle formation and their properties still needs to be explored. Also, there is a need to focus on the use of NPs in the agriculture field. This chapter provides a brief description of present research activities based on the synthesis of NPs and their implications for agriculture.

5.2 FEATURES OF NANOPARTICLES

Nanomaterials of less than 100 nm in size behave completely different and thus give way to the rules of quantum mechanics. NPs have unique functional properties due to various regions like:

- NPs have a large surface to volume ratio and are extremely reactive due to their nano size.
- NPs have more strength, heat resistance, a low melting point and different magnetic properties.
- NPs lead to variances in atomic distribution that affect the kinetics of the electron transfer rate between metal NPs and the corresponding adsorbed species (extract of plants and microbes).
- NPs have sharp edges and different structures, such as cubic or spherical, which has beeb recognized as improving chemical reactivity (Waleed and Abobatta, 2018).

In agriculture:

- The higher solubility of NPs in suspension.
- The higher surface area and particle size of the NPs, which facilitates penetration of seed coats and subsequently the emerging roots.

- Better bioavailability of molecules to the seed radicals.
- Providing the actual concentration and controlled release of fertilizers or pesticides.
- Improved targeted activity means specific plant protection activities, such as pesticides, fungicides, weedicides.
- Monitoring the quality of agricultural products.
- Monitoring plant growth.
- Lower harm to the environment with a safe and relaxed transport (Waleed and Abobatta, 2018).

5.3 CLASSIFICATION OF NANOPARTICLE

NPs are divided into many types based on their morphology (size and shape), and physical and chemical properties, as summarized in Table 5.1. There are carbon-based nanoparticles, metal nanoparticles, ceramic nanoparticles, nanoparticles with semiconductor properties, polymeric nanoparticles, and lipid-based nanoparticles each with their unique characteristics.

5.4 APPROACHES OF NANOPARTICLE SYNTHESIS

Numerous techniques are available for synthesizing metallic NPs, such as electrochemical, chemical, radiation, Langmuir-Blodgett, photochemical and biological methods, which also include the conversion of metal ions into metal NPs due to reduction (Dimitrijevic et al., 2001; Sun et al., 2002; Naik et al., 2002; Yin et al., 2003; Callegari et al., 2003; Swami et al., 2004; Zhang et al., 2008; Singh et al., 2020). NPs can be synthesized in an environmentally friendly fashion at low cost with little or low maintenance. NPs, which are synthesized from plants, are mostly stable and their synthesis rate is faster than other organisms (Ahmad and Sharma, 2012; Iravani and Zolfaghari, 2013). The synthesis of NPs follows two types of an alternative approach: the 'top-down approach' and the 'bottom-up approach'. Figure 5.1 summarizes these two top-down and bottom-up approaches for nanoparticle synthesis and the techniques employed in them.

5.4.1 Top-Down Approach

In this approach, starting materials of different sizes are reduced down using physical, mechanical, and chemical methods. There are several methods, but the most preferred method is pyrolysis, and attrition can be utilized in metallic NPs using the physical synthesis process (Khandelwal and Joshi, 2018).

5.4.2 Bottom-Up Approach

The bottom-up approach is used to construct the structure atom by atom, molecule by molecule, or cluster by cluster. By using a bottom-up approach, primarily nanostructures or NPs are synthesized, and through the chemical or biological procedures

TABLE 5.1
Classification of Nanoparticles

S. No.	Name of nanoparticles	Morphology	Properties	References
1	Carbon-based nanoparticles	elongated, tubular structure, 1–2 nm in diameter	Used as fillers and support medium for different inorganic and organic catalysts. Efficient gas adsorbents for environmental remediation	Mabena et al. (2011); Ibrahim (2013); Ngoy et al. (2014); Saeed and Khan, (2014, 2016)
2	Metal Nanoparticles	Possess unique optoelectrical properties.-2 0 to-50 nm particle size (Natural iron oxide ore) < 10 nm particle size (Transition metal dichalcogenide nanodots) 5.8 ± 1.1nm particle size (Co_3O_4)	Used in therapy	Dreaden et al. (2012); Priyadarshana et al. (2015); Zhang et al. (2015); Zhou et al. (2016)
3	Ceramic Nanoparticles	Amorphous, polycrystalline, dense, porous or hollow forms	Used in applications such as catalysis, photocatalysis, photodegradation of dyes, and imaging applications	Sigmund et al. (2006);Thomas et al. (2015)
4	Semiconductor Nanoparticles	Bandage gaps are present	Used in photocatalysis, photo optics, and electronic devices and also used in water splitting applications	Sun (2000); Hisatomi et al. (2014)
5	Polymeric Nanoparticles	Nanospheres or nanocapsules shaped. Other molecules are adsorbed at the outer boundary of the spherical surface	Used in tumour therapy	Mansha et al. (2017); Rao and Geckeler (2011)
6	Lipid-based Nanoparticles	Spherical with a diameter ranging from 10 to 1000 nm	drug carriers and delivery and RNA release in cancer therapy	Rawat et al., (2011); Puri et al. (2009); Gujrati et al. (2014)

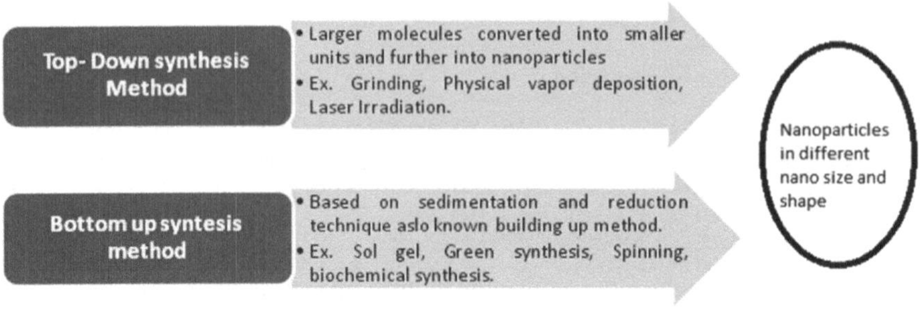

FIGURE 5.1 Preparation method of nanoparticles.

of synthesis, they can further be assembled into the final material (Khandelwal and Joshi, 2018).

5.5 NANOPARTICLE SYNTHESIS METHODS

NP synthesis can be done using three methods: physical, chemical, and green. The biosynthesis of NPs also depends on several experimental parameters, such as optimized temperature, pH, time of reaction, biomass concentration, and the type of metal ions (Khandel and Shahi, 2018). To obtain the desired size and shape of nanoparticle, it is necessary to establish a balance between all the above factors.

5.5.1 Physical Method

The various physical methods used in NPs are as follows:

5.5.1.1 Gas Condensation

Gas condensation was the first method used to synthesize metal NPs and alloys. In this method, thermal evaporation sources like Joule heated refractory crucibles and electron beam evaporation devices are used to vaporize metallic or inorganic material. Due to gas evaporation, a high gas pressure helps with the formation of ultrafine particles (100 nm) because of the gas-phase collision, but this is a slow method as gas pressure of greater than 3 mPa (10 Torr) gas pressure. This method has many limitations, such as the incompatibility of the source precursor, different ranges of temperature and variations in the evaporation rates of an alloy (Rajput, 2015).

5.5.1.2 Vaporization and Deposition of Vacuum

In this method, metals, alloys, elements, or compounds are vaporized and deposited under a vacuum (10 to 0.1 MPa). The complete process is carried out at a pressure of less than 1 m Torr or 0.1 Pa. The substrate temperature ranges up to 500°C. The advantages associated with the vacuum deposition process are that is has high deposition rates and is economic. However, the production of NPs from supersaturated vapour usually takes longer than the cluster, so the deposition of many compounds is difficult (Rajput, 2015).

5.5.1.3 Chemical Vapour Deposition (CVD) and Chemical Vapour Condensation (CVC)

CVC was developed in Germany in 1994. It involves the pyrolysis of vapours of metal-organic precursors in a reduced pressure atmosphere. In CVD the solid is deposited via a chemical reaction on a heated surface after the vapour or gas phase. CVD needs different temperatures, such as in thermal CVD where the activation of the reaction is achieved by a high temperature (above 900°C). In plasma CVD, the reaction is activated between 300 and 700°C using plasma. In laser CVD, pyrolysis takes place when laser thermal energy heats an absorbing substrate. Ultraviolet radiation has sufficient photon energy to break the chemical bond in the reactant molecules during the chemical reaction, and this technique is used in photo-laser CVD. In this process, the reaction is photon activated, and deposition occurs at room temperature. Unlike from CVD, a CVC reaction requires activation energy to proceed (Rajput, 2015).

5.5.2 Mechanical Method: There Are Various Following Mechanical Methods

5.5.2.1 Mechanical Attrition

Unlike many of the methods, mechanical attrition produces its nanostructured molecules through the structural decomposition of coarser-grained structures as a result of plastic deformation, not by cluster assembly (Rajput, 2015).

5.5.2.2 Chemical Precipitation

The basis of the chemical precipitation method is to synthesize nanoparticles by allowing the reaction among the constituents in a suitable medium enriched with dopant in a controlled manner. The main precaution to be taken in this technique is inhibiting the process of aggregation of the small molecules, and this can be achieved with the use of a surfactant. Temperature is kept low and non-aqueous solvents are used so as to control the thermal coagulation and Oswald ripening. The nanocrystals thus formed can be precipitated out by centrifugation, washing, and vacuum drying. UV curing is used on the dried nanocrystals for the possible polymerization of the surfactant capping film on the surface of the nanocluster for imparting true quantum confinement (Konrad et al., 2001; Sharma et al., 2009).

5.5.2.3 Sol-Gel Technique

The sol-gel technique involves the evolution of networks through the formation of a colloidal suspension (sol) and gelatine to form a network in the continuous liquid phase (gel). The precursor for synthesizing these colloids consists of ions of metal alkoxides and alkoxysilanes. Mutual solvent alcohol is used. A catalyst is used to start the reaction and control pH. Sol-gel formation occurs in four stages: hydrolysis, condensation, growth, and the agglomeration of particles (Rajput, 2015).

5.5.3 Biological Method

The biological synthesis of gold, silver, platinum, palladium, selenium, zinc, and copper particles uisng bacteria, actinomycetes, fungi, and yeasts have been reported worldwide. The green biosynthesis of NPs can be achieved by selecting an environmentally acceptable solvent with eco-friendly reducing and stabilizing agents. These biological systems can transform inorganic metal ions into metal NPs via the reductive capacities of the proteins and metabolites present in these organisms (Soni et al., 2018).

5.5.3.1 Synthesis of NPs Using Plant Extract

NPs can also be synthesized using various plant extracts. For this, plant parts like the root, leaf, and bark are washed, cut into small pieces under sterile conditions, and boiled to obtain the extract. The extract is further purified by filtration and centrifugation. The purified plant extract, metal salt, and water are incubated to obtain the NPs of the desired metal ion (Soni et al., 2018). Table 5.2 gives an overview of various plant parts that may be used for the synthesis of nanoparticles such as the

TABLE 5.2
Examples of Plant Parts Used for Nanoparticles Synthesis (Singh et al., 2016)

S.No.	Parts of plants	Name of plants	Morphology	Application
1	Thallus	*Sargassum algae*	Palladium NPs, 5–10 nm in size	Electrocatalytic activities towards hydrogen peroxide
2	Leaves	*Euphorbia prostrate, Ginkgo biloba, Cymbopogon citrates, Azadirachta indica, Nigella sativa, Cocos nucifera,* Catharanthus roseus, *Abutilon indicum*	Silver, gold, titanium dioxide (TiO2) and Copper NPs. Spherical, triangular, hexagonal, and rod, 10–50 nm in size	Biolarvicidal, cytotoxicity, antimicrobial
3	Root	Silver and gold NPs. Red ginseng, Panax ginseng	Spherical, 10–30 nm in size	Antibacterial
4	Fruits	*Artocarpus gomezianus, Citrus medica,* orange and pineapple	Copper and silver NPs. Spherical in shape, 10–300 nm in size	Antimicrobial
5	Cones	*Pinus densiflora*	Silver NPs. Oval in shape, few triangular-shaped, 30–80 nm in size	Antimicrobial
6	Peel	Banana	Cadmium sulphide NPs. 1.48 nm in size	–

thallus, leaves, roots, cones, peel, etc. Several plants have been used for the synthesis process, and nanoparticles of different sizes and properties are produced.

5.5.3.2 Synthesis of NPs Using Microorganism

Various microorganisms: bacteria, yeasts, fungi, and algae have been extensively studied for the synthesis of nanoparticles – both extracellular and intracellular. Due to their eco-friendliness and cost-effectiveness, using microorganisms as nanofactories have received much attention in recent years. The intracellular synthesis of nanoparticles has some limitations as it requires downstream processing for the recovery of NPs, including a number of steps such as breaking the cell wall with sonication, repeated steps of centrifugation, and washing for purifying the nanoparticles. Thus extracellular methodologies for the biosynthesis of nanoparticles have an advantage over the intracellular ones because the former lacks all the shortcomings of the latter. A number of other factors like metal-resistant genes, proteins, peptides, enzymes, reducing cofactors, and organic materials also play an important role as reducing agents. (Singh et al., 2016; Soni et al., 2018; and Yousaf et al., 2020).

In extracellular synthesis, biomass is harvested by centrifugation after culturing the microorganisms in a rotating shaker under optimum growth conditions. The supernatant recovered after centrifugation is incubated with a metal salt solution to synthesize the desired NPs of the metal ion. A colour change in the reaction mixture is the first qualitative indication that NPs are being synthesized. For example, for silver nanoparticles, the colour changes to deep brown, and for gold nanoparticles it changes from ruby red to a deep purple colour. The Tyndall effect, a property of colloidal particles in solution, is used to detect the presence of NPs in solution (Poinern, 2014). The nanoparticles can be synthesized by employing various raw materials, such as bacterial biomasses, supernatants, and derived components. These biological materials act as natural capping agents that prevent nanoparticles from agglomerating and thus this enhances the stability of the synthesized nanoparticles for a more extended period. Centrifugation at high speed helps precipitate out the desired NPs, which later, by washing in ethanol/methanol solvent, can be collected. In the intracellular synthesis of nanoparticles, biomass is harvested after culturing microorganisms under optimum growth conditions. The biomass is washed thoroughly with sterile water and incubated with a metal ion solution. A colour change is an indication of the NPs synthesis. After the incubation period, biomass is subjected to ultrasonication, washing, and centrifugation, which release the NPs after the breakdown of the cell wall. Finally, the released NPs are washed, centrifuged, and collected. The enzyme mainly responsible for the conversion of metal ions into metallic nanoparticles is the reductase enzyme (Singh et al., 2016).

As the reaction mechanism and behaviour of every microorganism towards metal ions differ, a thorough understanding of the nanoparticle synthesis mechanism is a must. The physical synthesis methods require a quite large sum of money and material, which poses a hindrance to the large-scale production of nanoparticles using these approaches. The chemical methods used in the mass production of nanoparticles cause harm to the environment by generating pollutants. Thus, the only choice that remains is the biological synthesis process, which makes up almost all the different life forms existing on the earth, for synthesizing nanoparticles in eco-friendly, safe, and biodegradable forms. The bio-components of living cells, such as enzymes, sugars, proteins, etc., act as capping and stabilizing agents for the nanoparticles. This green synthesis approach has significant advantages over other methods, such as it is simple, with an extensive range of samples, comprehensively available, easy to prepare and process, cost-effective, rapid, reproducible, stable in component production and environmentally friendly (Kharissova et al., 2013; Lade and Shanware, 2020).

Several microorganisms from bacterial species from fungi, actinomycetes, and yeasts are useful for the biological synthesis of nanoparticles, with potential applications in the agriculture sector. Table 5.3 gives an overview of such microbes.

5.5.4 Mechanism of Bio-Reduction

Plant extracts are rich in several biomolecules, like proteins, sugars, amino acids, enzymes, and other traces of metals. These components are actively involved in the

TABLE 5.3
Examples of Microbes Used in Nanoparticles Synthesis and Their Applications (Singh et al., 2016)

S. No.	Types	Name of Microorganism	Morphology	Application
1	Bacterial Species	*Pseudomonas deceptionensis, Bacillus methylotrophicus, Brevibacterium frigoritolerans, Bacillus amyloliquefaciens, Bacillus pumilus, Bacillus persicus, Bacillus licheniformis, Listeria monocytogenes, Bacillus subtilis, Streptomyces anulatus, Weissella oryzae,*	Extracellular and intracellular. Silver, gold and cadmium NPs. Spherical and Anisotropic in shape, mostly 10–30 nm in size	Antimicrobial, antibiofilm, mosquitocidal
2	Fungus	*Neurospora crassa*	Intra and extracellular NPs. Silver and gold. Quasi-spherical in shape >100 nm in size	–
3	Actinomycetes	*Streptomyces* sp. LK3	Silver NPs. Spherical in shape. 5 nm in size	Acaricidal
4	Yeast	*Yarrowia lipolytica* NCYC 789, *Rhodosporidium diobovatum, Candida utilis* NCIM 3469	Extracellular and intracellular Silver NPs Spherical in shape. 15 nm in size	Antibiofilm and Antibacterial

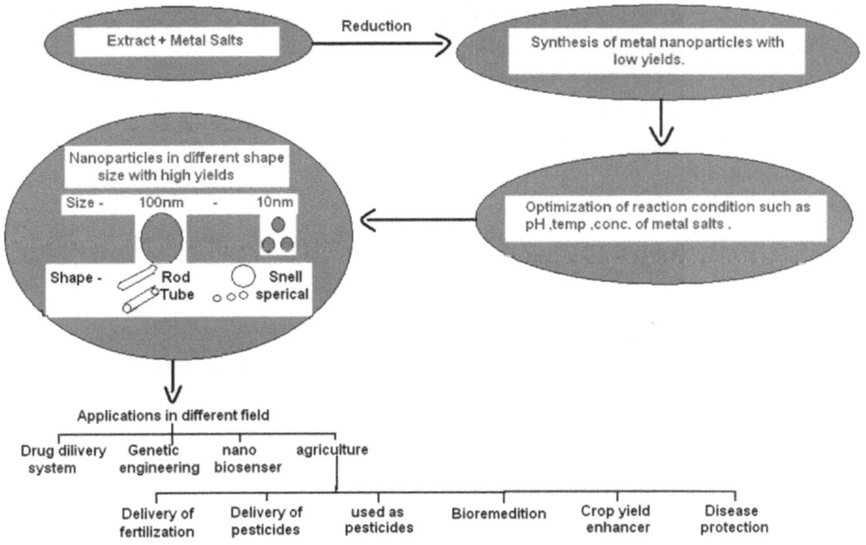

FIGURE 5.2 Synthesis of nanoparticles (NPs) and their application.

bio-reduction process. Many metals showed a reduction process during NP formation, such as silver nitrate ($AgNO_3$) and gold (Tripathy et al., 2010), platinum and copper (Ramanathan et al., 2013), and zinc nitrate ($ZnNO_3$) (Sangeetha et al., 2011).

Figure 5.2 explains the reduction of metal salts by plants and a nanoparticle synthesis process, which may be of variable sizes. Optimization can be done with different variable factors, such as pH, temperature, etc., to generate high yields and suitable-sized nanoparticles with great potential in agriculture along with other sectors.

The physical method involves the use of costly equipment, high temperatures and pressures, it needs a large area for the setting up of machines (Chandrasekaran et al., 2016), and the chemical method involves the use of toxic chemicals, which can be hazardous to the person and the environment (Agarwal et al., 2017).

Moreover, the toxicity of the chemicals used in the physical and chemical methods make it unsafe for use in the medical field. As compared to conventional methods, biological methods are considered safe and ecologically sound for nanomaterial fabrication. Therefore an environment and eco-friendly approach for the synthesis of NPs using microorganisms and different plants, commonly referred to as the green approach, is the most researched area for synthesis of NPs as it holds immense potential (Soni et al., 2018).

5.6 APPLICATIONS IN AGRICULTURE

5.6.1 Nanocapsules for Efficient Delivery of Pesticides, Fertilizers, and Other Agrochemicals

The process of encapsulating insecticides, desired chemicals, or DNA to control insect pests with the use of nanocapsules is currently the most promising technology for the protection of host plants against insect pests, and is termed nanoencapsulation. Nano-encapsulated chemicals are slowly but efficiently released into a particular host plant so as to properly absorb into the plant (Scrinis and Lyons, 2007; Suman et al., 2010). Diffusion, dissolution, biodegradation, and osmotic pressure with specific pH are the mechanisms involved in the release of the nano-encapsulated agents (Park et al., 2006).

5.6.2 Nanoparticles as Pesticides

Nanoparticles have been found to be novel and practical formulations against pests and insects (Barik et al., 2008; Gajbhiye et al., 2009). The DNA and other required chemical agents may also be transferred explicitly using nanoparticles to protect plants against their pathogens. The pesticide (validamycin) can be efficiently delivered and released in a controlled manner into the plant system using porous hollow silica NPs (PHSNs). These promising NPs act as target-specific nanocarriers and enhance the efficacy of commercial pesticides, herbicides, and fertilizers (Wanyika et al., 2012).

5.6.3 NPs for Remediation of Heavy Metal Toxicity

In a study conducted by different research groups, it was reported that silicon nanoparticles (Si NPs) could act as a good method to defend against heavy metal toxicity (Cui et al., 2017), UV-B stress (Tripathi et al., 2017), salinity stress (Haliem et al., 2017), dehydration (Jullok et al., 2016), etc. Rastogi et al. in 2019 reported that Si NPs could be used as a good fertilizer, herbicide, and effective agent against pests.

5.6.4 NPs as a Crop Yield Enhancer

It was demonstrated that tomato (*Lycopersicon esculenttum* Mill. cv. Super Strain B) seedlings grew faster after treatment with fumed silica nanoparticles. The application of the NPs was made before and after germination. The percentage of *Lycopersicon esculentum* seed germination was also improved.

5.6.5 NPs as Crop Protection and Disease Management

The research on nanoparticle use in plant protection is still in its early stages, and only a few reports are available which provide long-term trial information on this technology. Reports proposing the use of RNA interference technology and nanoparticles against viruses came out in 2017 (Mitter et al., 2017). Another study conducted by Zhao et al. (2017) focused on the post-application effects of nanoformulations on plant pesticide levels. Yang et al. in 2009 tested the insecticidal efficacy of their nanoformulation on stored grains. Similarly, research was carried out by Jenne et al. (2018) in which they infested groundnuts with groundnut bruchid insects and then studied the effect of the azadirachtin with either zinc oxide or chitosan NPs on it.

5.7 FUTURE OUTLOOK

The meticulous use of precision agricultural techniques and soil microorganisms should be useful to simultaneously enhance nutrients and crop yields without causing any damage to the natural environment, such as soil and water or reducing the nutrients due to leaching and emissions (Waleed and Abobatta, 2018).

Besides the many advantages of nanoparticles used in diverse areas for society's benefit, there are some negative impacts on living beings and nature, such as chemical reactions caused by the introduction of nanoparticles into the edible plants or free radical generation in some cases by these nanoparticles. So, it is highly important to carefully watch and scrutinize the use of these nanoparticles in biological systems (Waleed and Abobatta, 2018).

The field of nanobiotechnology is very dynamic, and the biosynthesis of NPs using different microorganisms and plants is the focus of current research. The biosynthesis of NPs is a cost-effective and environmentally friendly alternative to the chemical and physical methods. The plant-mediated synthesis of NPs is a green chemical approach that connects nanotechnology with plants and provides a sustainable, resource-efficient, and cheap method of synthesis. The biosynthesis of NPs has great potential as this method is free from the toxic and hazardous chemicals used in conventional physical and chemical methods and is less expensive. NPs find applications in biology, medical sciences, drug delivery, and electronics, to name a few.

Among all these available approaches, the biological method is the most preferred for synthesizing nanomaterials. The biological route is less toxic and environmentally friendly. Moreover, it also eliminates the use of external capping agents. Organic molecules are used to enhance the stability of synthesized NPs and to inhibit its self-agglomeration. There has been extensive research over the last two decades in the field of the synthesis of nanoparticles using a biological approach. The use of plant materials and their metabolites on a large scale makes the process of synthesis of metallic nanoparticles eco-friendly. In this manner, the nanoparticles synthesized are an excellent resource to be used in different areas, such as medicines, cosmetics, energy, other healthcare, and commercial products. As the nanoparticles are synthesized and derived from plant materials, they have minimal side effects compared to the other available diagnosis and treatment methods. Thus, in the future, the biologically synthesized nanoparticles can prove a boon to society, especially for the pharmaceutical, cosmetic, medical, food, and agriculture sectors. To maximally explore the production and application of biologically synthesized nanoparticles, the following points need to be considered:

i. Identification of the active component among the plant metabolites, which has the property to naturally reduce the metal ions and thus enhance the production of nanoparticles.
ii. Scale-up the production of nanoparticles from laboratory to industrial level and
iii. The mode of action and mechanism of synthesized nanoparticles against the target pathogens should be known for their better use in the treatment process.

5.8 CONCLUSION

Nanotechnology is a new and advanced approach to increase agricultural production by improving plant varieties, enhancing nutrients, diagnosing plant diseases, providing remedies for plant protection against pathogens, and monitoring plant overall growth. Nanotechnology offers real solutions to various crops related issues, such as the increased production volume needed for the increasing global population, and also provides many potential modern applications for development of the agricultural sector.

REFERENCES

Agarwal, H., Venkat, K.S., and Rajesh, K.S. 2017. A review on green synthesis of zinc oxide NPs–an eco-friendly approach. *Resource-Efficient Technologies* 3: 406–413.

Ahmad, N., and Sharma, S. 2012. Green synthesis of silver nano particles using extracts of *Ananas comosus*. *Green and Sustainable Chemistry* 2: 141–147.

Ahmed, S., Annu, S., and Yudha, S.S. 2016. Biosynthesis of gold nanoparticles: A green approach. *Journal of Photochemistry and Photobiology, Part B: Biology* 161: 141–153.

Aromal, A., and Philip, D. 2012. Green synthesis of gold nanoparticles using *Trigonella foenum-Graecum* and its size-dependent catalytic activity. *Spectrochimica Acta. Part A: Molecular and Biomolecular Spectroscopy*: 1–5.

Awwad, A.M., Salem, N.M., and Abdeen, A.O. 2013. Green synthesis of silver NPs using carob leaf extract and its antibacterial activity. *International Journal of Industrial Chemistry* 4(1): 29–35.

Barik, T.K., Sahu, B., and Swain, V. 2008. Nanosilica-from medicine to pest control. *Parasitology Research* 103(2): 253–258.

Callegari, A., Tonti, D., and Chergui, M. 2003. Photochemically grown silver NPs with wavelength-controlled size and shape. *Nano Letters* 3(11): 1565–1568.

Chandrasekaran, R., Gnanasekar, S., Seetharaman, P., Keppanan, R., Arockiaswamy, W., and Sivaperumal, S. 2016. Formulation of *Carica papaya* latex-functionalized silver NPs for its improved antibacterial and anticancer applications. *Journal of Molecular Liquids* 219: 232–238.

Cui, J., Liu, T., Li, F., Yi, J., Liu, C., and Yu, H. 2017. Silica NPs alleviate cadmium toxicity in rice cells: Mechanisms and size effects. *Environmental Pollution* 228: 363–369.

Daniel, M.C., and Astruc, D. 2004.Gold nanoparticles: Assembly, supramolecular chemistry, quantum-size-related properties, and applications toward biology, catalysis, and nanotechnology. *Chemical Reviews* 104(1): 293–346.

Dimitrijevic, N.M., Bartels, D.M., Jonah, C.D., Takahashi, K., and Rajh, T. 2001. Radio lytically induced formation and optical absorption spectra of colloidal silver NPs in supercritical ethane. *Journal of Physical Chemistry. Part B* 105(5): 954–959.

Dreaden, E.C., Alkilany, A.M., Huang, X., Murphy, C.J., and El-Sayed, M.A. 2012. The golden age: Gold NPs for biomedical. *Chemical Society Review* 41: 2740–2779.

Dubey, S.P., Lahtinen, M., Sarkka, H., and Sillanpa, M. 2010. Bioprospective of *Sorbus aucuparia* leaf extract in development of silver and gold nanocolloids. *Colloids and Surfaces, Part B: Biointerfaces* 80(1): 26–33.

Gajbhiye, M., Kesharwani, J., Ingle, A., Gade, A., and Rai, M. 2009. Fungus mediated synthesis of silver nanoparticles and its activity against pathogenic fungi in combination of fluconazole. *Nanomedicine* 5(4):282–286.

Gujrati, M., Malamas, A., Shin, T., Jin, E., Sun, Y., and Lu, Z.R. 2014. Multifunctional cationic lipid-based NPs facilitate endosomal escape and reduction triggered cytosolic siRNA release. *Molecular Pharmaceutics* 11(8): 2734–2744.

Haliem, M.E.F.A., Hegazy, H.S., Hassan, N.S., and Naguib, D.M. 2017. Effect of silica ions and nano silica on rice plants under salinity stress. *Ecological Engineering* 99: 282–289.

Hisatomi, T., Kubota, J., and Domen, K. 2014. Recent advances in semiconductors for photo catalytic and photo electrochemical water splitting. *Chemical Society Review* 43(22): 7520–7535.

Huang, H., and Yang, X. 2004. Synthesis of polysaccharide-stabilized gold and silver nanoparticles: A green method. *Carbohydrate Research* 339(15): 2627–2631.

Ibrahim, K.S. 2013. Carbon nanotubes-properties and applications: A review. *Carbon Letters* 14(3): 131–144.

Iravani, S., and Zolfaghari, B. 2013. Green synthesis of silver nanoparticles using *Pinus eldarica* bark extract. *Biotechnology and Green Chemistry*: 1–5.

Jenne, M., Kambham, M., Tollamadugu, N.P., Karanam, H.P., Tirupati, M.K., Reddy Balam, R., Shameer, S., and Yagireddy, M. 2018. The use of slow releasing nanoparticle encapsulated Azadirachtin formulations for the management of *Caryedon serratus* O (groundnut bruchid). *IET Nanobiotechnology* 12(7): 963–967.

Jha, A.K., Prasad, K., Kumar, V., and Prasad, K. 2009. Biosynthesis of silver NPs using *Eclipta* leaf. *Biotechnology Progress* 25: 1475–1477.

Jullok, N., Van Hooghten, R., Luis, P., Volodin, A., Van Haesendonck, C., Vermant, J., and Van, D.B. 2016. Effect of silica NPs in mixed matrix membranes for pervaporation dehydration of acetic acid aqueous solution: Plant-inspired dewatering systems. *Journal of Cleaner Production* 112: 4879–4889.

Kato, H. 2011. In vitro assays: Tracking NPs in/side cells. *Nature Nanotechnology* 6(3): 139–140.

Khandel, P., and Shahi, K.S. 2018. *Journal of Nanostructure in Chemistry* 8(4): 369–391.

Khandelwal, A., and Joshi, R. 2018. Synthesis of NPs and their application in agriculture. *Acta Scientific Agriculture* 2: 10–13.

Kharissova, O.V., Dias, H.V.R., Kharisov, B.I., Pérez, B.O., and Pérez, V.M.J. 2014 The greener synthesis of nanoparticles. *Trends in Biotechnology*: 240–248.

Kim, J.S., Kuk, E., Yu, K.N., Jong-Ho, K., Park, S.J., Lee, H.J., and Kim, S.H. 2007. Antimicrobial effects of silver nanoparticles. *Nanomedicine: Nanotechnology, Biology, and Medicine* 3(1): 95–101.

Konrad, A., Herr, U., Tidecks, R., Samwer, F., and Samwer, K. 2001. Luminescence of bulk and nanocrystalline cubic yttria. *Journal of Applied Physics* 90(7): 3516–3523.

Lade, B.D., and Shanware, A.S. 2020. Phytonanofabrication: Methodology and factors affecting biosynthesis of nanoparticles. In: *Nanosystems*. Intech Open. 1–17.

Liu, J., Qiao, S., Hu, Z., Q.H., and Lu, G.Q. 2011. Magnetic nanocomposites with mesoporous structures: Synthesis and applications. *Small* 7(4): 425–443.

Luechinger, N.A., Grass, R.N., Athanassiou, E.K., and Stark, W.J. 2010. Bottom-up fabrication of metal/metal nanocomposites from NPs of immiscible metals. *Chemistry of Materials* 22(1): 155–160.

Mabena, L.F., Sinha Ray, S., Mhlanga, S.D., and Coville, N.J. 2011. Nitrogen-doped carbon nanotubes as a metal catalyst support. *Applied Nanoscience* 1(2): 67–77.

Mani, P.K., Mondal, S., Kole, C., Sakthi Kumar, D., and Khodakovskaya, M.V. 2016. Agrinanotechniques for plant availability. In: *Plant Nanotechnology. Principles and Practices*. Switzerland: Springer International Publishing. 263–303.

Mansha, M., Khan, I., Ullah, N., and Qurashi, A. 2017. Synthesis, characterization and visible-light-driven photoelectrochemical hydrogen evolution reaction of carbazole-containing conjugated polymers. *International Journal of Hydrogen Energy* 42(16): 10952–10961.

Mitter, N., Worrall, E.A., Robinson, K.E., Li, P., Jain, R.G., Taochy, C., Fletcher, S.J., Carroll, B.J., Lu, G., and Xu, Z.P. 2017. Clay nanosheets for topical delivery of RNAi for sustained protection against plant viruses. *Nature Plants* 3: 16207.

Mohanpuria, P., Rana, N.K., and Yadav, S.K. 2008. Biosynthesis of nanoparticles: Technological concepts and future applications. *Journal of Nanoparticle Research* 10(3): 507–517.

Moulton, M.C., Braydich-Stolle, L.K., Nadagouda, M.N., Kunzelman, S., Hussaina, S.M., and Varma, R.S. 2010. Synthesis, characterization and biocompatibility of 'green' synthesized silver NPs using tea polyphenols. *Nanoscale* 2(5): 763–770.

Muhammad, O., Ali, T.K., Nazar, U.I., Irshad, A., Muhamamd, A., Muthupandian, S., Zabta, K.S., and Sudip, M. 2018. Role of plant phytochemicals and microbial enzymes in biosynthesis of metallic NPs. *Applied Microbiol Biotechnology* 102: 6799–6814.

Naik, R.R., Stringer, S.J., Agarwal, G., Jones, S., and Stone, M.O. 2002. Peptide templates for nanoparticle synthesis derived from polymerase chain reaction-driven phage display. *Advanced Functional Materials* 14: 25–30.

Ngoy, J.M., Wagner, N., Riboldi, L., and Bolland, O. 2014. A CO_2 capture technology using multi-walled carbon nanotubes with polyaspartamide surfactant. *Energy Procedia* 63: 2230–2248.

Park, H.J., Kim, S.H., Kim, H.J., and Choi, S.H. 2006. A new composition of nano sized silica-silver for control of various plant diseases. *The Plant Pathology Journal* 22: 25–34.

Parveen, K., Banse, V., and Ledwani, L. 2016. Green synthesis of nanoparticles: Their advantages and disadvantages. *AIP Conference Proceedings* 1724: 020048.

Poinern, G.E.J. 2014. *A Laboratory Course in Nanoscience and Nanotechnology*. Boca Raton, FL: CRC Press Taylor & Francis

Puri, A., Loomis, K., Smith, B., Lee, J.-H., Yavlovich, A., Heldman, E., and Blumenthal, R. 2009. Lipid-based NPsas pharmaceutical drug carriers: from concepts to clinic. *Critical Reviews in Therapeutic Drug Carrier Systems* 26(6): 523–580.

Rajput, N. 2015. Methods of preparation of nanoparticles – A review. *International Journal of Advances in Engineering and Technology* 7: 1806–1811.

Ramanathan, R., Field, M.R., O'Mullane, A.P., Smooker, P.M., Bhargava, S.K., and Bansal, V. 2013. Aqueous phase synthesis of copper nanoparticles: A link between heavy metal resistance and nanoparticle synthesis ability in bacterial systems. *Nanoscale* 21(5(6)): 2300–2306.

Rao, J.P., and Geckeler, K.E. 2011. Polymer nanoparticles: preparation techniques and size-control parameters. *Progress in Polymer Science* 36(7): 887–913.

Rastogi, A., Tripathi, D.K., Yadav, S., Chauhan, D.K., Živčák, M., Ghorbanpour, M., ElSheery, N.I., and Brestic, M. 2019. Application of silicon NPs in agriculture. *Biotechnology* 9: 90.

Rawat, M.K., Jain, A., Singh, S., Mehnert, W., Thunemann, A.F., Souto, E.B., Mehta, A., and Vyas, S.P. 2011. Studies on binary lipid matrix based solid lipid NPs of repaglinide: In vitro and in vivo evaluation. *Journal of Pharmaceutical Sciences* 100: 2366–2378.

Saeed, K., and Khan, I. 2014. Preparation and properties of single-walled carbon nanotubes/poly(butylene terephthalate) nanocomposites. *Iranian Polymer Journal* 23(1): 53–58.

Saeed, K., and Khan, I. 2016. Preparation and characterization of singlewalled carbon nanotube/nylon 6,6 nanocomposites. *Instrumentation Science and Technology* 44: 435–444.

Sangeetha, G., Rajeshwari, S., and Venckatesh, R. 2011. Green synthesis of zinc oxide NPs by *aloe barbadensis* miller leaf extract: Structure and optical properties. *Materials Research Bulletin* 46: 2560–2566.

Sanghi, R., and Verma, P. 2009. Biomimetic synthesis and characterization of protein capped silver nanoparticles. *Bioresource Technology* 100(1): 501–504.

Scrinis, G., and Lyons, K. 2007. The emerging nano-corporate paradigm: Nanotechnology and the transformation of nature, food and agri-food systems. *International Journal of Sociology of Agriculture and Food* 15: 22–44.

Sharma, A.B., Sharma, M., Pandey, R., and K. 2009. Synthesis, properties and potential applications of semiconductor quantum particles. *Asian Journal of Chemistry* 10: S033–038.

Sigmund, W., Yuh, J., Park, H., Maneeratana, V., Pyrgiotakis, G., Daga, A., Taylor, J., and Nino, J.C. 2006. Processing and structure relationships in electrospinning of ceramic fiber systems. *Journal of the American Ceramic Society* 89(2): 395–407.

Singh, P., Kim, Y.J., Zhang, D., and Yang, D.C. 2016. Biological synthesis of NPs from plants and microorganisms.*Trends in Biotechnology* 34(7): 588–599.

Singh, A., Gaud, B., and Jaybhaye, S. 2020. Optimization of synthesis parameters of silver nanoparticles and its antimicrobial activity. *Materials Science for Energy Technologies* 3: 232–236.

Soni, M., Mehta, P., Soni, A., Goswami, K., and Girish, K. 2018.Green nanoparticles: Synthesis and applications. *Journal of Biotechnology and Biochemistry* 4: 78–83.

Suman, P.R., Jain, V.K., and Varma, A. 2010. Role of nanomaterials in symbiotic fungus growth enhancement. *Current Science* 99: 1189–1191.

Sun, S. 2000. Monodisperse FePt NPsand ferromagnetic FePt nanocrystal superlattices. *Science* 80(287): 1989–1992.

Sun, Y., Yin, Y., Mayers, B.T., Herricks, T., and Xia, Y. 2002. Uniform silver nanowires synthesis by reducing $AgNO_3$ with ethylene glycol in the presence of seeds and poly (vinyl pyrrolidone). *Chemistry of Materials* 14(11): 4736–4745.

Swami, A., Selvakannan, P.R., Pasricha, R., and Sastry, M. 2004. One-step synthesis of ordered two-dimensional assemblies of silver NPs by the spontaneous reduction of silver ions by pentadecylphenol langmuir monolayers. *Journal of Physical Chemistry. Part B* 108(50): 19269–19275.

Thomas, S., Harshita, B.S.P., Mishra, P., and Talegaonkar, S. 2015. Ceramic nanoparticles: Fabrication methods and applications in drug delivery. *Current Pharmaceutical Design* 21(42): 6165–6188.

Tiwari, D., Behari, K.J., and Sen, P. 2008. Time and dose-dependent antimicrobial potential of Ag NPs synthesized by top-down approach. *Current Science* 95: 647– 655.

Tripathi, D.K., Singh, S., Singh, V.P., Prasad, S.M., Dubey, N.K., and Chauhan, D.K. 2017. Silicon NPs more effectively alleviated UV-B stress than silicon in wheat (*Triticum aestivum*) seedlings. *Plant Physiology and Biochemistry* 110: 70–81.

Tripathy, A., Raichur, A.M., Chandrasekaran, N., Prathna, T.C., and Mukherjee, A. 2010. Process variables in biomimetic synthesis of silver NPs by aqueous extract of *Azadirachta indica* (Neem) leaves. *Journal of Nanoparticle Research* 12: 237–246.

Waleed, F., and Abobatta 2008. Nanotechnology application in agriculture. *Acta Scientific Agriculture* 6: 99–102.

Wanyika, H., Gatebe, E., Kioni, P., Tang, Z., and Gao, Y. 2012. Mesoporous silica NPs carrier for urea: Potential applications in agrochemical delivery systems. *Journal of Nanoscience and Nanotechnology* 3: 2221–2228.

Wei, D., and Qian, W. 2008. Facile synthesis of Ag and Au nanoparticles utilizing chitosan as a mediator agent. *Colloids and Surfaces, Part B: Biointerfaces* 15(1): 136–142.

Yang, F.L., Li, X.G., Zhu, F., and Lei, C.L. 2009. Structural characterization of NPs loaded with garlic essential oil and their insecticidal activity against *Tribolium castaneum* (Herbst)(Coleoptera: Tenebrionidae). *The Journal of Agricultural and Food Chemistry* 57(21): 10156–10162.

Yin, B., Ma, H., Wang, S., and Chen, S. 2003. Electrochemical synthesis of silver NPs under protection of poly (n-vinylpyrrolidone). *Journal of Physical Chemistry. Part B* 107: 8898–8904.

Yousaf, H., Mehmood, A., Shafique, A., and Raffi, M. 2020. Green synthesis of silver nanoparticles and their applications as an alternative antibacterial and antioxidant agents. *Materials Science and Engineering: Part C* 112.

Zhang, L., Shen, Y.H., Xie, S.K., Li, S.K. and Wang, C. 2008. One-step synthesis of silver NPs in self-assembled multilayered films based on a Kegg in structure compound. *Journal of Materials Chemistry* 18: 1196–1203.

Zhang, X., Lai, Z., Liu, Z., Tan, C., Huang, Y., Li, B., Zhao, M., Xie, L., Huang, W., and Zhang, H. 2015. A facile and universal top-down method for preparation of monodisperse transition-metal dichalcogenide nanodots. *Angewandte Chemie International. Edition* 54(18): 5425–5428.

Zhao, P., Cao, L., Ma, D., Zhou, Z., Huang, Q., and Pan, C. 2017. Synthesis of pyrimethanil-loaded mesoporous silica NPs and its distribution and dissipation in cucumber plants. *Molecules* 22(5): 817.

Zhou, Y., Dong, C.K., Han, L., Yang, J., and Du, X.-W. 2016. Topdown preparation of active cobalt oxide catalyst. *ACS Catalysis* 6(10): 6699–6703.

Part 2

Application of Nanoparticles

6 Potential of Agro-Products in the Green Synthesis of Nano-Metallopharmaceuticals

Deepak Panchal, Rita S. Dhodapkar, and Sukdeb Pal

CONTENTS

6.1 INTRODUCTION

The versatile characteristics of metals successfully brought humans out of the Stone Age. It would not be wrong to say that metals are the most valuable class of material and have become fundamental to life (Medici et al., 2015). In the periodic table, more than two-thirds of the elements are metals. They are a multi-functional material in terms of their various uses, such as in weapons, decorations, ornamentals, industrialization, currency, etc., helping humans to progress and advance. However, when we think about noble metals, images of wealth, such as jewels, coins, and expensive works of art, flash in the back of our minds. The counterparts to these bulk metals, nano or colloidal states, make up another unique class of material exhibiting exceptional features that are widely accepted in healthcare, drug delivery, biological imaging, hydrogen evolution reactions, electronics, etc. (Dizaj et al., 2014; Wilczewska et al., 2012; Singh and Nalwa, 2011; Tran et al., 2018; Ko et al., 2007).

More recently, with nanotechnology's arrival, this class of element has gained a lot of attention in applications related to the life sciences as a potential tool for the treatment/cure of various medical conditions (Wani et al., 2016; Lippert, 2013; Muhammad and Guo, 2014). However, these metal-based remedies have been in use since ancient times. For example, silver is one of the well-known metals used for killing harmful bacteria. For instance, its biocidal activity has been long recognized and was even used by ancient Greek physicians for wound treatment, and silver vessels were also used to store liquids in their pure form. In the modern age, the discovery of cisplatin in 1965, which showed the inhibition to cellular division in *Escherichia coli*, opened the gates of the unexplored world of metal-based drugs (Rosenberg, Camp, and Krigas, 1965). Metal-based drugs, also known as 'metallopharmaceuticals' or 'metallodrugs', are examples of sophisticated compounds that contain metal as an active component. Besides the treatment applications, nanotechnology-enabled biosensors' can miniaturize with a higher specificity, which has simplified diagnosis and reduces the invasiveness (Boisseau and Loubaton, 2011). As the nanoparticles (NPs) operate within dimensions ranges of 1–100 nm, it can be anticipated that NPs can interact efficiently with biomolecules on surfaces and safely transport the drug molecules to target cells (Wani et al., 2016).

Hence, these metal-nanoparticle-based pharmaceuticals are valuable tools in the field of therapeutics and diagnosis. Various chemical, physical, and biological methods have been employed for synthesizing NPs. The biosynthesis or 'green synthesis' of these metallic NPs is highly preferred over others due to being less expensive, eco-friendly, and avoiding chemical contamination during administration in medical and biological applications. Higher biocompatibility to human cells, scalability, and agro-based nanoparticles' applicability to the medical world has opened up a new avenue for therapeutic use. Furthermore, the current and prospective use of biogenic nano-metallopharmaceuticals in tumour diagnosis, cellular visualization, anticancer, drug delivery systems, and bio-imaging have attracted progressive research. This chapter emphasizes the importance and potential of agro-products for the synthesis of nano-metallopharmaceuticals for biomedical applications. It also discusses different specific nano-metallopharmaceuticals for their potential uses for curing severe diseases.

6.2 NANO-METALLOPHARMACEUTICALS

Metallopharmaceuticals have a biologically active metal or are carriers of active pharmaceutical components that are promising candidates for the smart treatment of various diseases. The application of nanoparticles in therapeutic strategies and drug delivery systems has been investigated for more than two decades, bringing innovations in dosage forms with improved physicochemical effects. Compared to the conventional form of drugs, nano-metallopharmaceuticals deliver the drug not only in a controlled fashion but also with high specificity to the targeted cells (Wilczewska et al., 2012; Wani et al., 2016). Some of the well-known nano-metallopharmaceuticals include platinum (Pt), palladium (Pd), ruthenium (Ru), titanium (Ti), gold (Au), silver (Ag), cerium oxide (CeO), copper/copper oxide (CuO), and zinc oxide (ZnO) as anticancer, antitumor and antibacterial agents (Song, 2010; Bharti and Singh, 2009; Pal et al., 2009). Several mechanisms have been proposed for their actions, including the cleavage of DNA strands, intercalation with DNA/RNA, internalization into the cell, and disrupting many cellular processes that lead to cell death (Dizaj et al., 2014; Medici et al., 2015; Wani et al., 2016). Biocompatibility, high electron density, intense optical absorption, photoluminescence, and the magnetic properties of nanoparticles make them useful for imaging and therapy in biomedical systems. Basically, nano-metallopharmaceuticals are nanostructured, which helps with the diagnosis, prevention, and treatment of diseases, including injuries and pain, improving human health using nanotechnology and knowledge of the human body (Wani et al., 2016).

Furthermore, metallic nanoparticles' physicochemical and morphological characteristics have a significant impact on their activity and can be controlled by tuning their particle shape and size (Pal, Tak, and Song, 2007; Menon et al., 2012). For example, features like the transformation of absorbed light (excitation energy) into heat are exploited in photothermal cancer treatments and were found to be dependent on the architecture of AuNPs (Hirsch et al., 2003; Fazal et al., 2014).

6.3 WHAT MAKES THEM USEFUL FOR THERAPEUTICS/ POTENTIAL AS THERAPEUTIC AGENTS

After the synthesis of NPs as pharmaceutically active agents or carriers, their significant uptake by the targeted cell is the final step for the successful action of the drug. The shape, size, and surface functionalization of nano-metallopharmaceuticals are the chief reasons for the enhanced action and bioavailability after both systemic and local administration (Sara, Solmaz, and Ahmad, 2015). Therefore, metallic nanoparticles' physicochemical and morphological characteristics have a significant impact on their activity and can be controlled by tuning particle shape and size (Pal, Tak, and Song, 2007; Menon et al., 2012). A small change in the size and shape strongly affects plasmon resonance frequency (LSPR), shifting the band absorption into the visible and near-infrared region (NIR), which is of huge interest to the biomedical world (Kohout, Santi, and Polito, 2018). Interestingly, several inorganic nanoparticles such as silver, gold, platinum, and silica nanospheres are self-therapeutic and need no surface functionalization for their action (Rai et al., 2016). At the nanoscale, metallic particles can easily penetrate through biological barriers, which extend their bioavailability to the targeted cells, such as tumour cells (Kim et al., 2009). For example, various forms of silver (Ag and Ag_2O NPs) have been shown to have a bactericidal impact against different bacterial strains (Liu and Lal, 2014). However, surface functionalization also enhances specific delivery to targeted cells, such as surface-derivatized cadmium sulfide (CdS) nanocrystals, which were used as removable caps for mesoporous silica nanospheres, and showed a controlled release of the drug (Lai et al., 2003). Therefore, the synthesis of an appropriate, biocompatible, and bioavailable nanoparticle is essential for expediting drugs to the targeted cell.

6.4 SYNTHETIC ROUTES FOR NANO-METALLOPHARMACEUTICALS

As morphology principally determines a metallic nanoparticle's fundamental characteristics, one could control and fine-tune the various synthetic methods. Researchers around the globe have used numerous methods for the synthesis of NPs, as shown in Figure 6.1:

1. Chemical
2. Physical
3. Biological

The chemical and physical methods are well-known and familiar but pose threats to human health and the environment. Chemical routes require several reducing and capping agents for metal precursors, such as sodium borohydride, citrate, hydrazine, DMF, cetyltrimethylammonium bromide (CTAB), etc. Exposure to these chemical salts leads to pulmonary oedema, hypocalcaemia, affects the nervous system, and severely irritates the skin and eyes (Yugandhar, Haribabu, and Savithramma, 2015; Yugandhar et al., 2017). Physical methods like pulverization, laser ablation, thermochemical decomposition, arc-discharge, microwave-assisted, microemulsion, sonochemical, ball milling, etc., are complicated, expensive, consume high amounts of energy, and require a large space (Yugandhar et al., 2017; Capek, 2004; Tavakoli, Sohrabi, and Kargari, 2007). Therefore, a continuous effort is being made in search

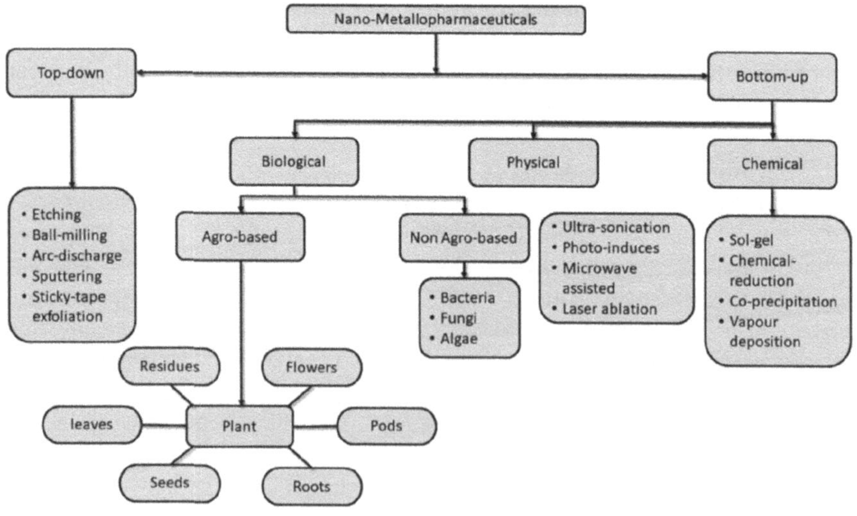

FIGURE 6.1 Various methods for the synthesis of nano-metallopharmaceuticals.

of suitable methods. Continuing in this quest, 'green' technology, such as using microbes, fungi, enzymes, plant/plant parts extracts, and phytochemicals, offers environment-friendly protocols and can synthesize biocompatible NPs for numerous biomedical applications without using toxic chemicals (Barabadi et al., 2017).

6.5 POTENTIAL OF AGRO-PRODUCTS – AN IDEAL APPROACH

Although chemical and physical methods are more popular, the use of toxic chemicals significantly limits their biomedical world, particularly in clinical fields. With these constraints, the biosynthesis of NPs has gained considerable attention due to their negligible toxicity towards humans and the environment. For instance, substantial work has been done on the biosynthesis of NPs, especially using microorganisms (Singaravelu et al., 2007; Castro-Longoria, Vilchis-Nestor, and Avalos-Borja, 2011; Bar et al., 2009; Li et al., 2011). Nevertheless, the use of bacteria, fungi, and enzymes are still under investigation. On the other hand, agricultural products such as leaves, seeds, roots, pods, fruits, and waste for nanoparticle synthesis are evolving as new branches but are still currently relatively unfamiliar. A list of different nano-metallopharmaceuticals derived from various agro-products is given in Table 6.1.

It has been shown that morphological features such as shape and size can be controlled by altering the ratio of the metal precursor to the plant extract. However, the bio-reduction mechanism of metal salts into nanoparticles is not very well known. It is recognized that plant metabolites comprising different functional groups, particularly the phenolic, carboxyl, and amine, are primarily involved in the reduction. Bioactive secondary metabolites and acids,, such as flavonoids, alkaloids, terpenoids, pentacyclic triterpene acids, and fatty acids present in the agro-products (plants parts), are suggested to be responsible for the reduction and capping of metal precursors (Arya et al., 2018; Barabadi et al., 2017; Mousavi, Tafvizi, and Zaker Bostanabad, 2018; Huo et al., 2018; Wang et al., 2017; Abdi et al., 2018; Kuppusamy, Yusoff, and Maniam, 2016). The potential of agro-products for the synthesis of metallic NPs, including Pt, Pd, Au, Ag, Cu, and other oxides like CeO, CuO, and ZnO NPs with medical applications are described in Figure 6.2.

6.6 BIOSYNTHESIS OF DIFFERENT NANO-METALLOPHARMACEUTICALS

6.6.1 Platinum (Pt)

Platinum was the starting point for use of metal complexes in medicine, which was chiefly represented by cisplatin and widely accepted for anticancer therapy. Being the noble metal, Pt has been continuously applied in treatments from cancer drugs to antibacterial applications. In the modern era, a lot of efforts have been made to enhance its specificity, availability, and biocompatibility using various agricultural plant species and nanotechnological approaches (Medici et al., 2015). The metal toxicity tolerance ability of *Medicago sativa* was exploited for the *in vivo* formation of Pt nanoparticles in roots ranging from 3 to 100 nm. At higher concentrations, Pt(II)

TABLE 6.1
Potential of Various Agro-Products in the Synthesis of Nano-Metallopharmaceuticals

Plants/plant extracts	Metal nanoparticle	Reducing and capping agent	Therapeutic application	References
Allium cepa	Ag	Extract (bulbs)	Antibacterial activity	(Saxena, Tripathi, and Singh, 2010)
Green tea	Ag	Leaf extract	Antibacterial activity	(Sun et al., 2014)
Brassica rapa L.	Ag	Leaf extract	Antifungal activity	(Narayanan and Park, 2014)
Vaccinium macrocarp	Ag	Fruit extract	Antibacterial activity	(El-Kamel et al., 2015)
Anthemis atropatana	Ag	Plant extract	Anticancer, Antimicrobial activity	(Dehghanizade, Arasteh, and Mirzaie, 2018)
Aloe vera	Ag	Leaf extract	Antifungal activity	(Medda et al., 2015)
Musa paradisiaca	Ag	Peel extract	Antimicrobial activity	(Ibrahim, 2015)
Chaenomeles sinensis	Ag/Au	Fruit extract	Anticancer, Antibacterial activity	(Oh et al., 2017)
Glycyrrhiza uralensis	AgCl/Au	Root extract	Anticancer, Antimicrobial activity	(Soshnikova et al., 2018)
Mentha piperita	Ag/Au	Leaf extract	Antibacterial activity	(Mubarakali et al., 2011)
Aloe vera	Ag/Au	Leaf extract	Anticancer	(Chandran et al., 2006)
Vitis vinifera L.	Au	Grape skin, stalk and seeds		(Krishnaswamy, Vali, and Orsat, 2014)
Allium sativum	Au	Clove extract	Anticancer	(Menon et al., 2012)
Theobromo cacao	Au	Seed extract	Anticancer	(Fazal et al., 2014)
Rabdosia rubescens L.	Au	Leaf extract	Anticancer	(Zhang et al., 2019)
Mangifera indica L.	Au	Peel extract	Anticancer	(N. Yang, Li, and Hao, 2014)
Hibiscus sabdariffa	Au	Leaf and stem extract	Anticancer	(Mishra et al., 2016)
Medicago sativa and *Brassica juncea*	Pt	Rooted seedlings		(Bali, Siegele, and Harris, 2010)

(Continued)

TABLE 6.1 (CONTINUED)
Potential of Various Agro-Products in the Synthesis of Nano-Metallopharmaceuticals

Plants/plant extracts	Metal nanoparticle	Reducing and capping agent	Therapeutic application	References
Diopyros kaki	Pt	Leaf extract		(Yong, Kwon, and Kim, 2010)
Ocimum sanctum	Pt	Leaf extract		(Soundarrajan et al., 2012)
Asparagus racemosus L.	Pt/Pd	Root extract		(Raut, 2013)
Gloriosa superba	Pt/Pd	Tubers extract	Anticancer	(Rokade et al., 2018)
Moringa oleifera	Pd	flower extract	Anticancer, Antimicrobial activity	(Anand et al., 2016)
Tea and coffee	Pd	Extract		(Nadagouda and Varma, 2008)
Hippophae rhamnoides L.	Pd	Leaf extract		(Nasrollahzadeh, Sajadi, and Maham, 2015)
Annona squamosa L.	Pd	Peel extract	Acaricidal, insecticidal, and larvicidal efficacy	(Roopan et al., 2013)
Cinnamomum camphora	Pd	Leaf extract	Biomedical application	(X. Yang et al., 2010)
Dioscorea bulbifera	Pt, Pd, Pt–Pd	Tuber extract	Anticancer	(Ghosh et al., 2015)
Gloriosa superba L.	CeO	Leaf extract	Antibacterial activity	(Arumugam et al., 2015)
Aloe Barbadensis	CeO	Leaf extract		(Sai Priya et al., 2014)
Hibiscus Sabdariffa	CeO	Flower extract		(Thovhogi et al., 2015)
Artemisia haussknechtii	Cu/TiO_2	leaf extract	antibacterial, total antioxidant, scavenging	(Alavi and Karimi, 2018)
Broccoli	Cu	Whole extract	anticancer	(Prasad, Kanchi, and Naidoo 2016)
Acalypha indica	Cu	Leaf extract	anticancer	(Sivaraj et al., 2014)
Punica granatum	Cu	Peel extract	Antibacterial Activity	(Kaur, Thakur, and Chaudhury, 2016)

(Continued)

TABLE 6.1 (CONTINUED)
Potential of Various Agro-Products in the Synthesis of Nano-Metallopharmaceuticals

Plants/plant extracts	Metal nanoparticle	Reducing and capping agent	Therapeutic application	References
Catha edulis	CuO	Leaf extract	Antibacterial Activity	(Gebremedhn, Kahsay, and Aklilu, 2019)
Gloriosa superba L.	CuO	Plant extract	Antibacterial Activity	(Naika et al., 2015)
Lavandula vera	Zn	Leaf extract	Anticancer and antioxidant activity	(Salari et al., 2017)
Solanum tuberosum	ZnO	Potato extract		(Buazar et al., 2016)
Matricaria chamomilla L., *Olea europaea and red Lycopersicon esculentum* M.	ZnO	Flower, leaf, fruit extract	Antibacterial activity	(Ogunyemi et al., 2019)
Mangifera indica	ZnO	Leaf extract	Anticancer	(Rajeshkumar et al., 2018)
Camellia sinensis	ZnO	Leaf extract	Antimicrobial activity	(Senthilkumar and Sivakumar, 2016)
Lycopersicon esculentum	ZnO	Fruit extract		(Sutradhar and Saha, 2016)
Corymbia citriodora	ZnO	Leaf extract		(Zheng et al., 2015)
Rosa canina	ZnO	Fruit extract	Anticancer, antibacterial activity	(Jafarirad et al., 2016)
Gloriosa superba L.	Ru	leaf extract	Antibacterial activity	(Gopinath et al., 2014)
Acalypha indica	RuO_2	plant extract	Antibacterial activity	(Kannan and Sundrarajan, 2015)
Couroupita guianensis Aubl.	Fe_3O_4	Fruit extract	Anticancer, antibacterial	(Sathishkumar et al., 2017)

was sequestered and reduced to Pt(0) due to the action of plant metabolites (Bali, Siegele, and Harris, 2010). The leaf extracts of *Diospyros kaki* and *Ocimum sanctum* are the two main plants cultivated for their medicinal value, and they have been found to reduce metal precursors, most likely due to the many functional groups present in their metabolites (Song, 2010; Soundarrajan et al., 2012).

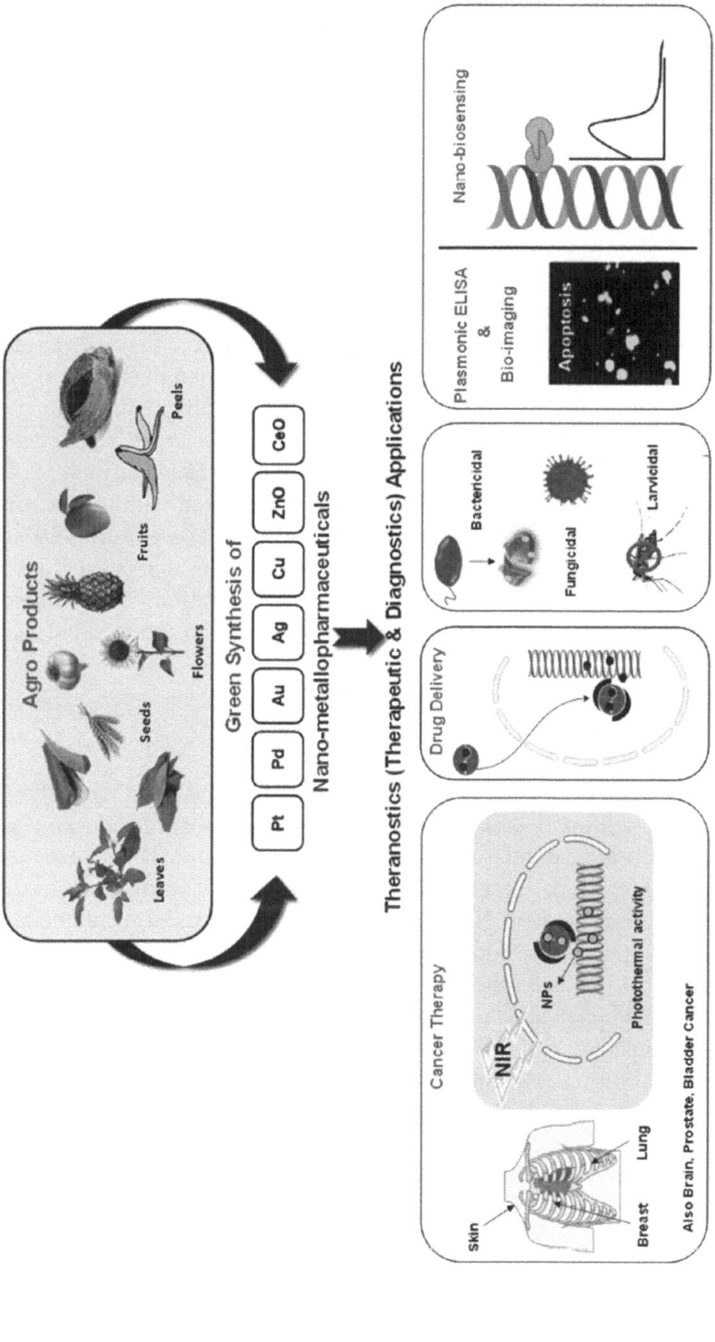

FIGURE 6.2 Potential of different agro-products for green synthesis of metal-based nanoparticles with useful applications in theranostics.

Interestingly, the root extract of *Asparagus racemosus* Linn. very finely controlled the size of monodispersed Pt NPs between 1 to 6 nm within a few minutes (Raut, 2013). Uniform, almost isodiametric (spherical), Pt NPs were produced using a medicinal plant *Gloriosa superba* tuber extract (GSTE). These monodispersed NPs showed strong anticancer action against MCF-7 cell lines by inducing apoptosis (Rokade et al., 2018). The release of ions, coated drugs from the surface of Pt NPs, induce the cytotoxicity towards cancerous cells, oxidative stress, and arrest the cell cycles (Thondavada, Chokkareddy, and Redhi, 2018).

6.6.2 Palladium (Pd)

Because of its similar coordination chemistry to Pt(II), Pd is a valid successor. However, the real interest in its therapeutics started in two decades ago (Medici et al., 2015). Moreover, during the evaluation of anticancer activity, Pd complexes were observed to be more liable and bind to DNA faster than Pt (Ray et al., 2007). Recently, biosynthesis of Pd NPs with *Gloriosa superba* tuber extract showed significant anticancer potential against the human breast adenocarcinoma cells that supported the use of Pd as nano-metallopharmaceuticals (Rokade et al., 2018). In a simple green approach, Pd nanocrystals were synthesized within a narrow size range of 20–60 nm using coffee and tea extract at room temperature (Nadagouda and Varma, 2008). This size range of Pd NPs was further narrowed from 3.2 to 6.0 nm and simply controlled by varying the concentrations of *Cinnamomum camphora* leaf broth, which may have an essential role in the medical field (X. Yang et al., 2010).

6.6.3 Gold (Au)

Among all noble metals, gold NPs are considered to be valuable in clinical research because of their non-toxic, ease of functionalization, polyvalent effects, and easy detection. Au NPs are essential antibacterial agents, and they is also used in drug formulations and for the delivery of drugs to targeted cells. Therefore, the biosynthesis of these multipurpose (Au) NPs using agro-products (extract) could be a potent metallodrug. Polyphenolic compounds available in grape skins, seeds, and stalks produced spherical to quasi-spherical biocompatible Au NPs (Krishnaswamy, Vali, and Orsat, 2014). Phytochemicals of garlic, a medicinal herb, were used to synthesize NPs with tunable optical characteristics due to the range of anisotropy with good *in vitro* stability, suggesting the potential application in imaging and nanomedicines (Fazal et al., 2014). Similarly, anisotropic gold nanoparticles showed good photothermal activity for cancer therapy, using *Theobroma cacao* (cocoa) extract for synthesis (Fazal et al., 2014). Particles synthesized from the seeds of cardamom fruits and the leaf extract of *Indigofera tinctoria* shows a cytotoxic effect on breast and lung cancer cells, respectively (Soshnikova et al., 2018; Vijayan, Joseph, and Mathew, 2018). Zhang et al. (2019) synthesized gold nanoparticles from *Rabdosia rubescens* and showed the potent anticancer and apoptotic effects against lung carcinoma A549 cell lines (Zhang et al., 2019).

6.6.4 Silver (Ag)

Ag nanoparticles have been extensively used for their popularly known antibacterial efficacy. Increasing antibiotic resistance and the potential bactericidal activity of Ag NPs have drawn interest towards Ag as nano-metallopharmaceuticals. Although the exact mechanism of the antibacterial activity is still unknown, most researchers still believe that Ag NPs induce pits into the cell membrane and disrupts the metabolic pathways, which ultimately causes the death of the cells (Pal, Tak, and Song, 2007). Use of agro-products such as green tea leaf extract, onion (*Allium cepa*) extract, turnip leaf extract (*Brassica rapa* L.), cranberry powder aqueous extract, etc., have been described as synthesizing Ag NPs with a significant antibacterial activity (Sun et al., 2014; Saxena, Tripathi, and Singh, 2010; Narayanan and Park, 2014; El-Kamel et al., 2015). Even after using the leading products in agriculture, the residues which were considered to be waste were found to have the potential for nanoparticle synthesis. For instance, low-cost coconut-coir and banana peel have been utilized for the production of Ag NPs, and they showed significant larvicidal and antibacterial activity, respectively (Roopan et al., 2013; Ibrahim, 2015). The fresh aerial part of the plant *Anthemis atropatana* extract was used to synthesize spherical Ag NPs which had good cytotoxic effects on colon cancer cell lines (HT29) (Dehghanizade, Arasteh, and Mirzaie, 2018). Huo and co-workers showed not only strong antibacterial activity against disease-causing pathogens (*Staphylococcus aureus*, *Escherichia coli*, and *Pseudomonas aeruginosa*) but significant cytotoxicity against breast cancer cell lines was also represented (MCF7) (Huo et al., 2018).

Silver particles are also used in bio-imaging as a fluorescent probe to detect tumour cells. For example, Ag NPs for cancer diagnosis through plasmonic ELISA (enzyme-linked immunosorbent assay) is of great use in therapeutic monitoring (Barabadi et al., 2017). Ag NPs were also used in wound dressings, catheters, dental composites, biosensing cardiovascular implants, and other therapeutic applications (Rafique et al., 2017).

6.6.5 Cerium Oxide (CeO)

Recently, cerium oxides have emerged as an essential metallodrug to protect from cellular damage in severe pathological conditions, brain or cardiac ischaemia, and neurological disorders from toxicants. To reduce their poor solubility, particles can be functionalized with polymers for enhanced solubility and biocompatibility (Rajeshkumar and Naik, 2018). Spherical nanoceria can be synthesized using the leaf extract of *Gloriosa superba* L., which showed strong antibacterial activity against the gram (+/–) bacteria (Arumugam et al., 2015). Different parts of the plant, such as the leaves of *Aloe barbadensis* and flower's extract of *Hibiscus sabdariffa* were used to synthesize CeO_2 NPs with sizes 63.6 and 3.9 nm, respectively (Sai Priya et al., 2014; Thovhogi et al., 2015).

6.6.6 Copper (Cu)

Copper NPs are of great interest in research because of their low-cost preparation and unique physicochemical, biological properties, and vigorous antibacterial activity. Recently, preparations using the various metabolites present in plants, such as amino acids, polysaccharides, enzymes, and alkaloids, have been used widely (Dizaj et al., 2014). The leaf extract of a Chinese medicinal plant, *Artemisia haussknechtii*, was used to produce Cu NPs of sizes 35.36 ± 44.4 nm and showed good antibacterial activity against many MDR (multiple drug resistance) species (*Staphylococcus epidermidis* ATCC 12258, *Staphylococcus aureus* ATCC 43300, *Escherichia coli* ATCC 25922, and *Serratia marcescens* ATTC13880) (Alavi and Karimi, 2018). In another study, *Punica granatum* peel extract reduced copper salt to spherical Cu NPs with sizes varying between 15–20 nm and showed bactericidal activity against opportunistic pathogens (Kaur, Thakur, and Chaudhury, 2016). Copper oxide NPs formed due to thermodynamic instability and exposure to air lead to an oxide layer over the NP surface. However, plant parts such as the leaf extract of *Gloriosa superba* L. have also been found to synthesize very narrowly ranged (5–10 nm) CuO NPs that exhibited significant antibacterial activity (Naika et al., 2015). Knowing the various medicinal use of *Azadirachta indica*, CuO NPs were synthesized using a leaf extract and were found to activate TNF-a and caspases signalling pathways to induce apoptosis against MCF-7 and Hela cells as well as in a Balb/C mice model (Dey et al., 2019).

6.6.7 Zinc Oxide (ZnO)

The biocompatibility of ZnO NPs with human cells makes them an appropriate additive for metallodrugs or in textiles for an improved therapeutic use. Besides the chemical synthesis, biogenic routes have emerged as environmentally sound techniques using various plant part extracts. Potato extract was used as a reducing and stabilizing agent for synthesizing ZnO NPs with promising potential biomedical applications (Buazar et al., 2016). The leaf extract of *Olea europaea* produced the smallest size range of between 40.5 to 124.0 nm NPs with the highest antibacterial efficacy when compared with the fruit extract of *Lycopersicon esculentum* M. (tomato) and flower extract of *Matricaria chamomilla* L. (Ogunyemi et al., 2019). The potential of a tomato extract (*Lycopersicon esculentum*) was also tested, which produced well-defined ZnO NPs with excellent reproducibility (Sutradhar and Saha, 2016). The leaf extract of *Mangifera indica* (mango) was used to synthesize ZnO NPs, which showed antioxidant properties equivalent to standard vitamin C and dose-dependent cytotoxic effects against lung cancer A549 cell lines (Rajeshkumar et al., 2018). Some of the higher plants, such as cones of *Pinus densiflora* and the leaf extract of *Corymbia citriodora* have also shown the potential to produce ZnO NPs for biomedical and other applications (Velmurugan et al., 2016; Zheng et al., 2015).

All these findings show how relevant and useful these agro-products are for synthesizing potential nano-metallopharmaceuticals. At the same time, these agro-products are environmentally benign, which allows for their unrestricted use and, thus, does not pose any threat to the environment.

6.7 CONCLUSIONS AND FUTURE OUTLOOK

The green synthesis of nano-metallopharmaceuticals is a reliable tool for the diagnosis and treatment of numerous medical implications. Some of the nanoparticles can directly impart cytotoxic effects as 'self-therapeutics' such as Ag and Au. In contrast, others such as Pt and Pd, are modified or functionalized and recognized as safe delivery vehicles for drugs. The use of agricultural products improves the biocompatibility of nanoparticles for biomedical applications and eliminates the need to use toxic chemicals. These naturally occurring agro-products from the whole plant to phytochemicals and metabolites are found to play an active role in the synthesis, which helps to design safe nano-metallopharmaceuticals. However, the insight into the green synthesis mechanism of such nano-metallopharmaceuticals using agro-products is not sufficient, and detailed research can improve their potential biomedical applications. It is expected that agro-product mediated green synthesis of nano-metallopharmaceuticals will emerge as therapeutics and diagnostics agents in the near future.

ACKNOWLEDGEMENTS

DP acknowledges the University Grant Commission, New Delhi, India, for providing Senior Research Fellowship. Director, CSIR-NEERI is thankfully acknowledged for giving the opportunity to pursue the work in CSIR-NEERI, Nagpur, India. The article is checked for plagiarism using the iThenticate software and recorded in the Knowledge Resource Center, CSIR-NEERI, Nagpur for anti-plagiarism (KRC No.: CSIR-NEERI/KRC/2020/FEB/WWT-DRC/1).

REFERENCES

Abdi, Vahideh, Iman Sourinejad, Morteza Yousefzadi, and Zahra Ghasemi. 2018. 'Mangrove-Mediated Synthesis of Silver Nanoparticles Using Native *Avicennia marina* Plant Extract from Southern Iran'. *Chemical Engineering Communications* 205(8): 1069–76. https://doi.org/10.1080/00986445.2018.1431624.

Alavi, Mehran, and Naser Karimi. 2018. 'Characterization, Antibacterial, Total Antioxidant, Scavenging, Reducing Power and Ion Chelating Activities of Green Synthesized Silver, Copper and Titanium Dioxide Nanoparticles Using *Artemisia haussknechtii* Leaf Extract'. *Artificial Cells, Nanomedicine and Biotechnology* 46(8): 2066–81. https://doi.org/10.1080/21691401.2017.1408121.

Anand, K., C. Tiloke, A. Phulukdaree, B. Ranjan, A. Chuturgoon, S. Singh, and R. M. Gengan. 2016. 'Biosynthesis of Palladium Nanoparticles by Using *Moringa oleifera* Flower Extract and Their Catalytic and Biological Properties'. *Journal of Photochemistry and Photobiology, Part B: Biology* 165: 87–95. https://doi.org/10.1016/j.jphotobiol.2016.09.039.

Arumugam, Ayyakannu, Chandrasekaran Karthikeyan, Abdulrahman Syedahamed Haja Hameed, Kasi Gopinath, Shanmugam Gowri, and Viswanathan Karthika. 2015. 'Synthesis of Cerium Oxide Nanoparticles Using *Gloriosa superba* L. Leaf Extract and Their Structural, Optical and Antibacterial Properties'. *Materials Science and Engineering C: Biomimetic Materials Sensors and Systems* 49: 408–15. https://doi.org/10.1016/j.msec.2015.01.042.

Arya, Geeta, R. Mankamna Kumari, Nidhi Gupta, Ajeet Kumar, Ramesh Chandra, and Surendra Nimesh. 2018. 'Green Synthesis of Silver Nanoparticles Using Prosopis Juliflora Bark Extract: Reaction Optimization, Antimicrobial and Catalytic Activities'. *Artificial Cells, Nanomedicine and Biotechnology* 46(5): 985–93. https://doi.org/10.1080/21691401.2017.1354302.

Bali, Roza, Rainer Siegele, and Andrew T. Harris. 2010. 'Biogenic Pt Uptake and Nanoparticle Formation in *Medicago sativa* and *Brassica juncea*'. *Journal of Nanoparticle Research* 12(8): 3087–95. https://doi.org/10.1007/s11051-010-9904-7.

Bar, Harekrishna, Dipak Kr. Bhui, Gobinda P. Sahoo, Priyanka Sarkar, Santanu Pyne, and Ajay Misra. 2009. 'Green Synthesis of Silver Nanoparticles Using Seed Extract of *Jatropha curcas*'. *Colloids and Surfaces. Part A: Physicochemical and Engineering Aspects* 348(1–3): 212–16. https://doi.org/10.1016/j.colsurfa.2009.07.021.

Barabadi, Hamed, Muhammad Ovais, Zabta Khan Shinwari, and Muthupandian Saravanan. 2017. 'Anti-Cancer Green Bionanomaterials: Present Status and Future Prospects'. *Green Chemistry Letters and Reviews* 10(4): 285–314. https://doi.org/10.1080/17518253.2017.1385856.

Bharti, Sanjay K., and Sushil K. Singh. 2009. 'Recent Developments in the Field of Anticancer Metallopharmaceuticals'. *International Journal of PharmTech Research* 1(4): 1406–20.

Boisseau, Patrick, and Bertrand Loubaton. 2011. 'Nanomedicine, Nanotechnology in Medicine'. *Comptes Rendus Physique* 12(7): 620–36. https://doi.org/10.1016/j.crhy.2011.06.001.

Buazar, F., M. Bavi, F. Kroushawi, M. Halvani, A. Khaledi-Nasab, and S. A. Hossieni. 2016. 'Potato Extract as Reducing Agent and Stabiliser in a Facile Green One-Step Synthesis of ZnO Nanoparticles'. *Journal of Experimental Nanoscience* 11(3): 175–84. https://doi.org/10.1080/17458080.2015.1039610.

Capek, I. 2004. 'Preparation of Metal Nanoparticles in Water-In-Oil (w/o) Microemulsions'. *Advances in Colloid and Interface Science* 110(1–2): 49–74. https://doi.org/10.1016/j.cis.2004.02.003.

Castro-Longoria, E., Alfredo R. Vilchis-Nestor, and M. Avalos-Borja. 2011. 'Biosynthesis of Silver, Gold and Bimetallic Nanoparticles Using the Filamentous Fungus *Neurospora crassa*'. *Colloids and Surfaces, Part B: Biointerfaces* 83(1): 42–8. https://doi.org/10.1016/j.colsurfb.2010.10.035.

Chandran, S. P., Minakshi Chaudhary, Renu Pasricha, Absar Ahmad, and Murali Sastry. 2006. 'Synthesis of Gold Nanotriangles and Silver Nanoparticles Using Aloe W Era Plant Extract'. *Biotechnology Progress* 22(2): 577–83.

Dehghanizade, Saeede, Javad Arasteh, and Amir Mirzaie. 2018. 'Green Synthesis of Silver Nanoparticles Using Anthemis Atropatana Extract: Characterization and In Vitro Biological Activities'. *Artificial Cells, Nanomedicine and Biotechnology* 46: 160–68. https://doi.org/10.1080/21691401.2017.1304402.

Dey, Aditi, Subhankar Manna, Sourav Chattopadhyay, Dipankar Mondal, Dipankar Chattopadhyay, Anupam Raj, Subhajit Das, Braja Gopal Bag, Somenath Roy. 2019. '*Azadirachta indica* Leaves Mediated Green Synthesized Copper Oxide Nanoparticles Induce Apoptosis through Activation of TNF- a and Caspases Signaling Pathway against Cancer Cells'. *Journal of Saudi Chemical Society* 23(2): 222–38. https://doi.org/10.1016/j.jscs.2018.06.011.

Dizaj, Solmaz Maleki, Farzaneh Lotfipour, Mohammad Barzegar-Jalali, Mohammad Hossein Zarrintan, and Khosro Adibkia. 2014. 'Antimicrobial Activity of the Metals and Metal Oxide Nanoparticles'. *Materials Science and Engineering C: Biomimetic Materials Sensors and Systems* 44: 278–84. https://doi.org/10.1016/j.msec.2014.08.031.

El-Kamel, Amal, Asmaa Ashour, Dina Raafat, and Hanan El-Goweily. 2015. 'Green Synthesis of Silver Nanoparticles Using Cranberry Powder Aqueous Extract: Characterization and Antimicrobial Properties'. *International Journal of Nanomedicine* 10: 7207–21. https://doi.org/10.2147/ijn.s87268.

Fazal, Sajid, Aswathy Jayasree, Sisini Sasidharan, Manzoor Koyakutty, Shantikumar V Nair, and Deepthy Menon. 2014. 'Green Synthesis of Anisotropic Gold Nanoparticles for Photothermal Therapy of Cancer'. *ACS Applied Materials & Interfaces* 6: 8080–89.

Sathishkumar, G., V. Logeshwaran, S. Sarathbabu, Pradeep K. Jha, M. Jeyaraj, C. Rajkuberan, N. Senthilkumar, and S. Sivaramakrishnan. 2017. 'Green Synthesis of Magnetic Fe 3 O 4 Nanoparticles Using *Couroupita guianensis* Aubl. Fruit Extract for Their Antibacterial and Cytotoxicity Activities'. *Artificial Cells, Nanomedicine, and Biotechnology* 46(0): 589–98. https://doi.org/10.1080/21691401.2017.1332635.

Gebremedhn, K., Mebrahtu Hagos Kahsay, and Muluken Aklilu. 2019. 'Green Synthesis of CuO Nanoparticles Using Leaf Extract of *Catha edulis* and Its Antibacterial Activity'. *Journal of Pharmacy and Pharmacology* 7(7): 327–42. https://doi.org/10.17265/2328-2150/2019.06.007.

Ghosh, Sougata, Rahul Nitnavare, Ankush Dewle, Geetanjali B Tomar, Rohan Chippalkatti, Piyush More, Rohini Kitture, Sangeeta Kale, Jayesh Bellare, and Balu A Chopade. 2015. 'Novel Platinum – Palladium Bimetallic Nanoparticles Synthesized by *Dioscorea bulbifera* lbiferaa inum – Palladium Bimetallic Na'. *International Journal of Nanomedicine* 10: 7477–90.

Gopinath, Kasi, Viswanathan Karthika, S. Gowri, V. Senthilkumar, S. Kumaresan, and A. Arumugam. 2014. 'Antibacterial Activity of Ruthenium Nanoparticles Synthesized Using *Gloriosa superba* L. Leaf Extract'. *Journal of Nanostructure in Chemistry* 4(1): 83. https://doi.org/10.1007/s40097-014-0083-4.

Hirsch, L. R., R. J. Stafford, J. A. Bankson, S. R. Sershen, B. Rivera, R. E. Price, J. D. Hazle, N. J. Halas, and J. L. West. 2003. 'Nanoshell-Mediated Near-Infrared Thermal Therapy of Tumors under Magnetic Resonance Guidance'. *Proceedings of the National Academy of Sciences of the United States of America* 100(23): 13549–54. https://doi.org/10.1073/pnas.2232479100.

Huo, Yue, Priyanka Singh, Yeon Ju Kim, Veronika Soshnikova, Jongpyo Kang, Josua Markus, Sungeun Ahn, et al. 2018. 'Biological Synthesis of Gold and Silver Chloride Nanoparticles by *Glycyrrhiza uralensis* and In Vitro Applications'. *Artificial Cells, Nanomedicine and Biotechnology* 46(2): 303–12. https://doi.org/10.1080/21691401.2017.1307213.

Ibrahim, Haytham M. M. 2015. 'Green Synthesis and Characterization of Silver Nanoparticles Using Banana Peel Extract and Their Antimicrobial Activity against Representative Microorganisms'. *Journal of Radiation Research and Applied Sciences* 8(3): 265–75. https://doi.org/10.1016/j.jrras.2015.01.007.

Jafarirad, Saeed, Meysam Mehrabi, Baharak Divband, and Morteza Kosari-nasab. 2016. 'Biofabrication of Zinc Oxide Nanoparticles Using Fruit Extract of *Rosa canina* and Their Toxic Potential against Bacteria : A Mechanistic Approach'. *Materials Science and Engineering C: Biomimetic Materials Sensors and Systems* 59: 296–302. https://doi.org/10.1016/j.msec.2015.09.089.

Kannan, S. K., and M. Sundrarajan. 2015. 'Green Synthesis of Ruthenium Oxide Nanoparticles : Characterization and Its Antibacterial Activity'. *Advanced Powder Technology* 26(6): 1505–11. https://doi.org/10.1016/j.apt.2015.08.009.

Kaur, Pawan, Rajesh Thakur, and Ashok Chaudhury. 2016. 'Biogenesis of Copper Nanoparticles Using Peel Extract of *Punica granatum* and Their Antimicrobial Activity against Opportunistic Pathogens'. *Green Chemistry Letters and Reviews* 9(1): 33–8. https://doi.org/10.1080/17518253.2016.1141238.

Kim, Jeong Hun, Jin Hyoung Kim, Kyu Won Kim, Myung Hun Kim, and Young Suk Yu. 2009. 'Intravenously Administered Gold Nanoparticles Pass through the Blood-Retinal

Barrier Depending on the Particle Size, and Induce No Retinal Toxicity'. *Nanotechnology* 20(50): 505101. https://doi.org/10.1088/0957-4484/20/50/505101.

Ko, Seung H., Inkyu Park, Heng Pan, Costas P. Grigoropoulos, Albert P. Pisano, Christine K. Luscombe, and Jean M. J. Fréchet. 2007. 'Direct Nanoimprinting of Metal Nanoparticles for Nanoscale Electronics Fabrication'. *Nano Letters* 7(7): 1869–77. https://doi.org/10.1021/nl070333v.

Kohout, Claudia, Cristina Santi, and Laura Polito. 2018. 'Anisotropic Gold Nanoparticles in Biomedical Applications'. *International Journal of Molecular Sciences* 19(11): 3385. https://doi.org/10.3390/ijms19113385.

Krishnaswamy, Kiruba, Hojatollah Vali, and Valérie Orsat. 2014. 'Value-Adding to Grape Waste: Green Synthesis of Gold Nanoparticles'. *Journal of Food Engineering* 142(June): 210–20. https://doi.org/10.1016/j.jfoodeng.2014.06.014.

Kuppusamy, Palaniselvam, Mashitah M. Yusoff, and Gaanty Pragas Maniam. 2016. 'Biosynthesis of Metallic Nanoparticles Using Plant Derivatives and Their New Avenues in Pharmacological Applications – An Updated Report'. *Saudi Pharmaceutical Journal: SPJ: The Official Publication of the Saudi Pharmaceutical Society* 24(4): 473–84. https://doi.org/10.1016/j.jsps.2014.11.013.

Lai, Cheng-yu, Brian G.Trewyn, Dusan M. Jeftinija, Ksenija Jeftinija, Shu Xu, Srdija Jeftinija, and Victor S. Lin. 2003. 'A Mesoporous Silica Nanosphere-Based Carrier System with Chemically Removable CdS Nanoparticle Caps for Stimuli-Responsive Controlled Release of Neurotransmitters and Drug Molecules'. *Journal of the American Chemical Society* 125(6): 4451–59.

Li, Xiangqian, Huizhong Xu, Zhe-Sheng Chen, and Guofang Chen. 2011. 'Biosynthesis of Nanoparticles by Microorganisms and Their Applications'. *Journal of Nanomaterials* 2011: 1–16. https://doi.org/10.1155/2011/270974.

Lippert, B. 2013. *Uses of Metal Compounds in Medicine. Reference Module in Chemistry, Molecular Sciences and Chemical Engineering.* Elsevier Inc. https://doi.org/10.1016/b978-0-12-409547-2.05375-0.

Liu, Ruiqiang, and Rattan Lal. 2014. 'Synthetic Apatite Nanoparticles as a Phosphorus Fertilizer for Soybean (*Glycine max*)'. *Scientific Reports* 4: 1–6. https://doi.org/10.1038/srep05686.

Medda, Shreya, Amita Hajra, Uttiya Dey, Paulomi Bose, and Naba Kumar Mondal. 2015. 'Biosynthesis of Silver Nanoparticles from Aloe vera Leaf Extract and Antifungal Activity against Rhizopus sp. and Aspergillus sp.'. *Applied Nanoscience* 5(7): 875–80. https://doi.org/10.1007/s13204-014-0387-1.

Medici, Serenella, Massimiliano Peana, Valeria Marina Nurchi, Joanna I. Lachowicz, Guido Crisponi, and Maria Antonietta Zoroddu. 2015. 'Noble Metals in Medicine: Latest Advances'. *Coordination Chemistry Reviews* 284: 329–50. https://doi.org/10.1016/j.ccr.2014.08.002.

Menon, Deepthy, Amritha Basanth, Archana Retnakumari, K. Manzoor, and Shantikumar V. Nair. 2012. 'Green Synthesis of Biocompatible Gold Nanocrystals with Tunable Surface Plasmon Resonance Using Garlic Phytochemicals'. *Journal of Biomedical Nanotechnology* 8(6): 901–11. https://doi.org/10.1166/jbn.2012.1455.

Mishra, Pratik, Sambit Ray, Sayantan Sinha, Bhaskar Das, Md. Imran Khan, Susant K. Behera, Suraj Kumar Tripathy Soon-IL Yun, Amrita Mishra, and A. Mishra. 2016. 'Facile Bio-Synthesis of Gold Nanoparticles by Using Extract of *Hibiscus sabdariffa* and Evaluation of Its Cytotoxicity against U87 Glioblastoma Cells under Hyperglycemic Condition'. *Biochemical Engineering Journal* 105: 264–72. https://doi.org/10.1016/j.bej.2015.09.021.

Mousavi, Bita, Farzaneh Tafvizi, and Saied Zaker Bostanabad. 2018. 'Green Synthesis of Silver Nanoparticles Using Artemisia Turcomanica Leaf Extract and the Study of

Anti-Cancer Effect and Apoptosis Induction on Gastric Cancer Cell Line (AGS)'. *Artificial Cells, Nanomedicine and Biotechnology* 46(supl): 499–510. https://doi.org/10.1080/21691401.2018.1430697.

Mubarakali, D., N. Thajuddin, K. Jeganathan, and M. Gunasekaran. 2011. 'Colloids and Surfaces B : Biointerfaces Plant Extract Mediated Synthesis of Silver and Gold Nanoparticles and Its Antibacterial Activity against Clinically Isolated Pathogens'. *Colloids and Surfaces, Part B: Biointerfaces* 85(2): 360–65. https://doi.org/10.1016/j.colsurfb.2011.03.009.

Muhammad, Nafees, and Zijian Guo. 2014. 'Metal-Based Anticancer Chemotherapeutic Agents'. *Current Opinion in Chemical Biology* 19(1): 144–53. https://doi.org/10.1016/j.cbpa.2014.02.003.

Nadagouda, Mallikarjuna N., and Rajender S. Varma. 2008. 'Green Synthesis of Silver and Palladium Nanoparticles at Room Temperature Using Coffee and Tea Extract'. *Green Chemistry* 10(8): 859–62. https://doi.org/10.1039/b804703k.

Naika, H. Raja, K. Lingaraju, K. Manjunath, Danith Kumar, G. Nagaraju, D. Suresh, and H. Nagabhushana. 2015. 'Green Synthesis of CuO Nanoparticles Using *Gloriosa superba* L. Extract and Their Antibacterial Activity'. *Journal of Taibah University for Science* 9(1): 7–12. https://doi.org/10.1016/j.jtusci.2014.04.006.

Narayanan, Kannan Badri, and Hyun Ho Park. 2014. 'Antifungal Activity of Silver Nanoparticles Synthesized Using Turnip Leaf Extract (*Brassica rapa* L.) against Wood Rotting Pathogens'. *European Journal of Plant Pathology* 140(2): 185–92. https://doi.org/10.1007/s10658-014-0399-4.

Nasrollahzadeh, Mahmoud, S. Mohammad Sajadi, and Mehdi Maham. 2015. 'Green Synthesis of Palladium Nanoparticles Using *Hippophae rhamnoides* Linn Leaf Extract and Their Catalytic Activity for the Suzuki – Miyaura Coupling in Water'. *Journal of Molecular Catalysis A: Chemical* 396: 297–303. https://doi.org/10.1016/j.molcata.2014.10.019.

Ogunyemi, Solabomi Olaitan, Yasmine Abdallah, Muchen Zhang, Hatem Fouad, Xianxian Hong, Ezzeldin Ibrahim, Md Mahidul Islam Masum, Afsana Hossain, Jianchu Mo, and Bin Li. 2019. 'Green Synthesis of Zinc Oxide Nanoparticles Using Different Plant Extracts and Their Antibacterial Activity against *Xanthomonas oryzae* Pv. Oryzae'. *Artificial Cells, Nanomedicine, and Biotechnology* 47(1): 341–52. https://doi.org/10.1080/21691401.2018.1557671.

Oh, Keun Hyun, Veronika Soshnikova, Josua Markus, Yeon Ju Kim, Sang Chul, Priyanka Singh, Verónica Castro-aceituno, et al. 2017. 'Biosynthesized Gold and Silver Nanoparticles by Aqueous Fruit Extract of Chaenomeles sinensis and Screening of Their Biomedical Activities'. *Artificial Cells, Nanomedicine, and Biotechnology* 46: 599–606. https://doi.org/10.1080/21691401.2017.1332636.

Pal, Sukdeb, Yu. Kyung Tak, and Joon Myong Song. 2007. 'Does the Antibacterial Activity of Silver Nanoparticles Depend on the Shape of the Nanoparticle? A Study of the Gram-Negative Bacterium *Escherichia coli*'. *Applied and Environmental Microbiology* 73(42): 1712–20. https://doi.org/10.1128/AEM.02218-06.

Pal, Sukdeb, Eun Jeong Yoon, Yu. Kyung Tak, Eung Chil Choi, and Joon Myong Song. 2009. 'Synthesis of Highly Antibacterial Nanocrystalline Trivalent Silver Polydiguanide'. *Journal of the American Chemical Society* 131(44): 16147–55. https://doi.org/10.1021/ja9051125.

Prasad, P.R., S. Kanchi, and E. B. Naidoo. 2016. 'In-Vitro Evaluation of Copper Nanoparticles Cytotoxicity on Prostate Cancer Cell Lines and Their Antioxidant, Sensing and Catalytic Activity: One-Pot Green Approach'. *Journal of Photochemistry and Photobiology, Part B: Biology* 161: 375–82. https://doi.org/10.1016/j.jphotobiol.2016.06.008.

Rafique, Muhammad, Iqra Sadaf, M. Shahid Rafique, and M. Bilal Tahir. 2017. 'A Review on Green Synthesis of Silver Nanoparticles and Their Applications'. *Artificial Cells, Nanomedicine and Biotechnology* 45(7): 1272–91. https://doi.org/10.1080/21691401.2016.1241792.

Rai, Mahendra, Avinash P. Ingle, Sonal Birla, Alka Yadav, and Carolina Alves Dos Santos. 2016. 'Strategic Role of Selected Noble Metal Nanoparticles in Medicine'. *Critical Reviews in Microbiology* 42(5): 696–719. https://doi.org/10.3109/1040841X.2015.1018131.

Rajeshkumar, S., S. Venkat Kumar, Arunachalam Ramaiah, Happy Agarwal, T. Lakshmi, and Selvaraj Mohana Roopan. 2018. 'Biosynthesis of Zinc Oxide Nanoparticles UsingMangifera Indica Leaves and Evaluation of Their Antioxidant and Cytotoxic Properties in Lung Cancer (A549) Cells'. *Enzyme and Microbial Technology* 117(July 2017): 91–5. https://doi.org/10.1016/j.enzmictec.2018.06.009.

Rajeshkumar, S., and Poonam Naik. 2018. 'Synthesis and Biomedical Applications of Cerium Oxide Nanoparticles – A Review'. *Biotechnology Reports* 17(October 2017): 1–5. https://doi.org/10.1016/j.btre.2017.11.008.

Raut, R. W. 2013. 'Rapid Biosynthesis of Platinum and Palladium Metal Nanoparticles Using Root Extract of *Asparagus racemosus* Linn'. *Advanced Materials Letters* 4(8): 650–54. https://doi.org/10.5185/amlett.2012.11470.

Ray, Sriparna, Renu Mohan, Jay K. Singh, Manoja K. Samantaray, Mobin M. Shaikh, Dulal Panda, and Prasenjit Ghosh. 2007. 'Anticancer and Antimicrobial Metallopharmaceutical Agents Based on Palladium, Gold, and Silver N-Heterocyclic Carbene Complexes'. *Journal of the American Chemical Society* 129(48): 15042–53.

Rokade, Shalaka S., Komal A. Joshi, Ketakee Mahajan, Saniya Patil, Geetanjali Tomar, Dnyanesh S. Dubal, Vijay Singh Parihar, Rohini Kitture, Jayesh R. Bellare, and Sougata Ghosh. 2018. 'Gloriosa superba Mediated Synthesis of Platinum and Palladium Nanoparticles for Induction of Apoptosis in Breast Cancer'. *Bioinorganic Chemistry and Applications* 2018: 1–9. https://doi.org/10.1155/2018/4924186.

Roopan, Selvaraj Mohana, G. Madhumitha, A. Abdul Rahuman, C. Kamaraj, A. Bharathi, and T. V. Surendra. 2013. 'Low-Cost and Eco-Friendly Phyto-Synthesis of Silver Nanoparticles Using *Cocos nucifera* Coir Extract and Its Larvicidal Activity'. *Industrial Crops and Products* 43(1): 631–35. https://doi.org/10.1016/j.indcrop.2012.08.013.

Rosenberg, Barnett, Loretta Van Camp, and Thomas Krigas. 1965. *Inhibition of Cell Division in Escherichia coli by Electrolysis Products from a Platinum Electrode.* Nature Publishing Group 205, pp. 698–9.

Sai Priya, G., Abhimanyu Kanneganti, K. Anil Kumar, K. Venkateswara Rao, and Satish Bykkam. 2014. 'Bio Synthesis of Cerium Oxide Nanoparticles Using Aloe barbadensis Miller Gel'. *International Journal of Scientific and Research Publications* 4(6): 2250–3153.

Salari, Zeinab, Atefeh Ameri, Hamid Forootanfar, Mahboubeh Adeli-sardou, Mandana Jafari, Mitra Mehrabani, and Mojtaba Shakibaie. 2017. 'Microwave-Assisted Biosynthesis of Zinc Nanoparticles and Their Cytotoxic and Antioxidant Activity'. *Journal of Trace Elements in Medicine and Biology* 39: 116–23. https://doi.org/10.1016/j.jtemb.2016.09.001.

Sara Salatin, Maleki Dizaj Solmaz, and Yari Khosroushahi Ahmad. 2015. 'Effect of the Surface Modification, Size, and Shape on Cellular Uptake of Nanoparticles'. *Cell Biology International* 39(8): 881–90. https://doi.org/doi:10.1002/cbin.10459.

Saxena, Antariksh, R. M. Tripathi, and R. P. Singh. 2010. 'Biological Synthesis of Silver Nanoparticles by Using Onion (*Allium cepa*) Extract and Their Antibacterial Activity'. *Digest Journal of Nanomaterials and Biostructures* 5(2): 427–32.

Senthilkumar, S. R., and T. Sivakumar. 2016. 'Green Tea (*Camellia sinensis*) Mediated Synthesis of Zinc Oxide (ZnO) Nanoparticles and Studies on Their Antimicrobial Activities Innovare Academic Sciences Green Tea (*Camellia sinensis*) Mediated Synthesis of Zinc Oxide (ZnO) Nanoparticles and Stud'. *International Journal of Pharmacy and Pharmaceutical Sciences* 6(January 2014): 461–65.

Singaravelu, G., J. S. Arockiamary, V. Ganesh Kumar, and K. Govindaraju. 2007. 'A Novel Extracellular Synthesis of Monodisperse Gold Nanoparticles Using Marine Alga, Sargassum wightii Greville'. *Colloids and Surfaces, Part B: Biointerfaces* 57(1): 97–101. https://doi.org/10.1016/j.colsurfb.2007.01.010.

Singh, Ravina, and Hari Singh Nalwa. 2011. 'Medical Applications of Nanoparticles in Biological Imaging, Cell Labeling, Antimicrobial Agents, and Anticancer Nanodrugs'. *Journal of Biomedical Nanotechnology* 7(4): 489–503. https://doi.org/10.1166/jbn.2011.1324.

Sivaraj, Rajeshwari, Pattanathu K. S. M. Rahman, P. Rajiv, S. Narendhran, and R. Venckatesh. 2014. 'Spectrochimica Acta Part A : Molecular and Biomolecular Spectroscopy Biosynthesis and Characterization of *Acalypha indica* Mediated Copper Oxide Nanoparticles and Evaluation of Its Antimicrobial and Anticancer Activity'. *Spectrochimica Acta. Part A: Molecular and Biomolecular Spectroscopy* 129: 255–58. https://doi.org/10.1016/j.saa.2014.03.027.

Song, Joon Myong. 2010. 'Silver as Antibacterial Agent: Metal Nanoparticles to Nanometallopharmaceuticals'. In *International Conference on Nano/Molecular Medicine and Engineering*, pp. 98–101.

Soshnikova, Veronika, Yeon Ju Kim, Priyanka Singh, Yue Huo, Josua Markus, Sungeun Ahn, Verónica Castro-Aceituno, J. Kang, M. Chokkalingam, R. Mathiyalagan, and D. C. Yang. 2018. 'Cardamom Fruits as a Green Resource for Facile Synthesis of Gold and Silver Nanoparticles and Their Biological Applications'. *Artificial Cells, Nanomedicine and Biotechnology* 46(1): 108–17. https://doi.org/10.1080/21691401.2017.1296849.

Soundarrajan, C., A. Sankari, P. Dhandapani, S. Maruthamuthu, S. Ravichandran, G. Sozhan, and N. Palaniswamy. 2012. 'Rapid Biological Synthesis of Platinum Nanoparticles Using Ocimum Sanctum for Water Electrolysis Applications'. *Bioprocess and Biosystems Engineering* 35(5): 827–33. https://doi.org/10.1007/s00449-011-0666-0.

Sun, Qian, Xiang Cai, Jiangwei Li, Min Zheng, Zuliang Chen, and Chang Ping Yu. 2014. 'Green Synthesis of Silver Nanoparticles Using Tea Leaf Extract and Evaluation of Their Stability and Antibacterial Activity'. *Colloids and Surfaces. Part A: Physicochemical and Engineering Aspects* 444: 226–31. https://doi.org/10.1016/j.colsurfa.2013.12.065.

Sutradhar, Prasanta, and Mitali Saha. 2016. 'Green Synthesis of Zinc Oxide Nanoparticles Using Tomato (*Lycopersicon esculentum*) Extract and Its Photovoltaic Application'. *Journal of Experimental Nanoscience* 11(5): 314–27. https://doi.org/10.1080/17458080.2015.1059504.

Tavakoli, A., M. Sohrabi, and A. Kargari. 2007. 'A Review of Methods for Synthesis of Nanostructured Metals with Emphasis on Iron Compounds'. *Chemical Papers* 61(3): 151–70. https://doi.org/10.2478/s11696-007-0014-7.

Thondavada, Niranjan, Rajasekar Chokkareddy, and Gan G. Redhi. 2018. Green synthesis of platinum nanoparticles and their biomedical applications. *Green Metal Nanoparticles: Synthesis, Characterization and Their Applications* 22: 603–27. Scrivener Publishing LLC.

Thovhogi, N., A. Diallo, A. Gurib-Fakim, and M. Maaza. 2015. 'Nanoparticles Green Synthesis by *Hibiscus sabdariffa* Flower Extract: Main Physical Properties'. *Journal of Alloys and Compounds* 647: 392–96. https://doi.org/10.1016/j.jallcom.2015.06.076.

Tran, Tien D., Mai T. T. Nguyen, Hoang V. Le, Duc N. Nguyen, Quang Duc Truong, and Phong D. Tran. 2018. 'Gold Nanoparticles as an Outstanding Catalyst for the Hydrogen Evolution Reaction'. *Chemical Communications* 54(27): 3363–66. https://doi.org/10.1039/c8cc00038g.

Velmurugan, Palanivel, Jung Hee Park, Sang Myeong Lee, Jum Suk Jang, Young Joo Yi, Han Sang Sub, Sang Hyun Lee, Kwang Min Cho, Min Cho, and Byung Taek Oh. 2016. 'Phytofabrication of Bioinspired Zinc Oxide Nanocrystals for Biomedical Application'. *Artificial Cells, Nanomedicine and Biotechnology* 44(6): 1529–36. https://doi.org/10.3109/21691401.2015.1058811.

Vijayan, Remya, Siby Joseph, and Beena Mathew. 2018. 'Indigofera tinctoria Leaf Extract Mediated Green Synthesis of Silver and Gold Nanoparticles and Assessment of Their Anticancer, Antimicrobial, Antioxidant and Catalytic Properties'. *Artificial Cells, Nanomedicine and Biotechnology* 46(4): 861–71. https://doi.org/10.1080/21691401.2017.1345930.

Wang, Dandan, Josua Markus, Chao Wang, Yeon Ju Kim, Ramya Mathiyalagan, Verónica Castro Aceituno, Sungeun Ahn, and Deok Chun Yang. 2017. 'Green Synthesis of Gold and Silver Nanoparticles Using Aqueous Extract of Cibotium Barometz Root'. *Artificial Cells, Nanomedicine and Biotechnology* 45(8): 1548–55. https://doi.org/10.1080/21691401.2016.1260580.

Wani, Waseem A., Sanjiv Prashar, Sheikh Shreaz, and Santiago Gómez-Ruiz. 2016. 'Nanostructured Materials Functionalized with Metal Complexes: In Search of Alternatives for Administering Anticancer Metallodrugs'. *Coordination Chemistry Reviews* 312: 67–98. https://doi.org/10.1016/j.ccr.2016.01.001.

Wilczewska, Agnieszka Z., Katarzyna Niemirowicz, Karolina H. Markiewicz, and Halina Car. 2012. 'Nanoparticles in Drug Delivery Systems'. *Pharmacological Reports* 64(5): 1020–37. https://doi.org/10.1201/b14802.

Yang, Ning, Weihong Li, and Lin Hao. 2014. 'Biosynthesis of Au Nanoparticles Using Agricultural Waste Mango Peel Extract and Its In Vitro Cytotoxic Effect on Two Normal Cells'. 134: 67–70. https://doi.org/10.1016/j.matlet.2014.07.025.

Yang, Xin, Qingbiao Li, Huixuan Wang, Jiale Huang, Liqin Lin, Wenta Wang, Daohua Sun, et al. 2010. 'Green Synthesis of Palladium Nanoparticles Using Broth of *Cinnamomum camphora* Leaf'. *Journal of Nanoparticle Research* 12(5): 1589–98. https://doi.org/10.1007/s11051-009-9675-1.

Yong, Jae, Song Æ. Eun-yeong Kwon, and B. S. Kim. 2010. 'Biological Synthesis of Platinum Nanoparticles Using *Diopyros kaki* Leaf Extract'. *Bioprocess and Biosystems Engineering* 33(1): 159–64. https://doi.org/10.1007/s00449-009-0373-2.

Yugandhar, Pulicherla, Reddla Haribabu, and Nataru Savithramma. 2015. 'Synthesis, Characterization and Antimicrobial Properties of Green-Synthesised Silver Nanoparticles from Stem Bark Extract of *Syzygium alternifolium* (Wt.) Walp'. *3 Biotech* 5(6): 1031–39. https://doi.org/10.1007/s13205-015-0307-4.

Yugandhar, Pulicherla, Thirumalanadhuni Vasavi, Palempalli Uma Maheswari Devi, and Nataru Savithramma. 2017. 'Bioinspired Green Synthesis of Copper Oxide Nanoparticles from *Syzygium alternifolium* (Wt.) Walp: Characterization and Evaluation of Its Synergistic Antimicrobial and Anticancer Activity'. *Applied Nanoscience (Switzerland)* 7(7): 417–27. https://doi.org/10.1007/s13204-017-0584-9.

Zhang, Xi, Zhenyue Tan, Kunjing Jia, Wenzhi Zhang, and Minyan Dang. 2019. '*Rabdosia rubescens* Linn: Green Synthesis of Gold Nanoparticles and Their Anticancer Effects against Human Lung Cancer Cells A549'. *Artificial Cells, Nanomedicine, and Biotechnology* 47(1): 2171–78. https://doi.org/10.1080/21691401.2019.1620249.

Zheng, Yuhong, Li Fu, Fuigui Han, Aiwu Wang, Wen Cai, Yu Jinping, Jun Yang, and Feng Peng. 2015. 'Green Biosynthesis and Characterization of Zinc Oxide Nanoparticles Using Corymbia Citriodora Leaf Extract and Their Photocatalytic Activity'. *Green Chemistry Letters and Reviews* 8(2): 59–63. https://doi.org/10.1080/17518253.2015.1075069.

7 Myconanoparticles Synthesis and Their Applications in the Agriculture Sector

Rohit Sam Ajee, Neha Sharma,
Rajesh Singh Tomar, and Shuchi Kaushik

CONTENTS

7.1 INTRODUCTION

Agriculture plays a vital role in human society and may even be the most essential sector when it comes to sustaining human life on this planet. Crops and other products of this industry play a crucial role in supporting human life; plants as food, and other by-products like jute as cloth material, resins, and gums for adhesives, medically active metabolites, etc. are some examples of the use of plant products in daily life (Pohl and Paul, 2018; Tavares et al., 2020). Agriculture provides food and satisfies other material needs. It also acts as one of the most outstanding employment providers, especially in agrarian countries like India and China. Hence, it is very evident that agriculture plays an integral part in our lives, and as such any threats to this field would directly affect human life to a great extent.

The significant losses observed in this sector are down to diseases, pests, infections spoiling the agricultural harvest, and the chemical means used to control these losses. Insects cause average annual damage of approximately 7.7% in crop production in the fields of Brazil, which in terms of money is an annual economic loss of around 14.7 billion US$ (Oliveira et al., 2014). On a global scale, insect pests cause losses as high as 50% in the case of wheat and 80% in the case of cotton production (Oerke, 2006). Maximum crop loss was observed as a result of the growth of weeds in agricultural fields (34%) followed by crop loss due to animal pests (18%), pathogenic losses (16%) (Oerke, 2006), and 12.3% losses due to parasitic nematodes (Kumar et al., 2020). These losses were observed even after the placement and usage of pest and disease management techniques, like pesticide and insecticide usage, and proper weed removal (Oerke, 2006). Of the total global loss, an estimated 30% occurs in countries in Asia, followed by 25% in Europe and 15% in the countries of Central and North America (James et al., 1991 revised). More significant crop loss was observed in the tropics, sub-tropics, and areas which showed higher crop productivity, as these regions allow for enhanced crop damage and more significant growth rates of microbes and viruses (Oerke, 2006, Millan et al., 2020). Despite all the reported losses, and the continued use of massive quantities of chemical agents to control the diseases, infections, and pests, the human population still continues to face rising losses when it comes to crop production. Moreover, chemical pesticides, insecticides, and weedicides are harmful to the soil and the environment, in addition to the killing/inhibitory action they exhibit on the target organism (Savary et al., 2019). Keeping all of the previous facts in mind, new ways and substances are needed to control pathogens and pests.

A suitable place to look for new and improved ways to control pathogens and pests would be in the field of nanoscience. The term nanotechnology was coined by Norio Taniguchi in 1974, though the underlying concepts of it were discussed for the first time in 1959 by Richard Feynman in his historical talk. 'There is Plenty of Room at the Bottom'. In his talk, Feynman spoke about the self-assembling systems that made themselves by arranging their own atoms and made the desired chemical compounds via mechanistic manipulations (Fanfair et al., 2007). Nanotechnology can be described as the study of compounds or particles present in nanoscale ranges, i.e. less than 100 nm (Pitkethly, 2008).

7.1.1 Nanotechnology

As it became clearer that conventional means of improving the crop yields were not meeting the growing requirements of the global population, in addition to causing seemingly irreparable damage to arable land, the world started shifting its attention to other areas that could fill the void, one of which was nanotechnology (Mukhopadhyay, 2014). Nanoparticles, especially metallic nanoparticles, have a wide range of properties, including antimicrobial properties, biocide properties, catalysis in chemical reactions, etc. One of the most widely accepted and debated nanotech instances being used in the context of agriculture, precisely pest control, is silver nanoparticles. It has been suggested that silver has medicinal properties since as early as 400 BC, and its applications in the agricultural sector as a pesticide can be traced back to 1945 (Bergerson, 2010a). The nanoscale nature of nanosilver particles increases its effectiveness, and it is an established nanopesticide with a well-transcribed record of safe use. Nanopesticides have the added benefit of being more efficient and therefore reduce dependency on more conventional pest control techniques. Other nanotechnology applications in pest control would be its use in smart control systems, which monitor the fields, locate and report on the pathogens, and then disperse the needed pesticide or control agents (Bergerson, 2010b). In addition to nano-formulation based insecticides and pesticides, nanomaterials can also be used for soil remediation. Soil that is contaminated can be returned to arable states through the actions of nanomaterials with specific properties like magnetic or catalytic nanomaterials (Usman et al., 2020). Keeping in mind that nanotechnology already seems to play quite a significant role in the pest control sector and other areas as well, there is still a need to establish reliable and economic sources of nanoproducts (Chippa, 2019). Several different nanoparticles have been synthesized and used in the plant sector; a few examples are quoted in Table 7.1 along with the salient features of these nanoparticles.

Nanoparticle synthesis can be carried out using either the bottom-up approach or the top-down approach. Under these broad approaches, several techniques, like the ball mill method, scotch tape synthesis, inert gas condensation, sol-gel fabrication, etc., have been used by different groups of researchers. The main problems associated with the classical methods of nanoparticle synthesis are:

a. Time is taken to synthesize the desired products,
b. Cost of production,
c. The efficiency of said synthesis method,
d. And the environmental effect of the synthesis method.

Keeping these factors in mind and the growing importance of resorting to more environmentally friendly approaches in nanoparticle synthesis, we need to look towards greener synthesis methods such as the biological synthesis routes. This chapter focuses on myconanotechnology, a type of biological synthesis for nanoparticles, and their possible agriculture applications.

7.1.2 Myconanotechnology

Myconanotechnology can be defined as the synthesis or production of nanoparticles using fungi and their subsequent utilization in the required fields. The discipline is an interface between mycology (the study of fungi) and nanotechnology and has considerable potential due to the diversity and variety of fungi and different particles that can be synthesized by them. Another factor working in favour of myconanotechnolgy is that since it relies on biological systems (fungi) to synthesize the desired products, any harmful effects to the environment will be minimized (Gupta et al., 2012).

Fungi belong to the eukaryotic domain and are mostly multicellular organisms. These are heterotrophs that play a significant role in the ecosystem. They are capable of multiplication by both asexual and sexual modes of reproduction. Fungi can live symbiotically with bacteria and plants. Examples include moulds, mildews, rusts,

TABLE 7.1
Different Nanoparticles, Their Uses, and Salient Features

Nanoparticle	Use	Salient Features	Reference
Carbon Nanoparticles (C NPs)	Plant growth promoter	C NPs are shown to enhance germination and plant growth	Shojaei et al., 2019
Graphene Nanosensors	Detecting the presence of *E. coli* in water	A graphene-based nanosensor was created to detect the presence of *E. coli* in water; same techniques could be useful for aquaculture techniques	Wibowo et al., 2018
Zinc Oxide Quantum Dots (ZnO QDs)	Detecting the presence of pesticides in water	The pesticides Aldrin and Tetradifon were easily detected by ZnO QDs while other pesticides like atrazine showed lesser levels of detection; The QDs can be used to detect the levels of pesticides in food crops before consumption	Sahoo et al., 2018
Carbon NanoOnions (CNOs)	Seed growth enhancer for increased yield	The CNOs were used for the germination of seeds which were then subsequently allowed to grow normally; it was observed that the yields of gram plants grown via this method were superior to normally grown crops	Tripathi et al., 2017
Nanofertilizers	Enhancing Plant Growth	Nanofertilizers allow for enhanced nutrient uptake, cost the same as conventional fertilizers and require less frequent application	Dimkpa et al., 2017
Mesoporous Silica Nanoparticles (MSN)	Transfer of Genetic Material	Cre proteins were loaded into the MSN and transferred into maize cells via biolistic method followed by recombination	Martin-Ortigosa et al., 2014
NP enhanced fertilizers	Enhancing Plant Growth	Although NP augmented fertilizers are generally non-specific and can be utilized for general uses, they lack several of the drawbacks of conventional fertilizers	Liu et al., 2015
Poly(epsilon-caprolactone) nanocapsules (PCL)	Carrier for the herbicide Atrazine	Mustard plants were treated with Atrazine loaded PCL capsules; Results indicated that the herbicide functioned efficiently and lower levels of herbicide would prevent environmental damage	Oliveira et al., 2015
Chitosan Nanoparticles	Transfer of Genetic Material, Triggering plant defence system	Not only can these NPs be used to efficiently transfer genetic material to produce recombinant crops, but they are also shown to induce anti-viral activity in plants	Iriti and Varoni, 2014

(*Continued*)

TABLE 7.1 (CONTINUED)
Different Nanoparticles, Their Uses, and Salient Features

Nanoparticle	Use	Salient Features	Reference
Zinc Oxide Nanoparticles (ZnO NPs)	Antifungal agent	The growth of several fungal species like *Aspergillus niger*, *F. culmorum*, etc. were inhibited by the action of the ZnO NPs	Rajiv et al., 2013
Titanium Dioxide Nanoparticles	Food additive	Titanium Dioxide NPs serve as resistant and robust colouring agent for a variety of food products like gum, cake, etc.	Weir et al., 2012
Nanocapsules	Site specific delivery of pesticides, insecticides and fertilizers	Surface functionalized nanocapsules can be loaded with the desired active substance and targeted release with such a system is possible	Nair et al., 2010
Amourphous Nanosilica	Biopesticide	Causes death by direct physical action; Is absorbed into the cuticular cells of the insect pests without affecting plant systems	Ulrichs et al., 2006
Porous Hollow Silica Nanoparticles (PHSNs)	Carrier for pesticide Validamycin	PHSNs were shown to act as efficient carrier agents	Liu et al., 2006
Copper-based nanocomposites	Antifungal activity	A polymer network acts as the host for storing the copper nanoparticles and gradually releases it to inhibit fungal growth	Cioffi et al., 2004

yeasts, and mushrooms, etc. (Duhan et al., 2017). Fungi are found to be much more efficient in terms of nanoparticle production when compared to bacteria. This may be because fungi naturally have the necessary metabolites and biomolecules that are utilized for nanoparticle synthesis and produce a large number of enzymes, mainly reductases, which can reduce and stabilize nanoparticles. Additionally, the fungal cell wall has a net negative charge and adhesive substances on its surface, which attracts metal ions and binds them via electrostatic interactions. The nanoparticles can be synthesized by both extracellular as well as intracellular processes of fungal cells. The composite film formation is more favourable to the intracellular synthesis process, while the extracellular synthesis process involves the immobilization of metal ions in a suitable carrier or support. The secretion of active components, which can convert toxic matters into the non-toxic matter, might also be involved in the nanoparticles synthesis process (Owaid and Ibraheem, 2017).

In addition to fungi, other microorganisms like bacteria and even plants have been studied for their nanoparticle synthesis and formation properties. A brief overview of microbes utilized in nanoparticle synthesis is summarized in Table 7.2.

TABLE 7.2
Overview of Various Microorganisms Involved in Nanoparticles Synthesis

S. No.	Microorganisms	Nanoparticles synthesized	References
1.	Bacteria		
	Bacillus amyloliquefaciens	CdS	Singh et al., 2011
	Bacillus brevis	Au	Saravanan et al., 2018
	Bacillus licheniformis	Ag	Elbeshehy et al., 2015
	Bacillus methylotrophicus	Ag	Wang et al., 2016
	Bacillus persicus	Ag	Elbeshehy et al., 2015
	Bacillus pumilus	Ag	Elbeshehy et al., 2015
	Bacillus subtilis	Ag	Soni and Prakash, 2015; Southam and Beveridge, 1994
	Bhargavaea indica	Ag and Au	Singh et al., 2015
	Brevibacterium frigoritolerans Listeria	Ag	Singh et al., 2015
	monocytogenes	Ag	Camas et al., 2018
	Mycobacterium sp.	Ag	Soni and Prakash 2015
	Pseudomonas deceptionensis	Ag	Lv et al., 2018
	Shewanella loihica	Cu	Jo et al., 2016
	Streptomyces anulatus	Ag	Vijayabharathi et al., 2018
	Streptomyces griseoplanus	Au	Soni and Prakash, 2015
	Weissella oryzae	Ag	Singh et al., 2016
	Lactobacillus strains	Tt	Prasad et al., 2007
2.	Cyanobacteria		
	Plectonema boryanum UTEX 485	Au	Lengke et al., 2006
	Plectonema boryanum Pd	Pd	Lengke et al., 2007
3.	Actinomycetes		
	Streptomyces sp. LK3	Ag	Singh et al., 2016
	Rhodococcus sp.	Au	Ahmad et al., 2003
	Streptomyces clavuligerus	Au	Kumar et al., 2015
	Streptomyces fulvissimus	Au	Kumar et al., 2015
	Thermomonospora sp.	Au	Sedaghat et al., 2016
4.	Algae		
	Aphanothece sp.	Ag	Soltani et al., 2015
	Oscillatoria sp.	Ag	Soltani et al., 2015
	Microcoleus sp.	Ag	Soltani et al., 2015
	Aphanocapsa sp.	Ag	Soltani et al., 2015
	Phormidium sp.	Ag	Soltani et al., 2015
	Lyngbya sp.	Ag	Soltani et al., 2015
	Gleocapsa sp.	Ag	Soltani et al., 2015
	Synechococcus sp.	Ag	Soltani et al., 2015
	Spirulina sp.	Ag	Soltani et al., 2015
	Chlorella pyrenoidosa	Ag	Ahmad et al., 2003
	Nostoc linkia	Ag	Vanlalveni et al., 2018

(Continued)

TABLE 7.2 (CONTINUED)
Overview of Various Microorganisms Involved in Nanoparticles Synthesis

S. No.	Microorganisms	Nanoparticles synthesized	References
5.	Fungi		
	Alternaria alternate	Au	Gajbhiye et al., 2009
	Amylomyces rouxii	Au	Gurunathan et al., 2009
	Aspergillus clavatus	Au	Verma et al., 2011
	Aspergillus flavus	Ag, MgO	Moharrer et al., 2012; Raliya et al., 2013
	Aspergillus fumigatus	Ag, ZnO	Kowshik et al., 2003
	Aspergillus nidulans	Co	Vijayanandan and Balakrishnan, 2018
	Aspergillus niger	Au, ZnO	Mukherjee et al., 2001; Salunke et al., 2016; Kalpana et al., 2018
	Aspergillus oryzae	Au	Binupriya et al., 2010
	Aspergillus oryzae TFR9	$FeCl_3$	Li et al., 2011
	Aspergillus sydowii	Au	Kelly et al., 2003
	Aspergillus terreus	Ag	Liet al., 2011
	Aspergillus tubingensis	$Ca_3P_2O_8$	Abdullaeva et al., 2012
	Aureobasidium pullulans	Au	Bhainsa et al., 2006
	Bipolaris nodulosa	Ag	Qian et al., 2013
	Candida albicans	Au	Qian et al., 2013
	Candida glabrata	CdS, Ag	Li et al., 2011; Jalal et al., 2018
	Cladosporium cladosporioides	Ag	Balaji et al., 2009
	Cochlibolus lunatus	Cu, Al	Ahmad et al., 2003
	Colletotrichum sp.	Au	Shivshankar et al., 2003
	Coriolus versicolor	Au	Sanghi et al., 2010
	Cylindrocladium floridanum	Au	Philip et al., 2010
	Epicoccum nigrum	Au	Ghiassi et al., 2018
	Fusarium oxysporum	Pt, Ag, Au, CdS, TiO_2, CdSe, $SrCO_3$, $BaTiO_3$, ZrO_2	Ahmad et al., 2003; Duran et al., 2005; Ahmad et al., 2002; Mittal et al., 2013
	Fusarium semitectum	Au	Gurunathan et al., 2013
	Ganoderma neo-japonicum Imazeki	Ag	Gurunathan et al., 2009
	Ganoderma sp.	Ag, Au	Balakumaran et al., 2015
	Guingnardia mangiferae	Ag	Amutha et al., 2011
	Hansenula anomala	Au	Kumar et al., 2008
	Helminthosporum solani	Au	Varshney et al., 2009
	Hormoconis resinae	Au	Salvadori et al., 2013
	Hypocrea lixii	Cu	Mondal et al., 2014
	Macrophomina phaseolina	Ag	El-Baz et al., 2016
	Monascus purpureus	Ag	Castro-Longoria et al., 2011

(*Continued*)

TABLE 7.2 (CONTINUED)
Overview of Various Microorganisms Involved in Nanoparticles Synthesis

S. No.	Microorganisms	Nanoparticles synthesized	References
	Neurospora crassa	Au, Ag	Owaid et al., 2015; Castro-Longoria et al., 2011
	Neurospora oryzae	Ag	Qian et al., 2013
	Pleurotus citrinopileatus	Ag	Raman et al., 2015
	Pleurotus djamor	Ag	Borovaya et al., 2015
	Penicillium fellutanum	Au	Kathiresan et al., 2009
	Penicillium chrysogenum	Pt	Subramaniyan et al., 2018
	Penicillium nalgiovense AJ12	Ag	Wysocki et al., 2010
	Pleurotus ostreatus	CdS	Shaligram et al., 2009
	Penicillium rugulosum	Au	El-Baz et al., 2016
	Pediococcus pentosaceus	Ag	Sanghi et al., 2010
	Penicillium brevicompactum	Au	Zhang et al., 2009
	Penicillium sp.	Au	Raheman et al., 2011
	Pestalotia sp.	Ag	Thirumurugan et al., 2009
	Phanerochaete chrysosporium	Au	Musarrat et al., 2010
	Phoma glomerata	Ag	Rajput et al., 2017
	Phoma sorghina	Ag	Rajput et al., 2017
	Phytopthora infestans	Ag	Ge et al., 2014
	Pichia jadinii	Au	Bhat et al., 2011
	Pichta capsulate	Ag	Mazumdar et al., 2017
	Pleurotus florida	Ag	Shi et al., 2015
	Pleurotus sajor caju	Ag	Das et al., 2009
	Pycnoporus sanguineus	Au	Afreen and Ranganath, 2011
	Rhizopus oryzae	Au	Cuevas et al., 2015
	Rhizopus stolonifer	Au, Ag	Kowshik et al., 2002; Abdel Rahim et al., 2017
	Saccharomyces boulardii	Ag	Lok et al., 2007
	Saccharomyces cerevisiae	Au	Cuevas et al., 2015
	Schizosaccharomyces pombe	CdS	Cuevas et al., 2015
	Sclerotium rolfsii	Au	Philip, 2009
	Stereum hirsutum	Cu, CuO	Abdullaeva et al., 2012
	Trichoderma crassum	Ag	Karbasian et al., 2008
	Trichoderma harzianum	Ag	Thirumurugan et al., 2009
	Trichoderma koningii	Au	Thirumurugan et al., 2009
	Trichoderma reesei	Ag	Philip, 2009
	Trichoderma viride	Ag	Thostenson et al., 2005
	Thraustochytrium sp.	Ag, Zn	Liu et al., 2010
	Trichoderma asperellum	Ag	Abdel-Hafez et al., 2016
	Verticillium luteoalbum	Au	Sarsar et al., 2016
	Verticillium sp.	Au, Ag	Khan et al., 2018; Mukherjee et al., 2001

(Continued)

TABLE 7.2 (CONTINUED)
Overview of Various Microorganisms Involved in Nanoparticles Synthesis

S. No.	Microorganisms	Nanoparticles synthesized	References
	Volvariella volvacea	Au-Ag	El-Rafie et al., 2010
	Yarrowia lipolytica NCIM 3589	Au	Agnihotri et al., 2009

Compared to bacteria, fungi have a better tolerance for metals and a greater capacity for intaking them by binding to metal ions when compared to bacteria. Additionally, the lower costs involved in fungal biomass cultivation and the ease of recovery makes them more useful for the large-scale production of nanoparticles (Yadav et al., 2015).

Compared to other biological means of nanoparticle synthesis, fungi have a few advantages, namely the fact that they are much easier to isolate and culture. They are also capable of secreting large amounts of enzymes and proteins quickly, making them capable of nanoparticle fabrication faster than some chemical methods, significantly more non-hazardous than most chemical methods, and with better yields. They are also readily manipulated and have more convenient downstream processing methods making them more preferable as compared to classical methods (Rai et al., 2009). Various metal nanoparticles have been fabricated using different fungal strains (Gade et al., 2010). The significant factors influencing nanoparticle properties are its shape and size, which are better manipulated and controlled by fungi due to their slower kinetic properties and thus they provide a long-term stability to the mycogenically synthesized nanoparticles (Zhao et al., 2018).

The nanoparticles are generally produced as a result of the interaction between the biomass and the metallic salt solutions. There are two modes of production for nanoparticles which are discussed in subsequent sections (intracellular and extracellular). The mycosynthesis of nanoparticles falls under the bottom-up approach, which is composed of redox reactions giving rise to colloidal substrates. Certain fungi produce metabolites or enzymes that have a reducing capability, which allows for the conversion of the metal particles into the corresponding metallic nanostructures. This process is considered to either be part of the organisms' survival mechanisms under high metal concentrations or a normal bioprocess that is needed for the survival of these organisms. Certain metals are needed to carry out average growth and development like sodium, potassium, copper, etc., while others have no relevant use in the microbes but are still accumulated within the cell, like aluminium, cadmium, lead, etc. In addition to the above-mentioned redox reactions that the microbial enzymes carry out on the metals, they can also bind the metals to their cell walls,

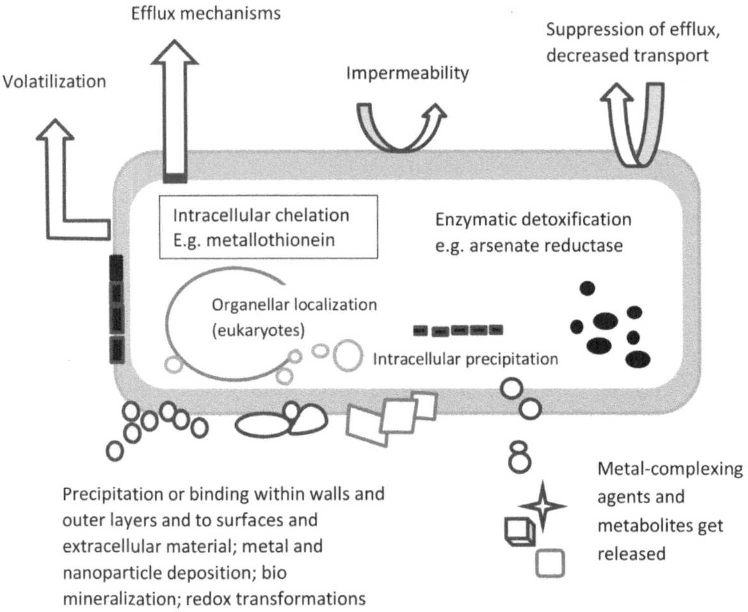

FIGURE 7.1 Representation of metal transformation and reduction processes in microbes used of nano synthesis. (Adapted from Gadd Metals, minerals, and microbes: Geomicrobiology and bioremediation, *Microbiology* (2010), 156, 609–643.)

internally compartmentalize them, carry out organic or inorganic precipitation, etc. as depicted in Figure 7.1 where different mechanisms of metal transformation adopted by the microbial cell are shown, which may be dependent or independent of microbial metabolic reactions, such as restricted entry into the cell, enzymatic detoxification by metal reductase, or intracellular chelation or precipitation, etc., by which the metals get reduced into nanoparticles. A variety of mechanisms may be involved in transport phenomena contributing to the decreased uptake and/or efflux. Biomineral formation (biomineralization) may be biologically induced, i.e. caused by physico-chemical environmental changes mediated by the microbes, or biologically controlled (Gadd, 2010).

7.2 MODES OF MYCONANOSYNTHESIS

Most biological methods of nanoparticle synthesis follow similar mechanisms even for different microbes and different metal nanoparticle synthesis. The necessary steps involved in the synthesis in the case of mycosynthesis are briefly illustrated below:

1. Conventionally available media must first be supplemented with sufficient concentrations of the metal. Sufficient in this context refers to a

concentration that is not excessively toxic to the fungi but, at the same time, causes the organism to activate its survival mechanisms, which are ultimately responsible for the metal reduction/oxidation process to form the desired nanoparticle.

2. The media must then be inoculated with the selected microbe (fungi) for the bioreduction of the target metal to the metal nanoparticle form.

The reduction process can occur either inside or outside the cell via intracellular or extracellular bioactive components. Furthermore, metal nanoparticles produced outside the cell by extracellular means may be taken up by the cell and stored inside it, making further processing necessary (Moghaddam, 2010). The process's reaction conditions can be optimized to diversify the results, such as changes in the shape, chemical composition, or size of the nanoparticles formed by altering the pH, temperature, light source, etc.

7.2.1 Extracellular Synthesis Process

Nanoparticle synthesis and accumulation can be of two distinct types, either intracellular or extracellular. The extracellular biosynthesis can be further classified into slow and rapid synthesis methods (Moghaddam, 2010). Speaking from an industrial point of view, the extracellular mode of synthesis is preferable as the metal nanoparticle formed is present outside the cells making their subsequent collection much more comfortable than in the case of intracellular synthesis, in which additional processing steps (cell lysis by detergents or other means) become necessary in order to retrieve the nanoparticles. Another advantage of the extracellular mode is that it is cheaper and has easier downstream processing methods.

The extracellular mechanism depends on enzymes that the microbes secrete out to reduce the metals into the nanophase. This method involves making use of the microbial biomass for the reduction, or using the supernatant of the cell culture (obtained after centrifugation to remove the cell mass) or using cell-free extracts. The supernatant of the culture can be further purified by passing it through a filter to remove all microbial cells to prevent the nanoparticles' intracellular accumulation. One mode of using the cell-free extract would be to keep the cells in water for a day and then remove the cells; the active biomaterials get released into the water, which can then be used for future reduction (Ovais et al., 2018; Moghaddam, 2010). The enzymes secreted are commonly argued to be nitrate reductases and NADH-dependent reductases. Protein assays were used to detect the presence of NADH-dependent reductase in the biosynthesis. The enzyme gains electrons from the NADH molecule and then reduces the metal by transferring the electrons it gained from NADH, performing a redox reaction (Senapati et al., 2005).

7.2.2 Intracellular Synthesis Process

The intracellular mechanism uses a combination of the ion transport system, which transports the metal cations into the cell, and the enzymes (located in the cell wall or other organelles), which reduces the metal particles into the corresponding nanoparticle form (Ovais et al., 2018). Most microbes also have some electrostatic means of attracting/trapping the metal ion. For example, bacterial cells have a negatively charged membrane which attracts the positive metal ions and allows for bioreduction. Fungal cells also trap the metal ions via electrostatic interactions (Hulkoti and Taranath, 2014). However, keeping in mind the limitations faced in this mode, research is generally focused on extracellular approaches.

7.3 MECHANISM OF SYNTHESIS

Fungi use various mechanisms (both intracellular and extracellular) to bioreduce the metals into the metal nanoparticle phase like changes in the solubility, transformation, metal complexion, sorption compartmentalization, redox, etc. (Khandel and Shahi, 2018). In the extracellular mode of reduction, the fungal cells secrete enzymes that reduce the metals. Of these, the most commonly observed enzyme is the nitrate reductase. In addition to the reductase, compounds with redox properties are also secreted by the cells (or may be present within the cell), which makes up the electron shuttle involved in the bioreduction process.

A generalized overview of the steps involved in mycogenic nanoparticle synthesis, characterization, and application is as follows:

1. Fungal extract (proteins, enzymes, nitrate reductase) + metal salt = low yield of heterogenous mycogenic nanoparticles obtained by reducing the metal ions.
2. Optimizing reaction conditions, such as pH, temperature, incubation time, the concentration of metal salt, and the concentration of fungal biomass, results in a high yield production of homogenous nanoparticles.
3. Characterization of mycogenically synthesized metal nanoparticles can be done with a number of techniques, such as UV-visible spectroscopy, transmission electron microscopy (TEM), scanning electron microscopy (SEM), Fourier transform infrared spectroscopy (FTIR), atomic force microscopy (AFM), differential light scattering (DLS), X-ray diffraction (XRD), energy-dispersive X-ray spectroscopy (EDX), etc.
4. The characterized nanoparticles may then be used in various fields like agriculture, environment, optical, medicine, cosmetics, etc.

7.3.1 Examples of Nanoparticles Synthesized Using Fungi

A number of studies and experiments have been conducted on various fungal species to characterize the formation of nanoparticles and to understand the mechanisms and compounds involved in the processes. One such study conducted on *Fusarium oxysporum* involved adding $AgNO_3$ to a flask containing water and a fungal biomass. It was concluded that the silver ions present in the solution were used for the formation of silver nanoparticles, as the yellowish colour of the solution (due to the fungal biomass) eventually faded to a brownish colour due to the accumulation of the nanoparticles. The formation of nanoparticles was confirmed via other spectroscopic and TEM readings. In order to verify that the reduction occurred outside the cells, the experiment was repeated with the changes that the biomass was first immersed in water for 72 hours, after which the aqueous solution was filtered out for the same experiment to be carried out. A similar colour change to the yellowish-brownish state was observed in the cell-free extract, indicating the reduction process (Ahmad et al., 2003).

When a similar study was done using *Fusarium moniliforme*, no nanoparticle synthesis was observed even though *F. moniliforme* contained similar reductases when compared to *F. oxysporum*. This could be attributed to the difference in the nature of the substrates that the reductases may target. It may also be due to the lack of the anthraquinone compounds in *F. moniliforme*, which are speculated to be a part of the electron shuttle in the bioreduction process. The presence of aromatic proteins, with tryptophan or tyrosine residues, was also observed from the solution's UV-vis spectra (Duran et al., 2005). Figure 7.2 represents a theorized mechanism

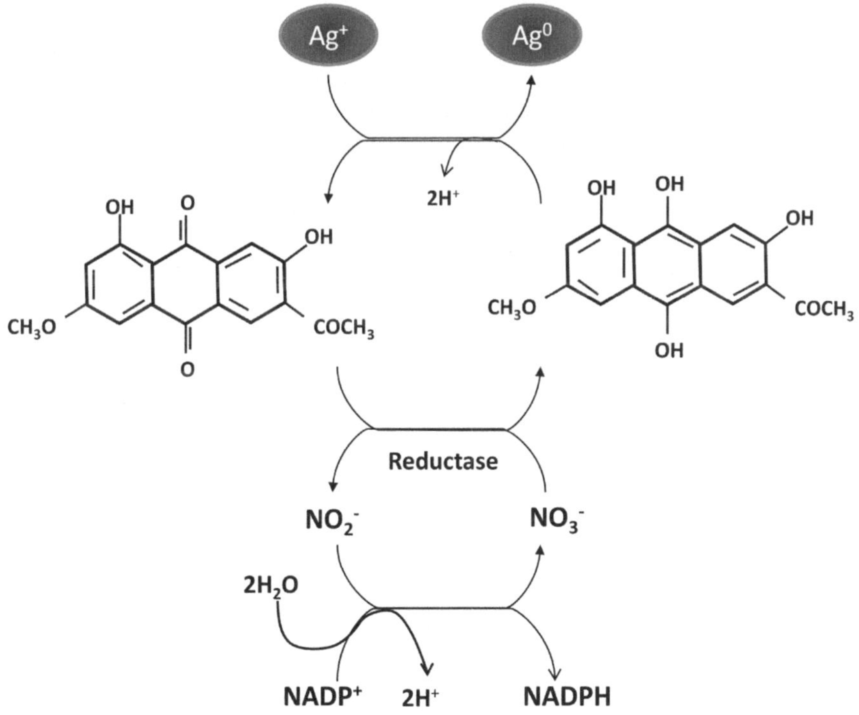

FIGURE 7.2 Hypothetical mechanism of Ag nanoparticle synthesis. (Adapted from Durán et al., Mechanistic aspects of biosynthesis of silver nanoparticles by several *Fusarium oxysporum* strains, *Journal of Nanobiotechnology*, 2005 3:8.)

via which the reductase enzymes may reduce the metal ions to the desired metal nanoparticles. The diagram shows how the enzyme gains an electron from NADH and then transports those electrons via the electron shuttle (which are comprised of unknown quinone compounds), which are then finally transferred to the metal ion to reduce them to the metal nanoparticle form.

A research group reported in 2007 that gold nanoparticles could be synthesized on the surface of *Escherichia coli* DH5α (Du et al., 2007). Similarly, another scientific group reported the synthesis of metal nanoparticles by microbes like the biosynthesis of silica nanoparticles assisted by *Fusarium oxysporum* and the synthesis of monodispersed peptide bound CdS nanoparticles by *Candida glubrata* (Narayanan and Sakthivel, 2010).

Another study elucidated the formation of cadmium sulphate (CdS) nanoparticles by *F. oxysporum* from Cd^{2+} and SO_4^{2-} ions, presented to the biomass via $CdSO_4$ solution. The test results were confirmed by the appearance of a yellow colour in the reaction mixture, which is indicative of the formation of nanoparticles, in this case CdS nanoparticles. The results were also confirmed by spectroscopic analysis, TEM image, and XRD analysis. Spectroscopic tests indicated the presence of proteins in the reaction mixture. These secreted proteins are thought to provide stability to the CdS nanoparticles and allow them to remain in a stable colloidal form for extended periods (up to one month) without accumulation (Ahmad et al., 2002). Silver nanoparticle synthesis was also observed in the case of *Verticillium* (isolated from *Taxus*). As with all the studies carried out, 10 mg of the fungal biomass (mycelia) was suspended in 100 ml of sterile distilled water, to which 100 ml of silver nitrate ($AgNO_3$) was added. The reaction was carried out over a period of 72 hours. The change in colour of the fungal biomass from pale yellow to brown indicates nanoparticle accumulation. The aqueous solution was observed to be colourless, indicating that the silver nanoparticle was present within the cells. The results were confirmed via UV-vis spectroscopy, TEM, and SEM. It was observed that the nanoparticles were present on the mycelial cell walls, with fewer particles present on the cytoplasmic membrane and within the cytoplasm. The cells were also observed to continue to grow in a usual manner even after exposure to $AgNO_3$ solution and formation of the silver nanoparticles, indicating that *Verticillium* cells aren't compromised on exposure to metallic ions (Mukherjee et al., 2001).

A similar experimental study was performed using the fungus *Penicillium* sp. where gold nanoparticles were synthesized by both intracellular and extracellular means. For intracellular biosynthesis, the fungal biomass itself was used, and for extracellular biosynthesis a cell-free filtrate was prepared to which $HAuCl_4$ was added to initiate nano synthesis. The colour of the filtrate changed from yellow to purple with the addition of $HAuCl_4$, which is characteristic of gold nanoparticles. The change in colour occurred almost instantaneously, within a minute, indicating rapid nanoparticle synthesis. UV-vis and other tests also confirmed the formation of the gold metal nanoparticles. For the intracellular biosynthesis, the colour of the culture was observed to change from milk-white to purple on the addition of $HAuCl_4$ solution, followed by an eight-hour incubation. SEM images confirmed the presence

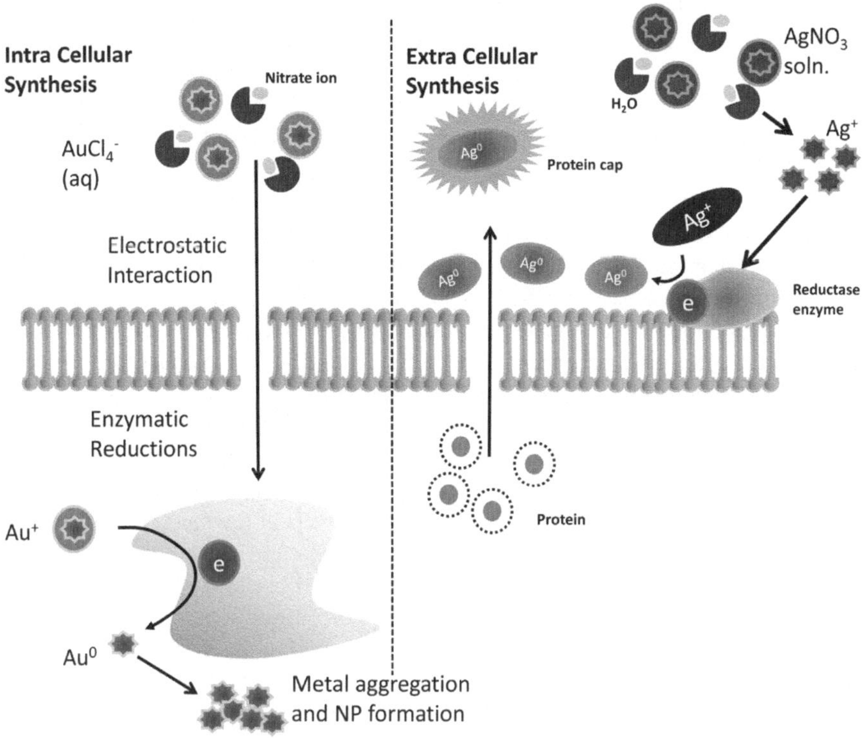

FIGURE 7.3 Mechanism for fungal bio-synthesis of nanoparticles (intra and extracellular). (Adapted from Kashyap et al., Myconanotechnology in agriculture: a perspective, World J Microbiol Biotechnol (2013) 29:191–207.)

of the particles within the cells. TEM images indicate the nanoparticle's presence on the cell walls and within the cells (Du et al., 2011).

Figure 7.3 explains the possible (a) intracellular and (b) extracellular mechanisms of nanoparticle synthesis using fungal biosystems. (a) Metal ions such as gold bind to the fungal cell surface via electrostatic interaction, which is later reduced by the intracellular enzymes and aggregated to form nanoparticles. In extracellular synthesis (b), when the metal ions get exposed to a fungus reductase enzyme (such as nitrate reductase), highly stable nanoparticles (e.g. silver) in solution are formed.

In 2015, a research group synthesized gold nanoparticles using an isolated endophytic fungus *Colletotrichum* sp. from *Geranium* leaf. They hypothesized that fungal polypeptides reducing actions and enzymes were the cause of stable different sized nanoparticles synthesized in their study (Kitching et al., 2015).

Cobalt oxide nanoparticles were observed to be synthesized by the action of the fungal species *Aspergillus nidulans*. Thc fungi were isolated and characterized by *Nothapodytes foetida*, and the mycelia of the fungal culture (grown on potato dextrose broth) were inoculated in a cobalt acetylacetone solution for a period of three days. Visual confirmation of nanoparticle formation was observed when the salt solution's pink colour changed to orange. Spectroscopic tests confirmed the same, Fourier transform infrared (FTIR) tests, X-ray diffraction, and transmission electron microscopy results. The FTIR spectrum also indicated the presence of proteins attached to the nanoparticles via amide and carboxylic bonds, which are speculated to act both as reducing as well as capping agents (Vijanandan et al., 2018).

Gold and silver nanoparticles were also observed to be synthesized within the cells of the fungal species *Neurospora crassa*. The fungal biomass was inoculated with 100 ml of $HAuCl_4$ and $AgNo_3$ solution in two different flasks, and colour changes were observed in the fungal biomass present in both flasks over an incubation period of 24 hours. The biomass incubated with $HAuCl_4$ became purple in colour after incubation, and the biomass incubated with $AgNO_3$ became reddish-brown after the incubation. Further analysis of the filtrate showed that silver nanoparticle formation also took place in the extracellular medium. Another set of flasks was also prepared, which contained a mix of gold and silver ions in different ratios to form gold/silver bimetallic nanoparticles. The formation of the bimetallic nanoparticles was confirmed by energy-dispersive X-ray spectroscopy. An undesirable outcome of the process was fungi cell death due to exposure to toxic levels of the metal ions (Langoria et al., 2011).

Zinc oxide (ZnO) nanoparticle synthesis was observed when the fungus *Aspergillus niger*'s cell filtrate was added to the zinc nitrate solution. The formation of white precipitate in a colourless filtrate, as compared to the light yellow colour of the filtrate before the addition of the salt solution, visually indicated the formation of the ZnO nanoparticles. The results were also confirmed by other means like FTIR, UV-vis spectroscopy, etc. Further studies were done to establish the antimicrobial nature of the produced nanoparticles (Kalpana et al., 2018).

Silver nanoparticle synthesis was also observed when the mycelial aqueous extract of *Rhizopus stolonifer* was subjected to incubation with silver nitrate. The yellow-coloured filtrate turned reddish-brown in colour, which served as a visual confirmation of silver nanoparticle presence. The same was confirmed with UV-vis tests, XRD, TEM, and FTIR. The FTIR test also indicated the presence of proteins that served as capping agents and some as reducing agents. The nanoparticle sizes were observed to vary with variations in physical parameters like the temperature (Rahim et al., 2017).

A study conducted by Kumar et al. (2007) tried to identify the proteins and enzymes involved in the extracellular pathway of fungal nanoparticle synthesis. They cultured *F. oxysporuym* and used it as the enzyme nitrate reductase source for the *in vitro* nanoparticle synthesis. The extracted enzyme was added to a reaction mixture containing silver nitrate, phytochelatin, 4-hydroxyquinoline, and α-NADPH, and incubated at an appropriate temperature. Reaction mixtures that lacked the enzyme, containing denatured versions of the enzyme, or lacking any one of the three other biomolecules were also created. Visual confirmation of silver nanoparticle formation was attained when the colourless reaction mixture became brown in colour after the incubation period, and spectroscopic tests also confirmed the same. The enzyme's role was confirmed when the UV-vis spectra of the reaction mixture lacking the enzyme were observed to not show a peak at 413 nm, which is the characteristic

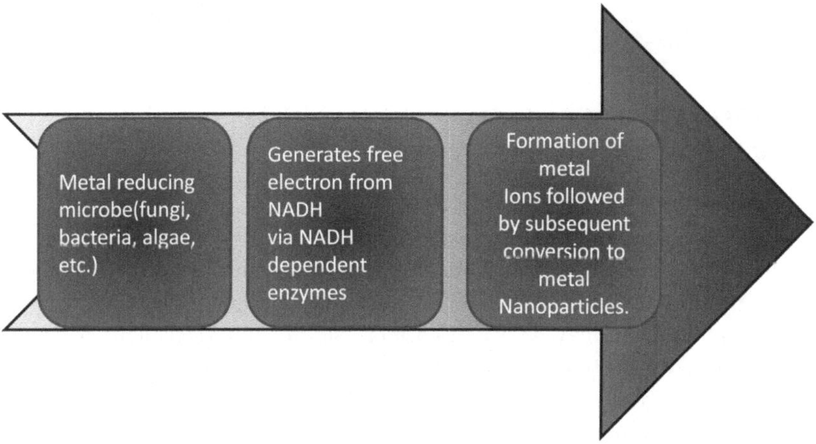

FIGURE 7.4 Flowchart depicting the stages/steps involved in NADH-dependent reduction of metals to metal nanoparticles.

surface plasmon resonance peak for silver nanoparticles. The peak was also absent in the reaction mixtures that didn't contain α-NADPH or hydroxyquinoline or phytochelatin, indicating that all these compounds are involved in some way in the bioreduction process. α-NADPH is speculated to be oxidized to α-NADP+, which is coupled with the reduction of the enzyme. The enzyme, in turn, gets oxidized and reduces the silver ion to the nanoparticle form. Hydroxyquinoline is speculated to act as an electron shuttle in the process. Phytochelatin is involved in the capping process, which stabilizes the nanoparticles and prevents their agglomeration.

The intracellular mode of synthesis is less widely explored due to the extra processing and downstream steps. Figure 7.4 explains the commonly accepted theory for the intracellular synthesis route. The metal ions, being positively charged, precipitate into the cell. The negative charges on the cell membrane or cell walls of certain microbes help in this precipitation process. Once the metals are within the cell, the enzymes present on the cell walls carry out the reduction process. The reduced nanoparticles accumulate within the cell or remain attached to the cell walls or membranes (Sastry et al., 2003). Another proposed theory of nanoparticle presence within the cells in the extracellular reduction of the metal ions is followed by their precipitation into the cells.

Further studies are needed to elucidate the exact mechanisms behind the mycosynthesis of nanoparticles and also to identify the precise roles of the various biomolecules and proteins present in the final product. The visual colour changes in all of the above synthesis steps are attributed to surface plasmon resonance. This is a type of non-radiative surface wave phenomenon attributed to the excitation of surface plasmons, which are free electrons present on the surface of the material, which also act as the interface between the substance and the media it is submerged in. Further microscopic and spectroscopic tests were conducted to confirm the nanoparticle's presence and the location of the synthesized nanoparticles with respect to the fungal biomass to identify intracellular or extracellular modes of synthesis in the above-summarized studies.

7.4 APPLICATION OF MYCONANOPARTICLES IN THE AGRICULTURE SECTOR

Despite using large amounts of conventional pesticides and other agrochemicals, worldwide crop losses have not decreased significantly over the last 40 years. According to a study by Oerke (2006), crop losses were most significant in the case of damage caused by weeds, followed by pests and pathogens. Nanotechnology can step into this field and may be used to a more significant effect than conventional means of managing pests and phytopathogens.

One of the most considerable drawbacks of using agrochemicals by routine spraying or injection methods is the lack of stability and solubility of the free compound. As a result, minimal quantities of the compound reach the ultimate target site, which is generally below the minimum required concentration to show productive activity. Due to the compound's ineffectiveness on its initial utilization, it would have to be applied to the plant crop multiple times to reach sufficient concentrations and show the desired activity (Nair et al., 2010). Nanoparticles can be used here to increase compound effectiveness by acting as delivery systems when the compounds are used in this state, compounds such as stability, solubility, and bio-distribution increase. Besides, we can also increase the compounds' effectiveness by either supplementing it with additives or increasing the permeability of the drug or some other means (Ingale et al., 2013; Nair et al., 2010).

In addition to the ineffectiveness of the chemicals used, annually, 2.5 metric tonnes of chemical pesticides are used to combat crop diseases and pests. A majority of these pesticides are harmful not just to the target organism but also to the surrounding environment, groundwater reservoirs (if the chemicals are capable of leaching into the underground reservoirs), the soil, and the air in addition to the toxicity they may show to non-target organisms and human beings (Yadav et al., 2018). Several studies have already been carried out, which confirm the effectiveness of the process and the safety afforded to the environment while making use of metal nanoparticles to control plant diseases. Several studies also show that compared to bulk-sized metal compounds, the nanosized compounds show higher activity inhibition inhibiting fungal and microbial growth.

Yet another drawback of using copious amounts of chemical pesticides is the increased resistance the pathogens and weeds may develop to the pesticide or chemical in use with time. A solution to this problem would be one of the most widely discussed and theorized applications of nanoparticles in the field of agriculture, which is the nanodelivery system. This system would make use of nanoparticles and a global positioning system. This smart delivery system would automatically detect the presence of pathogens or weeds and are self-regulated, removing the need for human control. The system would, after detection, disperse the required pesticide or agrochemical in a way that the targeted organism is efficiently suppressed or destroyed. Nanodelivery systems may also be designed so that they specifically target the infected tissues of plants without damaging other plant parts, thereby reducing the required quantities of the agrochemical and subsequently reducing the drug's phytotoxicity to the surroundings (Alghuthaymia et al., 2015). We could also design nanodelivery systems that would continuously circulate within the plant, via the circulatory systems, and liberate active compounds when a specific condition or set of conditions is met, like when exposed to some viral protein or other antigenic particle or so on. Such delivery systems would be analogous to the immune system found in vertebrates and would be able to remedy problems within the plants before they are even visible to the naked eye. Another possibility of disease mitigation lies in the triggering of defence responses within the plant by the smart delivery system via the system's action of genetic material. The DNA/RNA would activate plant defence mechanisms and prevent disease expression. Another way to prevent disease expression is to introduce small interfering RNA, which would disrupt specific pathogenic cycles and prevent the pathogen from fully expressing itself.

Nanotechnology in agriculture is limited to curative measures and can also be utilized in defensive capabilities. As such, myconanoparticles can be utilized for the formation of resistant and resilient varieties of plants. Such plant species would be comparatively more beneficial and economical in the long run as they would not depend on agrochemicals for protection against diseases. Nanoparticle mediated transformation of plant species could help create disease tolerant plant varieties and help establish new methods of combating phytopathogens (Alghuthaymia et al., 2015).

Nanotechnology can also be used to construct advanced sensor/biosensor systems. The nanoparticles in these sensors would play the role of acting as an attachment area for biomolecules like DNA or hormones. Using these systems, studies could be conducted on the role of various biomolecules within the plant. An example regarding the same would be the utilization of these sensors in auxin studies, which could explain the exact uses and mechanism that the hormone partakes within the plant (Ingale et al., 2013). Nanosensors can also be made that

a. Check for the presence of pathogens within the plants, which can be used in conjunction with the nanodelivery systems
b. Determine the stage of growth of the plants and hence help in identifying the harvest time of crops
c. Detect levels of nutrients within the plant, thereby providing an idea as to what should be provided to the plant for its proper growth (Sharon et al., 2010)
d. Detect viral DNA before the expression of the genetic material leads to a diseased state (which would be much harder to combat as compared to the detection process)

The post-harvest loss of crops is another phenomenon that can be resolved by using antimicrobial packaging or storage means, in which nanomaterials can contribute significantly. A study conducted by Marcato et al., in 2014, made use of silver nanoparticles that were synthesized through the extracellular mechanism of *F. oxysporum* and incorporated these particles into silk and cotton fabric to check for their bactericidal activity. It was observed that the cotton fabric combined with the nanoparticles inhibited the growth of *Staphylococcus aureus*. Such action was not seen in the case of silk garments, due to the low incorporation of the nanoparticles into the fabric. The study shows that silver nanoparticles inhibit microbial growth and can hence be used in various applications pertaining to agriculture, including storage of crops after their harvest and improved food packaging. Another way to protect fruits and vegetables after their harvest would be to make use of nano-laminates. These are defined as two or more layers of food-grade nanoparticles placed over the edible item meant to be protected. Nano-laminates are used by directly spraying them onto the substance, making them easy to use (Sharon et al., 2010).

Nanoparticles synthesized using biological approaches are eco-friendly and are found to be effective against several pathogens that cause diseases in the agricultural crops and thus act as biopesticides, in addition to their somewhat established role as biofertilizers (Singh et al., 2010). The nanoparticles could be effectively used for solar energy applications, as depicted in a study conducted by Mittal et al. in 2010 using several plant products to synthesize metal nanoparticles of FeO_3/Fe_3O_4 and CeO_2, which can provide a source of clean energy to farmlands (Mittal et al., 2013).

Nanoparticles of certain compounds show the properties of photocatalysis, i.e. when they are subjected to the light of specific wavelengths, they catalyze a unique process in certain chemical compounds to form an electron-hole (positive) and electron (which has been excited out of the valence shell). The pair of electron-hole and electron can work in conjunction or separately to perform various functions. The photocatalysis process in itself can be used for the decomposition or degradation of various harmful components, including pathogens. The negative electron is involved in multiple antimicrobial processes. Excessive amounts of pesticides or other agrochemicals that are part of the runoff from agricultural fields can be degraded via this process as well, which, if not treated, would eventually accumulate in various levels of the food chain and cause toxicity-related issues. The harmful compounds interact with positive electron-hole and are converted into non-toxic forms (Ditta, 2012).

A key strategy used for increasing crop yields, while keeping agricultural input minimum, is the precision farming technique. This farming model monitors the fields' localized conditions with the help of sensors, an integrated computer system, and global positioning to the system to ensure that the crops are growing efficiently. In case maximum efficiency is not achieved, the precision farming system scans the area and determines the problem and its source. The use of nanotechnology in this system is currently limited and can be augmented by increasing the system's technical capacity by providing progressive and tinier nanosensors and monitoring devices. Regardless of the challenges faced, the potential of this system is huge. Since the system can locate the area and source of the problem accurately, the farmers would be able to target the specific area and make use of the appropriate measures to remedy the problem, thereby saving time, money, and reducing the usage of agrochemicals, which would, in turn, reduce the harmful side effects they exhibit.

A pathogen that infects plants, *Bipolaris nodulosa*, was used by a group of researchers to synthesize anisotropic silver nanoparticles with excellent antimicrobial activity against harmful bacteria and fungi (Saha et al., 2010).

Nanoparticles can be an excellent source of biosensor development because of their unique optical and electronic characteristics. Yeast cells were employed to synthesize gold–silver alloy nanoparticles, which are five times more sensitive than electrochemical response sensors to the specific compound vanillin (Saha et al., 2010).

Nanofibrous mats were reported to be synthesized by Spasova et al., in 2011, by mixing chitosan and *Trichoderma viride*. These mats were manufactured faster and compete better with *Alternaria* sp. and *Fusarium* sp. for nutrients and space. Similarly, gold nanowires were synthesized by exposing *Aspergillus* and *Neurospora* species to gold nanoparticles for one week. The nanowires thus formed were then found to be surface functionalized with glutamate, aspartate, and polyethene glycol (Sugunan et al., 2007).

Figure 7.5 summarizes some of nanotechnology's potential applications in agriculture explained above in the text, such as genetic manipulation of plant species using nanoparticles as delivery systems, nanopesticides, insecticides, and herbicides

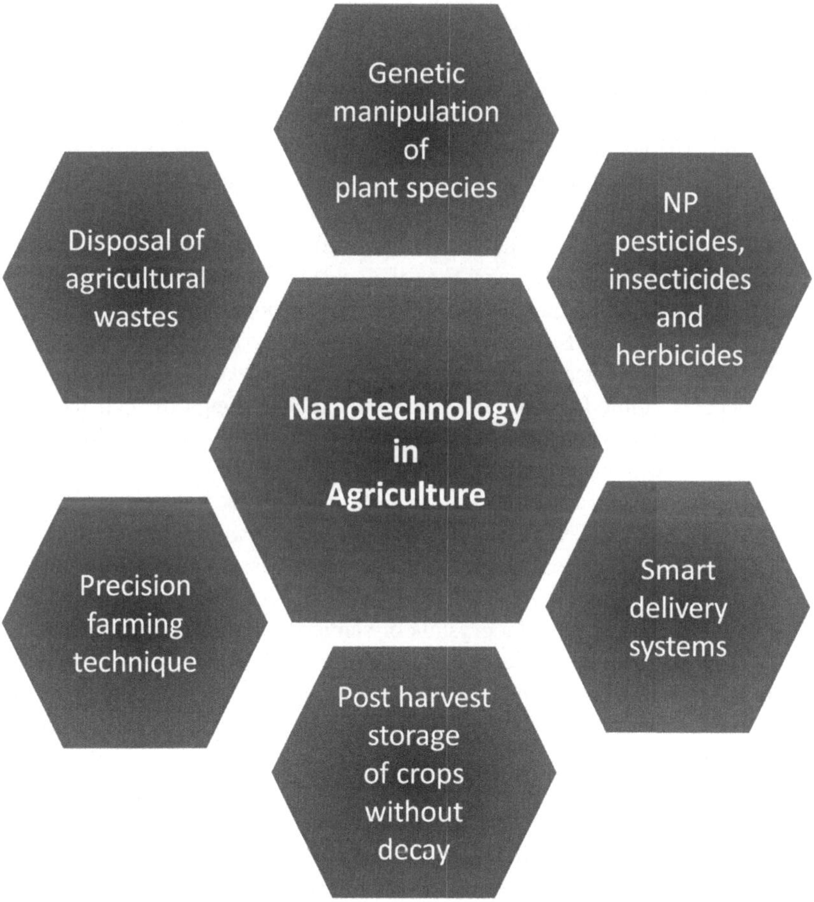

FIGURE 7.5 Various applications of nanoparticles in the agricultural sector.

for disease management in plants, post-harvest management of crops using nanotechnology, etc.

Agricultural waste and leftover substances after the processing of products can be put to good use using nanotechnology-enabled techniques. Nanofibres are the most prominent application of the same. Leftover fibres after cotton processing may be used for the synthesis of cotton nanofibres, by a technique known as electrospinning. These nanomaterials act as powerful absorbents, and they can be used to suck up the excess pesticides and other agrochemicals after they are sprayed onto the fields. The nanofibres can also be used for the dispersal of the same substances into the fields. As the organic fibres deteriorate over time, the chemicals absorbed into it are slowly released, allowing for a slower and more efficient dispersal system. Nanofibres can also be used to detect pathogen presence on a surface by simply wiping it over the surface. This works due to the interaction between specific antibodies immobilized on the nanofibres and the antigens specific to the immobilized antibodies. If the pathogen secretes the particular antigen, the interaction between the immunoglobulin and the antigen occurs, which would imply the pathogen (Mousavi and Rezaei, 2011).

Nanocapsules, a thin-shelled substance with a hollow internal area, can be used to deliver active content into the system and be used for agrochemical release into plant systems. Using nanocapsules for the delivery of active substances reduces the damage caused by the compounds while increasing their efficacy. The capsule bursting can be synchronized with the specific external parameters like pH, temperature, accumulation of certain compounds, etc. Furthermore, depending on the nature of the substance used for vesicle formation, the substance's penetration power can be altered. Liposome vesicles have been used for this purpose, but polymeric nanocapsules are observed to have a more excellent range of properties and are comparatively more mouldable and hence are being studied extensively. Liposomes, being amphiphilic molecules, allow for more effortless movement through the cell membrane. The polymeric nanocapsules can be modified by the addition of functional groups onto the polymeric chain, thereby changing the properties of the nanocapsule. These nanocapsules can also be used to absorb chemical signals secreted by plants that trigger parasite growth. The lack of a stimulating signal would prevent the parasite from initiating its life cycle and to avoid disease expression (P'erez-de-Luque and Rubiales, 2009).

Quantum dots (QDs) are nanocrystals of semiconductor materials and have the property to emit colour when exposed to UV light. This is due to the valence electrons breaking away from the valence band due to the absorption of energy, followed by the falling of the electron to the valence band associated with the liberation of the absorbed energy in the form of light of specific colour (depending on the difference in energy between the valence and conductance bands). QDs can serve to replace organic dyes as a stain to detect bacterial and other microbial cells. QDs are shown to be better than organic dyes due to their broad absorption spectra, narrow emission spectra, symmetry, and photostability (Ditta, 2012).

Hyperthermia is the process of transferring large amounts of heat; the heat transferred is used to kill microbial cells. Nanoparticles, specifically paramagnetic nanoparticles, can be used for the same. Such nanomaterials are made to accumulate within the target pathogen cells and then using electromagnetic waves, the temperature surrounding the nanoparticles is increased substantially, causing cell death. This cellular destruction mode can be made specific and allows for targeted cell destruction without causing damage to the host organism (P'erez-de-Luque and Rubiales, 2009).

An experimental study conducted by Liu et al. (2001(a)) showed that nanoparticle encapsulated delivery of fungicides increased the active ingredient's efficacy of action. Nanoparticle matrices were synthesized using polyvinyl pyridine (PVP), copolymers of PVP with styrene, and PVP with hyper-branched polymers (HBP). The fungicides used were tebuconazole or chlorothalonil, and median diameters of 100–200 nm were observed in the case of the synthesized nanoparticles. The compound tebuconazole is readily soluble in common organic solvents, while chlorothalonil shows lesser solubility. Biocides with lower solubility can only be used in the solution of hydrocarbon oils, thereby reducing the range of applications afforded to it. Using nano-encapsulated delivery methods would allow even biocides with lower solubility to be commercially used. The efficiency of a matrix-biocide nanoparticle preparation was observed to be relatively high due to the specific precipitation method used for synthesis. The effect of the biocide containing nanoparticle was tested against *Gloeophyllum trabeum* infected wood wafers. This fungal species is responsible for the brown rot wood decay. It was observed that the blank nanoparticles (lacking the biocide) showed little to no activity against the fungi, but nanoparticle containing the fungicides showed excellent efficiency in reducing the mass loss of the wood (southern yellow pine) caused by the fungi. It was observed that even minimal quantities of fungicide administered via nanocapsules showed optimal activity in reducing wood damage to tolerable levels. The reason nano-encapsulated fungicide shows such great antifungal activity could be attributed either due to the protection afforded to the fungicide by the matrix or due to the targeted delivery of fungicide to the infected areas. The study also showed that the matrix capsule had almost no effect on the nanoparticle's antifungal activity. Similar studies were conducted by varying the polymer matrix (polyvinyl chloride) and using additional biocides (KATHON 930, a fungicide, and chlorpyrifos, which is an insecticide). The active ingredient containing nanoparticles was synthesized similarly as in the previous study, and the particles were suspended in water to attain the desired levels of the drug. PVC was used as the polymer matrix in this study to demonstrate that the nanomatrix-encapsulated delivery of the active ingredient is practical even if we are to use 'difficult to work with' matrices. The wood used for this study was birch wood, which has smaller surface pit pores than southern yellow pine, thereby representing an extreme case. Birch treated with tebuconazole was shown to be protected from the fungal action of *Trametes versicolor*, which causes white-rot disease. The loss of wood mass was observed to be significantly lower in the cases of samples treated with tebuconazole than the control sample, even at low active ingredient loading concentrations. Similar results were obtained when chlorothalonil-containing nanoparticles were used for the antifungal treatment of the birch wood. The results showed that even biocides with lower solubility, like chlorothalonil, showed effective disease control when used in nanocapsule forms (Liu et al., 2002(b)). Further studies were also conducted by the same authors to further reduce the nanoparticle size, increase the suspension's stability, and simplify the procedure for synthesis (Liu et al., 2001(c)).

The most significant and most widespread application of nanomaterials is the use of silver nanoparticles (Ag NPs) as antimicrobial agents. Several studies have been conducted which show the efficiency of Ag NPs in controlling microbial growth (Choi et al., 2008; Bergerson, 2010a; Sotirou and Pratsinis, 2010; Kim et al., 2014; Marcato et al., 2014; Rizzello and Pompa, 2014; Singh et al., 2015; Yadav et al., 2018). Studies also show their role in controlling the growth of several fungal species like *Bipolaris sorokiniana*, which causes splotch blot disease in wheat (Mishra et al., 2014), *Candida* spp. which causes the disease candidiasis in humans (Panacek et al., 2009), *Trichosporona sahii*, which causes infections in humans (Xia et al., 2016).

Research on nano-based formulations has recently come to the rescue of agronomists and plant culturists from the common plant pathogens, mainly fungi (Singh and Arora, 2016). Silver nanoparticles are found to have antifungal activity against anthracnose, causing pathogen *Colletotrichum gloesporioides*, which is a common plant disease-causing pathogen in several fruits, such as apple, papaya, etc. (Aguilar-Mendez et al., 2011). Another study conducted on rose powdery mildew causing pathogen *Sphaerotheca pannosa* var. *rosae* reported that double capsulated silver nanoparticle antifungal activity diluted solution sprayed to the affected area caused 95% removal of the pathogen without reoccurrence for a period of one week (Kim et al., 2008). Hence, it could be concluded that nanosilver could be used as an alternative to chemical fungicide to manage plant diseases.

Keeping in mind the efficacy of these particles, their use can be extended to the agricultural field as pesticides. A collective count of all the pathogenic species that Ag NPs are effective against shows that they can kill or inhibit the action of approximately 650 different types of fungi, bacteria, viruses, etc. The exact mechanism that Ag NPs utilize to destroy microbial cells is not known. One theory speculates that the silver ions inhibit the enzyme's action used by bacterial and fungal cells to metabolize oxygen, thereby causing the cells to eventually die. Other theories that explain the inhibitory effect of Ag NPs, and metallic nanoparticles in general, are:

a. The greatest source of inhibitory action all metallic NPs have associated with them is their ionic properties. The ions liberated by these metallic nanoparticles may be genotoxic in nature, i.e. they may damage the microbial cell's genetic material, causing cell death.
b. Generation of reactive oxygen species by the NPs that damage and destroy the cells.
c. Ionic interaction with various microbial cells' proteins might render the cells non-functional and hamper cell functionality and may even cause death.
d. Metallic NPs may interfere with the cells' nutrient uptake (Yadav et al., 2018; Alghuthaymi et al., 2015).

Several studies have observed and characterized the formation of Ag NPs by fungal cells and also shown their efficacy in controlling microbial growth (Marcato et al., 2014; Fayaz et al., 2009). Fayaz et al. (2009) conducted an experimental study, and showed the extracellular formation of Ag NPs by the fungus *Trichoderma viride*. The formation of the nanoparticles was confirmed by all the usual means; observing the brown colour of the cell filtrate after the addition of silver nitrate was the visual confirmation for the formation of Ag NPs. Spectroscopic analysis, FTIR, TEM, and EDX results served to confirm the extracellular formation of Ag NPs. The study also tested the antimicrobial activity and efficiency of the so-formed Ag NPs using *E. coli* ATCC 8739 and *S. aureus* ATCC 6538. Cell cultures of both the bacteria showed decreased growth rates in the presence of Ag NP supplemented media. The Ag NPs were also incorporated into sodium alginate films, and a clear zone of inhibition was observed around the films, highlighting the antimicrobial activity of the Ag NPs. These films were used to coat carrots and pears, and it was noted that the sodium alginate incorporated with Ag NPs coated crop was edible even after ten days of storage as compared to sodium alginate lacking Ag NPs coated crop which lasted only for 6–8 days. The control carrots and pears, which didn't have any coat applied, could only be stored for shorter durations.

ZnO NPs also exhibit potent antimicrobial activity. The major mechanism attributed to their functionality is that once these NPs penetrate the cell membrane, they induce the formation of ROS, which causes cellular damage and destruction. Another point favouring the use of ZnO NPs in the agricultural sector is their lack of toxicity to non-target organisms and the surrounding ecosystem (Sabir et al., 2014).

The greatest application of nanoscale systems which can feasibly attain functionality in the near future is their use as a carrier of drugs, herbicides, pesticides, hormones, or other biologically active substances into the plant system. The carrier system would protect the substance from degradation until it reaches the target site and could be altered to have specific features like a timed release of the compounds or release only when certain conditions are met (like the presence of specific hormones). Usage of agrochemicals in such a manner would increase efficiency in the uptake of the active compounds and ultimately reduce the runoff and wastage of the agrochemical. Other applications like the smart delivery system, nano-laminates, nanoparticle-based genetic transformations to confer disease resistance to plants (Torney et al., 2007), nano-emulsions, etc. are all promising avenues to increase the productivity and decrease the losses faced in the agricultural sector.

7.5 FUTURE DEVELOPMENT AND NEEDS

The need of the hour is to find a way to provide food for the massive global population in a way that does not compromise the arable lands of the world or overburden the environment. Due to the ever-increasing nature of our population, there is massive pressure on the agricultural sector to keep up with food production. Diseases and post-harvest losses are another contributing factors to the need to develop new and innovative ways to combat the problems faced in this sector. Dependence on conventional pesticides and agrochemicals to boost plant growth and reduce disease rates is not a solution the human society can keep relying on forever, keeping in mind the detrimental effects most of these chemicals have on non-target organisms and the damage they may cause when they accumulate within the food chains. Hence we need to turn to ways that don't damage the ecosystem or the world we live in, that continues to produce sufficient edible and other agriculturally produced materials for the ever-growing population, is cost-effective, efficient and also reliable. Nanotechnology can be used to fulfil these roles, but the conventional (physical and chemical) methods of nanoparticle synthesis are neither reliable nor safe, environmentally speaking. They also cost a lot of money and take large amounts of time to produce nanoparticles. Natural alternatives are a greener approach to nanoparticle synthesis, especially synthesis using fungal species. Fungi are easy to handle and are shown to have greater nanoparticle yields due to the large amounts of proteins and extracellular enzymes they secrete, which are involved in the bioreduction process of the salt solution into the corresponding metal nanoparticle. Various studies and experiments elucidate the efficient and reliable production of myconanoparticles and their benefits to the agriculture sector. From acting as nano pesticides and herbicides to delivering agrochemicals and even genetic materials into the plants to evoke defence mechanisms, to increasing the yields of crops and protecting the farm produce from post-harvest fungal and microbial damage by making use of nanocoatings, my nanotechnology seems to provide a green, efficient and elegant solution to the current problems faced by the agriculture sector.

However, as with all solutions, myconanotechnology has its drawbacks, mainly the toxicity of the nanoparticles to the surrounding environment and non-target organisms. Nature and physical properties of the nanoparticles suggest that they may pose health-related problems to other organisms. As the levels of the NPs increase, the chances of accumulation of the particles within non-target organisms due to run off also increases. Accumulation of NPs may result in detrimental and maybe even lethal effects. They may produce ROS, induce genetic changes, cause conformational changes in proteins, thereby affecting their functionality, etc. To work beyond these problems, we need to completely characterize and understand the working and interactions of metallic NPs with the environment, target and non-target organisms. We also need to standardize their synthesis via fungal species (or other efficient green methods) and standardize the doses to be used while treating a disease or when they are being used for any other purpose in the fields. Another aspect that needs to be kept in mind is that the fungal secretions may be composed of substances that do not aid in bioreduction and may even cause changes in the final shape and sizes of the metallic NPs.

Thus it can be concluded that myconanotechnology holds immense potential for the safe and clean synthesis of NPs, and their applications in every sector of life are extremely diverse and abundant. Further studies and research needs to be done to fully utilize its untapped potential.

REFERENCES

Abdel-Hafez, S.I., Nafady, N.A., Abdel-Rahim, I.R., Shaltout, A.M., Mohamed, M.A.: Biogenesis and optimisation of silver nanoparticles by the endophytic fungus *Cladosporium sphaerospermum*. *Int. J. Nano Chem.* 2(1), 11–19 (2016).

Abdel Rahim, K., Mahmoud, S.Y., Ali, A.M., Almaary, K.S., Mustafa, A.E., Husseiny, S.M.: Extracellular biosynthesis of silver nanoparticles using *Rhizopus stolonifer. Saudi J. Biol. Sci.* 24(1), 208–216 (January 2017).

Abdullaeva, Z., Omurzak, E., Iwamoto, C., Ganapathy, H.S., Sulaimankulova, S., Liliang, C., Mashimo, T.: Onion-like carbon encapsulated Co, Ni, and Fe magnetic nanoparticles with low cytotoxicity synthesized by a pulsed plasma in a liquid. *Carbon* 50(5), 1776–1785 (2012).

Afreen, R.V., Ranganath, E.: Synthesis of monodispersed silver nanoparticles by *Rhizopus stolonifer* and its antibacterial activity against MDR strains of *Pseudomonas aeruginosa* from burnt patients. *Int. J. Environ. Sci.* 1(7), 1582–1592 (2011).

Afshar, P., Sedaghat, S.: Bio-Synthesis of silver nanoparticles using water extracts of *Satureja hortensis* L. and evaluation of the antibacterial properties. *Curr. Nanosci.* 12(1), 90–93 (2016).

Agnihotri, M., Joshi, S., Kumar, A.R., Zinjarde, S., Kulkarni, S.: Biosynthesis of gold nanoparticles by the tropical marine yeast *Yarrowia lipolytica* NCIM 3589. *Mater. Lett.* 63(15), 1231–1234 (2009).

Aguilar-Méndez, M.A., San Martín-Martínez, E., Ortega-Arroyo, L., Cobián-Portillo, G., Sánchez-Espíndola, E.: Synthesis and characterization of silver nanoparticles: Effect on phytopathogen *Colletotrichum gloesporioides. J. Nanopart. Res.* 13(6), 2525–2532 (2011).

Ahmad, A., Mukherjee, P., Senapati, S., Mandal, D., Khan, M.I., Kumar, R., Sastry, M.: Extracellular Biosynthesis of Silver Nanoparticles Using the Fungus *Fusarium oxysporum. Coll. Surf. B* 28, 313–318 (2003).

Ahmad, A., Mukherjee, P., Mandal, D., Senapati, Khan, M.I., Kumar, R., Sastry, M.: Enzyme mediated extracellular synthesis of CdS nanoparticles by the fungus, *Fusarium oxysporum. J. Am. Chem. Soc.* 124(41), 12108–12109 (2002).

Ahmad, A., Senapati, S., Khan, M.I., Kumar, R., Ramani, R., Srinivas, V., Sastry, M.: Intracellular synthesis of gold nanoparticles by a novel alkalotolerantactinomycete, *Rhodococcus species. Nanotechnology* 14(7), 824–828 (2003).

Alghuthaymi, M.A., Almoammar, H., Rai, M., Said-Galiev, E., Abd-Elsalam, K.A.: Myconanoparticles: Synthesis and their role in phytopathogens management. *Biotechnol. Biotechnol. Equip.* 29(2), 221–236 (2015).

Amutha, R., Muruganandham, M., Lee, G.J., Wu, J.J.: Facile microwave-combustion synthesis of wurtziteCdS nanoparticles. *J. Nanosci. Nanotechnol.* 11(9), 7940–7944 (2011).

Balaji, D.S., Basavaraja, S., Deshpande, R., Mahesh, D.B., Prabhakar, B.K., Venkataraman, A.: Extracellular biosynthesis of functionalized silver nanoparticles by strains of *Cladosporium cladosporioides* fungus. *Coll. Surf. B Biointerfaces* 68, 88–92 (2009).

Balakumaran, M.D., Ramachandran, R., Kalaichelvan, P.T.: Exploitation of endophytic fungus, *Guignardia mangiferae* for extracellular synthesis of silver nanoparticles and their in vitro biological activities. *Microbiol. Res.* 178, 9–17 (2015).

Bergeson, L.L.: Nanosilver pesticide products: What does the future hold? *Environ. Qual. Manag.* 19, 73–82 (2010b). https://doi.org/10.1002/tqem.20263.

Bergeson, L.L.: Nanosilver: Us EPA's pesticide office considers how best to proceed. *Environ. Qual. Manag.* 19, 70–85 (2010a). https://doi.org/10.1002/tqem.20255.

Bhainsa, K.C., D'Souza, S.F.: Extracellular biosynthesis of silver nanoparticles using the fungus *Aspergillus fumigatus. Coll. Surf. B: Biointerf.* 47, 160–164 (2006).

Bhat, R., Deshpande, R., Ganachari, S.V., Huh, D.S., Venkataraman, A.: Photo-irradiated biosynthesis of silver nanoparticles using edible mushroom *Pleurotus florida* and their antibacterial activity studies. *Bioinorg. Chem. Appl.* (2011).

Binupriya, A.R., Sathishkumar, M., Vijayaraghavan, K., Yun, S.I.: Bioreduction of trivalent aurum to nano-crystalline gold particles by active and inactive cells and cell-free extract of *Aspergillus oryzae* var. viridis. *J. Hazard. Mater.* 177(1–3), 539–545 (2010).

Borovaya, M., Pirko, Y., Krupodorova, T., Naumenko, A., Blume, Y., Yemets, A.: Biosynthesis of cadmium sulphide quantum dots by using *Pleurotus ostreatus* (Jacq.) P. Kumm. *Biotechnol. Biotechnol. Equip.* 29(6), 1156–1163 (2015).

Camas, M., Sazak Camas, A., Kyeremeh, K.: Extracellular synthesis and characterization of gold nanoparticles using Mycobacterium sp. BRS2A-AR2 isolated from the aerial roots of the Ghanaian mangrove plant, *Rhizophora racemosa. Indian J. Microbiol.* 58(2), 214–221 (2018).

Castro-Longoria, E., Vilchis-Nestor, A.R., Avalos-Borja, M.: Biosynthesis of silver, gold and bimetallic nanoparticles using the filamentous fungus *Neurospora crassa. Colloids Surf. B Biointerfaces* 83(1), 42–48 (2011).

Chippa, H.: Applications of nanotechnology in agriculture. *Methods Microbiol.* 46, 115–142 (2019).

Choi, O., Deng, K.K., Kim, N.J., Ross, L. Jr., Surampalli, R.Y., Hu, Z.: The inhibitory effects of silver nanoparticles, silver ions, and silver chloride colloids on microbial growth. *Water Res.* 42(12), 3066–3074 (2008).

Cioffi, N., Torsi, L., Ditaranto, N., Sabbatini, L., Zambonin, P.G., Tantillo, G., Ghibelli, L., D'Alessio, M., Bleve-Zacheo, T., Traversa, E.: Antifungal activity of polymer-based copper nanocomposite coatings. *Appl. Phys. Lett.* 85(12), 2417–2419 (2004).

Cuevas, R., Durán, N., Diez, M.C., Tortella, G.R., Rubilar, O.: Extracellular biosynthesis of copper and copper oxide nanoparticles by *Stereum hirsutum*, a native white-rot fungus from Chilean forests. *J. Nanomater.* 16(1), 1–7 (2015).

Das, R., Nath, S.S., Chakdar, D., Gope, G.R., Bhattacharjee, H.: Preparation of silver nanoparticles and their characterization. *J.Nanotech.* 5, 1–6 (2009).

Dimkpa, C.O., Bindraban, P.S.: Nanofertilizers: New products for the industry? *J. Agric. Food Chem.* 66(26), 6462–6473 (2017).

Ditta, A.: How helpful is nanotechnology in agriculture? *Adv. Nat. Sci. Nanosci. Nanotechnol.* 3(3), 033002 (2012). http://www.ncbi.nlm.nih.gov/pubmed/033002.

Du, L., Jiang, H., Liu, X., Wang, E.: Biosynthesis of gold nanoparticles assisted by *Escherichia coli* DH5α and its application on direct electrochemistry of hemoglobin. *Electrochem. Commun.* 9(5), 1165–1170 (2007).

Du, L., Xian, L., Feng, J.: Rapid extra-/intracellular biosynthesis of gold nanoparticles by the fungus *Penicillium* sp. *J. Nanopart. Res.* 13(3), 921–930 (2011).

Duhan, J.S., Kumar, R., Kumar, N., Kaur, P., Nehra, K., Duhan, S.: Nanotechnology: The new perspective in precision agriculture. *Biotechnol. Rep.* 15, 11–23 (2017).

Durán, N., Marcato, P.D., Alves, O.L., Souza, G.I., Esposito, E.: Mechanistic aspects of biosynthesis of silver nanoparticles by several *Fusarium oxysporum* strains. *J. Nanobiotechnology* 3, 8 (2005).

El-Baz, A.F., El-Batal, A.I., Abomosalam, F.M., Tayel, A.A., Shetaia, Y.M., Yang, S.T.: Extracellular biosynthesis of anti-Candida silver nanoparticles using *Monascus purpureus. J. Basic Microbiol.* 56(5), 531–540 (2016).

Elbeshehy, E.K., Elazzazy, A.M., Aggelis, G.: Silver nanoparticles synthesis mediated by new isolates of *Bacillus* spp., nanoparticle characterization and their activity against bean yellow mosaic virus and human pathogens. *Front. Microbiol.* 6, 453 (2015).

El-Rafie, M.H., Mohamed, A.A., Shaheen, T.I., Hebeish, A.: Antimicrobial effect of silver nanoparticles produced by fungal process on cotton fabrics. *Carbohydr. Polym.* 80(3), 779–782 (2010).

Fanfair, D., Desai, S., Kelty, C.: The Early History of Nanotechnology. *Connexions Module* 6, m14504 (2007).

Fayaz, M., Balaji, K., Girilal, M.,Kalaichelvan, P.T., Venkatesan,R.: Mycobased synthesis of silver nanoparticles and their incorporation into sodium alginate films for vegetable and fruit preservation. *J. Agric. Food Chem.* 57(14), 6246–6252 (2009).

Gadd, G.M.: Metals, minerals and microbes: Geomicrobiology and bioremediation. *Microbiology* 156(3), 609–643 (2010).

Gade, A., Ingle, A., Whiteley, C., Rai, M.: Mycogenic metal nanoparticles: Progress and applications. *Biotechnol. Lett.* 32(5), 593–600 (2010).

Gajbhiye, M., Kesharwani, J., Ingle, A., Gade, A., Rai, M.: Fungus-mediated synthesis of silver nanoparticles and their activity against pathogenic fungi in combination with fluconazole. *Nanomedicine* 5(4), 382–386 (2009).

Gardea-Torresdey, J.L., Gomez, E., Peralta-Videa, J.R., Parsons, J.G., Troiani, H., Jose-Yacaman, M.: Alfalfa sprouts: A natural source for the synthesis of silver nanoparticles. *Langmuir,* 194 1357 1361 (2003).

Ge, L., Li, Q., Wang, M., Ouyang, J., Li, X., Xing, M.M.: Nanosilver particles in medical applications: Synthesis, performance, and toxicity. *Int. J. Nanomed.* 9, 2399 (2014).

Ghiassi, S., Sedaghat, S., Mokhtary, M., Kefayati, H.: Plantmediated bio-synthesis of silver–montmorillonitenanocomposite and antibacterial effects on gram-positive and-negative bacteria. *J. Nanostruct. Chem.* 8(3), 353–357 (2018).

Gupta, S., Sharma, K., Sharma, R.: Myconanotechnology and application of nanoparticles in biology. *Recent Res. Sci. Technol.* 4(8), 36–38 (2012).

Gurunathan, S., Han, J.W., Kwon, D.N., Kim, J.H.: Enhanced antibacterial and anti-biofilm activities of silver nanoparticles against Gram-negative and Gram-positive bacteria. *Nanoscale Res. Lett.* 9(1), 373 (2014)

Gurunathan, S., Lee, K.J., Kalishwaralal, K., Sheikpranbabu, S., Vaidyanathan, R., Eom, S.H.: Antiangiogenic properties of silver nanoparticles. *Biomaterials* 30(31), 6341–6350 (2009).

Gurunathan, S., Raman, J., Malek, S.N.A., John, P.A., Vikineswary, S.: Green synthesis of silver nanoparticles using *Ganoderma neo-japonicum* Imazeki: A potential cytotoxic agent against breast cancer cells. *Int. J. Nanomed.* 8, 4399 (2013).

Huang, J., Li, Q., Sun, D., Lu, Y., Su, Y., Yang, X., Wang, H., Wang, Y., Shao, W., He, N., Chen, J.H.C.: Biosynthesis of silver and gold nanoparticles by novel sundried *Cinnamomum camphora* leaf. *Nanotechnology* 18, 105104 (2007).

Hulkoti, N.I., Taranath, T.C.: Biosynthesis of nanoparticles using microbes – A review. *Colloids Surf. B Biointerfaces* 121, 474–483 (2014).

Ingale, A.G., Chaudhari, A.N.: Biogenic synthesis of nanoparticles and potential applications an eco-friendly approach. *J Nanomed. Nanotechol.* 4, 165 (2013).

Iriti, M., Varoni, E.M.: Chitosan-induced antiviral activity and innate immunity in plants. *Environ. Sci. Pollut. Res.* 22(4), 2935–2944 (2015).

Jalal, M., Ansari, M.A., Alzohairy, M.A., Ali, S.G., Khan, H.M., Almatroudi, A., Raees, K.: Biosynthesis of silver nanoparticles from oropharyngeal *Candida glabrata* isolates and their antimicrobial activity against clinical strains of dacteria and fungi. *Nanomaterials* 8(8), 586 (2018).

James, W.C., Teng, P.S., Nutter, F.W.: *Estimated Losses of Crops from Plant Pathogens.* In *CRC Handbook of Pest Management in Agriculture*, 1st edition. CRC Press, Inc (1991), pp. 15–51.

Jo, J.H., Singh, P., Kim, Y.J., Wang, C., Mathiyalagan, R., Jin, C.G., Yang, D.C.: *Pseudomonas deceptionensis* DC5-mediated synthesis of extracellular silver nanoparticles. *Artif. Cells Nanomed. Biotechnol.* 44(6), 1576–1581 (2016).

Kalpana, N., Kataru, B.A.S., Sravani, N., Vigneshwari, T., Panneerselvam, A., Rajeswari, D.: Biosynthesis of zinc oxide nanoparticles using culture filtrates of *Aspergillus niger*: Antimicrobial textiles and dye degradation studies. *OpenNano* 3, 48–55 (2018).

Karbasian, M., Atyabi, S.M., Siadat, S.D., Momen, S.B., Norouzian, D.: Optimizing nanosilver formation by *Fusarium oxysporum* PTCC 5115 employing response surface methodology. *AJABS* 3(1), 433–437 (2008).

Kashyap, P.L., Kumar, S., Srivastava, A.K., Sharma, A.K.: Myconanotechnology in agriculture: A perspective. *World J. Microbiol.Biotechnol.* 29(2), 191–207 (2013).

Kathiresan, K., Manivannan, S., Nabeel, M.A., Dhivya, B.: Studies on silver nanoparticles synthesized by a marine fungus, *Penicillium fellutanum* isolated from coastal mangrove sediment. *Coll. Surf. B Biointerfaces* 71(1), 133–137 (2009).

Kelly, K.L., Coronado, E., Zhao, L.L., Schatz, G.C.: The optical properties of metal nanoparticles: The influence of size, shape, and dielectric environment. *J. Phys. Chem. B* 107(3), 668–677 (2003).

Khan, A.U., Malik, N., Khan, M., Cho, M.H., Khan, M.M.: Fungi-assisted silver nanoparticle synthesis and their applications. *Bioprocess Biosyst. Eng.* 41(1), 1–20 (2018).

Khandel, P., Shahi, S.K.: Mycogenic nanoparticles and their bioprospective applications: Current status and future challenges. *J. Nanostructure Chem.* 8(4), 369–391 (2018).

Kim, J.S., Kuk, E., Yu, K.N., Kim, J.H., Park, S.J., Lee, H.J., Kim, S.H., Park, Y.K., Park, Y.H., Hwang, C.Y., Kim, Y.K., Lee, Y.S., Jeong, D.H., Cho, M.H.: Corrigendum to 'antimicrobial effects of silver nanoparticles'. *Nanomed.Nanotechnol. Biol. Med.* 1, 95–101 (2007).

Kim, K.J., Sung, W.S., Moon, S.K., Choi, J.S., Kim, J.G., Dong, G.L.: Antifungal effect of silver nanoparticles on dermatophytes. *J. Microbiol. Biotechnol.* 18(8), 1482–1484 (2008).

Kitching, M., Ramani, M., Marsili, E.: Fungal biosynthesis of gold nanoparticles: Mechanism and scale up. *Microb. Biotechnol.* 8(6), 904–917 (2015).

Kowshik, M., Ashtaputre, S., Kharrazi, S., Vogel, W., Urban, J., Kulkarni, S.K., Paknikar, K.M.: Extracellular synthesis of silver nanoparticles by a silver-tolerant yeast strain MKY3. *Nanotechnology* 14(1), 95–100 (2003).

Kowshik, M., Vogel, W., Urban, J., Kulkarni, S.K., Paknikar, K.M.: Microbial synthesis of semiconductor PbS nanocrystallites. *Adv. Mater.* 14(11), 815 (2002).

Kumar, A., Vemula, P.K., Ajayan, P.M., John, G.: Silver-nanoparticle-embedded antimicrobial paints based on vegetable oil. *Nat. Mater.* 7(3), 236 (2008).

Kumar, C.G., Poornachandra, Y., Chandrasekhar, C.: Green synthesis of bacterial mediated anti-proliferative gold nanoparticles: Inducing mitotic arrest (G2/M phase) and apoptosis (intrinsic pathway). *Nanoscale* 7(44), 18738–18750 (2015).

Kumar, S.A., Abyaneh, M.K., Gosavi, S.W., Kulkarni, S.K., Pasricha, R., Ahmad, A., Khan, M.I.: Nitrate reductase-mediated synthesis of silver nanoparticles from $AgNO_3$. *Biotechnol.Lett.* 29, 439–445 (2007).

Kumar, V., Khan, M.R., Walia, R.K.: Crop loss estimations due to plant-parasitic nematodes in major crops in India. *Natl Acad. Sci. Lett.* (2020).

Lengke, M.F., Fleet, M.E., Southam, G.: Morphology of gold nanoparticles synthesized by filamentous cyanobacteria from gold (I)-thiosulfate and gold (III)-chloride complexes. *Langmuir* 22(6), 2780–2787 (2006).

Lengke, M.F., Fleet, M.E., Southam, G.: Synthesis of palladium nanoparticles by reaction of filamentous cyanobacterial biomass with a palladium (II) chloride complex. *Langmuir* 23(17), 8982–8987 (2007).

Li, G., He, D., Qian, Y., Guan, B., Gao, S., Cui, Y., Wang, L.: Fungus-mediated green synthesis of silver nanoparticles using *Aspergillus terreus*. *Int. J. Mol. Sci.* 13(1), 466–476 (2011).

Liu, F., Wen, L., Li, Z., Yu, W., Sun, H., Chen, J.: Porous hollow silica nanoparticles as controlled delivery system for water-soluble pesticide. *Mater. Res. Bull.* 41(12), 2268–2275 (2006).

Liu, R., Lal, R.: Potentials of engineered nanoparticles as fertilizers for increasing agronomic productions. *Sci. Total Environ.* 514, 131–139 (2015).

Liu, T., Liu, J., Sonshine, D.A., Shervani, S., Hurt, R.H.: Controlled release of biologically active silver from nanosilver surfaces. *ACS Nano* 4(11), 6903–6913 (2010).

Liu, Y., Laks, P., Heiden, P.: Controlled release of biocides in solid wood. I. Efficacy against brown rot wood decay fungus (*Gloeophyllum trabeum*). *J. Appl. Polym. Sci.* 86(3), 596–607 (2002a).

Liu, Y., Laks, P., Heiden, P.: Controlled release of biocides in solid wood. III. Preparation and characterization of surfactant-free nanoparticles. *J. Appl. Polym. Sci.* 86(3), 615–621 (2002b).

Liu, Y., Laks, P., Heiden, P.: Controlled release of biocides in solid wood. II. Efficacy against *Trametes versicolor* and *Gloeophyllum trabeum* wood decay fungi. *J. Appl. Polym. Sci.* 86(3), 608–614 (2002c).

Lok, C.N., Ho, C.M., Chen, R., He, Q.Y., Yu, W.Y., Sun, H., Tam, P.K.H., Chiu, J.F., Che, C.M.: Silver nanoparticles: Partial oxidation and antibacterial activities. *J. Biol. Inorg. Chem. JBIC Publ. Soc. Biol. Inorg. Chem.* 12(4), 527–534 (2007).

Lv, Q., Zhang, B., Xing, X., Zhao, Y., Cai, R., Wang, W., Gu, Q.: Biosynthesis of copper nanoparticles using *Shewanella loihica* PV-4 with antibacterial activity: Novel approach and mechanisms investigation. *J. Hazard. Mater.* 347, 141–149 (2018).

Marcato, P.D., De Souza, G.I.H., Alves, O.L., Esposito, E., Durán, N.: Antibacterial activity of silver nanoparticles synthesized by *Fusarium oxysporum* strain. *2nd Mercosur Congress on Chemical Engineering,4th Mercosur Congress* on *Process Systems Engineering, CEP 13083-970, SP/ Brazil* (2014).

Martin-Ortigosa, S., Peterson, D.J., Valenstein, J.S., Lin, V.S.Y., Trewyn, B.J., Lyznik, A., Wang, K.: Mesoporous silica nanoparticle-mediated intracellular Cre protein delivery for maize genome editing via loxP site excision. *Plant Physiol.* 164(2), 537–547 (2014).

Mazumdar, H., Haloi, N.: A study on biosynthesis of iron nanoparticles by *Pleurotus* sp. *Res. Rev. J. Microbiol. Biotechnol.* 1(3), 39–49 (2017).

Millan, V.E.G., Rankine, C., Sanchez-Azofeifa, G.A.: Crop loss evaluation using digital surface models from unmanned aerial vehicles data. *Remote Sens.* 12(6), 981 (2020).

Mishra, S., Singh, B.R., Singh, A., Keswani, C., Naqvi, A.H., Singh, H.B.: Biofabricated silver nanoparticles act as a strong fungicide against *Bipolaris ssorokiniana* causing spot blotch disease in wheat. *PLOS ONE* 9(5) (2014).

Mittal, A.K., Chisti, Y., Banerjee, U.C.: Synthesis of metallic nanoparticles using plant extracts. *Biotechnol. Adv.* 31(2), 346–356 (2013).

Moghaddam, K.M.: An introduction to microbial metal nanoparticle preparation method. *J. Young Investig.* 19(19) (2010).

Moharrer, S., Mohammadi, B., Gharamohammadi, R.A., Yargoli, M.: Biological synthesis of silver nanoparticles by *Aspergillus flavus*, isolated from soil of Ahar copper mine. *Indian J. Sci. Technol.* 5(S3), 2443–2444 (2012).

Mondal, N.K., Chowdhury, A., Dey, U., Mukhopadhya, P., Chatterjee, S., Das, K., Datta, J.K.: Green synthesis of silver nanoparticles and its application for mosquito control. *Asian Pac. J. Trop.* 4, S204–S210 (2014).

Mousavi, S.R., Rezaei, M.: Nanotechnology in agriculture and food production. *J. Appl. Environ. Biol. Sci.* 1(10), 414–419 (2011).

Mukherjee, P., Ahmad, A., Mandal, D., Senapati, S., Sainkar, S.R., Khan, M.I., Parishcha, R., Kumar, P.V.A., Alam, M., Kumar, R., Sastry, M.: Fungus-mediated synthesis of silver nanoparticles and their immobilization in the mycelial matrix: A novel biological approach to nanoparticle synthesis. *Nano Lett.* 1(10), 515–519 (2001).

Mukhopadhyay, S.S.: Nanotechnology in agriculture: Prospects and constraints. *Nanotechnol. Sci. Appl.* 7, 63–71 (2014).

Musarrat, J., Dwivedi, S., Singh, B.R., Al-Khedhairy, A.A., Azam, A., Naqvi, A.: Production of antimicrobial silver nanoparticles in water extracts of the fungus *Amylomyces rouxii* strain KSU-09. *Bioresour. Technol.* 101(22), 8772–8776 (2010).

Nair, R., Varghese, S.H., Nair, B.G., Maekawa, T., Yoshida, Y., Kumar, D.D.: Nanoparticulate material delivery to plants. *Plant Sci.* 179(3), 154–163 (2010).

Narayanan, K.B., Sakthivel, N.: Biological synthesis of metal nanoparticles by microbes. *Adv. Coll. Interface Sci.* 156(1–2), 1–13 (2010).

Oerke, E.C.: Crop losses to pests. *J. Agric. Sci.* 144(1), 31–43 (2006).

Oliveira, H.C., Stolf-Moreira, R., Martinez, C.B.R., Grillo, R., Jesus, M.B., Fraceto, L.F.: Nanoencapsulation enhances the post emergence herbicidal activity of atrazine against mustard plants. *PLOS ONE* 10(7) (2015).

Oliveira, C.M., Auad, A.M., Mendes, S.M., Frizzas, M.R.: Crop losses and the economic impact of insect pests on Brazilian agriculture. *Crop Prot.* 56, 50–54 (2014).

Ovais, M., Khalil, A.T., Ayaz, M., Ahmad, I., Nethi, S.K., Mukherjee, S.: Biosynthesis of metal nanoparticles via microbial enzymes: A mechanistic approach. *Int. J. Mol. Sci.* 19(12), 4100 (2018).

Owaid, M.N., Ibraheem, I.J.: Mycosynthesis of nanoparticles using edible and medicinal mushrooms. *Eur. J. Nanomed.* 9(1), 5–23 (2017).

Owaid, M.N., Raman, J., Lakshmanan, H., Al-Saeedi, S.S.S., Sabaratnam, V., Abed, I.A.: Mycosynthesis of silver nanoparticles by *Pleurotus cornucopiae* var. citrinopileatus and its inhibitory effects against *Candida* sp. *Mater. Lett.* 153, 186–190 (2015).

Pohl, F., Paul., K., T.L.: The potential use of plant natural products and plant extracts with antioxidant properties for the prevention/treatment of neurodegenerative diseases: In vitro, in vivo and clinical trials. *Molecules* 23(12), 3283 (2018).

P´erez-de-Luque, A., Rubiales, D.: Nanotechnology for parasitic plant control. *Pest Manag. Sci.* 65, 540–545 (2009).

Panacek, A., Kolar, M., Vecerova, R., Prucek, R., Soukupová., Krystof, J., PetrHamal, V., Zboril, R., Kvıtek, L.: Antifungal activity of silver nanoparticles against *Candida* spp. *Biomaterials* 30(31), 6333–6340 (2009).

Philip, D.: Biosynthesis of Au, Ag and Au–Ag nanoparticles using edible mushroom extract. *Spectrochim. Acta A* 73(2), 374–381 (2009).

Philip, D.: Green synthesis of gold and silver nanoparticles using *Hibiscus rosasinensis*. *Phys. E Low Dimens. Syst. Nanostruct.* 42(5), 1417–1424 (2010).

Pitkethly, M.: Nanotechnology: past, present, and future. *Nano Today* 3(3–4), 6 (2008). https://doi.org/10.1016/s1748-0132(08)70029-x

Prasad, K., Jha, A.K., Kulkarni, A.R.: *Lactobacillus* assisted synthesis of titanium nanoparticles. *Nanoscale Res. Lett.* 2(5), 248 (2007).

Qian, Y., Yu, H., He, D., Yang, H., Wang, W., Wan, X., Wang, L.: Biosynthesis of silver nanoparticles by the endophytic fungus *Epicoccum nigrum* and their activity against pathogenic fungi. *Bioprocess. Biosyst. Eng.* 36(11), 1613–1619 (2013).

Yadav, R.N., Chitara, M.K., Zaidi, N.W., Khan, A.I., Singh, U.S., Singh, H.B.: Novel facets and challenges in the management of phytopathogens using myconanoparticles. *Int. J. Curr. Microbiol. Appl. Sci.* 7(12) (2018).

Raheman, F., Deshmukh, S., Ingle, A., Gade, A., Rai, M.: Silver nanoparticles: Novel antimicrobial agent synthesized from an endophytic fungus *Pestalotia* sp. isolated from leaves of *Syzygium cumini* (L). *Nano Biomed. Eng.* 3(3), 174–178 (2011).

Rai, M., Yadav, A., Bridge, P., Gade, A.: Myconanotechnology: A new and emerging science. *Appl. Mycol.* 14, 258–267 (2009).

Rajiv, P., Rajeshwari, S., Venckatesh, R.: Bio-fabrication of zinc oxide nanoparticles using leaf extract of *Parthenium hysterophorus* L. and its size-dependent antifungal activity against plant fungal pathogens. *Spectrochim. Acta A Mol. Biomol. Spectrosc.* 112, 384–387 (2013).

Rajput, K., Raghuvanshi, S., Bhatt, A., Rai, S.K., Agrawal, P.K.: A review on synthesis silver nanoparticles. *Int. J. Curr. Microbiol. App. Sci.* 6(7), 1513–1528 (2017).

Raliya, R., Tarafdar, J.C., Gulecha, K., Choudhary, K., Ram, R., Mal, P.: Review article; scope of nanoscience and nanotechnology in agriculture. *J. Appl. Biol. Biotechnol.* 1, 41–44 (2013).

Raman, J., Reddy, G.R., Lakshmanan, H., Selvaraj, V., Gajendran, B., Nanjian, R., Sabaratnam, V.: Mycosynthesis and characterization of silver nanoparticles from *Pleurotus djamor* var. roseus and their in vitro cytotoxicity effect on PC3 cells. *Process. Biochem.* 50(1), 140–147 (2015).

Rizzello, L., Pompa, P.R.: Nanosilver-based antibacterial drugs and devices: Mechanisms, methodological drawbacks, and guidelines. *Chem. Soc. Rev.* 43(5), 1501 (2014).

Sabir, S., Arshad, M., Chaudhari, S.K.: Zinc oxide nanoparticles for revolutionizing agriculture: Synthesis and applications, the scientific world. *Sci. World J.* Article ID 925494 (2014).

Saha, S., Sarkar, J., Chattopadhyay, D., Patra, S., Chakraborty, A., Acharya, K.: Production of silver nanoparticles by a phytopathogenic fungus *Bipolaris nodulosa* and its antimicrobial activity. *Dig. J Nanomater. Biostruct* 5(4), 887–895 (2010).

Sahoo, D., Mandal, A., Mitra, T., Chakraborty, K., Bardhan, M., Dasgupta, A.K.: Nanosensing of pesticides by zinc oxide quantum dot: An optical and electro-chemical approach for the detection of pesticides in water. *J. Agric. Food Chem.* 66(2), 414–423 (2018).

Salunke, B.K., Sawant, S.S., Lee, S.I., Kim, B.S.: Microorganisms as efficient biosystem for the synthesis of metal nanoparticles: Current scenario and future possibilities. *World J. Microbiol.Biotechnol.* 32(5), 88 (2016).

Salvadori, M.R., Lepre, L.F., Ando, R.A., do Nascimento, C.A.O., Correa, B.: Biosynthesis and uptake of copper nanoparticles by dead biomass of *Hypocrea lixii* isolated from the metal mine in the Brazilian Amazon region. *PLOS ONE* 8(11), 1–8 (2013).

Sanghi, R., Verma, P.: pH dependant fungal proteins in the 'green' synthesis of gold nanoparticles. *Adv. Mater. Lett.* 1(3), 193–199 (2010).

Saravanan, M., Barik, S.K., MubarakAli, D., Prakash, P., Pugazhendhi, A.: Synthesis of silver nanoparticles from *Bacillus brevis* (NCIM 2533) and their antibacterial activity against pathogenic bacteria. *Microb. Pathog.* 116, 221–226 (2018).

Sarsar, V., Selwal, M.K., Selwal, K.K.: Biogenic synthesis, optimization and antibacterial efficacy of extracellular silver nanoparticles using novel fungal isolate *Aspergillus fumigatus* MA. *IET Nanobiotechnol.* 10(4), 215–221 (2016).

Sastry, M., Ahmad, A., Khan, M.I., Kumar, R.: Biosynthesis of metal nanoparticles using fungi and actinomycete. *Curr. Sci.* 85(2)(2003).

Savary, S., Willocquet, L., PethybridgeetS.J., EskerskP., McRobertscRN., Nelson, A.: The global burden of pathogens and pests on major food crops. *Nat. Ecol. Evol.* 3(3), 430–439 (2019).

Sedaghat, S., Afshar, P.: Green bio-synthesis of silver nanoparticles using *Ziziphora tenuior* L. water extract. *J. Appli. Chem. Res.* 10(1), 103–109 (2016).

Senapati, S., Ahmad, A., Khan, M.I., Sastry, M., Kumar, R. Extracellular biosynthesis of bimetallic Au–Ag alloy nanoparticles. *Small*, 1(5), 517–520 (2005).

Shaligram, N.S., Bule, M., Bhambure, R., Singhal, R.S., Singh, S.K., Szakacs, G., Pandey, A.: Biosynthesis of silver nanoparticles using aqueous extract from the compactin producing fungal strain. *Process. Biochem.* 44(8), 939–943 (2009).

Sharon, M., Choudhary, A.K., Kumar, R.: Nanotechnology in agricultural diseases and food safety. *J. Phytol.* 2(4), 83–92 (2010).

Shi, C., Zhu, N., Cao, Y., Wu, P.: Biosynthesis of gold nanoparticles assisted by the intracellular protein extract of *Pycnoporus sanguineus* and its catalysis in degradation of 4-nitroaniline. *Nanoscale Res. Lett.* 10(1), 147 (2015).

Shivshankar, S., Ahmad, A., Pasricha, R., Sastry, M.: Bioreduction of chloroaurate ions by Geranium leaves and its endophytic fungus yields gold nanoparticles of different shapes. *J. Mater. Chem.* 13(7), 1822–1826 (2003).

Shojaei, T.R., Salleh, M.A.M., Tabatabaei, M., Mobli, H., Aghbashlo, M., Rashid, S.A., Tan, T.: Applications of nanotechnology and carbon nanoparticles in agriculture. *Synth. Technol. Appl. Carbon Nanomater.* 11, 247–277 (2019).

Singh, B.R., Dwivedi, S., Al-Khedhairy, A.A., Musarrat, J.: Synthesis of stable cadmium sulfide nanoparticles using surfactin produced by *Bacillus amyloliquifaciens* strain KSU-109. *Coll. Surf. B* 85(2), 207–213 (2011).

Singh, N., Jenkins, G.J., Asadi, R., Doak, S.H.: Potential toxicity of super paramagnetic iron oxide nanoparticles (SPION). *Nano. Rev.* 1(1), 5358 (2010).

Singh, P., Kim, Y.J., Singh, H., Mathiyalagan, R., Wang, C., Yang, D.C.: Biosynthesis of anisotropic silver nanoparticles by *Bhargavaea indica* and their synergistic effect with antibiotics against pathogenic microorganisms. *J. Nanomater.* 4, 1–10 (2015).

Singh, P., Kim, Y.J., Singh, H., Wang, C., Hwang, K.H., Farh, M.E.A., Yang, D.C.: Biosynthesis, characterization, and antimicrobial applications of silver nanoparticles. *Int. J. Nanomed.* 10, 2567 (2015).

Singh, P., Kim, Y.J., Wang, C., Mathiyalagan, R., Yang, D.C.: *Weissella oryzae* DC6-facilitated green synthesis of silver nanoparticles and their antimicrobial potential. *Artif. Cells Nanomed.Biotechnol.* 44(6), 1569–1575 (2016).

Singh, R., Arora, N.K.: Bacterial formulations and delivery systems against pests in sustainable agrofood production. *Food Sci.* 1, 1–11 (2016).

Singh, R., Shedbalkar, U.U., Wadhwani, S.A., Chopade, B.A.: Bacteriagenic silver nanoparticles: Synthesis, mechanism, and applications. *Appl. Microbiol. Biotechnol.* 99(11), 4579–4593 (2015).

Soltani, N.M., Shahidi, B.G.H., Khaleghi, N.: Biosynthesis of gold nanoparticles using *Streptomyces fulvissimus* isolate. *Nanomed. J.* 2(2), 153–159 (2015).

Soni, N., Prakash, S.: Antimicrobial and mosquitocidal activity of microbial synthesized silver nanoparticles. *J. Parasitol. Res.* 114(3), 1023–1030 (2015).

Sotiriou, G.A.,Pratsinis, S.E.: Antibacterial activity of nanosilver ions and particles. *Environ. Sci. Technol.* 44(14), 5649–5654 (2010).

Southam, G., Beveridge, T.J.: The in vitro formation of placer gold by bacteria. *Geochim. Cosmochim. Acta* 58(20, October), 4527–4530 (1994).

Spasova, M., Manolova, N., Naydenov, M., Kuzmanova, J., Rashkov, I.: Electrospunbiohybrid materials for plant biocontrol containing chitosan and *Trichoderma viride* spores. *J. Bioact. Compat. Polym.* 26(1), 48–55 (2011).

Subramaniyan, S.A., Sheet, S., Vinothkannan, M., Yoo, D.J., Lee, Y.S., Belal, S.A., Shim, K.S.: One-Pot Facile Synthesis of Pt nanoparticles using cultural filtrate of microgravity simulated grown *P. chrysogenum* and their activity on bacteria and cancer cells. *J. Nanosci. Nanotechnol.* 18(5), 3110–3125 (2018).

Sugunan, A., Melin, P., Schnürer, J., Hilborn, J.G., Dutta, J.: Nutrition-driven assembly of colloidal nanoparticles: Growing fungi assemble gold nanoparticles as microwires. *Adv. Mater.* 19(1), 77–81 (2007)

Tavares, T.D., Joana, C.A., Ferreira., F., Felgueiras., H.P.: Biofunctionalization of natural fiber-reinforced biocomposites for biomedical applications. *Biomolecules* 10(1), 148 (2020)

Thirumurugan, G., Shaheedha, S.M., Dhanaraju, M.D.: In vitro evaluation of anti-bacterial activity of silver nanoparticles synthesized by using *Phytophthora infestans*. *Int. J. Chem. Tech. Res.* 1(3), 714–716 (2009).

Thostenson, E.T., Li, C., Chou, T.W.: Nanocomposites in context. *Compos. Sci. Technol.* 65(3–4), 491–516 (2005).

Torney, F., Trewyn, B.G., Lin, Victor S.-Y., Wang, K.: Mesoporous silica nanoparticles deliver DNA and chemicals into plants. *Nat. Nanotechnol.* 2(5) (2007).

Tripathi, K.M., Bhati, A., Singh, A., Sonker, A.K., Sarkar, S., Sonkar, S.K.: Sustainable changes in the contents of metallic micronutrients in first generation gram seeds imposed by carbon nano-onions: Life cycle seed to seed study. *ACS Sustain. Chem. Eng.* 5(4), 2906–2916 (2017).

Ulrichs, C., Krause, F., Rocksch, T., Goswami, A., Mewis, I.: Electrostatic application of inert silica dust based insecticides onto plant surfaces. *Commun. Agric. Appl. Biol. Sci.* 71(2 Pt A), 171–178 (2006).

Usman, M., Farooq, M., Wakeel, A., Nawaz, A., Cheem, S.A., Rehamn, H., Ashraf, I., Sanaullah, M.: Nanotechnology in agriculture: Current status, challenges and future opportunities. *Sci. Total Environ.* 721, 137778 (2020).

Vanlalveni, C., Rajkumari, K., Biswas, A., Adhikari, P.P., Lalfakzuala, R., Rokhum, L.: Green synthesis of silver nanoparticles using *Nostoc linckia* and its antimicrobial activity: A novel biological approach. *BioNanoScience* 8(2), 624–631 (2018).

Varshney, R., Mishra, A.N., Bhadauria, S., Gaura, M.S.: Novel microbial route to synthesize silver nanoparticles using fungus *Hormoconis resinae. Dig. J Nanomater. Biostruct* 4(2), 349–355 (2009).

Verma, V.C., Singh, S.K., Solanki, R., Prakash, S.: Biofabrication of anisotropic gold nanotriangles using extract of endophytic *Aspergillus clavatus* as a dual functional reductant and stabilizer. *Nanoscale Res. Lett.* 6(1), 16 (2011).

Vijayabharathi, R., Sathya, A., Gopalakrishnan, S.: Extracellular biosynthesis of silver nanoparticles using *Streptomyces griseoplanus* SAI25 anditsanti-fungal activity against *Macrophomina phaseolina*, the charcoal rot pathogen of sorghum. *Biocatal. Agric. Biotechnol.* 14, 166–171 (2018).

Vijayanandan, A.S., Balakrishnan, R.M.: Biosynthesis of cobalt oxide nanoparticles using endophytic fungus *Aspergillus nidulans. J. Environ. Manag.* 218, 442–450 (2018).

Wang, C., Kim, Y.J., Singh, P., Mathiyalagan, R., Jin, Y., Yang, D.C.: Green synthesis of silver nanoparticles by *Bacillus methylotrophicus*, and their antimicrobial activity. *Artif. Cells Nanomed. Biotechnol.* 44(4), 1127–1132 (2016).

Weir, A., Westerhoff, P., Fabricius, L., Hristovski, K., Goetz, N.: Titanium dioxide nanoparticles in food and personal care products. *Environ. Sci. Technol.* 46(4), 2242–2250 (2012).

Wibowo, K.M., Sahdan, M.Z., Ramli, N.I., Muslihati, A., Rosni, N., Tsen, V.H., Saim, H., Ahmad, S.A., Sari, Y., Mansor, Z.: Detection of *Escherichia coli* bacteria in wastewater by using graphene as a sensing material. *J. Phys. Conf. Ser.* 995, 012063 (2018).

Wysocki, R., Tamás, M.J.: How *Saccharomyces cerevisiae* copes with toxic metals and metalloids. *FEMS Microbiol. Rev.* 34(6), 925–951 (2010).

Xia, Z.K., Ma, Q.H., Li, S.Y., Zhang, D.Q., Cong, L., Tian, Y.L., Yang, R.Y.: The antifungal effect of silver nanoparticles on *Trichosporon asahii. J. Microbiol. Immunol. Infect.* 49(2), 182–188 (2016).

Yadav, A., Kon, K., Kratosova, G., Duran, N., Ingle, A.P., Rai, M.: Fungi as an efficient mycosystem for the synthesis of metal nanoparticles: Progress and key aspects of research. *Biotechnol. Lett.* 37(11), 2099–2120 (2015).

Zhang, X., He, X., Wang, K., Wang, Y., Li, H., Tan, W.: Biosynthesis of size-controlled gold nanoparticles using fungus, *Penicillium* sp. *J. Nanosci. Nanotechnol.* 9(10), 5738–5744 (2009).

Zhao, X., Zhou, L., Riaz Rajoka, M.S.R., Yan, L.: Jiang, C., Shao, D., Zhu, J., Shi, J., Huang, Q., Yang, H., Jin, M.: Fungal silver nanoparticles: Synthesis, application, and challenges. *Crit. Rev. Biotechnol.* 38(6), 817–835 (2018).

8 Green Synthesized Nanoparticles

A Way to Produce Novel Nano-Biosensor for Agricultural Application

Kumar Anand and Bijayananda Panigrahi

CONTENTS

8.1 INTRODUCTION

Nanotechnology is considered as one of the important branches of science, and it is used in many fields, including physics, chemistry, pharmaceutical science, material science, medicine, and agriculture. Globally, agriculture and its related sectors face huge challenges, due to the excessive use of agrochemicals which leads to environmental pollution, ecosystem imbalance, climate change, and other ecological impacts, including water pollution, soil fertility, and an uneven distribution of rainfall. To slow down the pollution and retain the natural environment, agricultural practices must employ the recently developed nano-based agricultural technologies to overcome the huge demands of agro-based products. At present, nanotechnology is widely implemented in agriculture to make increase accuracy in producrion. Nanotechnology can be used to boost agricultural production by improving nanoformulated fertilizers and agrochemicals for crop improvement (DeRosa et al., 2010). Nano-based biosensors can be used to detect plant–pathogen interactions and for the treatment of diseases. This technology can be used to remove contaminants from the soil and water. Nanomaterials are synthesized with one or more dimensions of different sizes in the range of 100 nm or less, which is used for the post-harvest management of vegetables, flowers, and the reclamation of salt-affected soils, etc. (Auffan et al., 2009). Well-known applications of nanomaterials include for plant protection, nutrition, and farm management. Due to their small size, high surface to volume ratio, and unique optical properties, nanoparticles are well appreciated in the scientific community (Ghormade et al., 2011). Quite a variety of materials are used to synthesize nanoparticles, including metallic nanoparticles, metal oxides, ceramics, magnetic materials, semiconductors, quantum dots (Petryayeva et al., 2013), lipids, peptides, polymers (synthetic or natural), dendrimers, and emulsions (Puoci et al., 2008). Metallic nanoparticles (Verma et al., 2019), including silver nanoparticles (Singh et al., 2018), gold nanoparticles (Kumari et al., 2019; Panigrahi et al., 2019), zinc oxide nanoparticles (Lin and Xing, 2008; Milani et al., 2015), titanium dioxide nanoparticles (Raliya et al., 2015), copper oxide nanoparticles (Kumari et al., 2018), etc. are commonly synthesized and implemented for different applications. The green synthesis of metallic nanoparticles is stabilized with different biological stabilizing agents; these could be plant extracts (Jena et al., 2016), bacterial extracst, fungal extracts, algal extracts, in addition to different biomacromolecules, including carbohydrate, cellulose, starch, heparin, chitosan (Kumar et al., 2015), dextran, glucose protein, peptides (Panigrahi et al., 2018), lipids, nucleic acids, as well as small phytochemicals (Katas et al., 2018). Similarly, chemical stabilizing agents, including ascorbate, sodium borohydride, elemental hydrogen, sodium citrate, polyol process, tollens reagent, N,N-dimethylformamide (DMF), and poly (ethylene glycol)-block copolymers are utilized for metallic nanoparticle synthesis (Deepak et al., 2019). The size, shape, and functional group on the exterior of the particle can be reflected in the application and utilization of the materials (Panigrahi et al., 2019). Green synthesized nanoparticles especially have a wide range of applications relating to medicine, particularly on the cellular and molecular level (Patel et al., 2019), including tissue engineering, energy and electronics, multicolour optical coding for biological assays, apoptosis of cancer cells (Verma et al., 2020), development of nanocarriers (Rout et al., 2018) for gene delivery applications (Panigrahi et al., 2019; Kerry et al., 2018) and biomolecules, detection of biomacromolecules, and the sensing of rare earth metals (Singh et al., 2018), etc.

The cellular uptake efficiency of the nanomaterials is induced with growth and metabolic functions that vary in different plant species. The concentration of nano-TiO_2 nanoparticles was reported as highly effective for enhancing processes like sprouting and development of the plant (Zheng et al., 2005). There are several uses, benefits, and potential applications of nanoparticles in agriculture. The most valuable use is for insect and pest management (Bansal et al., 2014). To increase agricultural productivity, Ragaei et al. (2014) developed a formulation of polymeric nanomaterial-based pesticides and insecticides. Using nanoparticles, encapsulated fertilizers are used for the sustained release of nutrients and water. To develop the insect and pest-resistant varieties, many positive charged nanomaterials are used for DNA or gene delivery into the plant (Rai et al., 2012). According to Rai et al., porous hollow silica nanoparticles can be applied as effective delivery systems for the application of pest control with validamycin as a pesticide agent. This delivery system can release validamycin with a controlled release (Rai et al., 2012). Many gene delivery systems were reported for siRNA delivery and the silencing of undesired gene expression; however, due to less efficiency and toxicity issues implementation is still in process. Recently Panigrahi et al. (2019) developed a cyclic peptide-based delivery system for siRNA delivery in mammalian cells. Similarly, gold nanoparticles were also developed using the green synthesis method for the transfer of siRNA (Panigrahi et al., 2019). These two peptide-generated nanomaterials could be applicable to deliver desired gene silencing to plant cells for the knockdown of undesired characters. Carbon nanotubes (Khodakovskaya et al., 2009) and metal-containing Au, SiO_2, ZnO, and TiO_2 nanoparticles contribute to ameliorating the development of plants by enhancing their elemental uptake and use of nutrients (Khot et al., 2012). However, the actual effect of nanomaterials was dependent on their configuration, concentration, size of the particles, surface charge on the materials, and physical-chemical properties. (Ma et al., 2010).

Only limited implementation of nanotechnology is reported in the area of food and agriculture, due to toxicity and hazardous chemical-based nanoparticles. Therefore, green synthesized nanomaterials were developed for eco-friendly and sustainable applications in the field of agriculture (Hossain et al., 2020). Presently, there is a need to shift towards green nanotechnology at a quicker speed for the implementation of its functions. The progress of the techno-driven agricultural system with the use of engineered smart green synthesized nanotools could be an outstanding approach to create a revolution in agricultural farming and thus moderate the influence of current agriculture on the surroundings, as well as to improve the value and quantity of yields (Liu and Lal, 2015). This technology is implemented in the agriculture and food industries for new nanomaterials designed for the monitoring of rapid disease diagnostics and improving the ability of plants to absorb nutrients, among others (Kerry et al., 2018; Kerry et al., 2019). The major benefits of using nano-based materials in agriculture include nanofertilizer and nanopesticides, which are used to increase productivity and protection against several insects, pests, and microbial diseases. Besides, it retains the fertility of soils, and is chemical contamination free to water in the agricultural field. Apart from these, in all the applications, nanosensors play important roles in the diagnosis of the plant pathogens, to maintain food quality estimation, environmental monitoring, metabolic complications detection of food spoilage, identification of soil contamination, detection as well as monitoring of soil quality in the agricultural field. Thus, they maintain the diversity

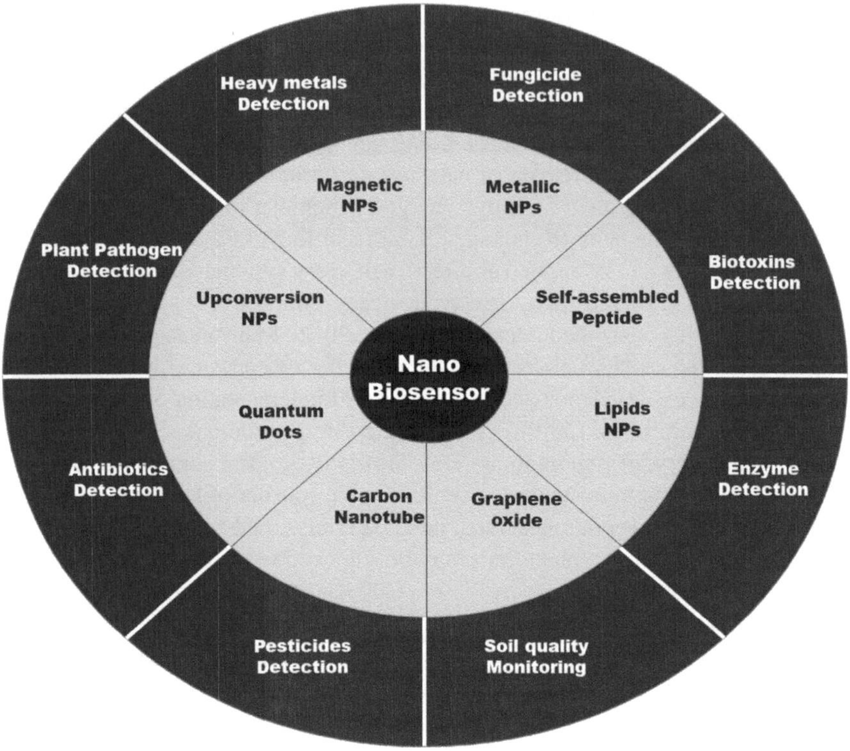

FIGURE 8.1 Different types of nano-biosensors and their potentials applications in agriculture.

of soil microorganisms as well help the healthy plants to grow and produce higher yields (Rout et al., 2019). A pictorial representation of a nanosensor and its potential applications is shown in Figure 8.1. Different nano-biosensors, including metallic nanoparticles, magnetic nanoparticles, upconversion nanoparticles, quantum dots, carbon nanotube, graphene oxide, lipids nanoparticles, self-assembled peptide nanostructure, are characterized for different biosensing applications in agriculture, such as monitoring of soil quality, detection of the pathogen, heavy metals, enzyme, antibiotics, fungicide, pesticides, biotoxins, and different agrochemicals which are harmful to mankind. Recently Mandal et al. (2018) reported on green synthesized nanoparticles and their potential nanosensor application with metal nanoparticles, including silver and gold. Green NPs have presented some highly exciting ingredients for the improvement of sensing phenomenon in the form of fluorescence and colourimetric detection using quantum dots and metal nanoparticles. This chapter is focused on green synthesis nanomaterials and their biosensor activities to increase the quality of farming, disease-free plant growth, diagnosis of diseased plants, and implementation of a suitable ecosystem for plant development to increase the yield without hostile ecological impact.

8.2 DIFFERENT TYPE OF NANO-BIOSENSOR

8.2.1 Quantum Dots

Semiconductor QDs are in the highest quantum yield with molar extinction coefficients, and have broad absorption spectra, symmetric fluorescence spectra spanning the UV to near-IR regions, great effective excitation, high stability to photoquenching, and a special resistance to photochemical deprivation. Thus, these have outstanding fluorescence, quantum confinement of charge transporter materials, and they possess size-tunable band energy. QDs are very small nanomaterials of less than 10 nm in size. QDs were reported to have rare spectral properties related to classical organic dyes; thus, recently, QDs were used as a new form of fluorophore in bioimaging and biosensing applications. They evolved as a commonly used tool in plant pathology. Indeed, they are successfully used for the regular monitoring of pathogens and beneficial organisms in a wide spectrum of microorganisms and substrates due to the gene profiling of wide microorganisms (Lévesque, 2001). Plant viruses are a major class of pathogen which can be diagnosed with QD-based biosensors. This technology has arisen as a novel class of sensors and is expected to open opportunities in plant virus detection; however, there have been very few practical applications. Some of the reported QDs, including CdSe-ZnS core, and PbS nanoparticles, were tested for the detection of CPMV2 and CaMV viruses, respectively. Other QDs, including InP, CdTe, Au NPs/QD, were reported for the detection of the CTV virus. ZnO films were implemented for the detection of the GVA virus. Similarly, Fe_3O_4/SiO_2 NPs and SiO2/UC NPs were investigated to detect ToRSV, BPMV, and ArMV plant viruses (Hong and Lee, 2018). Above all, the QD-based detection methods will be a great future tool to detect pesticides, herbicides, heavy metals, antibiotics, etc., in the new era of agriculture. Beside pathogen detection, QDs can detect harmful chemicals, such as pesticides, herbicides, enzymes, antibiotics, etc. The mechanisms of detection using QDs are represented in Figure 8.2.

8.2.2 Silver Nanoparticles

Silver nanoparticles have multiple uses, especially in the health sector, food storage, textile coatings, and several environmental applications. The extended limit of silver nanoparticle implementation and their toxicity has not been so clear. Several different accredited bodies have already approved products made from silver nanoparticles. Not only as a medicine but also silver nanoparticles to be used at home. Some of the most common applications of silver nanoparticles are seen for baby pacifiers, acne creams, computer keyboards, clothing (like socks and athletic wears) that protects against the emission of body odour in addition to deodorizing sprays (Calderón-Jiménez et al., 2017). Silver nanoparticles synthesized using green methods have been reported to have biomedical applications as well as for controlling pathogenic microbes. Silver nanoparticles were reported as antimicrobial agents against a wide range of over 650 microbes from diverse organisms, including Gram-positive, Gram-negative bacteria, fungi, or viruses. Recently, the selectivity of the clove-Ag NP-based sensor toward

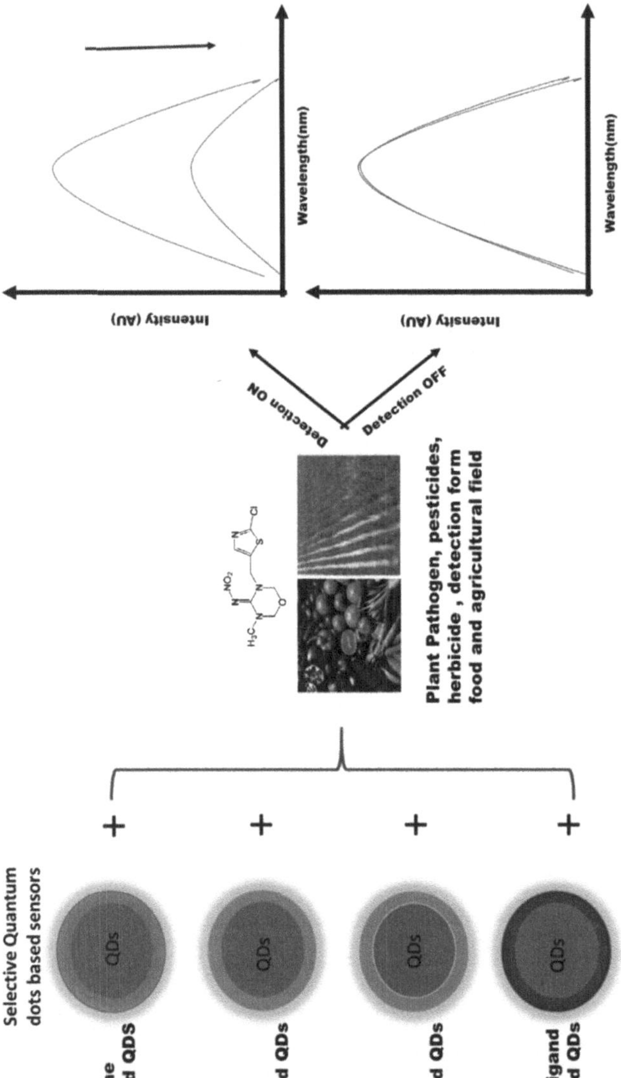

FIGURE 8.2 Pictorial representation of QDs based sensing.

vinclozolin (VIN), which is a common dicarboximide fungicide, was reported by Hussain et al. (2019) where a Ag NP-based nano-biosensor sensed VIN efficiently in the range of a 2–16 μM concentration with $R^2 = 0.997$. The AgNP-based sensor was green synthesized, simple, sensitive, selective, economical, and novel materials could detect trace amounts of VIN with a limit of detection (LOD) = 21 nM (Hussain et al., 2019). Therefore, this shows the potential for sensing many other small agrochemicals that are known as a major pollutants to water and the environment.

8.2.3 Gold Nanoparticles

Gold nanoparticles have a wide range of applications, including in biomedicine, the food industry, water purification, and agriculture. Gold nanoparticles are known to be biocompatible as compared to conventional reduction methods, which may compromise their advantages. Therefore, green synthesized Au NPs have much more potential in different fields because of their unique properties, including gold nanoparticle solution and colour absorption wavelength. Au NP solution will change when sensing a molecule or interacting with an agrochemical, which is related to the particle size and inter-particle spacing. These unique properties help with the exposure of hazardous and carcinogenic chemicals, including heavy metals, banned additives, pesticide residues, and biotoxins in food and agriculture. The gold nanoparticle has been reported as a potential biosensor for heavy metals, pesticides, veterinary drugs, and antibiotics, where different heavy metals, including Pd^{2+}, Pb^{2+}, Cd^{2+}, Hg^{2+}, Cr^{3+}, As^{3+}, Hg^{2+}, Pb^{2+}, were detected in a concentration of 4230 nM, 1.54 nM, 3.5 nM, 10 nM, 10 nM, 6.0 nM, and 2.38 nM, respectively. Different pesticides, such as atrazine, dithiocarbamate, triadimenol, imidacloprid, cyromazine, chlorsulfuron, and metsulfuron-methyl, were reported in minimum concentrations of 165 nM, 1.05 nM, 182 nM, 500 nM, 12 nM, 70 nM, and 131 nM, respectively. Similarly, veterinary drugs and antibiotics, including clenbuterol, ractopamine, salbutamol, sulfadimethoxine, ampicillin, oxytetracycline and melamine, were reported in minimum concentrations of 0.23 nM, 0.43 nM, 100 nM, 100nM, 179 nM, 4.9 nM, 25 nM, 5.1 nM, and 238 nM, respectively (Liu et al., 2018). The above data were all detected using colourimetric strategy-based sensing methods. Besides being used to detect small agrochemicals, Au NPs could be used as sensors for enzymes, pathogens, and many others in agricultural applications.

8.3 NANO-BIOSENSORS BASED NANOCHIP

Recently, researchers have become interested in a combined methodology of integrating nanoscience, electronics, and computers to create nano-biosensors with extraordinary sensing capabilities that show unprecedented spatial and temporal resolution and reliability. Nanosensors are designed with immobilized bioreceptor probes to detect or target a specific molecule for sensing. This is usually built on the nanoscale to obtain and investigate the data at the atomic level. Biosensing is one of the foremost modern topics and has been stimulating scientific studies for a long time. The reason for this is that biological entities are very complex and are directly related to the existence of a healthy environment. The enormous potential of biosensors have gained numerous applications, including in diverse food quality estimations, environmental monitoring, and the diagnosis of clinical and metabolic complications. Several nanomaterials, like nanoparticles, nanorods, nanotubes, and nanowires, have enabled faster detection (Gouda et al., 2019). Because of its unique properties, including high electrical conductivity, better shock bearing ability, and sensitive responses, such as piezoelectric and multipurpose colour-based detection mechanisms, biosensors are commonly called biological transducers. Additionally, the electrochemical character of nano-biosensors was reported as a wonderful asset for biosensor technology. Nano-biosensors were reported to sense biologically specific materials, including antibodies, proteins, enzymes, immunological molecules, and others. Nano-biosensors were measured as a biochemical or biological event using optical, electronic, or magnetic technology through a compact probe (You et al., 2009; Velasco-Garcia, 2009; Fan et al., 2008; Khanna, 2008). The probe was applied to detect analytes like urea, glucose, pesticides, etc., and various reports also indicate that they are used to monitor metabolites and detect various microorganisms/pathogens. Commercially available biosensors, including ethanol, glycerol, glucose, methanol, lactate for food quality, are available from the Analox Instruments company from the UK and USA. Other materials, including water-soluble vitamins, veterinary residues (chemical), and mycotoxins are marketed by Biacore AB, which is a Swedish-based company. Microorganisms are detected by biosensors which are provided by a French company, Biomerieux. Biochemical oxygen demand is provided by biosensors S.L., a Spanish-based company. To detect, alcohol, allergens, acids, and sulphites, Biosentec a French-based company has developed a new biosensor. Similarly, a Sweden based company Biotech IgG has developed a cheap based biosensor for allergens, vitamins, and microorganisms. To detect microorganisms and drug residue, different biosensors are available from the Eurofins company in Luxembourg. There are many other companies on the market for agriculture-based sensors to enable a better future for agriculture.

8.4 ROLE OF NANO-BIOSENSORS IN AGRICULTURE

Presently, nanomaterial-based biosensors exhibit fascinating prospects over traditional biosensors. Nano-biosensors have marked advantages, such as enhanced detection sensitivity/specificity and they possess great potential for applications in various different fields, including environmental and bioprocess control, the quality control of food, agriculture, biodefense, and, particularly, medical. However, this book chapter is concerned with the role of nanomaterial-based sensors in agriculture and agri-products, which is summarized in Table 8.1. The following are the key applications of nano-biosensors in the agriculture and food industry:

8.4.1 Detection and Sensing Application for the Agri-Food Industry

Nanosensors have potential applications in the food industry, especially in agriculture food processing, food packaging, food transport, and nutrition (Sekhon, 2014). They are merged with the packaging of food products for the monitoring of

TABLE 8.1
The Following Table Illustrates the Nanotechnology-Based Sensor and Its Applications in Agriculture

Nanomaterials	Synthesis	Shape and Size	Applications	Limit of Detection	References
aptasensorbased biosensor	Citrate-stabilized Au NPs with negative surface charges were synthesized *via* an easy one-pot reaction.	Spherical in shapc (36 nm)	Ametryn	0.15 µg/L	Qu et al., 2020
Au NPs	Au NPs-FMDV	12–13 nm in diameter	Detect foot and mouth disease virus (FMDV)	94.5% efficiency	Hamdy et al., 2018
Ag NPs	Green synthesized Ag NPs	Spherical (4 nm)	Detection of kanamycin and sulphide ions respectively	0.96 µM and 1.62 µM,	Singh et al., 2018
Au NPs	Citrate-stabilized Au NPs	Spherical shape (~8 nm)	Bacterial	~100 cells	Peng and Chen, 2018
QDs	Double QDs (ZnCdSe and CdSe)	–	Paraquat Dursban Dipterex Methyl thiophanate Cartap	20 nM	Fan et al., 2016
QDs	RFCdTe@ SiO2-Au NPs	Spherical in shape (13 nm)	Detection of Parathion-methyl	0.0684 nM	Yan et al., 2015
Au NPs	BSA conjugated gold nanoparticles	Spherical shape (32 nm)	aflatoxin M1 (AFM)	37.7 ng/L for AFM in PbS and 27.5 ng/L in skim milk	Zhang et al., 2013
Au NPs	Aptamer–polymeric microsphere–Au NPs	Spherical structure (40 nm)	Malathion	3.3 µg /mL	Barahona et al., 2013

(*Continued*)

TABLE 8.1 (CONTINUED)
The Following Table Illustrates the Nanotechnology-Based Sensor and Its Applications in Agriculture

Nanomaterials	Synthesis	Shape and Size	Applications	Limit of Detection	References
QDs	TGA-CdTe	Spherical	Fenamithion Acetamiprid	12 nM 34 nM	Qu et al., 2009
Lipids nanoparticles	liposome biosensors	Spherical in shape (300 ± 4 nm)	organophosphorus pesticides dichlorvos and paraoxon	2.0 pM 6.7 pM	Vamvakaki et al., 2007

chemicals released through food spoilage and serves as an electronic tongue (e.g. bitter, sweet, salty, umami, and sour detection), or nose (e.g. wine characterization). Portable nanosensors are used to detect chemicals, pathogens, and toxins in food. In addition to this, DNA biochips are used to detect harmful microbes and to determine the presence of different kinds of harmful pathogen in animal meat or sea fish, or fungus affected fruit (Bülbül et al., 2015). Nanosensors are used to detect food-borne contaminants or monitor environmental conditions on farms. These are used to selectively bind and remove chemicals or pathogens from food.

8.4.2 Aptasensor Based Detection in Various Agricultural Application

Nano-based biosensors are used for monitoring soil conditions (e.g. moisture, soil pH) and a wide variety of pesticides, herbicides, fertilizers, insecticides, pathogens, and crop growth as well (Kerry et al., 2017). Aptasensors are used for the determination of pesticides and insecticides (e.g. phorate, acetamiprid, isocarbophos). These are also used for the determination of antibiotics, drugs, and their residues (e.g. cocaine, oxytetracycline, tetracycline, kanamycin). In addition, to determine the heavy metals including Hg++, As3+, Cu++, aptasensors were reported as efficient nano-biosensors (Verdian, 2018). Similarly, aptasensors were also reported for the determination of microbial toxins (e.g. OTA, Fumonisin B1). They were also reported for the determination of microbial cells (e.g. *Salmonella typhimurium*, *Escherichia coli*, and *Listeria monocytogenes*).

8.4.3 Detection of Environmental Pollutants and Agrochemicals

The detection and removal of an increasing number of environmental pollutants are major challenges in day to day life, especially in the agrochemical industry. Sensing, diagnostic, and remediation devices for on-site implementation may permit close observation of environmental conditions. Therefore, they will increase agricultural productivity while reducing the use of agrochemicals from a precision farming perspective. Nanomaterials are promising materials for overall strategies to detect and remediate environmental contaminants (Iavicoli et al., 2017) which will help to improve positive health conditions for all ecosystems.

8.4.4 Electromechanical Nanosensors

Electromechanical nanosensors are used to detect ethylene which is prominently used for the ripening of raw fruits. Therefore smart-sensor technology is used to monitor the quality of grain, dairy products, fruit, and vegetables in a storage environment to detect the source and the type of spoilage (Popescu et al., 2010; Baruwati et al., 2009).

8.4.4.1 Nanosensors for Soil Quality and Disease Diagnosis

Nanosensor valuation is used to diagnose soil disease, which is caused by contaminating soil microorganisms, such as bacteria, viruses, and fungi, via the measurable quantity of differential oxygen consumption through the respiration of 'beneficial microbes' and 'harmful microbes' in the soil. Therefore, it is to be highlighted that the biosensor offers an inventive method of identifying soil conditions based on a semi-quantitative approach (Rai et al., 2012).

8.4.5 Detect Contaminants and Other Molecules

Numerous nanomaterials are designed to detect pollutants, toxins, pests, nutrient content, and plant stress caused due to drought, temperature, or pressure that are potentially helpful to farmers and agro-business persons for improving the quality and quantity of the yield. Organophosphorus pesticides, such as dichlorvos and paraoxon, at very low levels, were monitored by liposome-based biosensors (Vamvakaki and Chaniotakis, 2007).

8.4.6 Tool for Analysis in Food Products

Biosensor-based analysis is used in the food industry where it has several applications, including vitamin analysis, antibiotic detection, detection of food spoilage, detection of contaminant microbes, etc. Nano-based biosensors were used in surface plasmon resonance (SPR) biosensors for monitoring the interactions of a specific binding protein with the vitamin immobilized on a CM5 sensor chip. Similarly, the quality of food materials, especially milk, is very important in today's scenario, as milk is a good source of calcium and protein. However, many articles are published concerning cow's milk containing different types of antibiotics. The excessive use of antibiotics in cows for disease control leads to the secretion of antibiotics in the milk, which is harmful to mankind because long-term antibiotic consumption leads to multidrug resistance in the human body. Therefore to check the quality of the milk, Singh et al. (2018) recently reported a silver nanoparticle-based nanosensor for the detection kanamycin antibiotics in milk in trace amounts.

8.4.7 Detection of Food Spoilage

To detect histamine levels in tiger prawns (*Penaeus monodon*), an amperometric biosensor was developed using immobilized enzyme diamine oxidase (DAO) for the rapid monitoring of histamine (Keow et al., 2007). Similarly, a potentiometric biosensor could analyze isocitrate using a selective electrode and enzyme immobilization in a flow injection analysis (FIA).

8.4.8 Detection of a Plant Pathogen

Immunobiosensors were developed for the detection of different plant pathogens based on surface immobilization with monoclonal antibodies onto indium tin oxide (ITO) electrodes for *Escherichia coli* O157:H7 detection. Nanochips were developed based on the microarrays concept, which contains fluorescent-labelled oligo-attached probes where the hybridization principle is applied. The nanochips are known for their sensitivity and specificity for detecting single nucleotide changes in bacteria and viruses (López et al., 2009). Yao et al. (2009) utilized fluorescence attached to silica nanoparticles in a mixture of antibodies to detect *Xanthomonas axonopodis* pv. *vesicatoria* that causes bacterial spot disease in plants. Nanogold-based immunobiosensors are also being developed and could detect Karnal bunt disease in wheat (*Tillitia indica*) using SPR. Application of SPR sensors in seed certification and plant quarantines can prove highly effective and accurate for the detection of infection in plants.

8.5 CONCLUSION AND FUTURE PERSPECTIVES

Nanotechnology has a marked impact on health, energy, economy, and the environment by improving screening processes. Recently, a wide range of potential applications for nanotechnology have been proposed for use in agriculture, leading to intense research on both academic and industrial levels (Chen and Yada, 2011; Dasgupta et al., 2015). Nanotechnology can provide a green, efficient, and eco-friendly strategy for insect pest management in agriculture. Nanoparticles have great potential as 'magic bullets' loaded with fungicides, fertilizers, nutrients, herbicides, or nucleic acids for targeting specific plant tissues to release their charge to the desired part of the plant and achieve the desired results. Biotechnological advances to the speed and accuracy of diagnosis tools with the use of nanomaterials, including QDs, Au NPs, CDs, etc., have a great and bright future in the modern agri-biotech industry. Nanotechnology has proved to be a very significant blessing for the development of biosensors. It has revolutionized biological molecule detection. The future of nano-biosensors will come with quicker, smarter, cheaper, and more user friendly biomaterials.

REFERENCES

Auffan, M., Rose, J., Bottero, J. Y., Lowry, G. V., Jolivet, J. P., & Wiesner, M. R. (2009). Towards a definition of inorganic nanoparticles from an environmental, health and safety perspective. *Nature Nanotechnology*, 4(10), 634–641.

Bansal, P., Duhan, J. S., & Gahlawat, S. K. (2014). Biogenesis of nanoparticles: A review. *African Journal of Biotechnology*, 13(28), 2778–2785.

Barahona, F., Bardliving, C. L., Phifer, A., Bruno, J. G., & Batt, C. A. (2013). An aptasensor based on polymer-gold nanoparticle composite microspheres for the detection of Malathion using surface-enhanced Raman spectroscopy. *Industrial Biotechnology*, 9(1), 42–50.

Baruwati, B., Polshettiwar, V., & Varma, R. S. (2009). Glutathione promoted expeditious green synthesis of silver nanoparticles in water using microwaves. *Green Chemistry*, 11(7), 926–930.

Bülbül, G., Hayat, A., & Andreescu, S. (2015). Portable nanoparticle-based sensors for food safety assessment. *Sensors*, 15(12), 30736–30758.

Calderón-Jiménez, B., Johnson, M. E., Montoro Bustos, A. R., Murphy, K. E., Winchester, M. R., & Vega Baudrit, J. R. (2017). Silver nanoparticles: Technological advances, societal impacts, and metrological challenges. *Frontiers in Chemistry*, 5, 6.

Chen, H., & Yada, R. (2011). Nanotechnologies in agriculture: New tools for sustainable development. *Trends in Food Science and Technology*, 22(11), 585–594.

Dasgupta, N., Ranjan, S., Mundekkad, D., Ramalingam, C., Shanker, R., & Kumar, A. (2015). Nanotechnology in agro-food: from field to plate. *Food Research International*, 69, 381–400.

Deepak, P., Amutha, V., Kamaraj, C., Balasubramani, G., Aiswarya, D., & Perumal, P. (2019). Chemical and green synthesis of nanoparticles and their efficacy on cancer cells. In: *Green Synthesis, Characterization and Applications of Nanoparticles* (pp. 369–387). Elsevier.

DeRosa, M. C., Monreal, C., Schnitzer, M., Walsh, R., & Sultan, Y. (2010). Nanotechnology in fertilizers. *Nature Nanotechnology*, 5(2), 91–91.

Fan, X., White, I. M., Shopova, S. I., Zhu, H., Suter, J. D., & Sun, Y. (2008). Sensitive optical biosensors for unlabeled targets: A review. *Analytic Achimica Acta*, 620(1–2), 8–26.

Fan, Y., Liu, L., Sun, D., Lan, H., Fu, H., Yang, T., She, Y., & Ni, C. (2016). 'Turn-off' fluorescent data array sensor based on double quantum dots coupled with chemometrics for highly sensitive and selective detection of multicomponent pesticides. *Analytic Achimica Acta*, 916, 84–91.

Ghormade, V., Deshpande, M. V., & Paknikar, K. M. (2011). Perspectives for nano-biotechnology enabled protection and nutrition of plants. *Biotechnology Advances*, 29(6), 792–803.

Gouda, S., Kerry, R. G., Das, G., & Patra, J. K. (2019). Synthesis of nanoparticles utilizing sources from the mangrove environment and their potential applications: An overview. In: *Nanomaterials in Plants, Algae and Microorganisms* (pp. 219–235). Academic Press.

Hamdy, M. E., Del Carlo, M., Hussein, H. A., Salah, T. A., El-Deeb, A. H., Emara, M. M., Pezzoni, G., & Compagnone, D. (2018). Development of gold nanoparticles biosensor for ultrasensitive diagnosis of foot and mouth disease virus. *Journal of Nanobiotechnology*, 16(1), 48.

Hong, S., & Lee, C. (2018). The current status and future outlook of quantum dot-based biosensors for plant virus detection. *The Plant Pathology Journal*, 34(2), 85.

Hossain, A., Kerry, R. G., Farooq, M., Abdullah, N., & Islam, M. T. (2020). Application of nanotechnology for sustainable crop production systems. In: *Nanotechnology for Food, Agriculture, and Environment*. Springer, Cham, (135–159).

Hussain, M., Nafady, A., Avcı, A., Pehlivan, E., Nisar, J., Sherazi, S. T. H., Sherazi, S. T. H., Balouch, A., Shah, M. R., Almaghrabi, O. A., & Ul-Haq, M. A. (2019). Biogenic silver nanoparticles for trace colorimetric sensing of enzyme disrupter fungicide vinclozolin. *Nanomaterials*, 9(11), 1604.

Iavicoli, I., Leso, V., Beezhold, D. H., & Shvedova, A. A. (2017). Nanotechnology in agriculture: Opportunities, toxicological implications, and occupational risks. *Toxicology and Applied Pharmacology*, 329, 96–111.

Jena, S., Singh, R. K., Panigrahi, B., Suar, M., & Mandal, D. (2016). Photo-bioreduction of Ag+ ions towards the generation of multifunctional silver nanoparticles: Mechanistic perspective and therapeutic potential. *Journal of Photochemistry and Photobiology, Part B: Biology*, 164, 306–313.

Katas, H., Moden, N. Z., Lim, C. S., Celesistinus, T., Chan, J. Y., Ganasan, P., & Abdalla Suleman Ismail, S. (2018). Biosynthesis and potential applications of silver and gold nanoparticles and their chitosan-based nanocomposites in nanomedicine. *Journal of Nanotechnology*, 2018.

Keow, C. M., Bakar, F. A., Salleh, A. B., Heng, L. Y., Wagiran, R., & Bean, L. S. (2007). An amperometric biosensor for the rapid assessment of histamine level in tiger prawn (*Penaeus monodon*) spoilage. *Food Chemistry*, 105(4), 1636–1641.

Kerry, R. G., Gouda, S., Das, G., Vishnuprasad, C. N., & Patra, J. K. (2017). Agricultural nanotechnologies: Current applications and future prospects. In: *Microbial Biotechnology* (pp. 3–28). Springer, Singapore.

Kerry, R. G., Mahapatra, G. P., Patra, S., Sahoo, S. L., Pradhan, C., Padhi, B. K., & Rout, J. R. (2018). Proteomic and genomic responses of plants to nutritional stress. *Biometal: An International Journal on the Role of Metal Ions in Biology, Biochemistry, and Medicine*, 31(2), 161–187.

Kerry, R. G., Rout, J. R., Das, G., Fraceto, L. F., Paramithiotis, S., & Patra, J. K. (2019). Applications of nanotechnology in food and agriculture. In: *Food Molecular Microbiology* (pp. 223–249). CRC Press.

Khanna, V. K. (2008). New-generation nano-engineered biosensors, enabling nanotechnologies and nanomaterials. *Sensor Review*.

Khodakovskaya, M., Dervishi, E., Mahmood, M., Xu, Y., Li, Z., Watanabe, F., & Biris, A. S. (2009). Carbon nanotubes are able to penetrate plant seed coat and dramatically affect seed germination and plant growth. *ACS Nano*, 3(10), 3221–3227.

Khot, L. R., Sankaran, S., Maja, J. M., Ehsani, R., & Schuster, E. W. (2012). Applications of nanomaterials in agricultural production and crop protection: A review. *Crop Protection*, 35, 64–70.

Kumar, S., Chauhan, N., Gopal, M., Kumar, R., & Dilbaghi, N. (2015). Development and evaluation of alginate–chitosan nanocapsules for controlled release of acetamiprid. *International Journal of Biological Macromolecules*, 81, 631–637.

Kumari, P., Panda, P. K., Jha, E., Pramanik, N., Nisha, K., Kumari, K., Soni, N., Mallick, M. A., & Verma, S. K. (2018). Molecular insight to in vitro biocompatibility of phytofabricated copper oxide nanoparticles with human embryonic kidney cells. *Nanomedicine*, 13(19), 2415–2433.

Kumari, S., Panda, P. K., Pramanik, N., Verma, S. K., Mallick, M. A., & Mallick, M. A. (2019). Molecular aspect of phytofabrication of gold nanoparticle from Andrographispeniculata photosystem II and their in vivo biological effect on embryonic zebrafish (*Danio rerio*). *Environmental Nanotechnology, Monitoring and Management*, 11, 100201.

Lévesque, C. A. (2001). Molecular methods for detection of plant pathogens – What is the future? *Canadian Journal of Plant Pathology*, 23(4), 333–336.

Lin, D., & Xing, B. (2008). Root uptake and phytotoxicity of ZnO nanoparticles. *Environmental Science and Technology*, 42(15), 5580–5585.

Liu, G., Lu, M., Huang, X., Li, T., & Xu, D. (2018). Application of gold-nanoparticle colorimetric sensing to rapid food safety screening. *Sensors*, 18(12), 4166.

Liu, R., & Lal, R. (2015). Potentials of engineered nanoparticles as fertilizers for increasing agronomic productions. *Science of the Total Environment*, 514, 131–139.

López, M. M., Llop, P., Olmos, A., Marco-Noales, E., Cambra, M., & Bertolini, E. (2009). Are molecular tools solving the challenges posed by detection of plant pathogenic bacteria and viruses? *Current Issues in Molecular Biology*, 11(1), 13–45.

Ma, X., Geiser-Lee, J., Deng, Y., & Kolmakov, A. (2010). Interactions between engineered nanoparticles (ENPs) and plants: Phytotoxicity, uptake and accumulation. *Science of the Total Environment*, 408(16), 3053–3061.

Mandal, D., Mishra, S., & Singh, R. K. (2018). Green synthesized nanoparticles as potential nanosensors. In: *Environmental, Chemical and Medical Sensors* (pp. 137–164). Springer, Singapore.

Milani, N., Hettiarachchi, G. M., Kirby, J. K., Beak, D. G., Stacey, S. P., & McLaughlin, M. J. (2015). Fate of zinc oxide nanoparticles coated onto macronutrient fertilizers in an alkaline calcareous soil. *PLoS One*, 10(5), e0126275.

Panigrahi, B., Mishra, S., Singh, R. K., Siddiqui, N., Bal, R., & Mandal, D. (2019). Peptide generated anisotropic gold nanoparticles as efficient siRNA vectors. *International Journal of Pharmaceutics*, 563, 198–207.

Panigrahi, B., Singh, R. K., Mishra, S., & Mandal, D. (2018). Cyclic peptide-based nanostructures as efficient siRNA carriers. *Artificial Cells, Nanomedicine, and Biotechnology*, 46 (Suppl. 3), S763–S773.

Patel, P., Kumari, P., Verma, S. K., & Mallick, M. A. (2019). Cellular and molecular impact of green synthesized silver nanoparticles. In: *Silver Nanoparticles-Health and Safety*. IntechOpen.

Peng, H., & Chen, I. A. (2018). Rapid colorimetric detection of bacterial species through the capture of gold nanoparticles by chimeric phages. *ACS Nano*, 13(2), 1244–1252.

Petryayeva, E., Algar, W. R., & Medintz, I. L. (2013). Quantum dots in bioanalysis: A review of applications across various platforms for fluorescence spectroscopy and imaging. *Applied Spectroscopy*, 67(3), 215–252.

Popescu, M., Velea, A., & Lőrinczi, A. (2010). Biogenic production of nanoparticles. *Digest Journal of Nanomaterials and Biostructures (DJNB)*, 5(4).

Puoci, F., Iemma, F., Spizzirri, U. G., Cirillo, G., Curcio, M., & Picci, N. (2008). Polymer in agriculture: A review. *American Journal of Agricultural and Biological Sciences*, 3(1), 299–314.

Qu, F., Zhou, X., Xu, J., Li, H., & Xie, G. (2009). Luminescence switching of CdTe quantum dots in presence of p-sulfonatocalix [4] arene to detect pesticides in aqueous solution. *Talanta*, 78(4–5), 1359–1363.

Qu, Y., Qian, H., Mi, Y., He, J., Gao, H., Lu, R., Zhang, S., & Zhou, W. (2020). Rapid determination of the pesticide ametryn based on a colorimetric aptasensor of gold nanoparticles. *Analytical Methods*, 12(14), 1919–1925.

Ragaei, M., & Sabry, A. K. H. (2014). Nanotechnology for insect pest control. *International Journal of Science, Environment and Technology*, 3(2), 528–545.

Rai, M., & Ingle, A. (2012). Role of nanotechnology in agriculture with special reference to management of insect pests. *Applied Microbiology and Biotechnology*, 94(2), 287–293.

Rai, V., Acharya, S., & Dey, N. (2012). Implications of nanobiosensors in agriculture.

Raliya, R., Nair, R., Chavalmane, S., Wang, W. N., & Biswas, P. (2015). Mechanistic evaluation of translocation and physiological impact of titanium dioxide and zinc oxide nanoparticles on the tomato (*Solanum lycopersicum* L.) plant. *Metallomics: Integrated Biometal Science*, 7(12), 1584–1594.

Rout, G. K., Shin, H. S., Gouda, S., Sahoo, S., Das, G., Fraceto, L. F., & Patra, J. K. (2018). Current advances in nanocarriers for biomedical research and their applications. *Artificial Cells, Nanomedicine, and Biotechnology*, 46 (Suppl. 2), 1053–1062.

Rout, J. R., Kerry, R. G., Panigrahi, D., Sahoo, S. L., Pradhan, C., Ram, S. S., Chakraborty, A., & Sudarshan, M. (2019). Biochemical, molecular, and elemental profiling of Withaniasomnifera L. with response to zinc stress. *Environmental Science and Pollution Research International*, 26(4), 4116–4129.

Sekhon, B. S. (2014). Nanotechnology in agri-food production: An overview. *Nanotechnology, Science and Applications*, 7, 31.

Singh, R. K., Mishra, S., Jena, S., Panigrahi, B., Das, B., Jayabalan, R., Parhi, P. K., & Mandal, D. (2018). Rapid colorimetric sensing of gadolinium by EGCG-derived AgNPs: The development of a nanohybridbioimaging probe. *Chemical Communications*, 54(32), 3981–3984.

Singh, R. K., Panigrahi, B., Mishra, S., Das, B., Jayabalan, R., Parhi, P. K., & Mandal, D. (2018). pH triggered green synthesized silver nanoparticles toward selective colorimetric detection of kanamycin and hazardous sulfide ions. *Journal of Molecular Liquids*, 269, 269–277.

Vamvakaki, V., & Chaniotakis, N. A. (2007). Pesticide detection with a liposome-based nano-biosensor. *Biosensors and Bioelectronics*, 22(12), 2848–2853.

Velasco-Garcia, M. N. (2009, February). Optical biosensors for probing at the cellular level: A review of recent progress and future prospects. In: *Seminars in Cell & Developmental Biology* (Vol. 20, No. 1, pp. 27–33). Academic Press.

Verdian, A. (2018). Apta-nanosensors for detection and quantitative determination of acetamiprid–A pesticide residue in food and environment. *Talanta*, 176, 456–464.

Verma, S. K., Jha, E., Panda, P. K., Thirumurugan, A., & Suar, M. (2019). Biological effects of green-synthesized metal nanoparticles: A mechanistic view of antibacterial activity and cytotoxicity. In: *Advanced Nanostructured Materials for Environmental Remediation* (pp. 145–171). Springer, Cham.

Verma, S. K., Nisha, K., Panda, P. K., Patel, P., Kumari, P., Mallick, M. A., Sarkar, B., & Das, B. (2020). Green synthesized MgO nanoparticles infer biocompatibility by reducing in vivo molecular nanotoxicity in embryonic zebrafish through arginine interaction elicited apoptosis. *Science of the Total Environment*, 713, 136521.

Yan, X., Li, H., Han, X., & Su, X. (2015). A ratiometric fluorescent quantum dots based biosensor for organophosphorus pesticides detection by inner-filter effect. *Biosensors and Bioelectronics*, 74, 277–283.

Yao, K. S., Li, S. J., Tzeng, K. C., Cheng, T. C., Chang, C. Y., Chiu, C. Y., Liao, C. Y., Hsu, J. J., & Lin, Z. P. (2009). Fluorescence silica nanoprobe as a biomarker for rapid detection of plant pathogens. In: *Advanced Materials Research* (Vol. 79, pp. 513–516). Trans Tech Publications Ltd.

You, C., Bhagawati, M., Brecht, A., & Piehler, J. (2009). Affinity capturing for targeting proteins into micro and nanostructures. *Analytical and Bioanalytical Chemistry*, 393(6–7), 1563–1570.

Zhang, Z., Lin, M., Zhang, S., & Vardhanabhuti, B. (2013). Detection of aflatoxin M1 in milk by dynamic light scattering coupled with superparamagnetic beads and gold Nanoprobes. *Journal of Agricultural and Food Chemistry*, 61(19), 4520–4525.

Zheng, L., Hong, F., Lu, S., & Liu, C. (2005). Effect of nano-TiO 2 on strength of naturally aged seeds and growth of spinach. *Biological Trace Element Research*, 104(1), 83–91.

9 Sophorolipid
A New Path towards Greener Substitute in Nanotechnology

Priti Darne and Asmita Prabhune

CONTENTS

9.1 INTRODUCTION

Surfactants are surface-active agents that are widely used in detergents, emulsions, inks, herbicides, biocides, and even the shampoo.[1,2] Surfactants also serve as efficient emulsifiers, thus enabling liquids with low solubility to be miscible.

Due to their ability to manipulate and change the properties and dynamics of different surfaces and interfaces, the high number of possible uses of surfactants has led to the development of synthetic surfactants using chemicals. Chemical surfactants, such as petrochemicals or oleochemicals, are derived from non-renewable sources. Different surfactants have different properties, such as solubility dissimilarities, the ability to reduce surface and interfacial stress, and the ability to shape micelles when used with detergents at different critical micellar concentrations, thus allowing foaming and wetting to show application-specific utilities.[3–5] Such multiple applications contribute to the extensive use of surfactants, leading to an increase in the discharge of surfactants into the atmosphere through different mediums.

Chemical-based surfactants are not readily biodegradable, resulting in marine environments being polluted and destroyed and posing significant health threats to humans. The overall utilization volume in North America (the US and Canada) of major surfactants and detergent alcohols from 1990 to 2004 ranges from 719,000 metric tonnes in 1990 to 895,500 metric tonnes in 2004, with an average of around 787,000 metric tonnes. This indicates the amount of use of such goods and hence the cumulative harm to the environment.[6,12] With a direct proportionality found between the surfactant concentrations, chemical surfactants have several effects on the environment. Some of these harmful effects are described as follows.[7,13]

- Promotes and assists the growth of algae in the water, thus reducing the productivity of water sources and weakening the processes of the aquatic food chain.
- Increases the toxicity of aquatic organisms, inducing bioaccumulation, by feeding and skin penetration. A high concentration of toxins in the human body gradually builds up via the food chain, resulting in inhibited enzyme activity and loss of immunity.
- Owing to the emulsification and dispersion of contaminants, chemical surfactants facilitate insufficient filtering and purification of wastewater, decreasing the efficiency of sewage treatment plants.
- They are reported to cause skin damage, skin irritation, and liver damage and are also known to promote carcinogenic formation in some instances.
- Chemical surfactants can disturb fetal growth and are thus known as teratogens.

Due to the concentration of chemical surfactants in the atmosphere and the dependency on the use of surfactants, the multitude of harmful effects has led to an emphasis on the production of environmentally sustainable but productive surfactants using green chemistry principles. There are readily available sources of biosurfactants derived from animal or plant sources, but the cost of obtaining such surfactants via extraction, precipitation, or distillation methods is much higher compared to the production of chemical surfactants, rendering them economically unviable.

9.2 BIOSURFACTANTS

The use of fermentation techniques to tap the ability of microbial species for the production of natural surfactants of high yield was a better option. The highly efficient development of biosurfactants has become possible due to lower costs and sustainable production methods for bacteria, fungi, and yeast.[4] The use of glycolipids for the manufacture of biosurfactants using bacteria and yeasts has led to the production of three forms of surfactants: sophorolipids, rhamnolipids, and mannosylerythritol lipids; sophorolipids are simple to produce and show high yields, while the others need further studies and production to understand and improve their potential applications.[4,8]

A recent study published by Zion Research on the development of the market for microbial biosurfactants projected that by the end of 2020, demand for sophorolipids will rise to approximately USD 17.5 million globally, with significant requirements in the fields of detergents, industrial cleaners, chemicals for oilfield clearing, personal care, agricultural pesticides and herbicides, and food processing.[9]

Biosurfactants are currently in demand based on their increased applications in various industries, such as paint, detergents, cosmetics, agriculture, pharmaceuticals, textiles, food, etc. They have several properties that are better than their chemical equivalents, which are currently a big player in the market. These properties include biodegradability, low toxicity, broad pH range stability, temperatures, high salinity tolerance, low-foaming, detergent, and interfacial properties.[1,10,11] In the global market for surfactants, the presence of biosurfactants is estimated at 11.2 million metric tonnes. While very small, a recent market study projected a significant market penetration in the range of 8–25%, depending on the sector and applications in a year.[11] Biosurfactants are primarily hydrophilic with hydrophobic elements, making them amphiphilic. Biosurfactants are classified by their structure, unlike chemical surfactants that are classified based on their charge. Five types of biosurfactants exist: glycolipids, lipopeptides, fatty acids, biosurfactants of the polymer form, and particulates.

Glycolipids are the largest group of biosurfactants in the world. They are a mixture of the carbohydrate head group and the lipid tail. The most widely recognized glycolipids are rhamnolipids, sophorolipids, and trehalolipids.

Lipopeptides and lipoproteins include a variety of cyclic lipopeptides, such as decapeptide antibiotics (gramicidin), and polymyxin, and are known to possess surface-active properties. Many microorganisms may synthesize fatty acids and phospholipids when complemented with n-hexane.

Polymer biosurfactants are compounds such as emulsan, mannoproteins, liposan, etc.

Hydrocarbon-containing extracellular membrane vesicles conduct themselves as particulate biosurfactants.[12,13]

9.3 SOPHOROLIPID (SL)

This chapter is going to concentrate on sophorolipids (SLs). These are interesting molecules to study as they have demonstrated applications in many areas. Based on therapeutic potential, SL has been marketed by Soliance and MG Intobio Co. in the form of anti-dermatitis soaps, body washes, and creams.[14] These applications will be discussed briefly, but our main focus is to understand their role in the synthesis of nanoparticles. Sophorolipids were first synthesized using *Candida apicola* in the early 1960s by Gorin et al. Many SL generating strains were later identified, viz. *Rodotorula bogoriensis*, *Starmerella bombicola* (previously recognized as *Torulopsis bombicola* and *Candida bombicola*), and *Wickerhamiella domericqiae*. Among these, *Starmerella bombicola*, known as a non-pathogenic yeast, has emerged as the best candidate because of its high yielding ability (up to 400g/L) and higher conversion rate of the substrate to the product.[14–16]

Sophorolipids are made of a sophorose head and fatty acid chain, as the name implies. The two glucose units in sophorose are linked via a unique β,1–2 linkage that may or may not be acetylated with one or two acetyl groups. With varying degrees of saturation, the fatty acid chain is usually C16–18 long. During synthesis, organisms form *de novo* fatty acids through the acetyl-CoA pathway if a hydrophobic carbon source is not given. The fatty acids are then oxidized at the terminal (ω) or subterminal ($\omega - 1$) position by a cytochrome P450 mono-oxygenase enzyme. These result in the formation of hydroxylated fatty acids that are β-glycosidically connected by glycosyltransferase I to the first glucose molecule at the 1' position. A second glucose molecule is bound by glycosyltransferase II to the 2' position of the first one, yielding sophorolipid acid.[1,4,15] Two basic sophorolipid structures, the open acidic chain SL and the closed lactonic chain SL exist. In crude SL, the proportion of acid and lactonic depends on many factors, such as the organism used for production, medium sources of carbon, nitrogen, and salt, environmental conditions, temperature, agitation, pH, aeration, and dissolved oxygen, etc.[1]

9.3.1 Properties of Sophorolipids

Sophorolipids, being a class of biosurfactants, possess all the basic properties, like stability in a wide range of pH, temperatures, and salinity. They have low-foaming and detergent properties; they can sustain water hardness (high concentration of divalent cations) without affecting their interfacial properties. The increased proportion of the acidic form of SL increases its surfactant activity. They are highly biodegradable and have low critical micelle concentration (CMC) values with good surface tension and emulsification behaviour. Also, SL can be synthesized in large quantities using renewable resources, agro-industrial by-products, and residues with simplified product recovery.[1,16,17] Sophorolipids do not irritate skin and eyes and have an oral safety level which is greater than or equal to 5 mL per kg body weight and cause no allergic reactions; hence, they are preferred for food, cosmetic, and pharmaceutical applications.[15]

Sophorolipids display properties and advantages which are comparable to or even better than other surfactants (chemical or biosynthesized). Some of these advantages are listed as follows:[1,5]

- Environmental compatibility.
- High biodegradability.
- Low toxicity.
- Higher tolerance to a broad range of temperature, pH, and salinity conditions.
- Eco-friendly.
- Follow the approach of green chemistry.

9.3.2 Production

Sophorolipid is produced extracellularly when grown on a certain hydrophilic and lipophilic carbon source. The lipophilic substrate is usually in the form of free fatty acid, methyl ester (biodiesel), triglyceride, or alkane and the hydrophilic source is glucose.[18,19] Sophorolipid synthesis starts when yeast cells enter the stationary phase after the exhaustion of nitrogen, in the presence of suitable carbon and energy sources.[20] Declined levels of phosphate or nitrogen lead to a reduction in the specific activities of NAD+ and NADP+ dependent isocitrate dehydrogenase, thereby causing accumulation of isocitrate as well as citrate in the mitochondria. A transport of isocitrate and citrate occurs into the cytosol, where it is cleaved by adenosine triphosphate (ATP): citrate synthase to give rise to acetyl-CoA, the precursor for fatty acid synthesis.[21] As sophorolipid producing organisms are usually found in areas of high osmotic levels (high sugar content), the production of sophorolipid is a means of dealing with high levels of sugar.[18,22]

9.3.3 Producer Organisms

Gorin, Spencer, and Tulloch were the first to describe an extracellular glycolipid produced by yeast, *Torulopsis magnolia* in 1961, later confirming it to be *Torulopsis apicola* in 1968 (*Candida apicola*). In the same year, a new sophorolipid was identified from *Rhodotorula bogoriensis*.[23,24] A third SL yielding strain was identified as *Candida bombicola*.[25] However, the production of sophorolipids is not restricted to one single yeast species. SLs are also produced by several microorganisms, such as *Wickerhamiella domericqiae*, *Candida batistae* (CBS 8550), *Candida floricola* (TM1502), *Candida riodocensis*, *Candida stellata*, *Candida rugosa*, *Candida kuoi*, *Candida tropicalis*, and *Rhodotorula muciliginosa*.[26–31] Among these species, *Candida bombicola* (ATCC 22214) is the most studied yeast that exhibits the highest productivity. Furthermore, *Candida bombicola* comes under the title of being GRAS (generally recognized as safe).[32]

9.3.4 The Structure of a Sophorolipid

Sophorolipids, being surfactant molecules, are amphiphilic and interact with the phase boundary in heterogeneous systems and reduce the repulsive forces between

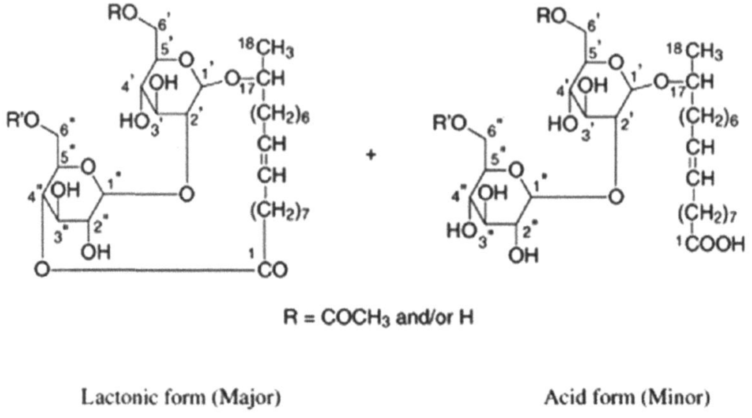

FIGURE 9.1 Structure of sophorolipid (a) acidic SL; (b) lactonic SL. (Image Ref. Parul Dubey et al., 2013.[35])

phases of different polarities (air–water/oil–water), thereby helping the phases to mix easily. They comprise of the hydrophobic fatty acid tail of 16–18 carbon atoms and sophorose, a hydrophilic carbohydrate head.[18,23,33] Sophorose (a dimeric sugar) and a hydroxyl fatty acid are linked together with a glycosidic bond.[34] Two different types of SLs are synthesized by the yeast, the non-lactonic (or acidic) SLs, and the lactonic SLs. The hydroxyl fatty acid moiety of the acidic SLs has a free carboxylic acid functional group. On the other hand, the lactonic SL forms a macrocyclic lactone ring with the 4''-hydroxyl group of the sophorose by intramolecular esterification. In some rare cases, 17-hydroxy oleic acid is found at the 6' or 6'' position. This is the most common lipophilic moiety used for the production of sophorolipids by *Candida bombicola*.[22] Stearic, linoleic, and palmitic acid are some of the commonly found fatty acids, all of which are hydroxylated on the ultimate or penultimate carbon that is used as a lipophilic carbon source (Figure 9.1).[34]

9.4 APPLICATIONS OF SL (FIGURE 9.2)

9.4.1 Agriculture

Due to their anti-microbial and anti-fungal properties, sophorolipids have proven to be highly effective in the field of agriculture. They can also work with their impressive surfactant properties as an adjuvant in herbicide formulation. SL is also effective against plant pathogenic fungi, *Phytophthora* sp., and *Pythium* sp. These fungi cause 'damping-off', a soil-borne root disease, and are capable of spreading rapidly through the soil in the form of zoospores with the aid of water for migration, infecting the seeds, leaves, and stems of water culturing plants. The addition of SL has proven to inhibit the mycelial growth and motility of the zoospore while causing lysis of the zoospore. They have great potential as an environmentally compatible bio-control agent,[22,36] since SL has a low toxicity and is biodegradable. They are also helpful in inhibiting the bloom of algals that affect the aquatic environment.[37]

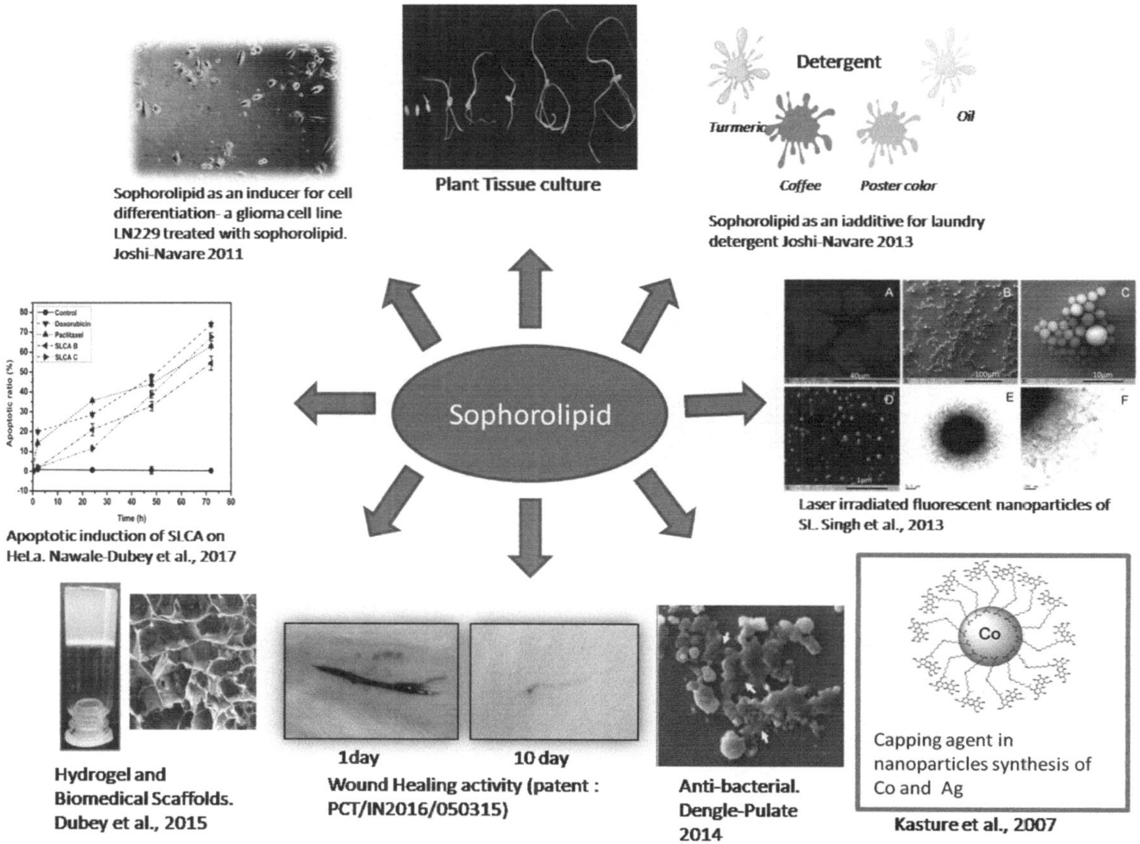

FIGURE 9.2 Various applications of SL.

9.4.2 Food

The addition of surfactants leads to the improvement of the creaminess, appearance, palatability, texture, consistency, rheology, and overall food quality.[38] When SL is added to wheat flour or a product constituting wheat flour, it improves the quality of the final product. When used for making bread, SL containing wheat flour gives the bread a better volume, improving the appearance and extending its shelf-life.[1] Microbial and chemical contamination of foodstuffs may have a significant effect on public and economic health. Agricultural workers involved in the handling of fruits and vegetables during transport or distribution create opportunities for microbial hazard contamination, such as *Salmonella* spp., or *Escherichia coli* (VTEC) developing verotoxin, and norovirus (NoV).[39] In 30 seconds, germicide formulations containing 1% SL could kill 100% of *Escherichia coli*, *Salmonella typhi*, and *Shigella dysenteriae*.[40] Owing to the presence of biofilm, food contamination may also be induced. Such biofilms are key sources of contamination that lead to food spoilage and the transmission of diseases.[1] Biofilm is a defensive matrix consisting of an extracellular polysaccharide or protein material that bacteria secrete to help them adhere to the surface.[41,42] Preventing the adherence of pathogenic microorganisms to food surfaces is a significant step in providing consumers with contamination-free and high-quality products.[38,43] Because of its anti-microbial properties against these organisms, the biofilm produced by Gram-positive (*Ralstonia eutropha* ATCC 17699) and Gram-negative microorganisms (*Bacillus subtilis* BBK006) is effectively disrupted by SL.[43]

9.4.3 Cosmetics

As an active ingredient in the cosmetics sector, SL biosurfactants have gained significance. SLs are known to be biocompatible as they show low cytotoxicity towards human keratinocytes and fibroblasts.[44,45] The incorporation of SL into creams and lotions promotes the metabolism of fibroblasts and collagen neo-synthesis in the dermis layer of the skin, thereby aiding to skin repair and skin toning.[46] Skin ageing is caused by an increase in dermal fibroblast elastase activity, contributing to the development of freckles and wrinkles. By reducing the production of free radicals responsible for skin ageing, SL inhibits the development of elastases.[47] SL helps to activate macrophages, fibrinolytic agents, desquamating agents, and, through partial inhibition of melanogenesis, also serves as a de-pigmenting agent to help reduce brown spots.[48] The French company, Soliance, uses sophorolipid in their cosmetic formulations,[22] as the bactericidal and bacteriostatic property of SL is useful for the control of dandruff, acne treatment, and for reducing body odour in the form of deodorants.[1] There is a product from Soliance (France, http://www.groupesoliance.com) called Sopholiance S, that uses rapeseed oil SL against body odour and acne.

9.4.4 Anti-Microbial Activity

SLs are anti-microbial. They are capable of destabilizing and altering the permeability of the cellular membrane causing cell rupture, which leads to extrusion of the cell contents.[37] Besides this, SL can also act as an anti-algal, anti-fungal, and antiviral agent.[36,37,49,50] The effect of SL on Gram-positive and Gram-negative organisms differs due to the differences in their cell wall structure and osmotic level.[1] SL produced by *Candida bombicola* using lauryl–myristyl alcohol as its hydrophobic moiety, shows potent activity against Gram-positive and Gram-negative organisms. It shows complete inhibition of Gram-negative bacteria *Escherichia coli* and *Pseudomonas aeruginosa*, which is mainly responsible for causing cross infections in hospitals and clinics. It also exhibits substantial activity against Gram-positive bacteria, such as *Staphylococcus aureus* and *Bacillus subtilis*, which are highly involved in food poisoning. Gram-positive bacteria are killed by cell rupturing (lysis) while the shrinking of the cells occurs in the case of Gram-negative bacteria.[51,52] Lactonic sophorolipid inhibited cell growth more effectively than acidic for both *B. subtilis* and *P. acne*.[53]

9.4.5 SL as a Cleaning Agent

Due to its potential as an anti-microbial agent and properties such as emulsifiers and low–high foaming surfactants, sophorolipids are the chosen biosurfactants for cleaning purposes. 'Sophoron', a low-foam dishwasher detergent based on sophorolipid, has been marketed by a Japanese firm Saraya. 'Sophogreen' is a bio-solubilizer recently developed using sophorolipids by Soliance that serves as a better alternative to solubilizers based on petroleum. Ecover (http:/www.ecover.com), based in Belgium, and MG Intobio, based in South Korea, both commercialize goods made using sophorolipids. Ecover's skin irritability test (Episkin test) on an SL-based product confirmed that SL does not irritate the skin. The toxicity and biodegradability test carried out on the Ecover product SL18 showed that the SL-based product was less toxic than non-ionic surfactants, alkyl polyglucosides (APG), and 100% biodegradable after 28 days in order of magnitude. Sophorolipids mild nature, low aquatic toxicity, and considerable surface activity have resulted in the launching of SL-based products by Ecover, such as an all-purpose cleaner, window cleaner, floor soap, and wax cleaner.[19,54] Germany-based Henkel has also been employing sophorolipid for its glass cleaning products, such as Sidolin, Tenn, Instanet, and Sonasol. A study conducted by Navare et al. (2013) also demonstrated that SL synthesized using jatropha oil, to curtail the production cost by using cheap sources, had the potential to be used as a fabric cleaner.[8]

9.4.6 Anti-Cancer

Sophorolipids induced the differentiation of human acute promyelocytic leukaemia cell line HL60 at a concentration of 10 μg mL^{-1}. Kinase C activity was observed to be inhibited after two days as the differentiation into monocytes took place. For two other leukaemic cell lines, the human myelogenous leukaemia cell line K562 and the human basophilic leukaemia cell line KU812, differentiation into megakaryocytes was also induced at a sophorolipid concentration of 15 μg mL–^{1}.[15] Joshi-Navare et al. have shown the anti-cancer activity of SL by inducing cell differentiation in human glioma cell LN-229.[16] A recent study conducted by Nawale et al. showed the anti-proliferative activity of purified forms of a novel SL derived from cetyl alcohol (SLCA) and their anti-cancer mechanism in human cervical cancer cells.[55,61]

9.4.7 Self-Assembly

A recent study conducted by Singh et al. demonstrated the spherical mesoscale molecular assembly of sophorolipid when irradiated by a pulse laser. These assemblies had fluorescent properties and a probable use for bio-imaging and drug delivery. The very first attempt to understand acidic SL assembly behaviour at a pH of 2.0 to 4.1 resulted in giant twisted helical ribbons with a 5–11 μm width and several hundred micrometres.[56,57] The self-assembly of the molecules depends on the concentration, wherein a study conducted by Baccile et al.,[58] showed a correlation between concentration and shape of the micelle. Spherical micelles were formed at < 1 wt% concentration and the shape elongated to a cylindrical shape with increasing concentration from > 0.5 wt% to 5 wt%. Apart from factors like pH and concentration, changes in structure at the molecular level can also influence the self-assembling behaviour of sophorolipids.[59] This was clearly illustrated using three different sophorolipids of oleic acid (OASL), stearic acid (SASL), and eladic acid (EASL) by Dhasaiyan et al.[60] The OASL formed a ribbon-like structure similar to those reported by Zhou et al.[57] in the case of SASL the self-assembled structures appeared like several open sheets lying on top of each other, and EASL (with a trans double bond) formed twisted ribbon structures, which was attributed to a favourable π overlap, leading to a squeeze in the middle of the molecule. The effect of unsaturation was evident from the study conducted by Dhasaiyan et al.[61] With the help of experimental and simulation studies, the author reported that the presence of a double bond in the hydrophobic core of the SL molecule greatly impacts the morphology of the self-assembled structures. They showed that oleic acid SL with one unsaturated bond forms ribbon-like structures (also reported previously), whereas linolenic acid SL with three double bonds assembles into vesicles.[61] Supporting this, recently Pandey et al.[62] also performed simulation studies and reported that linolenic acid sophorolipid with three double bonds, assembled into vesicles and aggregates depending upon the SL to water ratio.

9.4.8 Theranostic Application

Quantum dots (QDs) are well known for their optical and electronic properties, applicable in the biomedical field as labelling and imaging tools, but due to their toxicity and biocompatibility issues, they are not widely used. To introduce biocompatibility, without the loss of photoluminescence properties, Singh et al.[63] synthesized non-toxic cadmium telluride (CdTe) QDs using acidic SL as a surface-functionalizing agent. SL-CdTe QDs potential as a diagnostic tool was demonstrated by cellular internalization studies in ThP-1 (human acute monocytic leukaemia) cells. The cytotoxicity studies confirmed their biocompatibility and hence applicability.[63]

9.4.9 Wound Healing and Biomedical Scaffolds

A recent world patent filed by Prabhune et al. in 2017 (PCT/IN2016/050315), disclosed a study where SL in combination with silk sericin works on a par with commercialized wound healing creams. Rather they observed a distinct lack of scarring on the wound site when applied with SL-sericin-based cream. This has again opened a new avenue where SL can be explored further, and commercial products can be launched. Another study was conducted by Dubey et al.[35] where they used SL with silk fibroin for the rapid gelation and synthesis of scaffolds that could be used for biomedical applications.[64]

9.4.10 Sophorolipid in Nanoparticles Synthesis

The synthesis of nanoparticles mediated via surfactants is an emerging area in material science due to their potential for nanoparticle stability. Though chemical and physical processes are currently used, stable and size-controlled synthesis remains a challenge. Other disadvantages are that the process for synthesis is highly expensive and leads to harmful waste materials for humans and the environment. The size and shape of the nanoparticles determine their biological applicability. As a viable alternative, biosurfactants play a role in stabilizing and deciding the shape of the nanoparticles during synthesis.[65] The synthesis processes of nanoparticles based on biologically produced surfactants are rapidly gaining interest in ranges of 10–1000 nm.[13,19] In most cases, a bottom-up approach is used for nanoparticle synthesis. The following are examples of sophorolipids used as either reducing/capping agents or nanoparticles both reduced and capped. For a deeper understanding of the synthesis methods introduced in the articles, readers are encouraged to read these interesting articles.

9.5 COBALT NANOPARTICLES

A decade ago, when the synthesis of cobalt nanoparticles was first attempted, the earliest work on nanoparticles based on the sophorolipid system began. Oleic acid was usually used as a capping agent in early strategies for the synthesis of cobalt nanoparticles, but it did cause stability issues. In addition to this, oleic acid is also insoluble in water, so the mixture of methanol/ethanol–water was used during the cobalt nanoparticle preparation. Reports regarding oleic acid capped silver and nickel nanoparticles explain that polar solvents expose oleic acids –COOH group in the solvent and help with its binding to the nanoparticles surface via a double bond. But this also contributes to the exposed –CH3 group in the solvent, causing poor capping and instability of nanoparticles during dispersion. Kasture et al. used a sophorolipid biosurfactant to overcome this barrier, which is essentially a microbial alteration of oleic acid and soluble in water due to the attachment of the head group of polar sophorose. The superparamagnetic cobalt nanoparticles obtained by SL capping were stable in a powder form and readily dispersible in an aqueous medium without substantial morphological modifications.[66]

9.5.1 Silver Nanoparticles

The authors began to explore certain molecules that had a dual nature of both reducing and capping the nanoparticles in one go after a study with cobalt nanoparticles. These are preferred as they enable greater control over the reaction and avoid the use of an external reducing agent. This can help in scaling the process. The reduction of silver ions (Ag^+ ions) to Ag nanoparticles and capping with the SL is shown in this study. The difference in particle size was studied in this research as a function of time and temperature. For the synthesis of Ag NPs, viz., oleic acid SL (OASL)

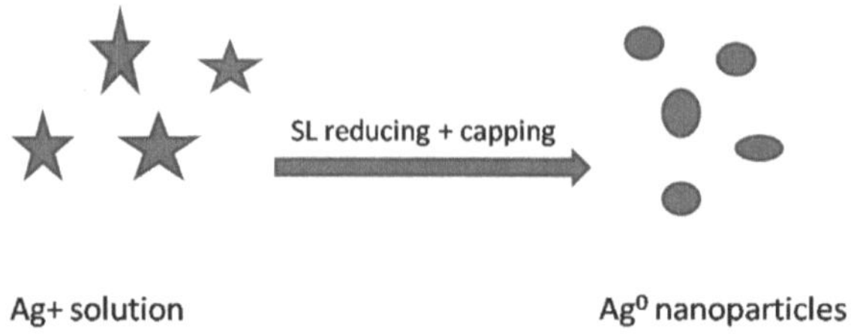

FIGURE 9.3 Schematic of Silver nanoparticles synthesis using SL as reducing and capping agent as a function of time and temperature.

and linoleic acid SL (LSL), two different SLs synthesized using two different fatty acids were used. At 40°C, OASL reduced-capped Ag NPs had a mean size of 20 nm and the particle size reduced to 5.5 nm with an almost spherical form at 90°C. In the case of LSL, the mean size obtained was 22 nm at 40°C, and a similar decrease in size to 11 nm was observed at 90°C. This can be attributed to lower temperatures as the formation of the particles becomes slow, but the synthesis becomes rapid as the temperature is elevated. From this study, it was concluded that in terms of size and morphology, the OASL derived Ag NPs were better than LSL derived (Figure 9.3).[67]

9.5.2 Silver Nanoparticles

Kumar et al. later took this study forward for a continuous synthesis using a microreactor. A batch process is a good approach for understanding the mechanism of nanoparticle synthesis, but up-scaling becomes a limiting factor. Limitations viz., the inefficient mixing, non-uniformity of temperature throughout the reactor, non-idealities in flow rates, usually result in a wider range of particle size distribution. Also, the thorough cleaning of the vessel after every batch is required to avoid contamination. To tackle this, micro-fluidic methods have been applied that have many advantages over batch processes, like reduced consumption of reagents, safety, and stoichiometric yield that satisfy the large-scale synthesis. But mostly, a high surface to volume ratio in the micro-channels leads to better heat and mass transfer, which gives better control over the reaction rates. Silver nanoparticles have well established anti-microbial activity and are considered a model system for microreactors. Therefore, in this study, a continuous flow synthesis of silver nanoparticles in a water medium via the use of sophorolipid as a reducing and capping agent was demonstrated. Any use of an external reducing agent was avoided, reducing the number of reactants in the reaction.[68]

Based on results obtained from previous batch process studies, stearic acid SL was chosen for continuous flow experiments. The particle size distributions of Ag NPs as shown by DLS were between 5–9 nm that were synthesized using a stainless tube model.

9.6 IRON NANOPARTICLES

The formation of sophorolipid-capped iron oxide nanoparticles was executed by Baccile et al. The magnetic iron nanoparticles were synthesized from iron (II) chloride and iron (III) chloride in the presence of ammonia at room temperature and at 80°C. Both two-step and one-step procedures were applied, which resulted in nanoparticles with maghemite (γ-Fe_2O_3) or two-line ferrihydrite ($Fe_5HO_8{\cdot}4H_2O$) structures, respectively. A very broad particle size distribution was obtained for the one-way synthesis pathway at room temperature, while monodispersed nanoparticles were obtained in all other synthesis processes mentioned above. A much higher increase in the average particle size for the nanoparticles synthesized at room temperature indicated that sophorolipids were more loosely bonded to the nanoparticle surface. FTIR analysis revealed that the sophorolipids mainly attached to the nanoparticle surface via the carboxylic acid function. Adsorption studies were performed for the sophorolipid-capped iron oxide nanoparticles with two lectins, namely concanavalin A and lectin from *Bandeiraea simplicifolia*. No specific interaction occurred with the lectin from *Bandeiraea simplicifolia*, but an affinity for concanavalin A for these capped NPs was observed. Sophorolipid-capped iron oxide nanoparticles also proved to be good selective targets for sugar/protein interactions.[69]

9.7 SILVER NANOPARTICLES FOR CYTOTOXIC AND GENOTOXIC STUDY

After identifying that sophorolipids could be used for nanoparticle synthesis, another set of experiments was carried out to study the cytotoxic and genotoxic effects of glycolipid synthesized Au NPs and Ag NPs; hepatic cell line HepG2 was selected and exposed to different concentrations of the metal nanoparticles. Oleic acid SL was used as a reducing and capping agent for nanoparticle synthesis. Silver NPs were synthesized by the previously mentioned method from the same group. These prepared OA-SL molecules, Au NPs, and Ag NPs were tested at different concentrations with human hepatic cells and were found to be neither cytotoxic nor genotoxic up to concentrations of 100 mM.[70]

9.8 ZINC NANOPARTICLES

Nanostructures that are functionalized with biomolecules can gain anti-microbial properties. Although there are some reports related to the basic guidelines for the functionalization and application of specialized nanomaterials, active research on functionalizing nanomaterials with biomolecules and their specific anti-microbial activity is poor. Scanty reports are available on the application of functionalized nanoparticles against pathogenic microbes. Therefore, in this study, an attempt was made to employ the sophorolipid (produced by yeast *Cryptococcus* sp. VITGBN2) as a bio-stabilizer for the synthesis of ZnO nanoparticles (ZnO). Also, ZnO NPs biofunctionalized by sophorolipids, were tested as anti-microbial agents. The activity of acidic diacetate sophorolipid by *Cryptococcus* sp. VITGBN2 against pathogens like *S. enterica* and *C. albicans* was established. Sophorolipid biofunctionalized ZnO showed higher anti-microbial activity compared to naked ZnO NP and other anti-microbial agents. Enhanced bioactivity of these NPs was attributed to the increased permeability through the microbial cells that led to the loss of cellular transport through the plasma membrane, resulting in cell death.[71]

9.9 GOLD NANOPARTICLES SURFACE FUNCTIONALIZED BY SL

The surface functionalization of gellan gum reduced gold nanoparticles with SL was proposed by Dhar et al. Taking advantage of SLs anti-cancer activity, these nanoparticles were then conjugated and loaded with the drug doxorubicin and tested against human glioma cells LN-229 as well as human glioma stem cells HNGC-2. Cytotoxicity was carried out without any labelling. A MTT assay was performed on a 24-hour-old plate. The incubation time was four hours, after which formazan crystals were dissolved in DMSO and read at 570 nm. TEM analysis and particle size distribution confirmed SL-GG-Au NPs of an average particle size of ~17 nm, which was slightly increased from that of pristine GG-Au NPs (~13 nm). In a cellular uptake study, gold nanoparticles were observed inside the cells as red dots and were localized mostly in the cytoplasm and perinuclear space of cells. From this study, it can be concluded that SL enhances the cell cytotoxicity of gellan gum nanoparticles against glioma cell lines (Figure 9.4).[72]

9.10 CADMIUM TELLURIDE (CDTE) QUANTUM DOTS

Singh et al. demonstrated the use of SL as a stabilizer for CdTe quantum dots that were synthesized using 3-mercaptopropionic acid (MPA) and amine-terminated by a coupling reaction between MPA and ethylenediamine (EDA). For the synthesis, the authors used NaHTe that was prepared by reacting $NaBH_4$ and Te powder. The reaction was considered completed once the black Te powder turned to a clear pink solution. NaHTe was then refluxed under nitrogen for two hours at 100°C with 36 mM MPA and $CdCl_2$ 15 mM at a pH of 9. Once cooled the CdTe solution was precipitated by acetone, centrifuged, and washed to remove un-reacted salts. The dried powder was then again re-dispersed for conjugation with EDC to achieve amine termination via sulfo-NHS (N-hydroxysulfo-succinimide) and EDC coupling. The

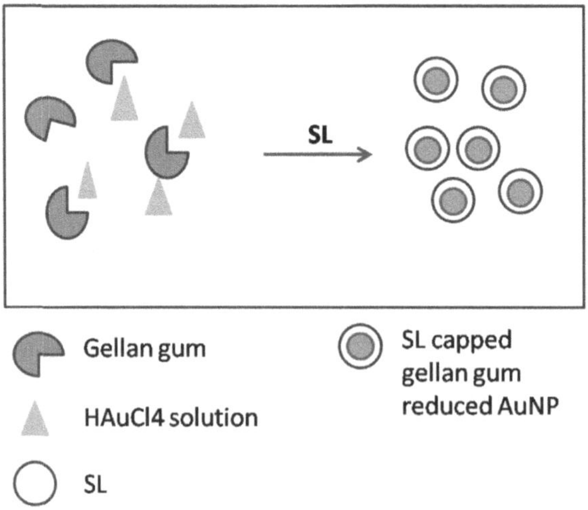

FIGURE 9.4 Surface functionalization of gellan gum reduced gold nanoparticles by SL.

amine-terminated CdTe QDs were then mixed with acidic SL (100 mg each) under stirring for 30 mins.

The QDs obtained via MPA were of 8 nm, as observed in TEM and DLS, whereas SL capped CdTe QDs were of 118 nm indicating encapsulation via SL. They were tested for their biocompatibility in both normal and cancer cell lines. It was shown that the SL capped CdTe QDs were biocompatible in normal cell lines and effective against cancer cell lines with enhanced cellular association and uptake, which was detected by the fluorescence of the QDs. Almost 100% viability in the case of normal NIH_3T_3 cell lines and 65% viability in the case of cancer MCF-7 cell line was observed, clearly indicating that composites provided a mode to diagnose cancer cells using their fluorescence properties as well as the cytotoxicity towards them. Thus authors have made a composite particle that had a high potential as a theranostic weapon.[63]

9.11 MESOPOROUS STRUCTURES OF SOPHOROLIPID

Singh et al. used a pulsed UV laser approach for self-assembly of SL molecules in water and achieved mesoporous structures. The structures thus formed were found to be fluorescent. They also demonstrated a simple loading of magnetic nanoparticles on these mesoporous structures during their synthesis for application in hyperthermia. The time evolution of the self-assembly was observed via SEM imaging at an interval of 10 mins. Both SEM/TEM analyses showed that the defined particles were formed after 40 mins of laser irradiation. These particles fluoresced when excited at 330 nm. This property was attributed to the addition of an extra double bond due to laser irradiation and π–π stacking. A cytotoxicity test was carried out on the HeLa cell line by MTT assay, which revealed that a 50 $\mu g\ ml^{-1}$ concentration was not detrimental for the cells. The hyperthermia experiment was carried out using a radio frequency source. Magnetite-loaded SL mesostructures (5 $mg\ ml^{-1}$ Fe_3O_4 nanoparticle) were used for this experiment. The data was collected at 0 mins, 4 mins, 10 mins, and 15 mins for both the simple SL solution and the magnetite-loaded SL mesostructures. It was observed that the magnetite-loaded SL mesoparticles were effective against hyperthermia.[56]

9.12 SOLID LIPID NANOPARTICLES

A recent example includes the use of SL for the synthesis of solid lipid nanoparticles (SLN) used to encapsulate the drugs rifampicin and dapsone. In this report, lactonic SL with a pluronic copolymer was used to obtain stable SLNs. These SLNs had a high entrapment efficiency and loading capacity. The release study conducted revealed a high release rate. Therefore, through this study, new areas were explored for solid lipid nanoparticles, and there is scope for development.[73]

9.13 CONCLUSION

In the previous literature on sophorolipid, various applications come to focus, and it imparts with the knowledge that SL is an amazing molecule with varied properties that can be applied to many fields. The biomedical aspect has been extensively explored to date, and a few reports cover nanoparticles synthesis, yet this seemingly interesting avenue remains a lesser-explored subject. The aim of this chapter was to bring readers attention to this wonderful molecule (SL) and to the new possibilities of applying SL to the multi-applied field of theranostics using nanoparticles.

REFERENCES

1. Oliveira MR De, Magri A, Baldo C, Camilios-neto D. Review : Sophorolipids a promising biosurfactant and it's applications. *Int. J. Adv Biotechnol Res.* 2015;6:161–174.
2. Vijayakuma S, Saravanan V. Biosurfactants-types, sources and applications. *Res J Microbiol.* 2015;10(5):181–192. doi: 10.3923/jm.2015.181.192.
3. Desai JD, Banat IM. Microbial production of surfactants and their commercial potential. *Microbiol Mol Biol Rev.* 1997;61(1):47–64. doi: 10.1016/S0140-6701(97)84559-6.
4. Marchant R, Banat IM. Biosurfactants: A sustainable replacement for chemical surfactants? *Biotechnol Lett.* 2012;34(9):1597–1605. doi: 10.1007/s10529-012-0956-x.
5. Banat IM, Franzetti A, Gandolfi I, et al. Microbial biosurfactants production, applications and future potential. *Appl Microbiol Biotechnol.* 2010;87(2):427–444. doi: 10.1007/s00253-010-2589-0.
6. Cowan-Ellsberry C, Belanger S, Dorn P, et al. Environmental safety of the use of major surfactant classes in North America. *Crit Rev Environ Sci Technol.* 2014;44(17):1893–1993. doi: 10.1080/10739149.2013.803777.
7. Yuan CL, Xu ZZ, Fan MX, Liu HY, Xie YH, Zhu T. Study on characteristics and harm of surfactants. *J Chem Pharm Res.* 2014;6(7): 2233–2237. doi: 10.1023/A:1011908323811.
8. Joshi-Navare K, Khanvilkar P, Prabhune A. Jatropha oil derived sophorolipids: production and characterization as laundry detergent additive. *Biochem Res Int.* 2013;2013:169797. doi: 10.1155/2013/169797.
9. Zion. Microbial Biosurfactants Market (Rhamnolipids, Sophorolipids, Mannosylerythritol Lipids (MEL) and Other) for Household Detergents, Industrial & Institutional Cleaners, Personal Care, Oilfield Chemicals, Agricultural Chemicals, Food Processing, Textile and Other Applications – Global Industry Perspective, Comprehensive Analysis and Forecast 2014–2020. 2014;2020:2018–2021. http://www.reportlinker.com/.
10. Kim K, Yoo D, Kim Y, Lee B, Shin D, Kim EK. Characteristics of sophorolipid as an anti-microbial agent. *J Microbiol Biotechnol.* 2002;12(2):235–241.
11. Solaiman DKY, Ashby RD, Foglia TA, Nuñez A, Marmer WN. Sophorolipids – Emerging microbial biosurfactants. *Inf - Int News Fats, Oils Relat Mater.* 2004;15(4).
12. Gharaei-Fathabad E. Biosurfactants in pharmaceutical industry: A mini-review. *Am J Drug Discov Dev.* 2011;1(1):58–69. doi: 10.3923/ajdd.2011.58.69.
13. Rodrigues LR. Microbial surfactants: Fundamentals and applicability in the formulation of nano-sized drug delivery vectors. *J Colloid Interface Sci.* 2015;449:304–316. doi: 10.1016/j.jcis.2015.01.022.
14. Dolman BM, Kaisermann C, Martin PJ, Winterburn JB. Integrated sophorolipid production and gravity separation. *Process Biochem.* 2017;54:162–171. doi: 10.1016/j.procbio.2016.12.021.
15. Delbeke EIP, Movsisyan M, Van Geem KM, Stevens CV. Chemical and enzymatic modification of sophorolipids. *Green Chem.* 2016;18(1):76–104. doi: 10.1039/C5GC02187A.
16. Darne P, Mehta M, Dubey P, Prabhune A. Bauhinia seed oil, a novel substrate for sophorolipid production. *World J Pharm Pharm Sci.* 2014;3(11).
17. Bogaert INA Van, Zhang J, Soetaert W. Microbial synthesis of sophorolipids. *Process Biochem.* 2011;46(4):821–833. doi: 10.1016/j.procbio.2011.01.010.

18. Fleurackers SJJ. On the use of waste frying oil in the synthesis of sophorolipids. *Eur J Lipid Sci Technol.* 2006;108(1):5–12. doi: 10.1002/ejlt.200500237.
19. Fleurackers SJJ, Van Bogaert INA, Develter D. On the production and identification of medium-chained sophorolipids. *Eur J Lipid Sci Technol.* 2010;112(6):655–662. doi: 10.1002/ejlt.200900158.
20. Davila A-M, Marchal R, Vandecasteele J-P. Sophorose lipid fermentation with differentiated substrate supply for growth and production phases. *Appl Microbiol Biotechnol.* 1997;47(5):496–501. doi: 10.1007/s002530050962.
21. Albrecht A, Rau U, Wagner F. Initial steps of sophoroselipid biosynthesis by *Candida bombicola* ATCC 22214 grown on glucose. *Appl Microbiol Biotechnol.* 1996;46(1):67–73. doi: 10.1007/s002530050784.
22. Van Bogaert IN, Saerens K, De Muynck C, Develter D, Soetaert W, Vandamme EJ. Microbial production and application of sophorolipids. *Appl Microbiol Biotechnol.* 2007;76(1):23–34. doi: 10.1007/s00253-007-0988-7.
23. Gorin PAJ, Spencer JFT, Tulloch AP. Hydroxy fatty acid glycosides of sophorose from *Torulopsis magnoliae. Can J Chem.* 1961;39(6199):846–855. doi: 10.1139/v61-104.
24. Tulloch AP and Spence JFT. A new hydroxy fatty acid sophoroside from Candida bogoriensis'. *Can J Chem.* 1968;46(21):3291–3300. doi: 10.1139/v68-545.
25. Gorin PAJ, Spencer JFT, Tulloch AP. Torulopsis bombicola sp. n. *Antonie Van Leeuwenhoek.* 1970;36(11234):129–133. doi: 10.1007/BF02069014.
26. Chandran P, Das N. Characterization of sophorolipid biosurfactants produced by yeast species grown in diesel oil. *Int J Eng Sci Technol* . 2011;2(1):63–71.
27. Chandran P, Das N. Biosurfactant production and diesel oil degradation by yeast species *Trichosporon asahii* isolated from petroleum hydrocarbon contaminated soil. *Int J Eng Sci Technol.* 2010;2(12):6942–6953. http://www.ijest.info/docs/IJEST10-02-12-045.pdf. Accessed 28 September 2014.
28. Kitamoto D, Morita T, Fukuoka T, Konishi M, Imura T. Self-assembling properties of glycolipid biosurfactants and their potential applications. *Curr Opin Colloid Interface Sci.* 2009;14(5):315–328. doi: 10.1016/j.cocis.2009.05.009.
29. Price NPJ, Ray KJ, Vermillion KE, Dunlap C, Kurtzman CP. Structural characterization of novel sophorolipid biosurfactants from a newly identified species of Candida yeast. *Carbohydr Res.* 2012;348:33–41. doi: 10.1016/j.carres.2011.07.016.
30. Kurtzman CP, Price NPJ, Ray KJ, Kuo T-M. Production of sophorolipid biosurfactants by multiple species of the Starmerella (Candida) bombicola yeast clade. *FEMS Microbiol Lett.* 2010;311(2):140–146. doi: 10.1111/j.1574-6968.2010.02082.x.
31. Chen J, Song X, Zhang H, Qu Y. Production, structure elucidation and anti-cancer properties of sophorolipid from *Wickerhamiella domercqiae. Enzyme Microb Technol.* 2006;39(3):501–506. doi: 10.1016/j.enzmictec.2005.12.022.
32. Bourdichon F, Casaregola S, Farrokh C, et al. Food fermentations: Microorganisms with technological beneficial use. *Int J Food Microbiol.* 2012;154(3):87–97. doi: 10.1016/j.ijfoodmicro.2011.12.030.
33. Claus S, Van Bogaert INA. Sophorolipid production by yeasts: A critical review of the literature and suggestions for future research. *Appl Microbiol Biotechnol.* 2017;101(21):7811–7821. doi: 10.1007/s00253-017-8519-7.
34. Hans-Joachim Asme, Siegmund Langa FW and VW. Microbial production, structure elucidation and bioconversion of sophorose lipids. *jaocs.* 1988;65(9):1460–1466. doi: 10.1007/BF02898308.
35. Dubey P, Sevaraj K, Prabhune A. Sophorolipids: in self assembly and nanomaterial synthesis. *World J Pharm.* 2013:1107–1133. http://www.wjpps.com/admin/assets/article_issue/137517738329 WJPPS 540.pdf. Accessed 28 September 2014.

36. Yoo DS, Lee BS, Kim EK. Characteristics of microbial biosurfactant as an antifungal agent against plant pathogenic fungus. *J Microbiol Biotechnol.* 2005;15(6): 1164–1169.
37. Baek SH, Sun XX, Lee YJ, Wang SY, Han KN, Choi JK, Noh JH, Kim EK. Mitigation of harmful algal bloom by sophorolipid. *J Microbial Biotechnol.* 2003;13(5): 651–659.
38. Kralova I, Sjöblom J. Surfactants used in food industry: A review. *J Dispers Sci Technol.* 2009;30(9):1363–1383. doi: 10.1080/01932690902735561.
39. Van Boxstael, S, Habib, I, Jacxsens, L, De Vocht, M, Baert, L, Van De Perre, E, Rajkovic, A, Lopez-Galvez, F, Sampers, I, Spanoghe, P, De Meulenaer, B, Uyttendaele, M. Food safety issues in fresh produce: Bacterial pathogens, viruses and pesticide residues indicated as major concerns by stakeholders in the fresh produce chain, *Food Control*, 2013, 32(1), 190–197. https://doi.org/10.1016/j.foodcont.2012.11.038
40. Pierce et al. United States Patent US 6,262,038 B1. 2001;2(12). doi: 10.1038/incomms1464.
41. Mukherji R, Prabhune A. Novel glycolipids synthesized using plant essential oils and their application in quorum sensing inhibition and as antibiofilm agents. *Scientific World J.* 2014;2014:890709. doi: 10.1155/2014/890709.
42. Patil A, Joshi-Navre K, Mukherji R, Prabhune A. Biosynthesis of glycomonoterpenes to attenuate quorum sensing associated virulence in bacteria. *Appl Biochem Biotechnol.* 2017;181(4):1533–1548. doi: 10.1007/s12010-016-2300-8.
43. Banat IM, De Rienzo MAD, Quinn GA. Microbial biofilms: biosurfactants as antibiofilm agents. *Appl Microbiol Biotechnol.* 2014;98(24):9915–9929. doi: 10.1007/s00253-014-6169-6.
44. Otto RT, Daniel HJ, Pekin G, et al. Production of sophorolipids from whey: II. Product composition, surface active properties, cytotoxicity and stability against hydrolases by enzymatic treatment. *Appl Microbiol Biotechnol.* 1999;52(4):495–501. doi: 10.1007/s002530051551.
45. Hirata Y, Ryu M, Oda Y, et al. Novel characteristics of sophorolipids, yeast glycolipid biosurfactants, as biodegradable low-foaming surfactants. *J Biosci Bioeng.* 2009;108(2):142–146. doi: 10.1016/j.jbiosc.2009.03.012.
46. Borzeix Frédérique. United States Patent (US006057302A). 2000.
47. Hillion et al. United States Patent (US005756471A). 1998.
48. Martine Maingault. WO1997001343A2.
49. Shah V, Jurjevic M, Badia D. Utilization of restaurant waste oil as a precursor for sophorolipid production. *Biotechnol Prog.* 2007;23(2):512–515. doi: 10.1021/bp0602909.
50. Shah V, Doncel GF, Seyoum T, et al. Sophorolipids, microbial glycolipids with anti-human immunodeficiency virus and sperm-immobilizing activities sophorolipids, microbial glycolipids with anti-human immunodeficiency virus and sperm-immobilizing activities. 2005;49(10):4093–4100. doi: 10.1128/AAC.49.10.4093.
51. Dengle-Pulate V, Chandorkar P, Bhagwat S, Prabhune AA. Antimicrobial and SEM studies of sophorolipids synthesized using lauryl alcohol. *J Surfactants Deterg.* 2013;17(3):543–552. doi: 10.1007/s11743-013-1495-8.
52. Dengle-Pulate V, Joshi J, Bhagwat S, Prabhune A. Application of sophorolipids synthesized using lauryl alcohol as a germicide and fruit-vegetable wash. 2014;3(7):1630–1643. http://www.wjpps.com/download/article/1404219095.pdf. Accessed 28 September 2014.
53. Lang S, Katsiwela E, Wagner F. Antimicrobial effects of biosurfactants. *Fat Sci Technol.* 1989;91(9):363–366. doi: 10.1002/lipi.19890910908.
54. Develter DWG, Lauryssen LML. Properties and industrial applications of sophorolipids. *Eur J Lipid Sci Technol.* 2010;112(6):628–638. doi: 10.1002/ejlt.200900153.

55. Nawale L, Dubey P, Chaudhari B, Sarkar D. Anti-proliferative effect of novel primary cetyl alcohol derived sophorolipids against human cervical cancer cells HeLa. *PLoS One* 2017;12:1–14.
56. Singh PK, Mukherji R, Joshi-Navare K, et al. Fluorescent sophorolipid molecular assembly and its magnetic nanoparticle loading: a pulsed laser process. *Green Chem.* 2013;15(4):943. doi: 10.1039/c3gc40108a.
57. Zhou S, Xu C, Wang J, et al. Supramolecular assemblies of a naturally derived sophorolipid. *Langmuir* 2004;(11):7926–7932.
58. Baccile N, Babonneau F, Jestin J, Pehau-arnaudet G, Bogaert I Van. Unusual, pH-Induced, self-assembly of sophorolipid biosurfactants. *ACS Nano* 2012;(6): 4763–4776.
59. Baccile N, Selmane M, Le Griel P, et al. PH-driven self-assembly of acidic microbial glycolipids. *Langmuir.* 2016;32(25):6343–6359. doi: 10.1021/acs.langmuir.6b00488.
60. Dhasaiyan P, Banerjee A, Visaveliya N, Prasad BLV. Influence of the sophorolipid molecular geometry on their self-assembled structures. 2013;8:369–372. doi: 10.1002/asia.201200935.
61. Dhasaiyan P, Pandey PR, Visaveliya N, Roy S, Prasad BLV. Vesicle structures from bolaamphiphilic biosurfactants: Experimental and molecular dynamics simulation studies on the effect of unsaturation on sophorolipid self-assemblies. *Chem - A Eur J.* 2014;20(21):6246–6250. doi: 10.1002/chem.201304719.
62. Pandey PR, Dhasaiyan P, Prasad BLV, Roy S. Structural insight into self assembly of sophorolipids: A molecular dynamics simulation study. *Zeitschrift fur Phys Chemie.* 2016;230(5–7):819–836. doi: 10.1515/zpch-2015-0719.
63. Singh P, Joshi K, Guin D, Prabhune AA. Chemically conjugated sophorolipids on CdTe QDs: A biocompatible photoluminescence nanocomposite for theranostic applications. *RSC Adv.* 2013;3(44):22319–22325. doi: 10.1039/c3ra44535f.
64. Advances RSC, Csir PD, Csir LN, Sarkar D. Sophorolipid assisted tunable and rapid gelation of silk fibroin to form porous biomedical scaffolds. *RSC Adv.* 2015;(May 2016). doi: 10.1039/C5RA04317D.
65. Kiran GS, Selvin J, Manilal A, Sujith S. Biosurfactants as green stabilizers for the biological synthesis of nanoparticles. *Crit Rev Biotechnol.* 2011;31(December 2009):354–364. doi: 10.3109/07388551.2010.539971.
66. Kasture M, Singh S, Patel P, et al. Multiutility sophorolipids as nanoparticle capping agents: Synthesis of stable and water dispersible co nanoparticles. *Langmuir.* 2007;23(23):11409–11412. doi: 10.1021/la702931j.
67. Kasture MB, Patel P, Prabhune AA, Ramana CV, Kulkarni AA, Prasad BLV. Synthesis of silver nanoparticles by sophorolipids : Effect of temperature and sophorolipid structure on the size of particles. *J Chem Sci.* 2008;120(6):515–520.
68. Kumar DVR, Kasture M, Prabhune AA, Ramana CV, Prasad BLV, Kulkarni AA. Continuous flow synthesis of functionalized silver nanoparticles using bifunctional biosurfactants. *Green Chem.* 2010;12(4):609. doi: 10.1039/b919550e.
69. Baccile N, Noiville R, Stievanod L, Van Bogaert I. Sophorolipids-functionalized iron oxide nanoparticles. *PhysChemChemPhys.* 2013;15:1606–1620. doi: 10.1039/c2cp41977g.
70. Singh S, D'Britto V, Prabhune AA, Ramana CV., Dhawan A, Prasad BLV. Cytotoxic and genotoxic assessment of glycolipid-reduced and -capped gold and silver nanoparticles. *New J Chem.* 2010;34(2):294. doi:10.1039/b9nj00277d.
71. Basak G, Das D, Das N. Dual role of acidic diacetate sophorolipid as biostabilizer for ZnO nanoparticle synthesis and biofunctionalizing agent against Salmonella enterica and *Candida albicans. J Microbiol Biotechnol.* 2014;24(1):87–96. doi:10.4014/jmb.1307.07081.

72. Dhar S, Reddy EM, Prabhune A, Pokharkar V, Shiras A, Prasad BLV. Cytotoxicity of sophorolipid-gellan gum-gold nanoparticle conjugates and their doxorubicin loaded derivatives towards human glioma and human glioma stem cell lines. *Nanoscale.* 2011;3(2):575–580. doi: 10.1039/C0NR00598C.
73. Kanwar R, Gradzielski M, Mehta SK. Biomimetic solid lipid nanoparticles of sophorolipids designed for antileprosy drugs. *J. Phys. Chem. B*, 2018, 122(26): 6837–6845. doi:10.1021/acs.jpcb.8b03081.

10 Nanoformulations in Soil Health and Agriculture

A.H. Sneharaniand and M. PavanKumar

CONTENTS

10.1 INTRODUCTION

Nanoparticles are becoming mainstream with their utilization in nourishment, agriculture, and medicine as well as in hardware, building, and manufacturing. Nanotechnology can comprehensively control matter within the scope of 1 to 100 nanometers. Mass materials when decreased to the nanoscale gain unique properties (liquefying point, physical quality, surface region, infiltration power, electric conductance, optical impact, attraction, and so forth), which differ from what they display on a full scale, empowering its application. These nanomaterials are either regular or designed and fabricated. At the nanoscale, gravity is less significant, while surface area and van der Waal strength are significantly increased. In present-day agribusiness, feasible creation and productivity are inconceivable without the utilization of agrochemicals, for example, pesticides and fertilizers. Nevertheless, having mentioned this, every agrochemical has potential issues, including the defilement of the soil and water or build-ups of toxic components that endanger the biotic and abiotic factors thereby unbalancing ecological well-being. These potential issues can be surpassed when the exact administration and control of information sources are recognized. The improvement of cutting-edge agrarian frameworks with the utilization of shrewdly designed nano instruments could be magnificent to achieve a revolution in agricultural/horticultural practices and the up-gradation of both quality and quantity of yields. Nano-enabled agriculture or agricultural nanotechnology enables control during farming by providing the nanostructured tools. The use of nanomaterials in horticulture works to lessen the excessive diffusion of synthetic chemicals and additives (fertilizers, pesticides, herbicides), thus preserving the nutrients and balancing ecological interactions. Nanotechnology imminently improves farming and nourishment industries with novel nanotools for controlling disease. The critical interest in utilizing nanotechnology for farming incorporates explicit applications like nanomanures and nanopesticides to increase supplement availability and decrease pests and pathogens. Nanotechnology sensors are used to examine and diagnose soil quality, which helps sustain some of the indigenous varieties of plants. Nanoparticle-intervened genetic engineering in plant care and maintenance for tackling pests and weeds, and also their diagnosis, can be carried out using biosensors. The utilization of a nano-earth composite superabsorbent for higher maintenance of soil dampness is under examination. Working within the plant and soil framework, nanomaterials aid the development of nanomaterial-based pesticides for pest management. To improve farming profitability bioconjugated nanoparticles are designed to provide sustained water and nutrient availability. Adding plant supplements through nanofeed substances helps with the nourishment of seeds and plants. Nanosensors are generally applied in agriculture because of their sensitivity and robust diagnosis of contamination in the soil and water. Specialists recommend the potential advantages of nanotechnology for horticulture, plant nourishment, and this should also be extended to fisheries and aquaculture. Nanomaterials are used to catalyze the conversion of waste, pollutants, and poisonous materials into non-toxic products concurrently to enhance beneficial soil microorganisms involved in bioremediation. Bioremediation utilizes living life forms to separate or expel poisons and harmful substances from agricultural soil and water. Depending on the life forms, the term bioremediation is replaced by phytoremediation when phytoplanktons are used, and mycoremediation implies the usage of fungi for the same process. With the bioremediation, toxic inorganic elements and organic compounds can be removed from soil and water naturally by microorganisms.

'Nano' signifies one-billionth; consequently nanotechnology manages materials estimated at around a billionth of a metre. A nanometre is 1/80,000 the distance across a human hair or roughly ten hydrogen particles wide. Nanotechnology brings together individual particles or atoms at a sub-atomic level into structures to make materials and gadgets with new or incomprehensibly advanced properties. Nanotechnology is an umbrella term covering physical science, chemical science, biological science, material science, agricultural science, and many more. Farming has consistently been the foundation of the sustainability of mankind. The increases in population, demand for energy consumption, and changes in the environment are bound to put increasing stress on agricultural production. Without putting much strain on the environment, the goal of food production can be achieved with a sustainable agricultural practice. Nano-enabled agriculture is an upcoming technology that has great potential to address the issues and needs for sustainable agricultural food and consumer goods production. Sustainable agriculture aims to increase agricultural yield without adverse environmental impact. This works within the framework of identifying and employing a combination of best agricultural practices. In addition to meeting the needs of hunger, agricultural practices fuel the economy. Providing food to fulfil the hunger of a huge global population is a gigantic task that needs to be approached with innovation, resulting in qualitative and quantitative yields. Factors influencing agriculture are inadequacies in supplement content, variable soil conditions, population blast, industrialization, water sources, and disintegration of the topsoil. In agriculture, there is a need to utilize compost to to provide supplements to maintain the quality of the soil. Since manures are a fundamental concern, creating nano-based compost could be a solution. Composts provide three basic elements: nitrogen, phosphorous, and potassium in equivalent proportions (Corradini et al., 2010), thus enriching the soil. Nanocomposts build the nutrient use effectiveness (NUE) of the soil by multiple times, and they likewise give a pressure enduring capacity. Nanomanures contain supplements; these are embodied in a nanoscale polymer, enabling the effective discharge of nutrients into the soil. Nanotechnology offers gigantic potential for development through its applications in farming (Ditta, 2012). Nanotechnology-based gadgets are being investigated in the field of plant propagation and for improving genetic traits (Torney et al., 2007). Further, the agrarian produce or its waste, like wheat straw and soy structures, could be successfully used by changing to bio-nanocomposites, with upgraded physical and mechanical properties, for bio-modern purposes (Alemdar and Sain, 2008).

10.2 NANOTECHNOLOGY IN AGRICULTURE

In December 2002, the United States Department of Agriculture (USDA) drafted the world's first 'guide' for applying nanotechnology to farming and nourishment. A huge gathering of strategy creators, policymakers, and researchers met at Cornell University (New York, USA) to share their vision on the most proficient method of nanoscale advances to revamp farming. Farming, as indicated by the new nanovision, should be uniform, progressively mechanized, industrialized, and yet fulfil the farming community. Nano-agrotechnology will expand agribusiness' capability to gather feedstocks with modern procedures. Small-scale farmers growing tropical products, such as rubber, cocoa, coffee, and cotton, will adapt to the new technology and materialize a nano-enabled economy. Nanotechnology is an incredible advancement in managing agricultural practice when implemented along with the conventional agricultural practice leading to precise cultivation.(Figure 10.1). Precision development is a strategy used for crop yield improvement by checking biological factors and applying controlled target action as demonstrated by each condition (Chen and Yada, 2011). The use of nanoformulation, nanoencapsulation, and functionalized nanomaterial, developed using cutting-edge technology, is applied for synthesizing fertilizers and pesticides to protect plants (Nair et al., 2010; Ghormade et al., 2011; Khot et al., 2012). Nano-agritechnology is an emerging frontier and has been opened up new strategies for agricultural practices by developing

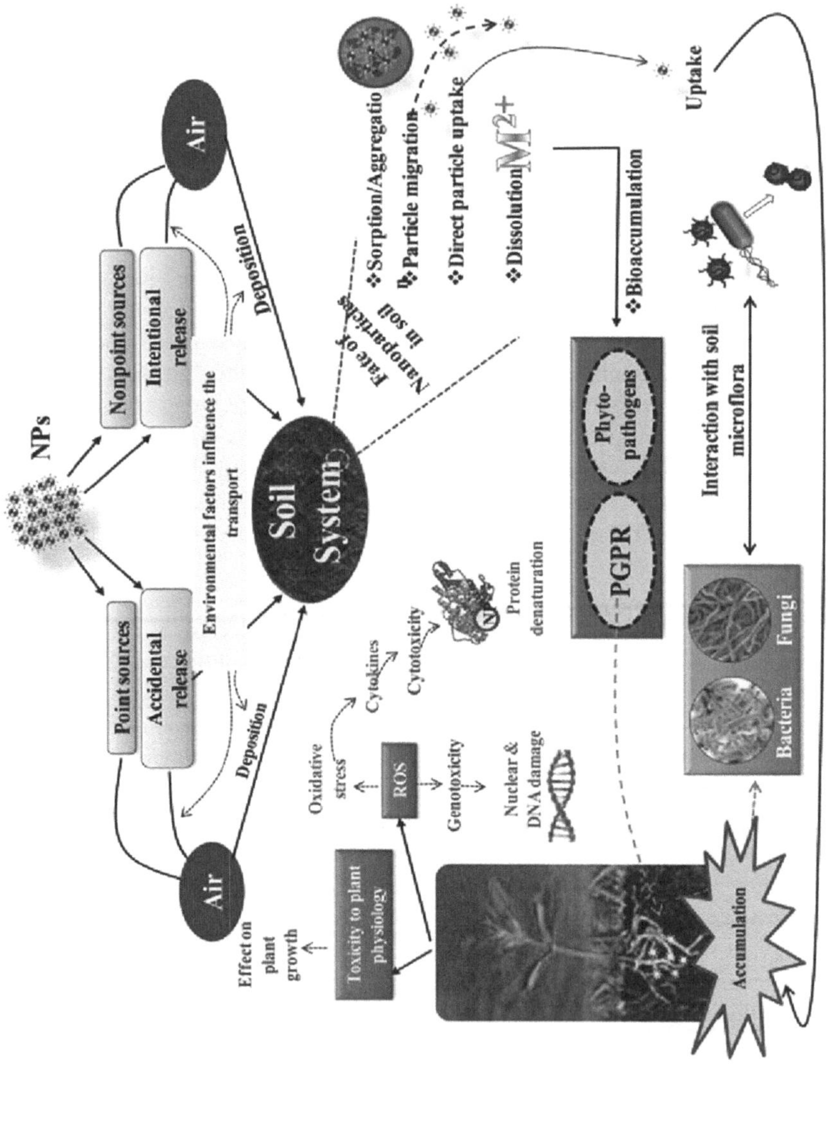

FIGURE 10.1 The course and impact of nanoparticles in the agro-environment.

nanofertilizers, nanopesticides, nanoherbicides, and nanosensors (Scott and Chen, 2013). Nanofertilzers provide the right amount of bioavailable nutrient for vigorous plant growth and development; nanopesticides and nanoherbicides help protect plants and seedlings against pest and weed infestations; nano-biosensors are smart devices that monitor soil and plant health conditions.

10.2.1 Nanofertilizers

Nanofertilizers are nanomaterials that enhance plant growth and support sustainability (Liu and Lal, 2015). Novel nanomaterials based on polymeric, inorganic, and lipid nanostructures have been developed to increase productivity. Their use is translated into developing intelligent nanosystems for the immobilization of nutrients and their release into the soil. Such systems reduce the leaching of minerals while improving the uptake of nutrients and minimizing eutrophication. After the green revolution, a tremendous increase in the required crop yield was observed for meeting the needs of the world's nutritional requirement. The vast usage of chemical fertilizers was recommended for boosting crops. In due course, through the processes of volatilization and leaching, the efficiency of the fertilizers was lost, adding lots of contaminants to the environment. A strategy used to face such a situation and to ensure sustainability was by developing nanofertilizers, which are slow-release fertilizers that improve soil quality by making nutrients available. Considering plant supplemental needs, nanofertilizers are classified into the following: (a) full-scale nanofertilizers or macronutrient nanofertilizers; (b) micronutrient nanofertilizers; and (c) nanoparticulate fertilizers. Nanofertilizers contain zinc, silica, iron and titanium dioxide, Mn/ZnSe quantum dots, ZnCdSe/ZnS core-shell quantum dots gold nanorods, core-shell quantum dots, and recently these nanofertilizers have become available commercially. They help with the controlled release of nutrients, thus improving the quality of crops (Lowry et al., 2019, Kah et al., 2019).

The growth and healthy development of plants is dependent on an ample supply of nutrients. Nitrogen (N), phosphorus (P), potassium (K), magnesium (Mg), sulphide (S), and calcium (Ca) are consolidated as macronutrients for plant growth. Urea encapsulated in zeolite chips and urea-balanced hydroxyapatite nanoparticles have demonstrated their capacities for providing a moderate and controlled availability of nitrogen to plants over a long period (Millan et al., 2008; Kottegoda et al., 2011). Additionally, Liu, and Lal (2014) prepared Ca and P hydroxyapatite nanoparticles and demonstrated a 20% and 33% growth in *Glycine max* seed yield, which was assessed against standard phosphorus. Previously, Liu et al. (2004) used calcium nanofertilizers as the supplement and evaluated a 15% increase in the biomass of *Arachishypogeae*. Further, Delfani et al. (2014) made magnesium nanofertilizers for supplementing Mg and evaluated a 7% expansion in *Vigna unguiculata* seed weight.

Micronutrients are required in minute amounts (B100 ppm), but yet they are central for different metabolic processes in plants. The nanoforms of micronutrients improve their bioavailable capacity to the plant and shows a colossal improvement in plant development and sustenance ability. Iron nanoparticles used as a micronutrient nanofertilizer on black-eyed peas were evaluated with a 10% increment in chlorophyll content in the leaves. In *G. max* chlorophyll content was extended using iron nanoparticles at 30–60 ppm. Ghafariyan et al. (2013) showed that the application of manganese nanoparticles to *Vigna radiata* extended root length by 52%, shoot length by 38%, rootlet by 71%, and biomass by 38% at 0.05 ppm compared with the use of bulk MnSO4. Zinc is a basic micronutrient and essentially required for different enzymatic activities in plants. ZnO nanoparticle showed an essential improvement in biomass, shoot length, root, chlorophyll and protein content, and phosphatate for development in *Vigna exude, Cicer ariatium, Cucumis sativas, Raphanus sativus, Brassica napus*, and *Cluster bean* (Lin and Xing, 2007; Zhao et al., 2013). Copper NPs improved photosynthesis in *Elodea despatch* by 35% at low concentrations, and seedling advancement up to 40% in lettuce. Molybdenum nanoparticles also exhibited improved microbial activity and seed improvement in chickpeas (Taran et al., 2014).

Diverse nanoparticles, for instance, TiO_2, SiO_2, and carbon nanotubes, (CNTs) demonstrated plant improvement. A TiO_2 and SiO_2 mixture extended nitrogen fixation in *Glycine max* and improved seed germination. CNT was supplemented in vegetables (tomato, cabbage, carrot, ambush, onion, and cucumber) and seeds (soybean, ryegrass, and corn) with an improvement in yield (Lin and Xing, 2007). Developing seeds to efficiently absorb carbon nanotubes boosted water uptake and improved plant growth (Srinivasan and Saraswathi, 2010).

10.2.2 Nanopesticide

The nanoformulation of conventional pesticides with polymers or metal nanoparticles is significantly valuable in the pesticide industry. The nanoencapsulation of pesticides is ideal for the controlled release of pesticides, thus dynamically fixing and controlling the pests with the slow release of pesticides within a set time frame. Nanoemulsions with water or oil improve the dissolution and potency of pesticides against different pests (Wang et al., 2007; Liu et al., 2006). Silver nanoparticles (Ag) have a broad range of action against phytopathogens: *Biploarisso rokinniana, Botrytis cinerea*, and *Colletotrichum gloeosporioides, Fusarium culmorum, Phythiumultimum, Phoma, Megnaporthegrisea, Trichoderma* sp. *Scalerotiniasclerotiorum, Sphaerotheca pannasa*, and *Rhizoctoniasolani* (Gajbhiye et al., 2009; Gopal et al., 2011). Silica nanoparticles with silver (Si–Ag NP) have been represented as very potent against development disorder in cucurbits. Porous silica nanomaterial was exploited for its use as a transporter of valinomycin and showed a controlled release of the pesticide (Liu et al., 2006). Likewise, Cu nanoparticles were reported to be compelling against Gram-positive and Gram-negative microorganisms and parasitic pathogens like *Fusarium* sp., *Phytophthora infestance, Xanthomonas oryzae*, and *Xanthomonas campestris* at low concentrations (Esteban-Tejeda et al., 2009). Zn nanoparticles were presented as nanofertilizers with an antifungal activity against *Penicillium expanalout, B. cinerea, Aspergillus flavus*, and *Aspergillus niger* (He et al., 2011). ZnO nanoparticles were less perilous to plants in assessment with Ag NP and could be an alternative to pesticide. Similarly, nano sulphur and a nanoformulation of hexaconazole were significantly toxic against plant pathogens like

Rhizopus solani, *Erysiphecichoracea*, and red little bug *Tetranychusutricae*. The impact of these nanomaterials on the soil and ecosystem should be dissected before their wide use and commercialization. To avoid hazardous impacts of their usage, a Nanopesticide Act should be formulated and considered by government associations.

10.2.3 Nano-Antimicrobials for Phytopathogens

Nanoscale materials are increasingly being used as novel antimicrobial agents due to their high surface area, and hence their ability to penetrate cells (Morones et al., 2005). Nanosilver is one such model, which is known to attack a wide range of microorganisms by distorting the cell bilayer membrane structure. This impedes the receptor proteins involved in ATP synthesis (Yamanaka et al., 2005; Lewinson et al., 2009). Polymer-based copper nanoparticles have been investigated for their antifungal action against plant pathogens. Silica silver nanoparticles have been presented as being convincingly antimicrobial against the plant pathogens *Rhizoctonia solani*, *Pythium ultimum*, *Botrytis cinerea*, *Magna porthegrisea*, and *Colletotrichum gloeosporioides* (Park et al., 2006). Antifungal and antibacterial nanoparticles are effective against plant pathogens such as Raffaelea sp., *Bipolaris sorokiniana*, *Magnaporthe*, *Grisea*, *Fusarium*, *Phoma*, and various other Gram-negative and Gram-positive microorganisms (Gajbhiye et al., 2009; Esteban-Tejeda et al., 2010). Nano-based formulations were used for the control of pumpkin illness. The development of parasitic hyphae and conidial germination could be significantly limited by nano-based methods, especially silver and copper nanoparticles. Thus, nanoherbicides, nano-fungicides, and nanopesticides have a great potential for augmentation in agriculture. Nanoemulsions are used for the preservation of pre- and post-harvest agricultural produce (Rickman et al., 1999; Zahir et al., 2012).

10.3 NANOTECHNOLOGY AND INTEGRATED PEST MANAGEMENT

10.3.1 Nanocarriers

Nanotechnology has an extended application in integrated pest management (IPM) because of the pest controlling capacity of nanomaterials. Nanocarriers were demonstrated to be feasible against a wide range of plant pathogens and pests (Esteban-Tejeda et al., 2010). The nanocarriers that have been used for the controlled release of agrochemicals are nanospheres, nanogels, nanotubes, and micelles. For example, permeable silica nanoparticles encapsulated within validamycin, sustainably diffused validamycin under controlled conditions (Liu et al., 2006). A wide range of agrarian pest infestations could be restrained with the utilization of nano-silica as a carrier for pesticides. Nanoparticles crosslinked with polyethylene glycol carrying garlic oil were exhibited to display biocidal action against *Tribolium castaneum*, a red flour scarab (Yang et al., 2009). Nanoemulsions are also efficient pesticides which deliver water-insoluble pesticides (Gao et al., 2007).

10.3.2 Nanospheres

Nanospheres are aggregates in which active ingredients are homogeneously mixed into polymeric structures. They are spherical particles with a size range between 10–200 nm. Due to their nano size, they exhibit novel physical, chemical, and biological properties comparable to their native element/compound. They can be synthesized by sequential dissociation, capturing, and encapsulation by fabricating active compounds with polymers (Sneharani and Bryappa, 2020). Nanospheres can be in suspension form or crystalline in nature (Singh et al., 2010).

10.3.3 Nanogels

Nanogels are considered favourable carriers over nanospheres, as they are insoluble in water and likewise less inclined to expansion or contraction with changes in moisture and environmental conditions (Bhagat et al., 2013). They significantly improve binding and release profiles (Paula et al., 2011). Nanogels have been proposed for the encapsulation and release of pheromones and major oils. Pheromones are seen as significantly unequivocal and eco-obliging natural control administrators, yet their course of action requires a moderate release. Bhagat et al. (2013) proposed the immobilization of pheromones inside a nanogel without using any cross-linkers (Bhagat et al., 2013). Scattering the pheromones in the nanogel was significantly decreased, and the efficacy could be extended up to 33 weeks when compared with only three weeks. The efficacy of a nanogel of the essential oil lippiasidoides was furthermore reported to be better than free oil. The use of unadulterated chitosan nanogels to improve the execution of antifungal prescriptions is reliant on copper (Brunel et al., 2013). Development of a copper (II)–chitosan complex is pH-dependent. Since most parasites will as a rule decrease the pH of their surroundings, copper (II) can be adequately applied during the advancement of the pathogen. A strong synergistic effect between chitosan and copper for preventing the development of *Fusarium graminea* was reported (Brunel et al., 2013). In tea leaves, chitosan nanoparticles showed promising phytoimmunogenic properties exhibited with the upregulation of defensive enzymes, like polyphenol oxidase, peroxidase, and nitric oxide. The results were compared with bulk chitosan (Chandra et al., 2015).

10.3.4 Nanotubes

Organic matter bound to carbon-based nanostructures, like fullerene-multiwalled coated nanotubes (MWCNT), was reported to be efficiently absorbed by seeds and seedlings. Carbon nanotubes were reported to generate new pores or regulate aquaporins leading to increased water permeation. This resulted in rapid seed germination and root growth, along with the enhancement of biomass. In seedlings of bitter gourd (*Momordica charantia*), the accumulation of fullerol led to the increased expression of anti-cancer and anti-diabetic compounds in fruits (Miralles et al., 2012). Despite the above positive effects of carbon nanotubes, the generation of free radicals and their facilitation in the uptake of heavy metals in plants are a few contrasting effects reported by Vithanage et al. (2017).

10.3.5 Nano-Biosensors

Nanotechnology finds its application in analytical devices due to its unique surface chemistry, electrical, thermal, and physical properties. Such unifying properties of nanoparticles are useful and employed to design analytical devices with enhanced sensitivity in their detection limits, cost-effective, and less response time, which can be embedded in multiplexed sensor systems. Nano-biosensor devices combine a biological recognition component, i.e. bioreceptor, with physical or chemical and electronic principles to assess the bioreliable changes in electrical signals (Gronow, 1984). Biosensors offer coherent mechanical assemblies with critical applications in the environmental, clinical diagnostic, food, beverage, and drug industries, and defence security, astronomy, and agriculture. In agriculture, nano-biosensors can be suitably used for distinguishing between a wide scope of fertilizers, herbicides, pesticides, pathogens, moisture content, physiology of seeds, seedlings and plants, and soil pH, and their controlled use can reinforce reasonable agriculture for improving yield (Rai et al., 2012). Nanosensors help farmers to maintain harvests with precision through the health monitoring of crops, by analyzing the nutrient requirements, pest infestations and management, and maximizing the efficiency of water absorption. To address these issues in plant health and growth, fabrication of nanomaterials, such as gold, silver, cobalt, carbon nanotubes, magnetic nanoparticles, and quantum dots, are developed for their use as biosensors. The fundamental guideline in soil investigation with a biosensor is to survey the general development of good and bad microorganisms in the soil. Field utilization of miniaturized biosensors can lessen the time required for microbial testing and immunoassays, and acknowledge the presence of contaminants in water supplies. Biosensors give high specificity and sensitivity, sharp responses, and are economic to use. Biosensors are used to detect Asian rust or soybean rust and aflatoxin in olive oil using the surface plasmon resonance (SPR) technique (Mendes et al., 2009).

Significant advancements have been made towards the association of colloidal semiconductor quantum spots (QDs), particularly using CdSe, CdS, and CdTe in developing biosensors (Park et al., 2007; Reiss et al., 2009). Semiconductor QDs have broad absorption spectra, with sharp fluorescence spectra spanning the UV to IR range along with a high quantum yield and molar extinction coefficient and high resistance to photobleaching and photochemical degradation. The exceptional spectral and fluorescent properties of QDs have been applied in bioimaging and biosensing. At low concentrations, QDs showed no cytotoxicity during seed germination and growth. This approach can be used to study the physiological process as a live imaging system. Pesticides/herbicides and growth hormones have been commonly used in agrarian cultivation. Indiscriminate use of these chemicals mean they find their way into the soil and from the soil will enter plants and microorganisms, which thereby also enters into different trophic levels. The tissues of animals feeding on such plants absorb these chemicals leading to the deposition and accumulation of toxic chemicals; a process termed bioaccumulation. Bioaccumulation hampers the metabolism and physiology of an organism leading to disorder and disease. Bioaccumulation can be prevented by checking that the presence of toxic chemicals in the soil and plants is beyond tolerable concentration. Semiconductor quantum dots in the form of biosensors are implemented to detect the for presence of such toxic pesticides and hormones in the soil. The optical properties of QDs basically and in conjugation with various substances have been comprehensively examined for their activity on the growth of plants. The real-time monitoring of cultivated lands is performed using a nanosensor-based global positioning system (GPS). Such real-time wireless monitoring gives comprehensive data on crop growth and provides opportunities for crop management. The automation of irrigation systems by nano-biosensors provide an excellent opportunity to maximize water use efficiency during irrigation. The application of nanosensors in agriculture is still developing and provides hope for a sustainable agriculture; however, the field requires much more input for successful implementation (Figure 10.1).

10.4 ECOTOXICOLOGICAL IMPLICATIONS OF NANOMATERIALS IN SOIL HEALTH

The use of nanoparticles in various forms and structures has raised concerns about the disadvantages and harmful effects on living organisms. Hence, why the relationships between nanoparticles and natural components like plants, microorganisms, and soil have always been considered (Bakshi et al., 2014). Once the nanoparticles find their way into the soil, their transport, bioavailable capacities and resulting risks are all considered as they could impact the earth's core properties (Shoults-Wilson et al., 2011; Cornelis et al., 2012). The earth's pH and cation exchange limit impacts the bioavailability of silver nanoparticles and silver metals in the soil. At higher pH, the soil shows an increasingly significant cation exchange limit due to which silver particles are absorbed into the earth's surface, which reduces their bioavailability. The physical properties of the soil and nanoparticles like size, shape, and surface charge are known to contribute to the stability of nanoparticles.

Various agencies involved in food security strongly promote 'sustainable crop intensification', which aims to meet the needs of the growing demand for food all over the world. Under the present scenario of climate change, nanotechnology is among the emerging technologies to address the needs of sustainable crop production (Leistritz et al., 2007). Nanomaterials, such as nanopesticides, nanofertilizers, nano-biosensors, nanocarriers, nanomaterials in packaging, and more, have revolutionized the agricultural sector, during pre- and post-harvesting.

The advancement of nanotechnology has led to extensive research on evaluating the engineered nanomaterials and evaluating the undesirable and toxic effects on the environment. Metal nanoparticles are antibacterial and antifungal possessing cytotoxic effects. The cytotoxic effects of metal nanoparticles are due to the electrostatic interactions between the charge on the cell membrane and the nanoparticles, which eventually gets accumulated into the cytoplasm. Several other types of nanoparticles (TiO_2, ZnO, and carbon-based nanomaterials) are photochemically active and when exposed to light they get excited. The excited electrons generate superoxide radicals that harm the cells. The broad range of antimicrobial properties of nanoparticles, particularly nanosilver, against human and plant pathogens have been generally reported (Shahverdi et al., 2007). Investigations have warned of the looming effect of silver nanoparticles on denitrifying microorganisms, which agitates and imbalances the denitrification in the soil (VandeVoort and Arai, 2012). The effect of nanoparticles on *Pseudomonas stutzeri* (denitrifier), *Azotobacter vinelandii* (nitrogen fixer), and *Nitrosomonas europaea* (nitrifier) have been studied (Yang et al., 2013). Zhang et al. (2014) reported no significant impact on the long-term use of nanosilver at a concentration of 0.10 μg/mL on the nitrifying bacterial system. At low doses, the nanosilver had no adversarial effect on nitrate reductase activity of *Rhizobium* and *Azotobacter.* Microorganisms play a vital role in maintaining soil health and thereby crop productivity. Nanoparticles may interact with other pollutants or toxic chemicals in the soil and enter and accumulate in different parts of the plant, leading to bioaccumulation. By entering through various trophic levels, bioaccumulated toxins cycle in the ecosystem. Hence, the introduction of nanoparticles in any form into the agricultural field should be followed by a routine check-up to maintain a friendly ecosystem. Ecotoxicological research needs extensive exploration to study the course and consequences of the materials gaining entry into different levels of the natural ecosystem (**Figure 10.1**).

10.5 FUTURE PERSPECTIVES

In this era of global growth and global warming, smart agriculture is key to meeting hunger. This can be realized by implementing nanotechnology, which is an emerging technology in sustainable agriculture. Sustainable agriculture should yield a balanced ecosystem which considers both abiotic factors and biotic living organisms, leading to a stabilized food chain and energy balance. Nevertheless, toxicological effects on living organisms, abiotic factors, and ecological cycle should be considered. International co-operation with defined regulation and legislation similar to genetically modified crops and organisms is necessary for the successful implementation of this technology.

10.6 CONCLUSION

Nanotechnology has an innovative and progressive growth in agribusiness, giving manageable devices to customary cultivating practices. It has many agricultural applications such as nanofertilizers, nano-biosensors, and nanopesticides. Broad research on the utilization of nanomaterials in agriculture is becoming normal, yet the threat to ecological security and sanitation should not be neglected. Understanding the impact of nanomaterials on the environment and ecological cycle is still undeveloped. Multifunctional, cost-effective, viable, and environment-friendly devices can be developed by combining expertise from material science, agronomy, chemistry, plant and soil physiology, and ecotoxicology. Sustainable and successful implementation is possible by government agencies aided by the knowledge of data scientists, artificial intelligence, and agribiotechnologists.

REFERENCES

Alemdar A, Sain M (2008) Isolation and characterizationof nanofibers from agricultural residues -Wheat strawand soy hulls. *Bioresour Technol* 99:1664–1671.

Bhagat D, Samanta SK, Bhattacharya S (2013) Efficient management of fruit pests by pheromone nanogels. *Sci Rep* 3:1–8.

Brunel F, El Gueddari NE, Moerschbacher BM (2013) Complexation of copper (II) with chitosan nanogels: Toward control of microbial growth. *Carbohydr Polym* 92(2): 1348–1356.

Chandra S, Chakraborty N, Dasgupta A, Sarkar J, Panda K, Acharya K (2015) Chitosan nanoparticles: A positive modulator of innate immune responses in plants. *Sci Rep* 5(15195):1–14.

Chen H, Yada R (2011) Nanotechnologies in agriculture: New tools for sustainable development. *Trends Food Sci Technol* 22:585–594. doi: 10.1016/j.tifs.2011.09.004

Cornelis G, Doolette C, Thomas M, McLaughlin MJ, Kirby JK, Beak DG, Chittle Borough D (2012) Retention and dissolution of engineered silver nanoparticles in natural soils. *Soils Sci Soc Am J* 76:891–902.

Corradini E, Moura MR, Mattoso LHC (2010) A preliminary study of the incorporation of NPK fertilizer into chitosan nanoparticles express. *Polymer Lett* 4: 509–515.

Ditta A (2012) How helpful is nanotechnology in agriculture. *Adv Nat Sci Nanosci Nanotechnol* 3:033002.

Delfani M, Firouzabadi MB, Farrokhi N, Makarian H (2014) Some physiological responses of black-eyed pea to iron and magnesium nanofertilizers. *Commun Soil Sci Plant Anal* 45:11. doi: 10.1080/00103624.2013.863911.

Esteban-Tejeda L, Malpartida F, Esteban-Cubillo A, Pecharromán C, Moya J (2009) Antibacterial and antifungal activity of a soda lime glass containing copper nanoparticles. *Nanotechnology* 20:505701.

Esteban-Tejeda L, Malpartida F, Pecharroman C, Moya JS (2010) High antibacterial and antifungal activity of silver monodispersed nanoparticles embedded in a glassy matrix. *Adv Eng Mater* 12(7): B292–B297.

Gajbhiye M, Kesharwani J, Ingle A, Gade A, Rai M (2009) Fungus mediated synthesis of silver nanoparticles and its activity against pathogenic fungi in combination with fluconazole. *Nanomedicine* 5(4):282–286.

Gao L, Zhuang J, Nie L, Zhang J, Zhang Y, Gu N, Yan X (2007) Intrinsic peroxidase-like activity of ferromagnetic nanoparticles. *Nat Nanotechnol* 2(9):577–583.

Ghormade V, Deshpande MV, Paknikar KM (2011) Perspectives for nano-biotechnology enabled protection and nutrition of plants. *Biotechnol Adv* 29:792–803. doi: 10.1016/j.biotechadv.2011.06.007.

Ghafariyan MH, Malakouti MJ, Dadpour MR, Stroeve P, Mahmoudi M (2013) Effects of magnetite nanoparticles on soybean chlorophyll. *Environ Sci Technol* 47:10645–10652. doi: 10.1021/es402249b.

Gopal M, Gogoi R, Srivastava C, Kumar R, Singh PK, Nair KK et al. (2011) Nanotechnology and its application in plant protection. In: Thind TS et al. (eds.), *Plant Pathology in India: Vision 2030*. pp. 224–230.

Gronow M (1984) Biosensors. *Trends Biochem Sci* 9:336–340.

He L, Liu Y, Mustapha A, Lin M (2011) Antifungal activity of zinc oxide nanoparticles against *Botrytis cinerea* and *Penicillium expansum*. *Microbiol Res* 166:207–215.

Kah M, Tufenhji N, White JC (2019) Nano-enabled strategies to enhance crop nutrition and protection. *Nat Nanotechnol* 14:532–540.

Khot LR, Sankaran S, Maja JM, Ehsani R, Schuster EW (2012) Applications of nanomaterials in agricultural production and crop protection: A review. *Crop Prot* 35:64–70. doi: 10.1016/j. cropro.2012.01.007.

Kottegoda N, Munaweera I, Madusanka N, Karunaratne V (2011) A green slow-release fertilizer composition based on urea-modified hydroxyapatite nanoparticles encapsulated wood. *Curr Sci* 101:73–78.

Leistritz FL, Hodur NM, Senechal DM, Stowers DM, Mc Calla D, Saffron CM et al. (2007) Biorefineries using Agri-cultural Residue Feedstock in the Great Plains. AAEReport 07001 Working Paper.

Lewinson O, Lee AT, Rees DC (2009) A P-type ATPase importer that discriminates between essential and toxic transition metals. *Proc Natl Acad Sci* 106:4677–4682.

Liu F, Wen L-X, Li Z-Z, Yu W, Sun H-Y (2006) Porous hollow silica nanoparticles as controlled delivery system for water-soluble pesticide. *Mater Res Bull* 41(12): 2268–2275.

Liu R, Lal R (2014) Synthetic apatite nanoparticles as a phosphorus fertilizer for soybean (*Glycine max*). *Sci Rep* 4:6. doi: 10.1038/ srep05686.

Liu R, Lal R (2015) Potentials of engineered nanoparticles as fertilizers for increasing agronomic productions. *Sci Total Environ* 514:131–139. doi: 10.1016/j.scitotenv.2015.01.104.

Liu X, Zhang F, Zhang S, He X, Wang R, Fei Z et al. (2004) Responses of peanut to nanocalcium carbonate. *Plant Nutr Fertil Sci* 11:385–389.

Lin D, Xing B (2007) Phytotoxicity of nanoparticles: Inhibition of seed germination and root growth. *Environ Pollut* 150:243–250. doi: 10.1016/j.envpol.2007.01.016

Liu X-M, Feng Z-B, Zhang S-Q, Zhang F-D, Zhang J-F, Xiao Q, Wang Y-J (2006) Preparation and testing of cementing sub-nano composites of slower controlled release of fertilizers. *Sci Agric Sin* 39:7.

Lowry G, Avellan A, Gilbertson LM (2019). Oppurtunities and challenges for nanotechnology in agri-tech revolution. *Nat Nanotechnol* 14:517–522.

Mendes RK, Ferreira DCM, Carvalhal RF, Peroni LA, Stach-Machado DR, Kubot LT (2009) Development of an electrochemical immunosensor for *Phakopsorapachy rhizi*. detection in the early diagnosis of soybean rust. *J Braz Chem Soc* 20:795–801.

Millan G, Agosto F, Vazquez M (2008) Use of clinoptilolite as a carrier for nitrogen fertilizers in soils of the Pampean regions of Argentina. *Cienc Investig Agrar* 35:293–302.

Miralles P, Johnson E, Church TL, Harris AT (2012) Multiwalled carbon nanotubes in alfalfa and wheat: Toxicology and uptake. *J R Soc Interface* 9:3514–3527.

Morones JR, Elechiguerra JL, Camacho A, Holt K, Kouri JB, Ramirez JT, Yacaman MJ (2005) The bactericidal effect of silver nanoparticles. *Nanotechnology* 16: 2346–2353.

Nair R, Varghese SH, Nair BG, Maekawa T, Yoshida Y, Kumar DS (2010) Nanoparticulate material delivery to plants. *Plant Sci* 179:154–163. doi: 10.1016/j.plantsci.2010.04.012.

Park HJ, Kim SH, Kim HJ, Choi SH (2006) A new composition of nanosized silica-silver for control of various plant diseases. *Plant Pathol J* 22:295–302.

Park J, Joo J, Kwon SG, Jang Y, Hyeon T (2007) Synthesis of monodisperse spherical nanocrystals. *Angew Chem Int Ed* 46:4630–4660.

Paula HCB, Sombra FM, Cavalcante RF, Abreu FOMS, de Paula RCM (2011) Preparation and characterization of chitosan/cashew gum beads loaded with *Lippiasidoides* essential oil. *Mater Sci Eng C Mater Biol Appl* 31(2):173–178.

Rai V, Acharya S, Dey N (2012) Implications of nanobiosensors in agriculture. *J Biomater Nanobiotechnol* 3:315–324.

Reiss P, Protiere M, Li L (2009) Core/shell semiconductor nanocrystals. *Small* 5:154–168.

Rickman D, Luvall JC, Shaw J, Mask P, Kissel D, Sullivan D (1999) Precision agriculture: Changing the face of farming. Geotimes feature article. www.ghcc.msfc.nasa.gove/precisionag Accessed 19 Nov 2011.

Scott N, Chen H (2013) Nanoscale science and engineering for agriculture and food systems. *Ind Biotechnol* 9:17–18. doi: 10. 1089/ind.2013.1555.

Shahverdi RA, Minaeian S, Shahverdi RH, Jamalifar H, Nohi AA (2007) Rapid synthesis of silver nanoparticles using culture supernatants of Enterobacteria: A novel biological approach. *Process Biochem* 42:919–923.

Shoults-Wilson W, Zhurbich O, McNear D, Tsyusko O, Bertsch P, Unrine J (2011) Evidence for avoidance of Ag nanoparticles by earthworms (*Eisenia fetida*). *Ecotoxicology* 20:385–396.

Singh A, Garg G, Sharma PK (2010) Nanospheres: A novel approach for targeted drug delivery system. *Int J Pharm Sci Rev Res* 5:84–88.

Sneharani AH, Byrappa K (2020) Facile chemical fabrication of designer biofunctionalised nanomaterials, chapter 2. In *Functionalized Nanomaterials I*, Eds. Kumar V, Guleria P, Dasgupta N, Rajan S. CRC Press. doi: 10.1201/9781351021623.

Srinivasan C, Saraswathi R (2010) Nano-agriculture – Carbon nanotubes enhance tomato seed germination and plant growth. *Curr Sci* 99:2

Taran NY, Gonchar OM, Lopatko KG, Batsmanova LM, Patyka MV, Volkogon MV (2014) The effect of colloidal solution of molybdenum nanoparticles on the microbial composition in rhizosphere of *Cicer arietinum* L. *Nanoscale Res Lett* 9:289. doi: 10.1186/1556-276X-9-289.

Torney F, Trewyn BG, Lin VSY, Wang K (2007) Mesoporous silica nanoparticles deliver DNA and chemicals into plants. *Nat Nanotechnol* 2:295–300.

VandeVoort AR, Arai Y (2012) Effect of silver nanoparticles on soil denitrification kineics. *Ind Biotechnol* 8:358–364.

Vithanage M, Seneviratne M, Ahmad, M, Sarkar B, Ok YS (2017) Contrasting effects of engineered carbon nanotubes on plants: A review. *Environ Geochem Health* 39(6):1421–1439.

Wang L, Li X, Zhang G, Dong J, Eastoe J (2007) Oil-in-water nanoemulsions for pesticide formulations. *J Colloid Interface Sci* 314:230–235. doi: 10.1016/j.jcis.2007.04.079.

Yamanaka M, Hara K, Kudo J (2005) Bactericidal actions of a silver ion solution on *Escherichia coli*, studied by energy filtering transmission electron microscopy and proteomic analysis. *Appl Environ Microbiol* 71:7589–7593.

Yang F, L Li, XG, Zhu F, Lei CL (2009) Structural characterization of nanoparticles loaded with garlic essential oil and their insecticidal activity against *Triboliumcastaneum* (Herbst) (Coleoptera: Tenebrionidae). *J Agri Food Chem* 57(21):10156–10162.

Yang Y, Wang J, Xiu Z, Alvarez PJJ (2013) Impact of silver nanoparticles on cellular and transcriptional activity of nitrogen-cycling bacteria. *Environ Toxicol Chem* 32:1488–1494.

Zahir AA, Bagavan A, Kamaraj C, Elango G, Rahuman AA (2012) Efficacy of plant-mediated synthesized silver nanoparticles against *Sitophilus oryzae*. *J Biopest* 288:95–102.

Zhao L, Sun Y, Hernandez-Viezcas JA, Servin AD, Hong J, Niu G et al. (2013) Influence of CeO_2 and ZnO nanoparticles on cucumber physiological markers and bioaccumulation of Ce and Zn: A life cycle study. *J Agri Food Chem* 61:11945–11951. doi: 10.1021/jf404328e.

Zhang C, Liang Z, Hu Z (2014) Bacterial response to a continuous long term exposure of silver nanoparticles at sub-ppm silver concentrations in a membrane bioreactor activated sludge system. *Water Res* 50:350–358.

11 Emerging Role of Biosynthesized Silver Nanoparticles for Agricultural Applications

Shaheen Husain, Suresh K Verma, Aditya Nandi, Shumaila Akram, Durdana Yasin, and Hemlata

CONTENTS

11.1 INTRODUCTION

Nanoscience and nanotechnology have been the cynosure over the last few years due to their potential effects on many scientific fields, such as oil, medicine, pharmaceutical, electronics, and space industries. Nanoscience and nanotechnology have attracted a great deal of interest over the past few years. The area focuses on small structures and low dimensional materials ranging from a few nanometres to less than 100 nm. Due to their wide surface-to-volume ratio, nanoparticles (NPs) exhibit special and significantly modified chemical, physical, and biological properties compared to bulk materials with the chemical composition. The most effectively studied nanoparticles today are those made from noble metals, in particular, silver (Ag), platinum (Pt), gold (Au), and palladium (Pd). Silver has especially drawn the attention of scientists because of its extensive application in the development of new technology in the areas of electronics, material science, and medicine at the nanoscale. New applications of nanoparticles and nanomaterials are emerging rapidly. Nanosilver particles have found tremendous applications in the field of high sensitivity biomolecular detection and diagnostics, antimicrobials, therapeutics, catalysis, and microelectronics. Therefore there is still a need for an economic and commercially viable as well as an environmentally friendly method for the synthesis of Ag NP.

Agricultural nutrient balances have differed noticeably with economic growth, and especially in developing countries the development of soil fertility is very important (Campbell et al., 2014). The development of agriculture is a compulsory for ending poverty and hunger, which is a critical situation. Therefore, we will have to take bold steps for agriculture development. New technology should be adopted that decidedly focuses on getting better agricultural production (Yunlong and Smit, 1994). The issue of food and nutritional security is a hot topic nowadays. Therefore, sustainable agriculture is strengthening the practical opportunities for ending poverty and hunger in poorer populations. Henceforth, a lot of new instrumentation techniques have been introduced and make use of synthesis, purification and characterisation of nanomaterials (Bonnell and Huey, 2001; Gibney, 2015). These materials have a role in the enhancement of agricultural productivity involving nanoporous materials for the slow release and efficient dosage of pesticides and fertilizers, leading to pest management. Nanoparticles have distinct properties, whether it they are physical, chemical, or biological when compared to the larger particles of bulk materials. It has been well established that metallic nanoparticles have unique catalytic activities for the removal of the contaminants, both organic and inorganic. The application of nanotechnological methods in plant pathology is still in its early stages concerning various agricultural purposes, such as the development of nanoformulations for pesticides, fungicides, herbicides, fertilizers, insecticides, etc. Efficient nanodelivery systems help to minimize the highly toxic fungicides and pesticides used in the agricultural sector. Similarly, nanoparticles help with gene transfer, which is an ideal tool for growing crops and plants resistant to pathogens and pests. Nanomaterials can be used as carriers for agrochemicals or used as biopesticides or biofertilizers. Nanoparticles have been proven to be efficient agrochemical agents for improving crop productivity and reducing pests. They have boosted agricultural applications by increasing the nutrient uptake by plants and also inhibiting the pathogens infecting the plants. Nanoparticles act as 'magic bullets' serving as herbicides, pesticides, and fertilizers, etc. (Figure 11.1). They efficiently penetrate the cuticle and tissues of the plants, thus allowing their slow and effective release into the target area. Nanotechnology has provided solutions to many agriculture-related problems, such as insect pest management and nutrient uptake efficiency, etc. Nanomaterials in different forms can be used for efficient pest management and crop production. Biosynthesized silver nanoparticles (Ag NPs) have proven their application for two major tasks of plant growth enhancement and plant disease management. Therefore, it can also be concluded that silver nanoparticles are an efficient tool for providing

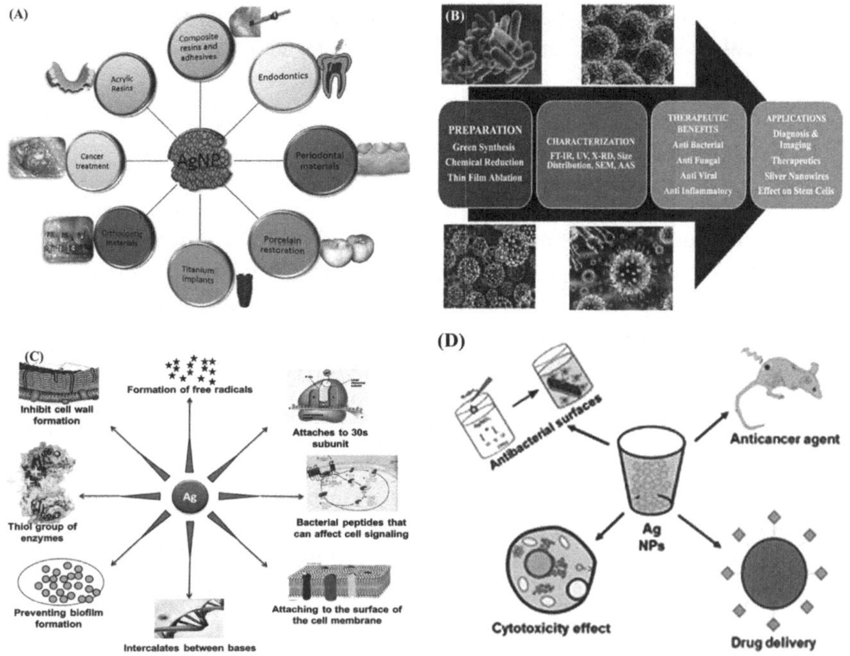

FIGURE 11.1 (A) Various applications of Ag NPs. (Figure adapted from Bapat et al., 2018.) (B) Preparation, characterization, therapeutic applications and applications of Ag NPs. (Figure adapted from Midha et al., 2016.) (C) Silver nanoparticles showing multiple bactericidal actions. (Figure adapted from Chen et al., 2014.) (D) (Figure adapted from Fahmy et al., 2019.)

green and eco-friendly alternatives. They have an antimicrobial activity against various plant pathogens and help the plants absorb important nutrients from the soil. The application of chemically synthesised nanomaterials is considered toxic; in this case, nanomaterials can be synthesised from the plant system and considered as green nanotechnology (Prasad, 2014a). Green nanotechnology is a safe process, which is energy-efficient, reduces waste, and lessens greenhouse gas emissions. The use of renewable materials in the production of such products is beneficial; thus, these processes have a low influence on the environment (Prasad et al., 2014b, 2016). Nanomaterials are eco-environmentally sustainable, and significant advances have been made in green nanotechnology. In the present decade, there has been more of a shift toward green nanotechnology for the implementation of its enhanced functions. In modern agriculture, sustainable production and efficiency are unimaginable without the use of agrochemicals such as pesticides, and fertilizers, etc. However, every agrochemical has potential issues, including the contamination of water or residues on food products that threaten human beings and environmental health; thus, precise management and control of inputs could reduce the risks (Kah, 2015). The development of the high-tech agricultural system with the use of engineered smart nanotools could be an excellent strategy for a revolution in agricultural practices, and thus reduce and/or eliminate the influence of modern agriculture on the environment as well as to enhance both the quality and quantity of yields (Sekhon, 2014; Liu and Lal, 2015).

11.2 CONVENTIONAL TECHNIQUES FOR SILVER NANOPARTICLE SYNTHESIS

Nanoparticles are often synthesized either through top-down or bottom-up approaches. The top-down approach is based on mechanical size reduction methods by slowly breaking down the bulk materials into nanostructures. Bottom-up methods are focused on the assembly of atoms or molecules into nanoscale molecular structures in nanoscale range (Narayanan and Sakthivel, 2010). Figure 11.2 indicates the different methods by which silver nanoparticles are synthesized. In top-down synthesis, nanoparticles are produced from a suitable starting material through a size reduction. Specific physical and chemical treatments can achieve a reduction in scale. There will be an imperfection in the surface structure of the substance in top-down approaches, and this is considered a significant limitation, due to the surface chemistry, and other physical properties of the nanoparticles, depending on the structure of the surface (Thakkar et al., 2011). Nanoparticles are composed of smaller molecules in the bottom-up method, such as atoms, molecules, and smaller structures (Hurst et al., 2006). Through this, the nanostructure, the nanoparticle building blocks, are first formulated and then assembled to create the final particle (Thakkar et al., 2011). Both chemical and biological approaches to metal nanoparticles rely on approaches from the bottom up (Narayanan and Sakthivel, 2010). The synthesis of metal nanoparticles is accomplished in bottom-up approaches with the chemical reduction process (Hurst et al., 2006). The chemicals involved are toxic during this form of synthesis process and have also contributed to non-eco-friendly by-products. Because of all these drawbacks these synthesis methods are not usually used for metal nanoparticle synthesis.

11.3 CHARACTERIZATION OF SILVER NANOPARTICLES

Silver nanoparticle characterization is also essential after the synthesis of nanoparticles. In general, nanoparticles are characterized by their scale, shape, dispersity, and surface area (Jiang et al., 2009). The developing techniques for nanoparticular characterization are as follows: UV–visible spectrophotometry, powder X-ray diffraction (XRD), dynamic light scattering (DLS), Fourier transform infrared spectroscopy (FTIR), scanning electron microscopy (SEM), transmission electron microscopy (TEM), and energy-dispersive spectroscopy (EDS) (Husain et al., 2015, 2019; Shahverdi et al., 2007) (Figure 11.3). UV–visible spectroscopy is a commonly used technique in which a light wavelength in the range 300–800 nm is generally used for the characterization of metal nanoparticles in the size range of 2–100 nm (Feldheim et al., 2002). For the characterization of silver nanoparticles, spectrophotometric absorption measurements in the size range of 400–450 nm is used. Similarly, DLS is used for the characterization of surface charge and the size distribution of the

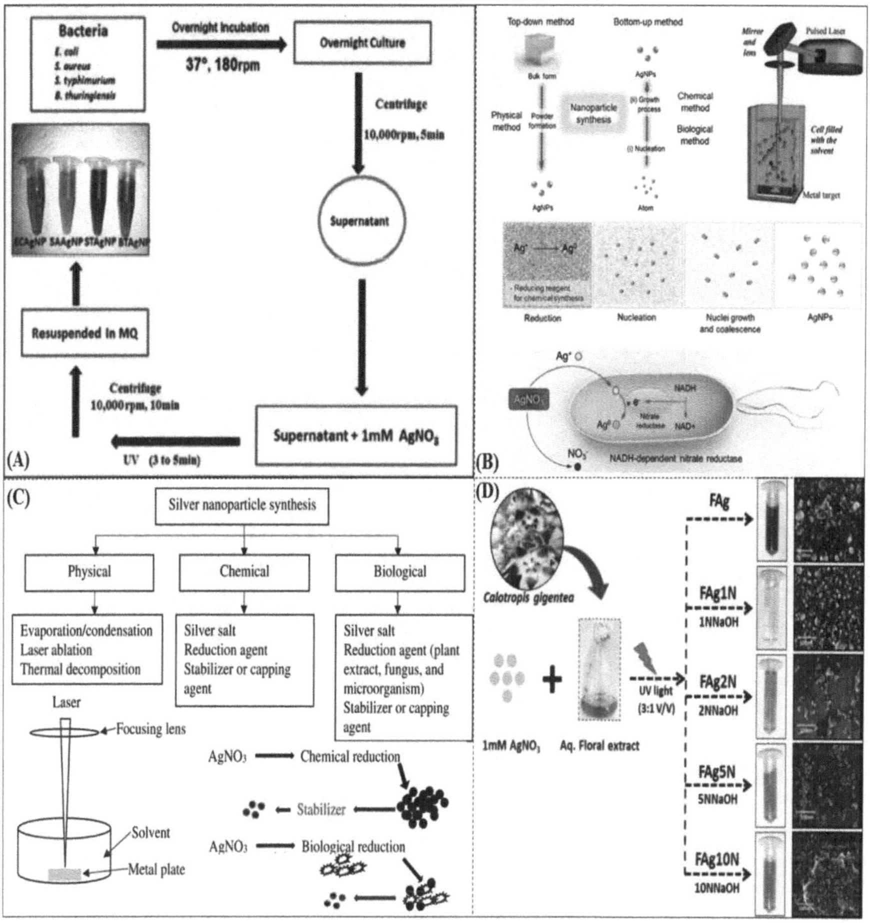

FIGURE 11.2 (A) Schematic representation of the synthesis of silver nanoparticles (ECAgNP, SAAgNP, STAgNP, and BTAgNP) from culture supernatant of *E. coli*, *S. aureus*, *Salmonella typhimurium*, and *B. thuringiensis*. (Figure adapted from Verma et al., 2018a.) (B) Diverse synthesis routes of silver nanoparticles (Ag NPs). Top-down and bottom-up methods, physical synthesis method, chemical synthesis method, plausible synthesis mechanisms of green chemistry. The electron transfer through nicotinamide adenine dinucleotide (NADH)-dependent reductase as an electron carrier to form NAD+, initiates the bioreduction. The resulting electrons are obtained by Ag+ ions which are reduced to elemental Ag NPs. (Figure adapted from Lee and Jun, 2019.) (C) General procedures to synthesize silver nanoparticles and schematic of the synthesis of silver nanoparticles using physical, chemical, and biological approaches. (Figure adapted from Syafiuddin et al., 2017.) (D) Schematic diagram showing the green synthesis of Ag NPs using *Calotropis gigantean* floral extract. (Figure adapted from Verma et al., 2018b.)

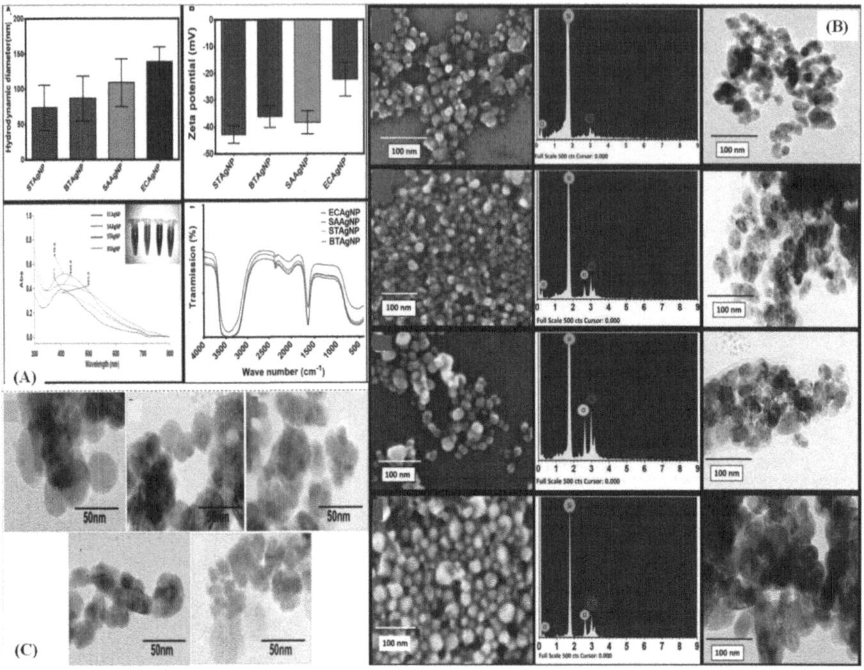

FIGURE 11.3 (A) Physiochemical characterizations of biogenic silver nanoparticles (ECAgNP, SAAgNP, STAgNP, and BTAgNP): hydrodynamic diameter; zeta potential; UV-vis spectra; and D, Fourier transform infrared spectra. Values are presented in Mean6SD. (B) Size determination and EDS analysis of silver nanoparticles by FESEM and TEM ECAgNP, SAAgNP, STAgNP, and BTAgNP. Right row presents FESEM images while middle and left rows present EDS and TEM images. ((A) and (B) are adapted from Verma et al., 2018a.) (C) Size determination of green synthesized Ag NPs from Calotropis gigantea by TEM – Fag, FAg1N, FAg2N, FAg5N, FAg10N. (Figure adapted from Verma et al., 2018b.)

nanoparticles, which are suspended in a liquid medium (Gunalan et al., 2011). The XRD pattern is also important for the phase identification and characterization of the crystallinity of nanoparticles and the phase distribution of the biosynthesized nanoparticles (Strasser et al., 2010). For structural information about the nanoparticles, X-rays penetrate the nanomaterials, and the resulting diffraction pattern is compared with the standard pattern to get the final information. Verma et al reported the formation of nanocrystals through the inference of the result obtained by Bragg's peak in XRD analysis. The characteristic peak of silver at 2Θ values found at 38.0, 45.45, 66.119, and 75.53 degrees corresponding to (111), (200), (220), and (311) planes exhibited the crystalline property of the silver nanoparticles. The elemental composition of nanoparticles is commonly determined using energy-dispersive spectroscopy (Sun et al., 2000). For the morphological characterization of nanoparticles at the nanometre and nanoscale ranges, SEM and TEM normally refers (Eppler et al., 2000). It has been found that TEM has a 1000-fold higher resolution than SEM. FTIR is also used for the characterization and identification of surface chemistry.

Organic functional groups (like carbonyls and hydroxyls) are found to attach to the surface of nanoparticles, and other surface chemical residues are generally determined using FTIR. Sankar et al. (2014) reported that FTIR spectroscopy is also used to determine the nature of the functional groups present on the surface of the nanoparticles and might be responsible for the synthesis and stabilization of metal nanoparticles.

11.4 ROLE OF SILVER NANOPARTICLES IN CROP PRODUCTION

Seed germination inhibition is a common toxic effect of metal-based nanoparticles. El-Temsah and Joner (2012) studied the toxic effect of zero-valent iron NPs and Ag NPs differing in average particle size from 1 to 20 nm on ryegrass, barley, and flaxseed germination exposed to 0–5000 mg L^{-1} zero-valent iron NPs or 0–100 mg L^{-1}Ag. Their study showed that inhibitory effects were observed in aqueous suspensions at 250 mg L^{-1} for zero-valent iron nanoparticles. Complete inhibition of germination was observed at 1000–2000 mg L^{-1} for zero-valent iron nanoparticles. AgNPs inhibited seed germination at lower concentrations, but showed no clear size-dependent effects, and never completely impeded germination. However, the positive effect of Ag NPs on seed germination has also been observed. Studies suggest that metal Ag NPs increase plant growth and development. Applications of Ag NPs increased plant height (Nejatzadeh et al., 2014), leaf number (Byczyńska, 2017), the weight of roots (Thuesombat et al., 2014), the biomass of the plant (Nejatzadeh et al., 2014; Byczyńska, 2017), seed germination (Mehrian et al., 2016), seed yield (Nejatzadeh et al., 2014; Byczyńska, 2017), quality of fruit (Shams et al., 2013), stem length (Gusev et al., 2016), the diameter of the canopy area (Aghajani and Ekhtiyari, 2013), root length (Gusev et al., 2016; Thangavelu et al., 2018; Sharma et al., 2012; Mehrian et al., 2016), and growth and development of explants under *in vitro* culture conditions (Homaee and Ehsanpour, 2015). The addition of an appropriate nanosilver concentration increased chlorophyll content (Sharma et al., 2012), carotenoids (Gusev et al., 2016), flavonoids (Homaee and Ehsanpour, 2015), photosynthetic quantum efficiency (Borase et al., 2014), enzymatic activity, the content of para hydroxybenzoic acid (Yasur and Rani, 2013), α-terpinyl acetate (Aghajani and Ekhtiyari, 2013), as well as antioxidative enzyme activities (Hatami and Ghorbanpour, 2014).

11.4.1 Ag NPs as Herbicides

Pesticides and herbicides are usually used in agriculture to get better crop yield and efficiency. Presently, the negative aspects of conventional pesticides and herbicides on the environment are contentious. The main disadvantages of pesticides are the development of pathogens and pest resistance, decrease in nitrogen fixation, reduction in soil biodiversity, the bioaccumulation of pesticides, pollinator decline, and destruction of the habitat for birds (Ghormade et al., 2011). Therefore, the use of NPs greatly put an end to these problems; its application with herbicides reduces the amount of herbicides required for weed eradication. With the active ingredient and smart delivery system, herbicides are released into the soil according to the soil conditions (Gruère et al., 2011). Ag NPs have a pesticidal activity against pathogenic fungi, and they are reported to have inhibitory effects on the conidial germination of the genus *Raffaelea*, which causes the mortality of oak trees (Nair et al., 2010). The use of nanoformulations may offer new ways to enhance the stability of biological agents. The role of NPs in insects and their potential for insect pest management has been reported (Bhattacharyya et al., 2010). The green synthesis of the silver nanoparticles of *Tinospora cordifolia* showed maximum cidal effect against the head louse *Pediculus humanus* and fourth instar larvae of *Anopheles subpictus* and *Culex quinquefasciatus* (Jayaseelan et al., 2011). Recently, research has concluded that nanoagrochemicals are generally nanoreformations of existing pesticides and fungicides. Probably the nanoformulations are expected to enhance the solubility of poorly soluble active constituents, releasing the active constituents in a targeted fashion that resists premature degradation (Kah et al., 2013). Nanopseticides significantly control the delivery of pesticides, with low chemical doses, and high positive results. Silver NPs at 100 mg/kg reduced mycelial growth and conidial germination on cucurbits and pumpkins against powdery mildew (Lamsal et al., 2011).

11.4.1.1 Ag NPs as Fungicides

Several studies have reported that Ag NPs can be used to suppress pathogens and increase crop growth. Jo et al. (2009) reported that Ag NPs in a 200 mg/l concentration reduced 50% of the colony formation of the pathogenic fungi that causes disease in ryegrass. Lamsal et al. (2011) reported that the application of Ag NPs enhanced disease suppression. The combined activities of Ag NPs with the fungicide fluconazole were found to be effective against *Candida albicans*, followed by *Phoma glomerata* and *Trichoderma* sp. (Gajbhiye et al., 2009). Jo et al. (2009) showed that effective concentrations for the inhibition of colony formation by silver compounds on *Bipolaris sorokiniana* was greater than that on *Magnaporthe grisea*. Silver compounds inhibited the colony formation of *B. sorokiniana* by 50% at the optimum concentration. A significant reduction in mycelial growth was observed from spores incubated with silver nanoparticles (Kasprowic et al., 2010). Silver nanoparticles greatly reduced the number of germinating fragments relative to the control in a 24-hour incubation of spores with a 2.5 ppm solution of nanoparticles. Field tests were also done with silver nanoparticles at various concentrations to determine antifungal activity. The highest inhibition rate for the growth of fungal pathogens on cucumber and pumpkins exposed to 100 ppm silver nanoparticles (Lamsal et al., 2011). Silver nanoparticles inhibited hyphal growth of the fungal phytopathogens *Colletotrichum coccodes*, *Monilinia* sp., and *Pyricularia* sp. at 2 mM and 4 mM. Other silver nanoparticles inhibited the growth of *Fusarium oxysporumat* at 8 g/ml (Gopinath and Velusamy, 2013). In another study, silver nanoparticles inhibited *Fusarium oxysporum* growth at ~6g/well (Musarrat et al., 2010). In a related study, silver nanoparticles exhibited dose-dependent growth inhibition against *Aspergillus* sp. and *Penicillium* sp. between 0.2–2 mg/ml and against *Pseudomonas aeruginosa* at 100 g/ml (Kanmani and Lim, 2013).

11.4.1.2 Ag NPs as Bacteriocides

Silver nanoparticles are used as antimicrobial agents because of their economical production and multiple inhibitory actions against microorganisms. Silver nanoparticles are the most studied and utilized nanoparticles in the bio-system because of their strong inhibitory and antimicrobial activities (Jo et al., 2009). Nanosilver colloid is more adhesive to the bacterial and fungal cell surface; hence, it acts as a better fungicide because of its well dispersed and stabilized silver nanoparticles solution. Ag NP exposure causes toxicity to bacteria. An Ag NP treatment can prevent replication and protein synthesis (Chaloupka et al., 2010). Tea extracts were used to prepare Ag NPs, which exhibited antibacterial activity against *Vibrio harveyi* (Vaseeharan et al., 2010). Silver was used a disinfectant and sterilizing agent by NASA and the Russian Space station for water (Vermeiren et al., 2002), silver zeolite, and silver. Gold has a high-temperature stability, low volatility and good anti-fungal and antimicrobial effects against 150 different bacteria (Kumar and Münstedt, 2005). The FDA in 2009 approved the direct use of silver as a disinfectant for commercial water, since they had proven to have effective result against microorganisms. Silver nanoparticles were shown to have an antimicrobial effect against *E. coli*, L. *mono cytogenes*, and *Staphylococcus aureus* (Necula et al., 2010), and nanosilver particles coated with cellulose acetate phthalate also provided similar results (Necula et al., 2010). Some nanoparticles have shown anti-fungal activity. These fungi include *Candida albicans*, *Aspergillus niger* (Sánchez et al., 2009) and yeast (Kim et al., 2003). Ag NPs are also found to be effective against methicillin-resistant *Staphylococcus aureus* (Mariadoss et al., 2019).

Antibacterial activity was also demonstrated with other types of silver nanoparticles. For example, on *Bacillus megaterium*, *Pseudomonas syringae*, *Burkholderia glumae*, *Xanthomonas oryzae*, and *Bacillus thuringiensis* (Velmurugan et al., 2015). Silver nanoparticles between 0.3–0.75 ppm (20 nm) or between 0.5–3 ppm (80 nm) exhibited dose-dependent toxicity toward *Nitrosomonas europaea*a bacterium and removed ammonia derived fertilizers from the soil, oxidizing the ammonia into nitrite. A toxic effect against *Nitrosomonas Europaea* was attributed to silver ions (Radnieck et al., 2011). The silver nanoparticles exhibit the bacteriocidal property by releasing silver ions through oxidation. The process is interrupted in aerobic condition leading to their non-toxicity to the bacteria like E.coli. It was found that in aerobic condition, the nanoparticles didn't exhibit toxicity to E.coli at concentration thousand times higher than the lethal concentration of silver ions (158mg/L and 195mg/L). However, under aerobic conditions the toxicity of nanoparticles to *E. coli* was exactly equal to that of $AgNO_3$ solutions, with the same concentrations of silver ions (~50–300 g/litre) (Xiu et al., 2012).

11.4.1.3 Ag NPs as Antivirals

Nanotechnology for controlling plant viruses, particularly spherical viruses, is considered to be naturally occurring in nanomaterials. The smallest plant virus known to date is the *satellite tobacco necrosis* virus, measuring only 18 nm in diameter (Höglund, 1968). Plant viruses are made up of single or double-stranded RNA/DNA genomes which are encapsulated in a protein coating. Their ability to infect, deliver the nucleic acid genome to a specific site in the host cell, replicate, package nucleic acid, and come out of the host cell precisely in an orderly manner has necessitated them to be used in nanotechnology. Devendra and Kothari (2014) showed a simple, rapid, and economical route to synthesize silver nanoparticles, and its application in plant virus inhibition was standardized. Biological synthesis provides particles with good control over size distribution and shape due to slower reaction kinetics. Kalimuthu et al. (2008) reported that maximum silver nanoparticle synthesis occurs at the stationary phase of *Bacillus licheniformis*. Pugazhenthiran et al. (2009) reported that the silver resistant *Bacillus* sp. when cultured with silver nitrate solution synthesizes silver nanoparticles in the periplasmic space of the cell. Further, an extract of alga was involved in the shape-controlled synthesis of Ag nanoplates because of different proteins present in the extract (Xie et al., 2007). In one additional study, silver nanoparticles exhibited strong inhibitory activity against *Alternaria alternata*, *Botrytis cinerea*, *Curvularialunata*, *Macrophomina phaseolina*, *Rhizoctonia solani*, and *Sclerotinia sclerotiorum* between 5–15 mg/plate of potato dextrose agar (Krishnaraj et al., 2012). The homemade silver nanoparticles at 3 ppm were more active against *Alternaria* sp., *Aspergillus niger*, *Botrytis cinerea*, *Penicillium expansum*, and *Rhizopus* sp. than the commercially available ones (Fu et al., 2020).

11.4.2 Role of Silver Nanoparticles in the Improvement of Soil Quality

Damage to the soil includes excessive tilling, which leads to erosion and irrigation without the required drainage, and this will lead to salinization. This is for the safety of food, animal feed and fibres requirements. Long-term experiments are required to show the effect of different practices on soil properties which are essential to sustainability and to provide important data on this objective. In the United States, the development of nanochemicals has produced promising agents for plant growth and pest control. Fertilizers are required for plant growth. Nanomaterials act as fertilizers that could have properties such as crop improvement and reduced ecotoxicity. Recent developments in agriculture cover the applications of nanoparticles for the more effective and safe use of chemicals on plants. The effects of different nanoparticles on plant growth and phytotoxicity were reported by several studies, including magnetite (Fe_3O_4) nanoparticles and plant growth (Shankramma et al., 2016), alumina, zinc, and zinc oxide on seed germination and root growth of five higher plant species; radish, rape, lettuce, corn, and cucumber, silver nanoparticles and seedling growth in wheat (Ling and Xing, 2007), sulfur nanoparticles on tomato (Salem et al., 2016), zinc oxide in mungbean, nanoparticles of AlO, CuO, FeO, MnO, NiO, and ZnO (Ghidan et al., 2018). Silver nanoparticles can stimulate wheat growth and yield. Soil applied with 25 ppm SNPs had highly favourable growth-promoting effects on wheat growth and yield.

11.4.3 Silver Nanoparticles as Pesticides

Silver nanoparticles caused the death of adult hematophagous flies *Hippobosca maculate* of between 5–25 mg/litres, and of cattle ticks *Rhipicephalus (Boophilus) microplus* of between 5–25 mg/litres (Santhoshkumar et al., 2012). In another study, it was also reported that silver nanoparticles had a dose-dependent activity against the larvae of *Rhipicephalus (Boophilus) microplus* between 1.25–20 mg/litre. Other silver nanoparticles exhibited dose-dependent activity against *Hippobosca maculate* and *Haema physalisbispinosa* between 2–10 mg/litre.

11.4.4 Phytotoxicity of Nanoparticles

11.4.4.1 Uptake of Ag NPs in the Soil by Plants

Considering the agricultural applications of nanoparticles, one should have in mind the possibility of adverse effects because of their phytotoxicity. Therefore, nanoparticles must be used carefully to prevent overdosage, and preliminary experiments in this sense may be required. The main parameters that are registered are root and shoot lengths compared to those of untreated plants. Potential eco-toxicity, a more general problem, should also be considered. The nanoparticles are expected to have negative effects on soils. Their intrinsic anti-fungal and antimicrobial activity may seriously threaten free-living nitrogen-fixing bacteria and disturb natural symbiotic relationships. Some nanoparticles, due to their long residence time in cells, may be transferred through the food chain. It is possible to imagine nanoparticles as miniaturized toxic delivery systems, as they release compounds or react against biological molecules at each trophic level without a remarkable loss of toxicity. Nanoparticles can also interact with natural organic matter, affecting their aggregation state, mobility, and bioavailability (Figure 11.4).

Silver nanoparticles displayed a nontoxic or poor phytotoxic effect on *Vigna unguiculata*, *Vigna mungo*, *Vigna radiata*, *Sorghum Vulgare*, and *Macrotyloma uniflorum*. The parameters investigated were germination time, shoot length, total chlorophyll, chlorophyll a, b, and carotene content (Namasivayam and Chitrakala, 2011). The treatment of *Triticum aestivum* seeds with silver nanoparticles at 10 mg/litre caused the diminishing of the shoot and root length. Silver uptake by *Triticum aestivum* at 1 and 10 mg/litre nanoparticles were observed to be dose-dependent. Because the nanoparticles found in roots were smaller than those originally applied, it was concluded that the phytotoxic effect was mediated by silver ions, and even by the complete dissolution of nanoparticles followed by a secondary nucleation (Vannini et al., 2014). The formation of silver nanoparticles was indeed shown in *Allium cepa*. 'Crystal white wax' bulbs after treatment with 40 mg/litre of $AgNO_3$ by foliar spray or after administration of 80 mg/litre of $AgNO_3$ dissolved in Douglas nutritive solution. The shape of the nanoparticles formed was close to spherical or irregular; the average size was about 500 nm, and the recorded size varied from ~300 nm to ~1000 nm. Silver nanoparticles showed *Zea mays* seed germination inhibition at 3.7 and 36.7 mg/litre compared with 500 mg [Ag]/litre for $AgNO_3$. Silver nanoparticles at 10 mg/litre inhibited *Oryza sativa* germination and the growth of germinated seedlings. The negative effect observed slightly depended on the size of the nanoparticles (Thuesomba et al., 2014). Between 0.5 and 1 mg/litre, silver nanoparticles caused a dose-dependent inhibitory effect on the growth of *Oryza sativa* seedlings. And oxidative stress was also observed. These nanoparticles at 1 mg/litre were partially

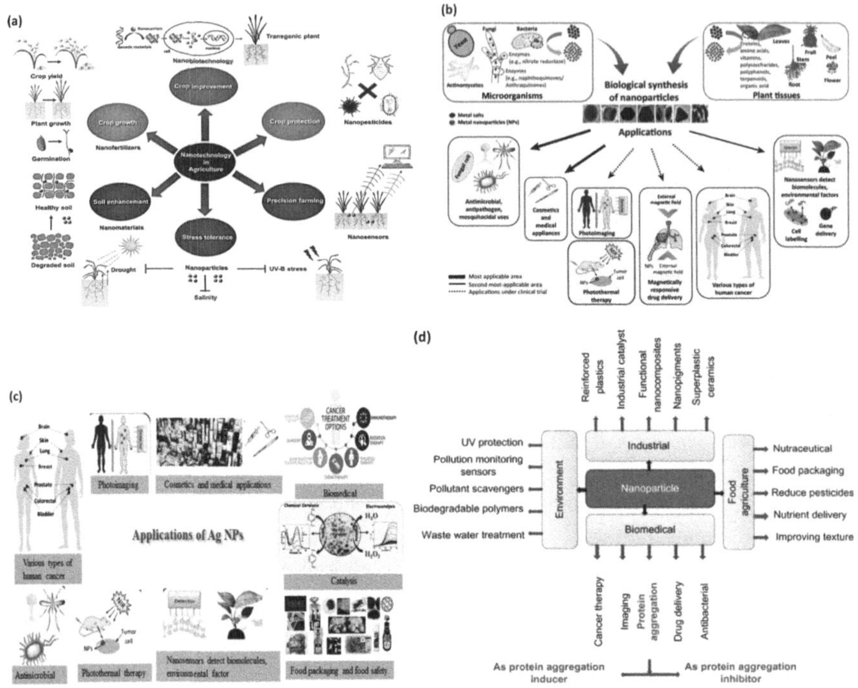

FIGURE 11.4 (A) Applications of nanotechnology in agriculture. Controlled released Nano fertilizers improve crop growth, yield and productivity. Nano-based target delivery approach (gene transfer) is used for crop improvement. Nano pesticides can be used for efficient crop protection. Uses of nanosensors and computerized controls greatly contribute to precision farming. Nanomaterials can also be used to promote plant stress tolerance and soil enhancement. (Figure adapted from Shang et al., 2019.) (B) Biological synthesis of nanoparticles and their applications. (Figure adapted from Haroon Anwar, 2018; Singh et al., 2016.) (C) Various applications of Ag NPs. (Figure adapted from Tarannum et al., 2019.) (D) Application of nanoparticles in various fields such as in the biomedical, environmental, industrial and food agriculture industries. (Figure adapted from Zaman et al., 2014.)

dissolved in the exposure medium used (without plants), releasing 5.815 g/litre of silver ions after 24 hours (Nair and Chung, 2014). A proteomic study of *Oryza sativa* irrigated with silver nanoparticles at 60 mg/litre revealed the expression of cellular detoxifying enzymes, as well as the expression of modification proteins as a response to oxidative stress. These nanoparticles also perturbed normal cell metabolic processes such as protein synthesis/degradation rate and the apoptosis process. Spherical silver nanoparticles with a size of 18.34 nm at 5 mg/litre caused fragmentation of the cell wall of rice rhizosphere bacteria, penetrated inside the bacterial cells, and transformed the bacteria into protoplasts. Some bacterial species were eliminated (Mirzajan et al., 2013). This bacterium can suppress *Lycopersicon esculentum* foot and root rot caused by *Fusarium oxysporum raditi lycopersici* (Validov et al., 2007) as well as root rot of *Lactuca sativa* caused by *Pythium aphanidermatum* and crown and root rot of *Cucumis sativus* caused by *Fusarium oxysporum* (Amer and Utkhede, 2000). Silver nitrate between 0.2–10 mg/litre caused the same bactericidal effect. In contrast, bulk Ag (size 44,000 nm) between 0.1–10 mg/litre did not have the same effect. Silver nanoparticles with a size of 10 nm enhanced poplar (*Populus deltoides×nigra*) evapotranspiration at 0.1 mg/litre and inhibited it at 1 mg/litre. The poplars were exposed to the nanoparticles via the roots. Silver ions exhibited the same phytotoxic or Phyto-stimulatory effects at these concentrations. In contrast, the exposition of these plants to PEG alone did not present this stimulatory effect. Silver nanoparticles under aerobic conditions continuously release silver ions, and 10 nm nanoparticles exerting higher phytotoxicity for poplars released more silver ions than the 25 nm ones; therefore, both effects were likely mediated by silver ions (Wang et al., 2013). *Lactuca Sativa* and *Cucumis sativus* seed treatments with silver nanoparticles at 100 mg/litre decreased their germination rate and inhibited root growth, but the effect was almost undistinguishable from that of the solvent alone.

11.4.4.2 Uptake of Ag NPs in the Soil by Insects

Previous studies in the literature have speculated that silver nanoparticles would be completely immobilized in sewage sludge due to the formation of insoluble silver sulfide (Cheng et al., 2013; Cross and Roberts, 2000; Shahverdi et al., 2007). It was theorized that the silver nanoparticles would remain immobile, and therefore unavailable for uptake after the sludge was applied as a land amendment. However, the present study showed that Ag NPs would be available for uptake by both insects and plants in high enough concentrations. It should be noted that this study used pristine nanoparticles rather than the silver that had interacted with sewage sludge. However, the widespread use of Ag NPs in detergents, textiles, etc. potentially serves as an anthropogenic source for the release of pristine silver nanoparticles into terrestrial ecosystems. Sewage sludge typically contains 0.3–2.3 wt% of sulfur (Wijnhoven et al., 2009). Hence, there may not be enough sulfur in the sludge in the event of the presence of a high concentration of Ag NPs. In the case of both the insects used in the study, a concentration-dependent increase in the uptake of silver from Ag NPs in the soil was observed. Insects exposed to the highest concentrations of Ag NPs used in the study (125 and 625 ppm) were observed to contain quantifiable concentrations of silver (> 0.1 ppm). Levels of silver in insects exposed to the lowest concentrations of Ag NPs used in the study (0 and 1 ppm) could not be quantified (below instrument detection limits or < 0.005 ppm). Lastly, insects exposed to intermediate concentrations of Ag NPs used in the study (5 and 25 ppm) were found to contain at least trace amounts of silver. In the case of *A. domesticus*, the levels of silver found in the highest treatment groups (125 and 625 ppm) were found to be significantly higher than the levels of silver in the control and 1 ppm treatment groups ($p < 0.05$). In the case of *T. molitor*, levels of silver in the insects from the 625 ppm treatment group were found to be significantly higher than the levels of silver in insects exposed to 1 ppm Ag NPs in the soil ($p < 0.05$). Past studies have found insects to uptake bulk metals from the soil (Elechiguerra et al., 2005; Naghsh et al., 2012). This result is in agreement with a previous study that has investigated the uptake of silver from Ag NPs in sandy loam soil. An analysis of the characteristics of the soil used in this study has determined that the soil is sandy loam in nature. The low organic matter content, low organic carbon content, low clay content, and low CEC of sandy loam soil usually result in more Ag ions being available for uptake. A similar result was observed in another study that dealt with the uptake of silver ions from Ag NPs by earthworms. The uptake of silver ions from Ag NPs can occur via ingestion or endocytosis. Either one of these processes could be responsible for the uptake of Ag NPs in *A. domesticus* and *T. molitor*. Nanoparticles are also known to adhere to the cell walls of damaged cells, a phenomenon that was previously observed with quantum dots. Another explanation could be the dermal exposure to silver ions by the test species. Studies have suggested that the uptake and accumulation of the whole particle are possible in cases of dermal exposure to nanoparticles (Davies, 2009). All *S. Vulgare* root samples were found to contain at least trace amounts of silver, regardless of the treatment group. No other plant tissue samples were found to contain silver except the leaf samples from the 625 ppm treatment group, which contained trace amounts. An increase in the root concentrations of Ag was observed as a function of increasing concentrations of Ag NPs in the soil. The *H. annuus* control samples were found to contain silver concentrations below detection limits, except one stem sample which was found to contain trace amounts of silver. Such an observation could be due to contamination and sample preparation and processing errors. The 1 ppm treatment group was found to contain no silver. All of the root and leaf samples of the 5, 25, 125, and 625 ppm treatment groups were found to contain at least trace amounts of silver. The stem samples of the highest treatment groups (125 and 625 ppm) were also found to contain at least trace amounts of silver. The root samples showed an overall trend of increasing concentrations of silver as a function of increasing concentrations of Ag NPs in the soil. There were no significant differences among any of the plant tissue samples. Additionally, in both plant species, the roots were found to contain the highest concentrations of silver, followed by the leaves, the stems, and finally the seeds in the *H. annuus* tissue samples. A similar result was observed by Reddy and Dunn (1984) who investigated the uptake of heavy metals like cadmium, nickle, and chromium by soybeans grown on sludge-amended soils (Reddy and Dunn, 1984). A major limitation that considerably affected the observed inferences was the limited sample size ($n = 2$). Such a small sample size considerably affected the power of the statistical tools used in analyzing the data. Uptake of silver from Ag NPs in the soil by the plants used in the studywa observed at the highest soil treatment groups. The *H. annuus* samples were found to take up more than the *S. vulgare*, although the amounts were not found to be significantly higher. These results were not unexpected as dicots are better able to take up metals from the soil than monocots due to the differences in root exudates (Jayalakshmi et al., 2006). The root exudates of dicots contain more organic acids than monocots. These organic acids include citric acid, maleic acid, ascorbic acid, and oxalic acid (Okafor et al., 2013; Abou El-Nouret al., 2010). The presence of organic acids in root exudates lowers the pH in the vicinity of the roots, thereby solubilizing metals and enabling their uptake (Bonsak et al., 2011; McFarland and Van Duyne, 2003). However, the uptake and accumulation of silver by *S. vulgare*, a monocot plant, may be explained by its root morphology. The presence of thin and numerous roots in monocots presents a high surface area for nanoparticles that facilitates penetration and accumulation into the system (Fauss, 2008). Additionally, recent studies have suggested that the presence of elevated levels of chlorine (Cl) in the soil, due to its salinity or when irrigated with water containing high amounts of Cl, may potentially increase the bioavailability and subsequent uptake of silver. This phenomenon could possibly be explained due to the formation of AgCl complexes insoluble or colloidal forms at elevated Cl concentrations. High Cl concentrations increase the mass transport of Ag and labile Ag, eventually resulting in the uptake of silver by plants (Roe et al., 2008). Additionally, *H. annuus* was also found to translocate silver to the shoot portion of the plant whereas *S. vulgare* did not. It has long been established that some plants can uptake metals from the soil via the roots and distribute them throughout the shoot system (Yan et al., 2002; Li et al., 2009, 2011). A possible explanation for the differences in translocation may be due to the differences in root exudates. If dicots are better at dissolving metals, a higher concentration of silver ions would be present around the roots; the ions would be able to travel further into a plant than a silver nanoparticle. Ions would also be able to diffuse across the same membrane or be actively transported across the lipid bilayer through transport proteins (Ma et al., 2008) whereas solid nanoparticles would not.

11.4.5 Ag NPs in the Food Industry

Post-harvest management, including the preservation of agricultural produce, is one of the important branches of agriculture. Previous studies have reported that Ag NP based antimicrobial packaging plays an important role in increasing the shelf life of fresh fruits and vegetables. An et al. (2008) evaluated the effect of an Ag NPs PVP coating on the preservation of green asparagus. They observed that the coating of Ag NPs PVP slowed down weight loss, ascorbic acid, and total chlorophyll, reduced the colour changes in the skin of asparagus, inhibited the increase of the tissue firmness, the growth of microorganism, and increased the shelf life of asparagus by about ten days at 2°C. Similarly, Mohammed Fayaz et al. (2009) also reported the preservation of vegetables and fruits using biosynthesized Ag NPs. They prepared Ag NP incorporated sodium alginate films and studied their antibacterial activity, measured the weight loss (%) of fruits and vegetables, and performed sensory analysis. They observed that Ag NP incorporated sodium alginate film retained their antibacterial activity against both Gram-positive and Gram-negative organisms. The minimum weight loss was observed in Ag NP incorporated sodium alginate film-coated carrots and pears compared to sodium alginate coated and uncoated control. The colour, appearance, texture, and taste of Ag NP incorporated sodium alginate film-coated carrots and pears were found to be acceptable with up to ten days of storage as compared to uncoated control and sodium alginate coated carrots and pears. A low-density polyethylene (LDPE) polymer matrix containing Ag NPs was studied to preserve and extend the shelf life of stored barberries (Valipoor et al., 2013). Nanopackaging with Ag NP-LDPE successfully maintained the sensory, physicochemical, and physiological qualities of barberry and strawberry fruits at a higher level than normal packaging with polyethylene bags. Similar effects were obtained in the case of Ag NP LDPE packages for the preservation of orange juice (Emamifar et al., 2010).

11.4.6 Conclusion

Silver nanoparticles have huge potential to enhance agricultural productivity, with remarkable crop growth biomass as well as adding value to nutrient content, signifying silver nanoparticle implementations as nanofertilizers and nanopesticides. It must be remembered that even certain nanoparticles synthesized using 'green' methods are not at all 'natural' in certain instances. Toxicity for humans must also be taken into consideration because in many cases the nanoparticles cannot be completely separated from the agricultural products eaten by humans. Until their introduction for real agricultural use, all these potential adverse effects must be examined for each particular form of nanoparticle. Most reports contain only preliminary data and do not take all these factors into account. Often the nanoparticles' characteristics are not provided so the data are more illustrative than realistic and reproducible for robust research field. In certain cases, the mechanisms of nanoparticles are poorly studied, for example, it is often uncertain if nanoparticles, ions emitted from them, or both are the actual cause of the effects observed. Besides, many beneficial effects of nanoparticles may be attributed to the influence of hormesis, which consists of a calming action at low doses of toxic substances. Silver nanoparticles assist with environmental conditions. Silver nanoparticles facilitate environmental remediation; however, the rate of remediation in various regions differs due to changes in soil type, salinity, temperature, humidity, water content, and accessible soil nutrients, and therefore more research is needed to improve the efficiency gains and compatibility of silver nanoparticles with appropriate plant species. Modern food packaging includes silver nanoparticles that express status information from within the food, and nanopackaging should be designed to release enzymes, flavours, antioxidants, nutraceuticals, and antimicrobials to improve the shelf life of the packaged food. Thus it is essential to conduct studies on the movement of silver nanoparticles in the human body and toxicological aspects to develop safe nanotechnology for food packaging. Another problem may arise from the difficult scalability of the synthetic techniques described. In any case, this is an interesting field of research that needs more deep investigation since the preliminary results seem very favourable.

11.5 REFERENCES

Aghajani, Z., and R. Ekhtiyari. 2013. Effect of nanosilver on stages of plant growth and yield and composition of essential oil of *Thymus kotschyanusBoiss*. Hohen. *African Journal of Agricultural Research*, 8(8), 707–710.

Amer, G. A., and R. S. Utkhede. 2000. Development of formulations of biological agents for management of root rot of lettuce and cucumber. *Canadian Journal of Microbiology*, 46(9), 809–816.

An, J., Zhang, M., Wang, S., and J. Tang. 2008. Physical, chemical and microbiological changes in stored green asparagus spears as affected by coating of silver nanoparticles-PVP. *LWT-Food Science and Technology*, 41(6), 1100–1107.

Bhattacharyya, A., Bhaumik, A., Rani, P. U., Mandal, S., and T. T. Epidi. 2010. Nanoparticles-A recent approach to insect pest control. *African Journal of Biotechnology*, 9(24), 3489–3493.

Bonnell, D. A., and B. D. Huey. 2001. 'Basic principles of scanning probe microscopy', in *Scanning Probe Microscopy and Spectroscopy: Theory, Techniques, and Applications*, ed. D. A. Bonnell (New York, NY: Wiley-VCH).

Bonsak, J., Mayandi, J., Thøgersen, A., Stensrud Marstein, E., and U. Mahalingam. 2011. Chemical synthesis of silver nanoparticles for solar cell applications. *Physica Status Solidi C*, 8(3), 924–927.

Borase, H. P., Salunke, B. K., Salunkhe, R. B., Patil, C. D., Hallsworth, J. E., Kim, B. S., and S. V. Patil. 2014. Plant extract: A promising biomatrix for ecofriendly, controlled synthesis of silver nanoparticles. *Applied Biochemistry and Biotechnology*, 173(1), 1–29.

Byczyńska, A. 2017. Nano-silver as a potential biostimulant for plant–A review. *World Scientific News*, 86(3), 180–192.

Campbell, B. M., Thornton, P., Zougmoré, R., van Asten, P., and L. Lipper. 2014. Sustainable intensification: What is its role in climate smart agriculture? *Current Opinion in Environmental Sustainability*, 8, 39–43. doi: 10.1016/j.cosust.2014.07.002

Chaloupka, K., Malam, Y., and A. M. Seifalian. 2010. Nanosilver as a new generation of nanoproduct in biomedical applications. *Trends in Biotechnology*, 28(11), 580–588.

Cheng, L., Zhang, K., Weir, M. D., Liu, H., Zhou, X., and H. H. Xu. 2013. Effects of antibacterial primers with quaternary ammonium and nano-silver on *Streptococcus mutans* impregnated in human dentin blocks. *Dental Materials*, 29(4), 462–472.

Cross, S. E., and M. S. Roberts. 2000. The effect of occlusion on epidermal penetration of parabens from a commercial allergy test ointment, acetone and ethanol vehicles. *Journal of Investigative Dermatology*, 115(5), 914–918.

Elechiguerra, J. L., Burt, J. L., Morones, J. R., Camacho-Bragado, A., Gao, X., Lara, H. H., and M. J. Yacaman. 2005. Interaction of silver nanoparticles with HIV-1. *Journal of Nanobiotechnology*, 3(1), 6.

El-Nour, K. M. A., Eftaiha, A. A., Al-Warthan, A., and R. A. Ammar. 2010. Synthesis and applications of silver nanoparticles. *Arabian Journal of Chemistry*, 3(3), 135–140.

El-Temsah, Y. S., and E. J. Joner. 2012. Impact of Fe and Ag nano-particles on seed germination and differences in bioavailability during exposure in aqueous suspension and soil. *Environmental Toxicology*, 27(1):42–49.

Emamifar A, Kadivar M, Shahedi M, and S. Soleimanian-Zad. 2010. Evaluation of nanocomposite packaging containing Ag and ZnO on shelf life of fresh orange juice. *Innovative Food Science and Emerging Technologies*, 11(4), 742–748.

Eppler, A. S., Rupprechter, G., Anderson, E. A., and G. A. Somorjai. 2000.Thermal and chemical stability and adhesion strength of Pt nanoparticle arrays supported on silica studied by transmission electron microscopy and atomic force microscopy. *Journal of Physical Chemistry B*. 104, 7286–7292.

Fauss, E. 2008. *The Silver Nanotechnology Commercial Inventory*. University of Virginia.

Feldheim, D. L. and C. A. Foss. 2002. *Metal Nanoparticles: Synthesis, Characterization, and Applications*. CRC, Boca Raton.

Fu, L., Wang, Z., Dhankher, O. P., and B. Xing. 2020. Nanotechnology as a new sustainable approach for controlling crop diseases and increasing agricultural production. *Journal of Experimental Botany*, 71(2), 507–519.

Gajbhiye, M., Kesharwani, J., Ingle, A., Gade, A., and M. Rai. 2009. Fungus-mediated synthesis of silver nanoparticles and their activity against pathogenic fungi in combination with *fluconazole*. *Nanomedicine: Nanotechnology, Biology and Medicine*, 5(4), 382–386.

Ghidan, A. Y., Al-Antary, T. M., Awwad, A. M., Ghidan, O. Y., Araj, S. E. A., and M. A. Ateyyat. 2018. Comparison of different green synthesized nanomaterials on green peach aphid as aphicidal potential. *Feb*-Fresenius Environmental Bulletin, 27, 7009–7016.

Ghormade, V., Deshpande, M. V., and K. M. Paknikar. 2011. Perspectives for nano-biotechnology enabled protection and nutrition of plants. *Biotechnology Advances*, 29(6), 792–803.

Gibney, E. 2015. Bucky balls in space solve 100-year-old riddle. *Nature News*. doi: 10.1038/nature.2015.17987.

Gopinath, V., and P. Velusamy. 2013. Extracellular biosynthesis of silver nanoparticles using *Bacillus sp.* GP-23 and evaluation of their anti-fungal activity towards *Fusariumoxysporum*. *Spectrochimica Acta Part A: Molecular and Biomolecular Spectroscopy*, 106, 170–174.

Gruère, G., Narrod, C., and L. Abbott. 2011. *Agricultural, Food, and Water Nanotechnologies for the Poor.* International Food Policy Research Institute, Washington, DC.

Gunalan, S., Sivaraj, R., and R. Venckatesh. 2011. Green synthesis of zinc oxide nanoparticles by Aloe barbadensis miller leaf extract: Structure and optical properties. *Materials Research Bulletin*, 46, 2560–2566.

Gusev, A. A., Kudrinsky, A. A., Zakharova, O. V., Klimov, A. I., Zherebin, P. M., Lisichkin, G. V., and Y. A. Krutyakov. 2016. Versatile synthesis of PHMB-stabilized silver nanoparticles and their significant stimulating effect on fodder beet (*Beta vulgaris* L.). *Materials Science and Engineering: C*, 62, 152–159.

Hatami, M., and M. Ghorbanpour. 2014. Defense enzyme activities and biochemical variations of Pelargonium zonale in response to nanosilver application and dark storage. *Turkish Journal of Biology*, 38(1), 130–139.

Höglund, S. 1968. Some electron microscopic studies on the satellite tobacco necrosis virus and its IgG-antibody. *Journal of General Virology*, 2(3), 427–436.

Homaee, M. B., and A. A. Ehsanpour. 2015. Physiological and biochemical responses of potato (*Solanum tuberosum*) to silver nanoparticles and silver nitrate treatments under in vitro conditions. *Indian Journal of Plant Physiology*, 20(4), 353–359.

Hurst, S. J., Lytton-Jean, A. K. R., and C. A. Mirkin. 2006. Maximizing DNA loading on a range of gold nanoparticle size. *Analytical Chemistry*, 78, 8313–8318.

Husain, S., Afreen, S., Yasin, D., Afzal, B., and T. Fatma. 2019. Cyanobacteria as a bioreactor for synthesis of silver nanoparticles-an effect of different reaction conditions on the size of nanoparticles and their dye decolorization ability. *Journal of Microbiological Methods*, 162, 77–82.

Husain, S., Sardar, M., and T. Fatma. 2015. Screening of cyanobacterial extracts for synthesis of silver nanoparticles. *World Journal of Microbiology and Biotechnology*, 31(8), 1279–1283.

Jayalakshmi, M., Mohan Rao, M., and K. Bum Kim. 2006. Effect of particle size on the electrochemical capacitance of a-Ni (OH) 2 in alkali solutions. *International Journal of Electrochemical Science*, 5.

Jayaseelan, C., Rahuman, A. A., Rajakumar, G., Kirthi, A. V., Santhoshkumar, T., Marimuthu, S., Bagavan, A., Kamaraj, C., Zahir, A. A., and G. Elango. 2011. Synthesis of pediculocidal and larvicidal silver nanoparticles by leaf extract from heartleaf moonseed plant, *TinosporacordifoliaMiers*. *Parasitology Research*, 109(1), 185–194.

Jiang, J., Oberdörster, G., and P. Biswas. 2009. Characterization of size, surface charge, and agglomeration state of nanoparticle dispersions for toxicological studies. *Journal of Nanoparticle Research*, 11, 77–89.

Jo, Y. K., Kim, B. H., and G. Jung. 2009. Anti-fungal activity of silver ions and nanoparticles on phytopathogenic fungi. *Plant Disease*, 93(10), 1037–1043.

Kah, M. 2015. Nanopesticides and nanofertilizers: Emerging contaminants or opportunities for risk mitigation? *Frontiers in Chemistry*, 3, 64. doi: 10.3389/fchem.2015.00064.

Kah, M., Beulke, S., Tiede, K., and T. Hofmann, 2013. Nanopesticides: State of knowledge, environmental fate, and exposure modeling. *Critical Reviews in Environmental Science and Technology*, 43(16), 1823–1867.

Kalimuthu, K., Babu, R. S., Venkataraman, D., Bilal, M., and S. Gurunathan. 2008. Biosynthesis of silver nanocrystals by *Bacillus licheniformis*. *Colloids and Surfaces B: Biointerfaces*, 65(1), 150–153.

Kanmani, P., and S. T. Lim. 2013. Synthesis and structural characterization of silver nanoparticles using bacterial exopolysaccharide and its antimicrobial activity against food and multidrug resistant pathogens. *Process Biochemistry*, 48(7), 1099–1106.

Kasprowicz, M. J., Kozioł, M., and A. Gorczyca, A. 2010. The effect of silver nanoparticles on phytopathogenic spores *of Fusariumculmorum*. *Canadian Journal of Microbiology*, 56(3), 247–253.

Kim, B., Kim, D., Cho, D., and S. Cho. 2003. Bactericidal effect of TiO_2photocatalyst on selected food-borne pathogenic bacteria. *Chemosphere*, 52(1), 277–281.

Krishnaraj, C., Ramachandran, R., Mohan, K., and P. T. Kalaichelvan. 2012. Optimization for rapid synthesis of silver nanoparticles and its effect on phytopathogenic fungi. *Spectrochimica Acta Part A: Molecular and Biomolecular Spectroscopy*, 93, 95–99.

Kumar, R., and H. Münstedt. 2005. Silver ion release from antimicrobial polyamide/silver composites. *Biomaterials*, 26(14), 2081–2088.

Lamsal, K., Kim, S. W., Jung, J. H., Kim, Y. S., Kim, K. S., and Y. S. Lee. 2011. Application of silver nanoparticles for the control of *Colletotrichum species* in vitro and pepper anthracnose disease in field. *Mycobiology*, 39(3), 194–199.

Li, L., Yi L., Jiashen L., Lei Y., Arthur F. T. M., Frank K., and Q. Ling. 2009. Antibacterial properties of nanosilver PLLA fibrous membranes. *Journal of Nanomaterials*.

Li, M., Zhu, L., and D. Lin. 2011. Toxicity of ZnO nanoparticles *to Escherichia coli*: Mechanism and the influence of medium components. *Environmental Science & Technology*, 45(5), 1977–1983.

Ling, D. H., and B. S. Xing. 2007. Phytotoxicity of nanoparticles: Inhibition of seed germination and root elongation. *Environmental Pollution*, 50, 243–250.

Liu, R., and R. Lal. 2015. Potentials of engineered nanoparticles as fertilizers for increasing agronomic productions. *Science of the Total Environment*, 514, 131–139. doi: 10.1016/j.scitotenv.2015.01.104

Ma, R. H., and Y. H. Yu. 2008. U.S. Patent No. 7,462,753. Washington, DC: U.S. Patent and Trademark Office.

Mariadoss, A. V. A., Ramachandran, V., Vijayakumar, S., Balupillai, A., Herbert, F. J., Kumar, S., et al. 2019. Green synthesis, characterization and antibacterial activity of silver nanoparticles by *Malusdomestica* and its cytotoxic effect on (MCF-7) cell line. *Microbial Pathogenesis*, 135, 1–9.

McFarland, A. D., and R. P. Van Duyne. 2003. Single silver nanoparticles as real-time optical sensors with zeptomole sensitivity. *Nano Letters*, 3(8), 1057–1062.

Mehrian, S. K., Heidari, R., Rahmani, F., and S. Najafi. 2016. Effect of chemical synthesis silver nanoparticles on germination indices and seedlings growth in seven varieties of *Lycopersiconesculentum Mill* (tomato) plants. *Journal of Cluster Science*, 27(1), 327–340.

Mirzajani, F., Askari, H., Hamzelou, S., Farzaneh, M., and A. Ghassempour. 2013. Effect of silver nanoparticles on *Oryza sativa* L. and its rhizosphere bacteria. *Ecotoxicology and Environmental Safety*, 88, 48–54.

Mohammed Fayaz, A., Balaji, K., Girilal, M., Kalaichelvan, P. T., and R. Venkatesan. 2009. Mycobased synthesis of silver nanoparticles and their incorporation into sodium alginate films for vegetable and fruit preservation. *Journal of Agricultural and Food Chemistry*, 57(14), 6246–6252.

Musarrat, J., Dwivedi, S., Singh, B. R., Al-Khedhairy, A. A., Azam, A., and A. Naqvi. 2010. Production of antimicrobial silver nanoparticles in water extracts of the fungus *Amylomycesrouxii* strain KSU-09. *Bioresource Technology*, 101(22), 8772–8776.

Naghsh, N., Safari, M., and P. Hajmehrabi. 2012. Comparison of nanosilver inhibitory effects growth between *Aspergillusniger* and *E. coli*. *Indian Journal of Science and Technology*, 5(3), 2448–2450.

Nair, P. M. G., and I. M. Chung. 2014. Physiological and molecular level effects of silver nanoparticles exposure in rice (*Oryza sativa* L.) seedlings. *Chemosphere*, 112, 105–113.

Nair, Remya, Saino Hanna Varghese, Baiju G. Nair, T. Maekawa, Y. Yoshida, and D. Sakthi Kumar. 2010. Nanoparticulate material delivery to plants. *Plant Science*, 179,154–163.

Namasivayam, S. K. R., and K. Chitrakala. 2011. Ecotoxicological effect of *Lecanicilliumlecanii* (*Ascomycota: Hypocreales*) based silver nanoparticles on growth parameters of economically important plants. *Journal of Biopesticides*, 4(1), 97.

Narayanan, K. B., and N. Sakthivel. 2010. Phytosynthesis of gold nanoparticles using leaf extract of *Coleus amboinicus* Lour. *Materials Characterization* 61(11), 1232–1238.

Necula, A. M., Dunca, S., Stoica, I., Olaru, N., Olaru, L., and S. Ioan. 2010. Morphological properties and antibacterial activity of nano-silver-containing cellulose acetate phthalate films. *International Journal of Polymer Analysis and Characterization*, 15(6), 341–350.

Nejatzadeh-Barandozi, F., Darvishzadeh, F., and A. Aminkhani. 2014. Effect of nano silver and silver nitrate on seed yield of (*Ocimumbasilicum* L.). *Organic and Medicinal Chemistry Letters*, 4(1), 11.

Okafor, F., Janen, A., Kukhtareva, T., Edwards, V., and M. Curley. 2013. Green synthesis of silver nanoparticles, their characterization, application and antibacterial activity. *International Journal of Environmental Research and Public Health*, 10(10), 5221–5238.

Prasad, R. 2014a. Synthesis of silver nanoparticles in photosynthetic plants. *Journal of Nanoparticles*, 2014, 963961. doi: 10.1155/2014/963961.

Prasad, R., Kumar, V., and K. S. Prasad. 2014b. Nanotechnology in sustainable agriculture: Present concerns and future aspects. *African Journal of Biotechnology*, 13, 705–713. doi: 10.5897/AJBX2013.13554.

Prasad, R., Pandey, R., and I. Barman. 2016. Engineering tailored nanoparticles with microbes: Quo vadis. *WIREs Nanomedicine and Nanobiotechnology*, 8, 316–330. doi: 10.1002/wnan.1363.

Radniecki, T. S., Stankus, D. P., Neigh, A., Nason, J. A., and L. Semprini. 2011. Influence of liberated silver from silver nanoparticles on nitrification inhibition of *Nitrosomonaseuropaea*. *Chemosphere*, 85(1), 43–49.

Reddy, M. R., and S. J. Dunn. 1984. Accumulation of heavy metals by soybean from sludge-amended soil. *Environmental Pollution Series B, Chemical and Physical*, 7(4), 281–295.

Roe, D., Karandikar, B., Bonn-Savage, N., Gibbins, B., and J. B. Roullet. 2008. Antimicrobial surface functionalization of plastic catheters by silver nanoparticles. *Journal of Antimicrobial Chemotherapy*, 61(4), 869–876.

Salem, N. M., Albanna, L. S., Abdeen, A. O., Ibrahim, Q. I., and A. M. Awwad. 2016. Sulfur nanoparticles improves root and shoot growth of tomato. *Journal of Agriculture Science* 8(4), 179.

Sánchez-Valdes, S., H. Ortega-Ortiz, L. F. Ramos-de Valle, F. J. Medellín-Rodríguez, and R. Guedea-Miranda. 2009. Mechanical and antimicrobial properties of multilayer films with a polyethylene/silver nanocomposite layer. *Journal of Applied Polymer Science*, 111(20), 953–962.

Sankar, R., Rizwana, K., Shivashangari, K. S., V. Ravikumar. 2014. Ultra-rapid photocatalytic activity of Azadirachta indica engineered colloidal titanium dioxide nanoparticles. *Appl. Nanosci.* 5, 731–736.

Santhoshkumar, T., Rahuman, A. A., Bagavan, A., Marimuthu, S., Jayaseelan, C., Kirthi, A. V., and K. Velayutham. 2012. Evaluation of stem aqueous extract and synthesized silver nanoparticles using *Cissusquadrangularis* against *Hippoboscamaculata* and *Rhipicephalus* (*Boophilus*) microplus. *Experimental Parasitology*, 132(2), 156–165.

Sekhon, B. S. 2014. Nanotechnology in agri-food production: An overview. *Nanotechnology, Science and Application*, 7, 31–53. doi: 10.2147/NSA.S39406.

Shahverdi, A. R., Fakhimi, A., Shahverdi, H. R., and S. Minaian. 2007. Synthesis and effect of silver nanoparticles on the antibacterial activity of different antibiotics against *Staphylococcus aureus* and *Escherichia coli*. *Nanomedicine: Nanotechnology, Biology and Medicine*, 3(2), 168–171.

Shahverdi, A. R., Minaeian, S., Shahverdi, H. R., Jamalifar, H., and A. A. Nohi. 2007. Rapid synthesis of silver nanoparticles using culture supernatants of Enterobacteria: A novel biological approach. *Pro-cessBiochem*, 42, 919–923.

Shams, G., Ranjbar, M., and A. Amiri. 2013. Effect of silver nanoparticles on concentration of silver heavy element and growth indexes in cucumber (*Cucumissativus L. negeen*). *Journal of Nanoparticle Research*, 15(5), 1630.

Shankramma, K., Yallappa, S., Shivanna, M. B., and J. Manjanna. 2016. Fe_2O_3 magnetic nanoparticles to enhance *S. lycopersicum* (tomato) plant growth and their biomineralization. *Applied Nanoscience*, 6(7), 983–990.

Sharma, P., Bhatt, D., Zaidi, M. G. H., Saradhi, P. P., Khanna, P. K., and S. Arora. 2012. Silver nanoparticle-mediated enhancement in growth and antioxidant status of *Brassica juncea*. *Applied Biochemistry and Biotechnology*, 167(8), 2225–2233.

Strasser, P., Koh, S., Anniyev, T., Greeley, J., More, K., and C. Yu. 2010. Lattice-strain control of the activity in dealloyed core–shell fuel cell catalysts. *Nature Chemistry* 2, 454–460.

Sun, S., Murray, C. B., Weller, D., Folks, L., A. Moser. 2000. Mono-disperse FePt nanoparticles and ferromagnetic FePtnanocrys-tal super lattices. *Science* 287, 1989–1992.

Thakkar, H. P., Patel, B. V., and S. P. Thakkar. 2011. Development and characterization of nanosuspensions of olmesartan medoxomil for bioavailability enhancement. *Journal of Pharmacy and Bioallied Sciences*, 3, 426.

Thangavelu, R. M., Gunasekaran, D., Jesse, M. I., SU, M. R., Sundarajan, D., and K. Krishnan. 2018. Nanobiotechnology approach using plant rooting hormone synthesized silver nanoparticle as 'nanobullets' for the dynamic applications in horticulture–an in vitro and ex vitro study. *Arabian Journal of Chemistry*, 11(1), 48–61.

Thuesombat, P., Hannongbua, S., Akasit, S., and S. Chadchawan. 2014. Effect of silver nanoparticles on rice (*Oryza sativa L. cv. KDML 105*) seed germination and seedling growth. *Ecotoxicology and Environmental Safety*, 104, 302–309.

Validov, S., Kamilova, F., Qi, S., Stephan, D., Wang, J. J., Makarova, N., and B. Lugtenberg. 2007. Selection of bacteria able to control *Fusariumoxysporum f. sp.*radicis-lycopersici in stonewool substrate. *Journal of Applied Microbiology*, 102(2), 461–471.

Valipoor, M. N., Hamed, M. M. T., and S. A. Mortazavi. 2013. Effect of polyethylene packaging modified with silver particles on the microbial, sensory and appearance of dried barberry. *Packaging Technology and Science*, 26(1), 39–49.

Vannini, C., Domingo, G., Onelli, E., De Mattia, F., Bruni, I., Marsoni, M., and M. Bracale. 2014. Phytotoxic and genotoxic effects of silver nanoparticles exposure on germinating wheat seedlings. *Journal of Plant Physiology*, 171(13), 1142–1148.

Vaseeharan, B., Ramasamy, P., and J. C. Chen. 2010. Antibacterial activity of silver nanoparticles (AgNps) synthesized by tea leaf extracts against pathogenic *Vibrio harveyi* and its protective efficacy on juvenile *Feneropenaeusindicus*. *Letters in Applied Microbiology*, 50(4), 352–356.

Velmurugan, P., Park, J. H., Lee, S. M., Jang, J. S., Lee, K. J., Han, S. S., and B. T. Oh. 2015. Synthesis and characterization of nanosilver with antibacterial properties using *Pinusdensiflora* young cone extract. *Journal of Photochemistry and Photobiology B: Biology*, 147, 63–68.

Vermeiren, L., Devlieghere, F., and J. Debevere. 2002. Effectiveness of some recent antimicrobial packaging concepts. *Food Additives and Contaminants*, 19(S1), 163–171.

Wang, J., Koo, Y., Alexander, A., Yang, Y., Westerhof, S., Zhang, Q., and P. J. Alvarez. 2013. Phytostimulation of poplars and Arabidopsis exposed to silver nanoparticles and Ag^+ at sublethal concentrations. *Environmental Science & Technology*, 47(10), 5442–5449.

Wijnhoven, S. W., Peijnenburg, W. J., Herberts, C. A., Hagens, W. I., Oomen, A. G., Heugens, E. H., and S. Dekkers. 2009. Nano-silver–A review of available data and knowledge gaps in human and environmental risk assessment. *Nanotoxicology*, 3(2), 109–138.

Xie, J., Lee, J. Y., Wang, D. I., and Y. P. Ting. 2007. Silver nanoplates: From biological to biomimetic synthesis. *ACS Nano*, 1(5), 429–439.

Xiu, Z. M., Zhang, Q. B., Puppala, H. L., Colvin, V. L., and P. J. Alvarez. 2012. Negligible particle-specific antibacterial activity of silver nanoparticles. *Nano letters*, 12(8), 4271–4275.

Yan, J., and Cheng, J., GloboAsia LLC. 2002. Nanosilver-containing antibacterial and antifungal granules and methods for preparing and using the same. U.S. Patent 6,379,712.

Yasur, J., and P. U. Rani. 2013. Environmental effects of nanosilver: Impact on castor seed germination, seedling growth, and plant physiology. *Environmental Science and Pollution Research*, 20(12), 8636–8648.

Yunlong, C., and B. Smit. 1994. Sustainability in agriculture: A general review. *Agriculture, Ecosystems & Environment*, 49, 299–307. doi: 10.1016/0167-8809(94)90059-0.

Part 3

Nanobioformulation and Plant Health

12 A Journey of Nano Iron Pyrite from the Chemosynthetic World of Hydrothermal Vents to the Photosynthetic World of Agricultural Fields

A New Class of Seed and Root Bio-Stimulant

Himanshi Jangir and Mainak Das

CONTENTS

12.1 INTRODUCTION

A series of serendipitous discoveries that took place during the 20th century marked the beginning of modern-day agriculture. The developmental story of 20th-century agriculture could be divided into four distinct quarters. The first quarter saw a breakthrough in ammonia synthesis and the beginning of an accessible source of nitrogenous fertilizers, viz., Haber's process. During the second quarter, while on one hand agricultural production started to rely heavily upon the use of synthetic fertilizers and pesticides; on the other hand, the discovery of dwarf grain varieties by Borlogue set the stage for the first green revolution. The green revolution was marked by the use of state of the art technologies for farming, viz., synthetic fertilizers, irrigation, pesticides, and high yielding dwarf varieties of major grain crops. The world was basking in the success of the green revolution during the third quarter of the 20th century, only to realize that excessive use of chemicals in the form of fertilizers and pesticides was severely damaging the ecosystem of Earth. Later on, further excessive dependence on the dwarf varieties led to a skewed gene pool for the major grain crops. Figure 12.1 illustrates the timeline of agricultural development

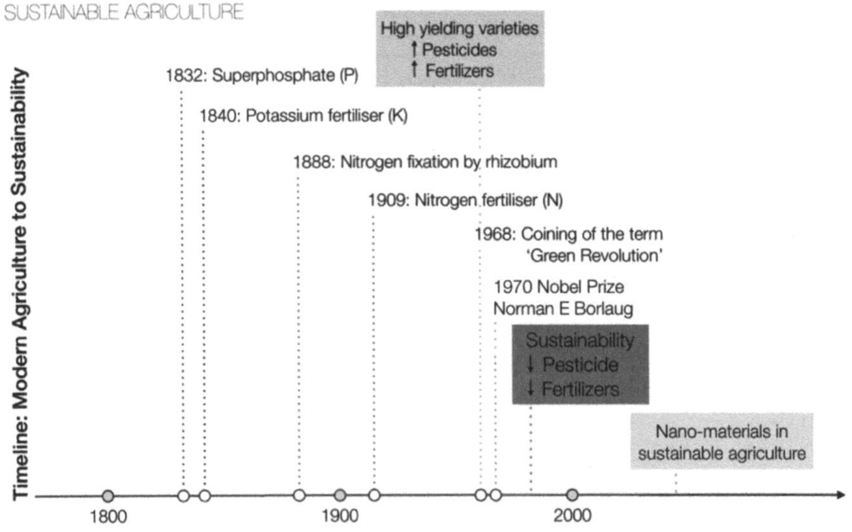

FIGURE 12.1 Major agricultural development took place during the 20th century.

which took place during the 20th century. The last quarter of the 20th century saw the emergence of the concept of sustainability, viz., in simple terms, this implies '*let development happen in harmony with nature*', requiring us 'to go back to our roots' and rediscover nature, from which we have evolved over billions of years. A journey back in time, to help us realize the true potential of this geological mass called planet Earth. Here, we wish to point out one such pioneer, whose philosophical work on agriculture in the form of a lesser-known book is worth reading: *One straw revolution* written by Ramon Magsaysay Award winner the late Masanobu Fukuoka (2 February 1913–16 August 2008), who was an early proponent of natural farming, and deeply influenced by Indian philosophy, the rural economy, and the age-old farming traditions of India (Jangir, 2020b; Srivastava, 2016).

Today, on this planet, we are standing at a cross-roads where we have to feed ourselves yet also have to ensure that the gift, which we have inherited in the form of 'planet Earth', remains liveable into the distant future. The immediate challenge that the agriculture community is facing is 'how to reduce the consumption of synthetic fertilizer, insecticides, and pesticides, yet maintaining a viable production level to feed the community'. While we desire to answer the call for sustainability, one has to realize a harsh truth: what mankind wishes to achieve in terms of sustainable agriculture is to topple the current inertia of using synthetic agrochemicals and redesign a new paradigm of agriculture. Such an approach will affect the large agrochemical sector, which has evolved during the best part of the 20th century and has been ruling the economy since then. So any approach to 'redesign a new sustainable agro paradigm' will have deep-rooted socio-economic and geo-political impacts (Jangir, 2020b).

This chapter is an attempt to 'redesign a new paradigm of sustainable agriculture', viz., *use of iron–sulphur based nanomaterial as seed and root priming agents.* Before we discuss our decade long journey with iron–sulphur nanomaterial and its emerging role in agriculture, we will first talk about what inspired us to explore this route. We drew our inspiration from the philosophical question of 'going back to the roots, rediscovering nature, from which we have evolved over billions of years'. In the search for an answer, we plunged back in time, to an ecosystem from where it all started, viz., the ancient iron–sulphur world and the basic energy principles governing the life forms that survived in such oxygen-free ecosystems (Das, 2015; Jangir, 2020b; Srivastava et al., 2014a, 2014b; Srivastava, 2016).

The present-day Earth thrives in an oxygen rich-environment, where the major source of energy is sunlight. The sunlight mediated reduction of carbon dioxide is assisted by an infinitesimally large electron donor, water, which leads to the formation of biomass on the floor of the Earth; a process commonly known as photosynthesis, in other words, the light-dependent synthesis of biomass by green plants and algae. Yet, going back into the distant past, where oxygen was not yet born (as part of the atmosphere), life evolved around a foul-smelling gas, hydrogen sulphide (H_2S), which is believed to be the predecessor of water as an electron donor. Species reminiscent of such oxygen-independent life forms are still observed in certain special ecosystems; namely, deep under the sea in hydrothermal vents, in coal mines, and possibly on the surface of our nearest planetary neighbour in the solar system, Mars, the red planet, with its martian environment (Das, 2015; Jangir, 2020b; Srivastava et al., 2014a, 2014b; Srivastava, 2016).

Throughout the 20th century, textbooks from primary schools to high schools mentioned photosynthesis as the only basic paradigm of biomass formation. However, it can be called serendipity or a twist of fate for humankind that a deep-sea search for deep ocean resources led to an amazing discovery. Ecosystems under high pressure, with an absence of oxygen and sunlight, where ancient black and white smokers emit gases like hydrogen sulphide (H_2S), and other transition metal sulphides (sulphides of Fe, Zn, Ni, Mn, Co, Mo, W), provide the necessary energy to support a myriad of life forms. Thus, a new model of biomass synthesis was discovered, viz., chemosynthesis (chemical+synthesis). A discovery that has led to the editing of textbooks and to pushing the limits of our indigenous genius to search for life forms independent of light and oxygen; to explore and investigate a past that is 4 billion years old (Das, 2015; Jangir, 2020b; Srivastava et al., 2014a, 2014b; Srivastava, 2016).

Later on, it was learned that chemosynthesis is another route for biomass synthesis in the iron–sulphur ecosystem in the absence of light and oxygen. Hence, this leads to the knowledge that plants, the autotrophs, are not the only avenue for biomass production. Thereafter, a tremendous amount of research has been going on to understand how life evolved in such harsh conditions; if we assume that life indeed evolved from these harsh environments of deep-sea hydrothermal vents, then it might be that we still carry some of these traits that have remained dormant. Figure 12.2 highlights the photosynthetic and chemosynthetic world and Figure 12.3 shows

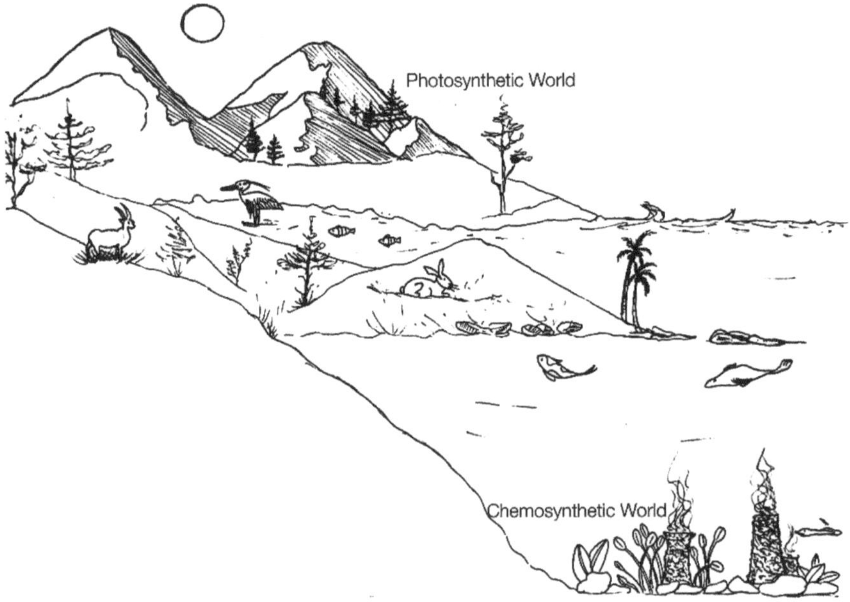

FIGURE 12.2 Photosynthetic and chemosynthetic world.

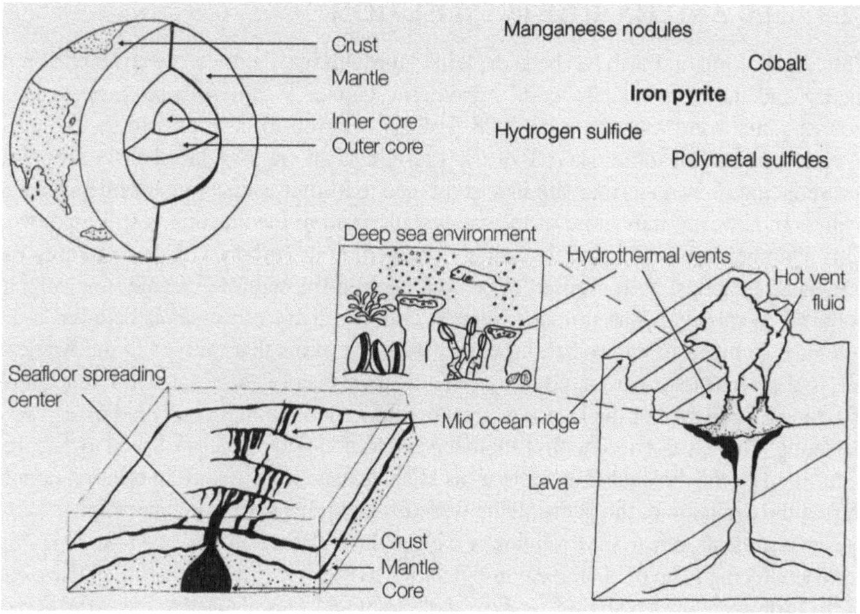

FIGURE 12.3 Deep-sea hydrothermal vents.

a schematic of the hydrothermal vents (Das, 2015; Jangir, 2020b; Srivastava et al., 2014a, 2014b; Srivastava, 2016).

This story is of one such agro-philosophical journey of the exploration of metal sulphide, which readily evolves into nanoforms in and around the deep-sea chimneys on the floor of the ocean. Hoping that we may find the solution to our ever-increasing demand for biomass in the form of food, shelter, etc., and by learning about how life survives and derives energy from these sulphides.

During the last decade, we have been exploring the effect of abundant metal sulphide on the Earth's crust, on crop plant life, viz., iron disulphide, FeS_2 in its nano form. This chapter describes the effects of nano FeS_2.

12.2 FES_2: A MULTI-FUNCTIONAL MOLECULE FOR ALL SEASONS

Iron pyrite (FeS_2) is believed to have evolved very early and finds its application in multiple domains. It has been documented as one of the important reagents for the preparation of vitriol and sulphuric acid. In 1876, Braun for the very first time reported iron pyrite's semiconductor properties. Following this, its rectifying property was exploited in radios to develop detectors. Interestingly, its non-homogenous nature, due to its stoichiometric defects, poses a problem for its use in solid-state electronics. It is still being explored as a photovoltaic material because of its absorption into the infrared region, and it is used as an electrode material for battery applications (Das, 2015; Dubey et al., 2016; Jangi, 2020b; Srivastava et al., 2014a, 2014b; Srivastava, 2016).

12.3 FES_2 AND ITS ROLE IN EVOLUTION

The origin of life on Earth has been explained through two theories: the pre-biotic broth theory and the iron–sulphur world hypothesis. Günter Wächtershäuser proposed the iron–sulphur world hypothesis in 1988. FeS_2 is believed to have existed in the harsh oxygen-less, sulphur-dominated T of the Pre-Archaean era. Wächtershäuser proposed that reducing H_2S and ferrous sulphide combined to form iron disulphide while releasing hydrogen; however, the nature of the catalyst involved in the reaction is still unknown. This hydrogen evolves during the reaction, helped by its carbon dioxide fixation in the presence of catalyst Ni (proposed to be contained in the semi-permeable iron–sulphur colloidal membrane). The primary electron donor in these processes is believed to be sulphide, so much so that some of the microbial life forms that evolved in the Archean era, and others existing even today, use FeS_2 as their major energy source. While evolving, the H_2S reserves of the Earth, which are required to maintain carbon dioxide, were depleting. There was a need to find another electron donor to replace H_2S, and to date that is fulfilled by the molecule known to us as H_2O. Because of this shift in electron donors during the evolution of the planet, there was also a parallel shift from anaerobic to aerobic environments, due to the evolution of oxygen through the splitting of water. This shift from a reducing to an oxidizing environment led to the oxidation of existing species, one of the prominent ones being Fe^{2+} to Fe^{3+} and S to SO_4^{2-}. This also led to the extinction of many anaerobic microbes, which failed to withstand the oxygen pressure, while others survived under the Earth's crust, where oxygen could not penetrate as freely. Existing anaerobic environments on the planet include coal mines, sewage ecosystems, and deep under the sea in sulphide-rich hydrothermal vents. These microbes, as discussed before, use sulphides as their primary source of energy via an exothermic reaction, leading to a release of 200 kilocalories per mole due to the oxidation of sulphide to sulphates. These series of events demonstrating that life could survive on sulphides brought us to a hypothesis: 'Could iron pyrite serve as a source of energy for photo-autotrophs that otherwise are known to depend on the sun for their energy needs'. The extreme case of this hypothesis: 'Can iron pyrite, as with case of thermopiles, serve as the only source of energy for autotrophs and influence its biomass production' (Das, 2015; Jangir, 2020b; Srivastava et al., 2014a, 2014b; Srivastava, 2016).

12.4 METHODOLOGIES AND CHALLENGES OF SYNTHESIZING NANO FES_2 IN LABORATORY CONDITIONS

Even though pyrite (FeS_2) is an abundant material, to unearth its role in biomass production it needs to be synthesized in a controlled environment. Further, to attain a nano-configuration, one has to opt for mechanical ball milling and other abrasive techniques. We developed a low temperature, ambient pressure chemical strategy to synthesize nano FeS_2. While developing the present chemical strategy, we considered the following four guidelines (a–c): a. The strategy should be economical and able to be scaled up or scaled down without major modifications; b. The technology should be translatable to large scale production; c. The energy expenditure for the synthesis must be within the affordable range for the developing and under-developed economies of the world; d. The technology should offer consistent, homogenous results, without any significant batch to batch variation in the physical and chemical properties of the nanomaterial. The basic requirement for chemical synthesis is a well ventilated wet laboratory.

The synthesis of nano iron pyrite is a two-step process, where all steps are carried out in a vacuumed environment. The first step is the preparation of poly-sulphide (preferably per-sulphide). Poly-sulphide is prepared by bubbling hydrogen sulphide gas through an aqueous solution of sodium hydroxide, to form sodium hydrosulphide. To which, fresh sodium hydroxide solution (of the same strength and volume to initiate the reaction) is added to obtain sodium sulphide. A known amount (one equivalent to the sodium sulphide) of elemental sulphur powder is then added into this sodium sulphide to get a solution predominantly consisting of sodium per-sulphide (disulphide).

In the second step of nano pyrite synthesis, an aqueous solution of ferric chloride is maintained at a pH of 5.6 using a buffer of sodium acetate and acetic acid to form a ferric acetate complex. Under continuous stirring, the prepared per-sulphide solution is added into this iron solution in a drop-wise manner leading to the reduction of ferric to ferrous with the formation of iron sulphide. At this point, the temperature of the oil bath is raised to 120°C, maintained for 2.5 hours, then further raised to 180°C, and allowed to stand for another 2.5 hours. The iron pyrite particles start to appear in the solvent medium that settles down as the stirring is discontinued. The overall schematic of the synthesis is shown in Figure 12.4.

The precipitate is centrifuged and washed with dilute hydrochloric acid to remove excess iron sulphide (FeS). The residue is washed with water to remove acid followed by acetone to clear the water. It is then washed with toluene to remove any excess of sulphur, and finally with acetone to remove the toluene. It is then dried in a desiccator. The nanoparticles thus obtained are characterized by the following properties: Structure and surface (scanning electron microscopy, X-ray diffraction, X-ray photoelectron spectroscopy), and bonding characteristics (Fourier transform infrared spectroscopy, Raman spectroscopy). Figure 12.5 shows an SEM image of the synthesized particles, XRD trace, FTIR, Raman spectra, and XPS (Das et al.,

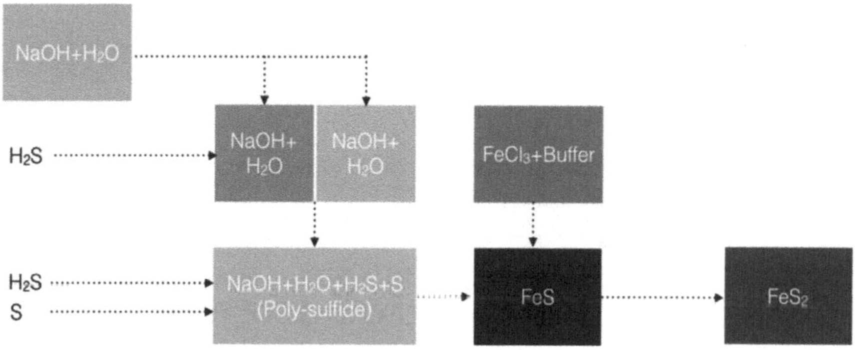

FIGURE 12.4 Chemical scheme for nano FeS_2 synthesis.

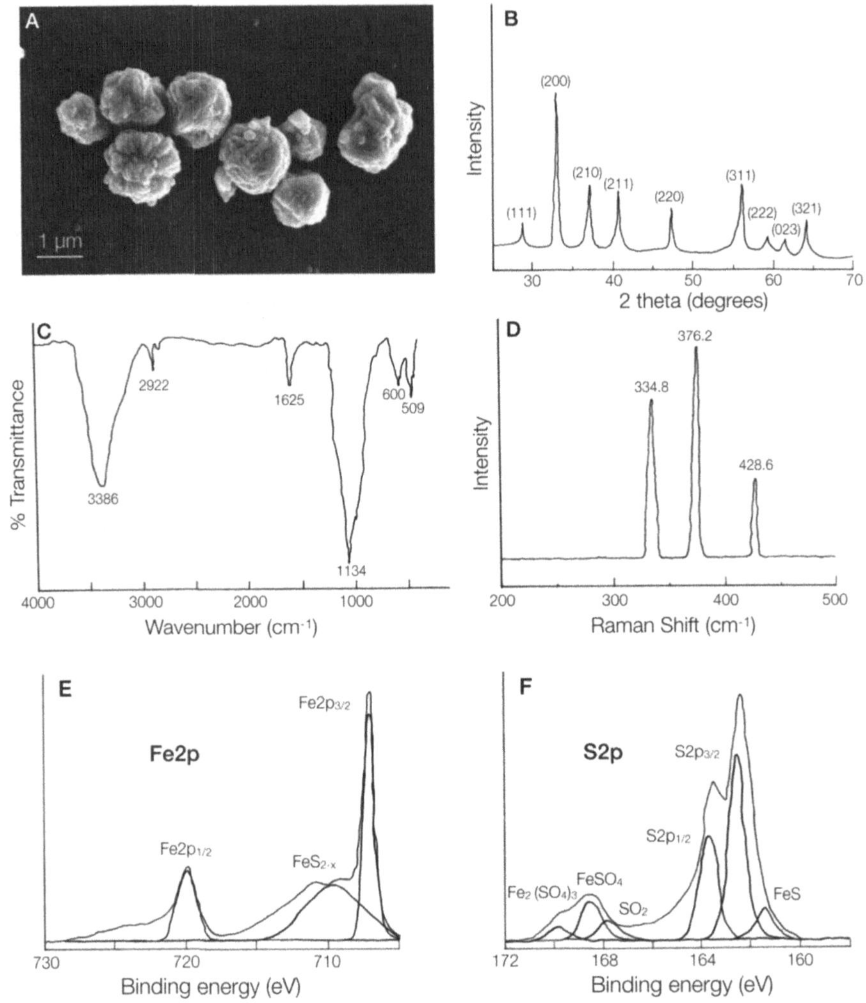

FIGURE 12.5 Characterization of the synthesized nano pyrite. A. Representative scanning electron microscopy picture of the particles. B. Representative XRD trace. C. Representative FTIR trace. D. Representative Raman spectra. E. Representative XPS of Fe2p. F. Representative XPS of S2p.

2016b; Dubey et al., 2016; Jangir et al., 2019a; Jangir, 2019b, 2019c, 2020a, 2020b; Srivastava et al., 2014b; Srivastava, 2016).

12.5 DISCOVERY OF NANO FES_2 AS A SEED STIMULANT AND WHY WE CHOSE A SEED TREATMENT STRATEGY

In the decades to come, nanomaterials will continue to make significant inroads into our ecosystem. While the raging debate will continue on the toxicity-related issues of nanomaterials, these debates will only fuel more research into the subject and will eventually help us to make a judicious call. In terms of nanomaterial applications in agriculture, there are serious apprehensions among scientists, who are concerned about the speculative detrimental effects of nanomaterials on soil and plant health as well as how they will affect human health. We, as a group, followed a different strategy, which relies solely on the surface properties of the nanomaterial, ensuring that the nanomaterial does not enter the soil or the plant directly. This aspect has been dealt with in detail in Section 12.6, where we talked about the mechanisms of nano pyrite. We used the nanomaterial as a seed treatment agent. So in our first set of experiments, we pre-treated chickpea seeds with an aqueous suspension of nano FeS_2 (@ 80μg/ml). After the treatment, the seeds were washed with water and then grown in test tubes of fresh water for seven days. While for the control the seeds were treated with water only. The seed pre-treatment was carried out for 12 hours. This brief seed pre-treatment yielded some interesting results, viz., as compared to the control plants, the treated seeds resulted in plants with significantly higher sulphur content and increased dry weight. In summary, we detected increased biomass and healthier plants following seed pre-treatment. This discovery provoked us to ask the following questions: a. Will the technology be successful under field conditions? b. What is the mechanism of action of nano FeS_2 on seeds? (Srivastava et al., 2014a; Srivastava, 2016).

12.6 TRANSLATING TECHNOLOGY FOR FIELD TRIALS ACROSS INDIA

During 2013–2014, we published our discovery that 'iron pyrite, a potential photovoltaic material, increases plant biomass upon seed pre-treatment'. The same year, 2014, we set a steep target to conduct a field trial at the institute nursery of the Indian Institute of Technology Kanpur. We chose spinach as our test crop, keeping in mind that it has a 50–80 day life cycle. We treated the control and the test seeds of spinach exactly as we did for the chickpeas; but instead of growing them in test tubes, this time we sowed them in the soil. Upon comparing the results at day 50, we observed that nano pyrite pre-treated seeds resulted in a greater number of leaves with a larger size, increased biomass, and had leaves with higher concentrations of manganese, zinc, and calcium. We published our first field trial titled 'Seed treatment with iron pyrite (FeS_2) nanoparticles increases the production of spinach'. We did not observe a similar effect on spinach when we pre-treated the seeds with other nanomaterials and their corresponding salts; signifying that the effect is neither oly because of the nano-configuration nor because of the constituting materials, but caused by the unique effect of nano pyrite (Srivastava et al., 2014b).

Next, we decided to see whether geographical locations had any effect on the activity of nano pyrite on seeds and how the other vegetables, oilseeds, and fodder crops performed with a nano pyrite seed treatment. Between 2014–2016, we performed a series of field trials in two different locations of India, viz., Meerut (Uttar Pradesh) India and Aurangabad (Maharashtra) India. These field trials were conducted by independent researchers, and they chose the crops as per their discretion. We supplied these researchers with the nanomaterial after thorough characterization. Investigators chose two vegetable crops: beetroot, carrot; a spice/vegetable crop: fenugreek; a fodder crop: alfalfa; and two oilseed crops: mustard and sesamum. While alfalfa and sesamum were grown in Aurangabad, the rest of the crops were grown in Meerut. While we observed an increase in fruit and biomass production in all the six crops under investigation, two interesting observations emerged out of this study. First, there was an increase in chlorophyll (a and b) content in the fenugreek leaves; these results indicate that nano pyrite treated seeds result in plants that can capture more sunlight and thus could produce more biomass. The second observation was made in the sesame crop. Sesame seeds were treated in three groups: control, nano FeS_2, and nano MoS_2. The idea was to evaluate if the effect could be mimicked by other sulphide nanomaterials and to evaluate the significance of Fe^{2+}. MoS_2 did not show any significant difference related to the control, in terms of production. These results further consolidated our working hypothesis that 'reactions happening on the surface of FeS_2 play a definitive role in boosting the metabolism of plant seeds' (Das, 2015; Das et al., 2016a, 2016b, 2018; Jangir, 2020a, 2020b; Srivastava et al., 2014a, 2014b; Srivastava, 2016). So the stage was set to explore the mechanism of action of nano FeS_2 on seeds. In Figure 12.6, a lab/field trial crop map is shown.

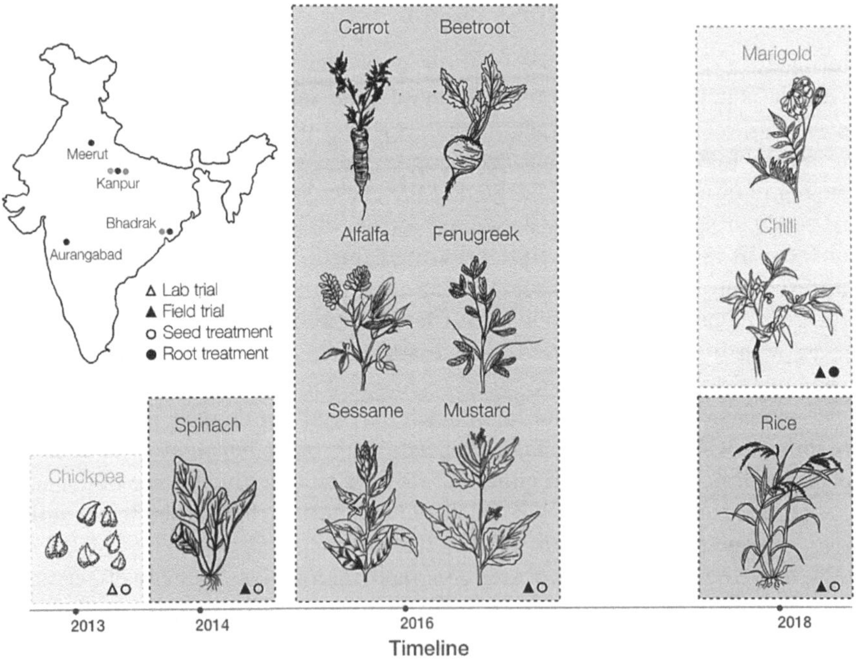

FIGURE 12.6 Crop map highlighting the lab/field trials performed with nano pyrite.

12.7 PARTIAL UNDERSTANDING OF THE MECHANISMS OF NANO FES_2 AS A SEED STIMULANT

We mentioned earlier that nano pyrite has stoichiometric defects on its surface; thus, it is customary to write the formula of nano pyrite as $FeS_{(2-X)}$. Unless nano pyrite synthesis is standardized thoroughly, it is exceptionally challenging to fix the amount of 'X'. This is one of the key reasons, why we suggest that the chemical synthesis of nano pyrite is a superior technique, as natural sources do not provide homogenous nano-range pyrite. During the first quarter of the 20th century, iron pyrite was used to fabricate solid-state electronic devices, but its use was discontinued due to its inherent stoichiometric defects. Interestingly stoichiometric defects on the surface of nano pyrite could be exploited for other sustainable applications. One such application is seed pre-treatment. In the subsequent paragraphs, we will discuss the putative reactions that happen in a milieu of FeS_2+H_2O+seed. We observed that in this milieu, after 14 hours, the amount of reducing sugar was significantly high. The next question was 'is there more starch breakdown from the seeds in the presence of FeS_2+H_2O? If it is so, then how?' The possible answer lies in the surface chemical properties of FeS_2.

FeS_2 is an exceptionally stable molecule. It does not dissolve in water, but rather forms a colloidal suspension. Due to the surface defects, FeS_2 in the presence of water oxidizes a small amount of water to form trace amounts of H_2O_2 (peroxide), along with the generation of trace amounts of Fe_2O_3, FeS, $FeSO_4$, $Fe(SO_4)_3$, SO_2, S, and H^+ ionic species. Each one of these species some way or another contributes to seed germination. Here, we will discuss the role of these ionic species in seed germination and long-term signalling.

First, we will explore the role of peroxide in seed germination. Endogenously seed germination is prevented by antioxidant enzymes, especially catalase. Catalase is an enzyme that breaks down peroxide. During the normal germination process, seeds produce a certain amount of endogenous peroxide, which initiates the breakdown of the stored starch, and later amylase takes over the job of breaking the rest of the starch to reducing sugar. If the seeds are incubated in a 'catalase inhibitor' solution, the germination will be faster; since the catalase inhibitor will block the endogenous catalase and will allow the endogenous peroxide to break down the starch without any stop signal, which otherwise would have been stopped by the catalase. Similarly, trace amounts of exogenous peroxide will also hasten the germination process. By the same analogy, if the seeds are incubated in catalase or a nano-analogue of catalase, viz., CeO_2, seed germination will be delayed. Endogenous peroxide activity followed by amylase action plays a critical role in seed germination and emergence.

In the case of nano pyrite seed treatment, we observe a higher level of reducing sugar in the milieu of FeS_2+H_2O+seed. This is due to the presence of trace amounts of peroxide in the milieu. Here it is noteworthy that the generation of peroxide in the presence of iron leads to the coining of terminology for FeS_2, viz., a 'pyrite only Fenton like reagent'. Peroxide can hydrolyze starch by attacking the glycoside linkages. Thus, a nano FeS_2+H_2O suspension acts similarly to amylase. So one can consider a nano FeS_2+H_2O suspension as an 'artificial enzyme system', which can break starch. Such starch breaking ability has a physiological significance in the life of the seed. The only source of energy for seed is starch. During germination, endogenous peroxide and amylase are activated, which breaks down the starch into reducing sugars. These reducing sugars provide the necessary metabolic energy at the early stages of the plant's life until it sends out its root to forage for nutrients. In the case of a nano pyrite treatment, the amylase activity is higher, since the seeds' endogenous amylase activity is supplemented by trace amounts of exogenous peroxide, which leads to more starch breakdown and offers the seeds a burst of energy for a limited period; thus significantly improving its germination potential, as observed by us. Quite interestingly, when seeds are incubated in nano CeO_2, an enzyme mimicking catalase activity, the effect of FeS_2 was nullified. CeO_2 is a potent antioxidant and is known to scavenge peroxide.

Apart from aiding in germination, trace amounts of peroxide functions as a chemical messenger. Peroxide acts as a secondary messenger in stimulating redox signalling, carbohydrate metabolism, and brassinosteroid mediated CO_2 assimilation. Later studies into spinach indicated that there is an increase in carbohydrate and chlorophyll in the leaves following nano pyrite seed pre-treatment.

Among the other ionic species, viz., Fe_2O_3, FeS, $FeSO_4$, $Fe(SO_4)_3$, SO_2, S, and H^+, H+ ions make the suspension slightly acidic, which helps the germinating seeds to ferry counter ions easily inside the seed. This effect has been dealt with in the root treatment section (Section 12.1). Fe_2O_3 is known to boost root growth. Other Fe and S ionic species together assist in the germination of the seeds. More research is needed for further understanding of the long-term effects of nano pyrite on seed physiology. In Figure 12.7, a schematic of the mechanism of action of nano pyrite on the seed is shown.

We can summarize this as follows: In nanoform, the iron pyrite can serve as a potential redox partner to provide electrochemical energy that affects the seed or the seeding at the initial stage. To tackle the seed coat, this nano iron pyrite slowly supplies the essential low potential oxidative environment. Iron pyrite has two vital roles to play, first, because of the presence of low potential S_2^{2-} (per-sulphide), it can smoothly operate as an oxidant to counter the antioxidant environment of the seed coat. By doing so, it generates FeS, which with the available amino acids quickly transforms into iron–sulphur proteins to be associated with the redox (reduction–oxidation) membrane reaction (Das et al., 2016a, 2016b, 2018; Jangir, 2020a, 2020b; Srivastava et al., 2014a, 2014b; Srivastava, 2016).

12.8 NANO FES_2 AND ITS INFLUENCE ON NPK FERTILIZER REQUIREMENTS

All previous discussions have concerned the non-utilization of NPK fertilizers. As discussed earlier, modern agriculture is heavily dependent on the use of fertilizers. It is not feasible for humankind to abruptly stop the use of synthetic fertilizers. Therefore, any approach to reduce the use of fertilizers is to be explored carefully. During 2017–2018, we designed an experiment to test the effect of seed pre-treatment on rice crops. This is the first study involving a grain crop and/or a monocot. The

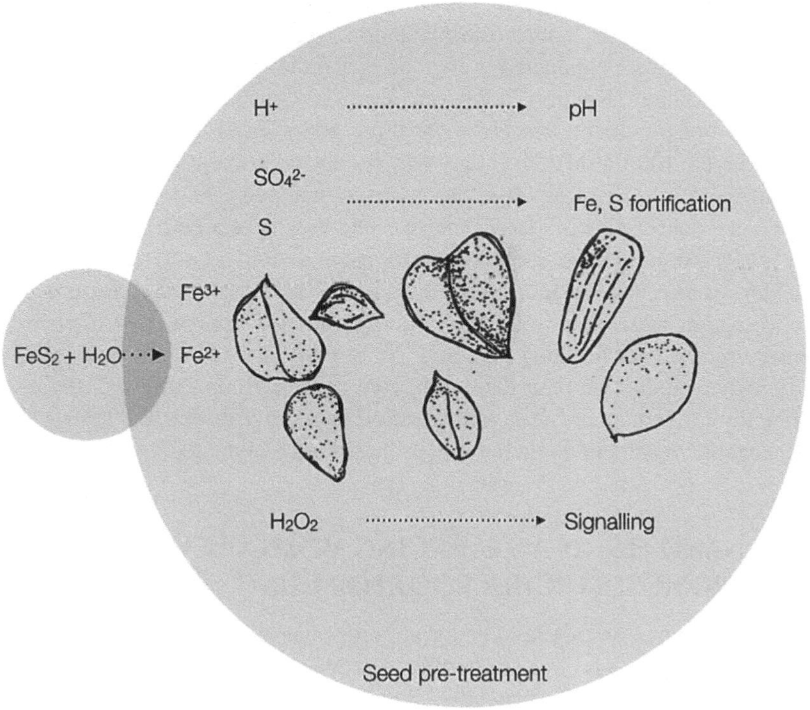

FIGURE 12.7 Mechanism of action of nano iron pyrite on seeds.

field experiments were conducted in Bhadrak, Odisha, India. An experiment was designed to test four groups, viz., group 1 (control), group 2 (nano pyrite seed treated but no NPK application), group 3 (NPK was applied but seeds were not treated with nano pyrite), and group 4 (seed pre-treated with nano pyrite, as well as NPK application). Here, a dose of 50 μg/ml nano pyrite aqueous suspension was optimized. We observed that group 2 and group 3 showed similar yield, while group 4 showed an increased yield. Therefore, we discovered that it is possible to achieve a NPK equivalent rice production, solely with nano pyrite seed dressing, and this can be regarded as a sustainable design for producing rice grain (Das et al., 2018). Wheat crop treated with nano pyrite has shown an increase in yield (Jangir, 2020a, 2020b).

12.9 DISCOVERY OF NANO FeS_2 AS A ROOT STIMULANT FOR CROPS REQUIRING TRANSPLANTATION

While we had explored seed treatment strategies quite extensively, we then turned our attention towards root treatments. The reason for exploring root treatment strategies was the following: For better crop quality, most of the vegetable growers and floriculturists grow their crops from transplanted seedlings. Therefore, they obtain the month-old seedlings from the nursery and transplant them into their field. So, keeping this in mind, during 2018–2019, we explored the possibility of a root treatment with nano pyrite, before transplantation. We choose two crops: chilli, and marigold. Trials were conducted in two locations: the nursery of Indian Institute of Technology Kanpur (Kanpur, India) and Bhadrak (Odisha, India). Unlike the seed treatment, root treatment was given for three hours in an aqueous suspension of nano pyrite (@ 100 µg/ml). The study was conducted without an NPK application. We made certain interesting observations from this study. The nano pyrite treated roots form a denser network under the soil, and they have a better spread. So there was improved root architecture, and they forage more into the soil in search of nutrition. This result is unique since all other efforts to improve root architecture involve genetics approaches. This is the first nano-agronomic approach to improve root architecture and root foraging potential. Nano pyrite treated roots showed significantly higher production than the non-treated control group (Jangir, 2019a, 2019b, 2020b). Root treatment of the samplings of cabbage, cauliflower, and tomato during transplantation results in an increase in profit (Jangir 2019b, 2020b).

12.10 NANO FES_2 AS AN EMERGING MOLECULE WHICH PROMOTES HIGHER ROOT FORAGING

Better root foraging among food and grain crops has a global significance. A large amount of the land-mass on the floor of the Earth has nutrient-deficient soil. So strategies have to be developed to bring these lands under cultivation. Currently, breeders are looking forward to being able to select and raise varieties with better root foraging potential; where such varieties perform better in nutrient-deficient soils, through their ability to draw upon nutrients in the soil through foraging. We are proposing a different strategy to meet this challenge; we propose that if we could tweak the metabolism of the root during transplantation, we may get the desired result without genetic manipulation. Our data on chilli, marigold, cabbage, cauliflower, and tomato shows this trend (Jangir, 2019a, 2019b, 2020b).

12.11 DISCOVERING THE INITIAL REACTIONS OF NANO FES_2 UPON ROOT TREATMENT

In Section 12.7, we discussed the mechanism of action of nano pyrite on seeds. Here, we will be discussing the putative mode of action on the root during the brief exposure. The root system, while interacting with the surroundings, maintains an electro-neutrality by pumping H^+ and OH^- ions into the environment; in this process, it ferries the growth limiting anions and cations from the surroundings to inside the root. When the root is exposed to a FeS_2+H_2O suspension for three hours, the root system is exposed to H^+, H_2O_2, Fe^{2+}, Fe^{3+}, and SO_4^{2-} ionic species. Iron is one of the rate-limiting species which has a long-term influence on plant growth. Due to a mildly acidic environment for three hours during the root treatment, the root took up Fe^{2+} and Fe^{3+} without any additional energy expense. The movement of OH^- ions outside the root maintains the electro-neutrality, and results in the counter-movement of

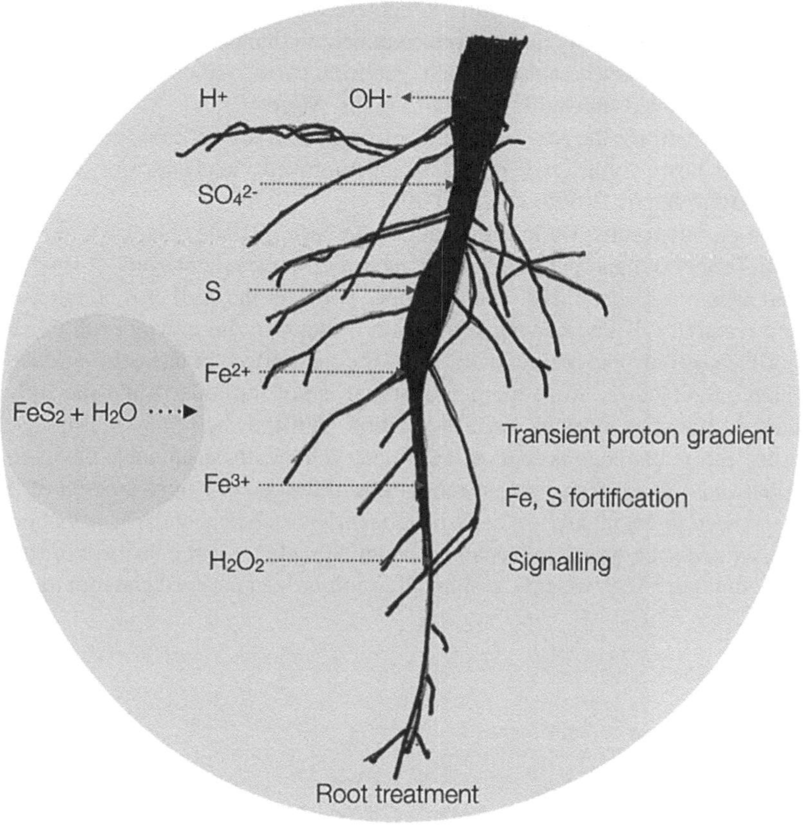

FIGURE 12.8 Mechanism of action of nano iron pyrite on root during brief 3 hours of root treatment.

SO_4^{2-} ionic species inside the root. During three hours of root treatment, the root is fortified with Fe and S, and peroxide signalling is initiated within the plant. Once the plant sapling is transplanted into the soil, the trace amount of H^+ ions and OH^- ionic species on the root surface result in the entry of other necessary cations (NH_4^+) and anions into the root system. We are currently exploring the long-term effect of such a treatment. In Figure 12.8, a schematic of the mechanism of action of nano pyrite on the root is shown (Jangir, 2019a, 2019b, 2020b).

12.12 NANO FES_2, A NON-GENETIC, AGRONOMIC TOOL FOR MARGINAL FARMERS

Nano pyrite (FeS_2) is an interesting molecule, which has been explored extensively for photovoltaic applications. Interestingly, in the year 2013, we discovered that this nanomaterial is a profound seed pre-treatment agent and results in a several-fold increase in the yield of different crops. Thus, ushering nano pyrite into the world of sustainable nano-agriculture to achieve precision farming (Jangir, 2020b). Since then, we discovered that seed pre-treatment with nano pyrite results in higher yields in grain (rice), vegetable (spinach, beetroot, carrot), spice (fenugreek), fodder (alfalfa), and oilseed (mustard, sesamum) crops. Moreover, we discovered that, for transplantation crops, the root treatments of vegetables (chilli, cabbage, cauliflower, tomato) and flower (marigold) results in an improved yield and root architecture (Jangir, 2019a; Jangir, 2019b; Jangir, 2020b).

Based on our results, we have explained the basic surface chemistry behind the beneficial effect of nano pyrite as a seed and root pre-treatment agent. It is noteworthy that nano pyrite, itself as a molecule, does not enter the seed or root, but a subtle surface reactivity of an aqueous suspension of nano pyrite enhances the metabolism of the seed or the root, to bring about the final effect in terms of productivity and plant architecture. Since the nanomaterial is not introduced into the soil, and it executes its action through surface reaction with the seed or root, during brief exposure, this technology is economically and ecologically sustainable. It targets the marginal and organic farmers who cannot afford the use of a huge amount of fertilizers and need to get maximum output from nutrient-deficient soils. This technology could eventually be translated as an agronomic practice for the farmers of the 21st century. In Figure 12.9, we have graphically summarized the seed and root treatment strategies.

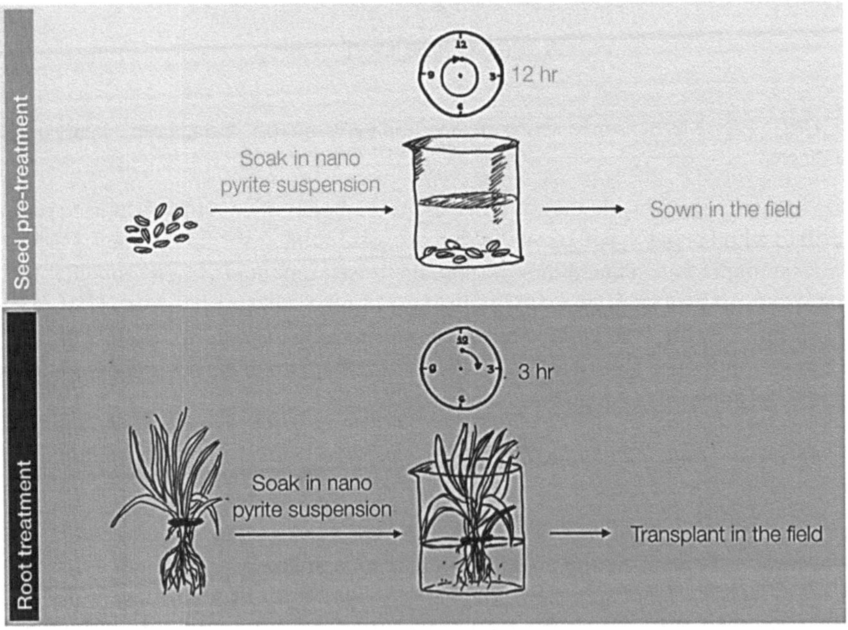

FIGURE 12.9 Two different agronomic strategies were suggested for nano pyrite treatment.

12.13 WHY NANO FES_2 AND NOT MACRO OR BULK FES_2?

In bulk FeS_2, which is mostly obtained from natural sources, it is challenging to control the surface reaction. Different studies have indicated that the surface reactivity of nano FeS_2 is many-fold higher than in bulk FeS_2. Since the long-term objective is to design a cost-effective, sustainable solution, instead of purifying the FeS_2 obtained from natural sources and then converting it into a nanoformulation, it is economical to generate the nanoformulation directly through an ambient chemical route. One of the prime characteristics of our synthetic approach (as described in Section 12.4) informing nano iron pyrite is its modularity. By varying the dilution and reaction times, the desired particle size can be attained. Moreover, this technique produces a homogenous form of iron pyrite, contrary to other known solid-state reactions where such controls are difficult to attain (Jangir, 2019a, 2019b, 2019c, 2020a, 2020b).

12.14 THE ECONOMICS FOR SYNTHESIS OF NANO FES_2?

The basic reagents which are needed for the synthesis of nano pyrite are the following: Iron (III) chloride ($FeCl_3$), sodium hydroxide (NaOH), sulphur powder (S), sodium acetate buffer (NaOAc), acetic acid (CH_3COOH), acetone ($(CH_3)_2CO$), toluene (C_7H_8), hydrochloric acid (HCl), iron sulphide (FeS), and argon gas cylinders. All these chemicals are easily available across the developing nations of the world. A small scale production unit needs 2–3 skilled labourers, who have a basic knowledge of working in a chemical laboratory. In one of our earlier works, we worked out the detailed economics of treating seeds of rice with nano pyrite (@ 50 μg/ml). Based on our findings, we concluded that the effective cost of treating seeds for planting a hectare of land is INR 26.00. We believe that if the process is scaled up, there will further cutting down on the synthesis costs (Das et al., 2018; Jangir, 2020a, 2020b).

12.15 PROPOSED MODEL TO TRANSLATE THE NANO FES_2 MOLECULES TO THE MARGINAL FARMERS OF ASIA AND AFRICA

We are proposing three different models for translating this technology to the marginal farmers of Asia and Africa (Jangir, 2020b). The major consideration which we have taken into account while proposing the strategies are the following: a. Most of the locations in Africa are fairly remote, and fertilizer supply is cost-prohibitive. b. Considering that nano pyrite is needed in a small amounts, it may be a better idea to commercialize it along with the seed.

Following the three approaches we have been proposing, viz., a. the factory in a suitcase; b. packaging them in small bags similar to tea bags; and c. formulating water-soluble capsules or gels while encapsulating the nanomaterial inside it. The approach of the factory in a suitcase demands on the spot synthesis and scaling down the production such that the demand can be met in the remote places while cutting down on the cost of transportation. Currently, we are working towards this approach. The other two approaches can be coupled with seed bags, viz., when the

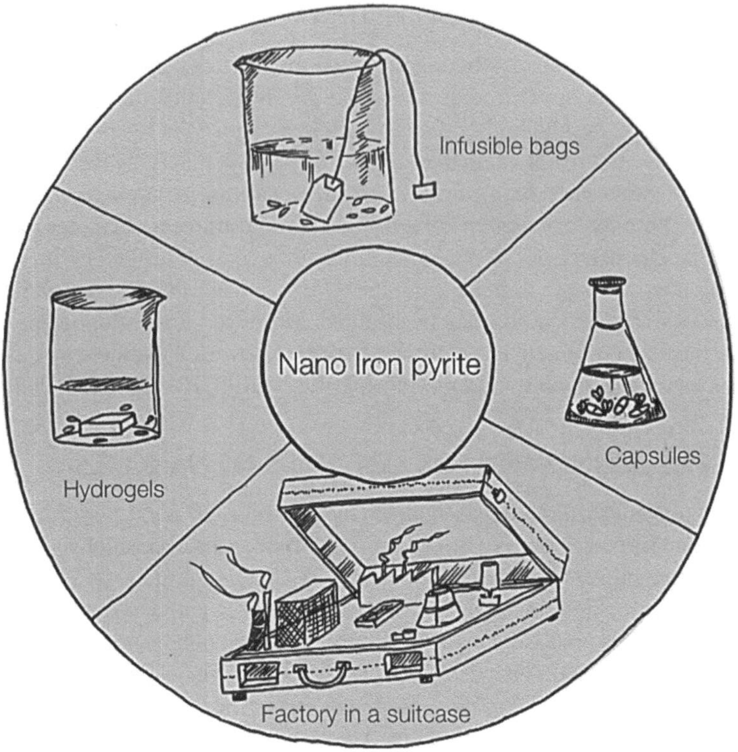

FIGURE 12.10 Schematic highlighting the different strategies which re being proposed here for translating the nano pyrite technology for the farmers.

farmer is buying the seeds, the nano pyrite pouches can be sold along with it. These approaches will further cut down on the heavy transportation charges incurred while shipping shipments in remote locations. Figure 12.10 describes a schematic of different strategies for translating the nano pyrite technology to the farmers.

12.16 GOING BACK TO OUR ROOTS: THE JOURNEY OF NANO FES_2 FROM DARK (CHEMOSYNTHESIS) TO LIGHT (PHOTOSYNTHESIS) AND A SUSTAINABLE FUTURE FOR POSTERITY

According to the theory of the chemical evolution of life, our present-day biosphere has its genesis in the iron–sulphur world. The iron–sulphur proteins as electron transfer carriers conserved their role from anaerobic primitive hyperthermophiles to present-day aerobic mesophiles. In our present effort, we explore the association of iron and sulphur in the form of nano iron pyrite, which is available in abundance in nature, albeit heterogeneously and in larger aggregates. In a laboratory setting, we produced homogenous nano iron pyrite. The reaction parameters can be uniquely varied to control the size of the nanoparticles. It is well known that iron–sulphur aggregate is mostly stabilized by thiolate groups associated with apoproteins in relevance to iron–sulphur proteins. The nano iron pyrite, once treated with thiolates, can readily be transferred to iron–sulphur–thiolate containing cores similar to active iron–sulphur proteins. We have discovered that this iron pyrite, which has sustained through evolution, has a vital role to play in seed germination and plant growth. Seed priming enhances seed germination, plant parameters, and yield; while root priming improves yield and modulates plant parameters and root architecture. This nanomaterial also has an application as an asymmetric charge storage device in an aqueous biocompatible electrolyte (Dubey et al., 2016). So we are gradually realizing that a three atom molecule, which is believed to be the major source of energy for the chemosynthesis dependent life forms from a distant past, is emerging as a ray of hope for the sustainable food production of the photosynthetic world.

12.17 CONCLUSION

We proposed a sustainable nano-agronomic design strategy for futuristic farming. We have demonstrated that higher productivity is feasible without using synthetic fertilizers in nine different crops-grains (wheat, rice), pulse (chickpea), vegetables (cauliflower, cabbage, tomato), spice (chilli, onion), and flower (marigold) (Figure 12.11).

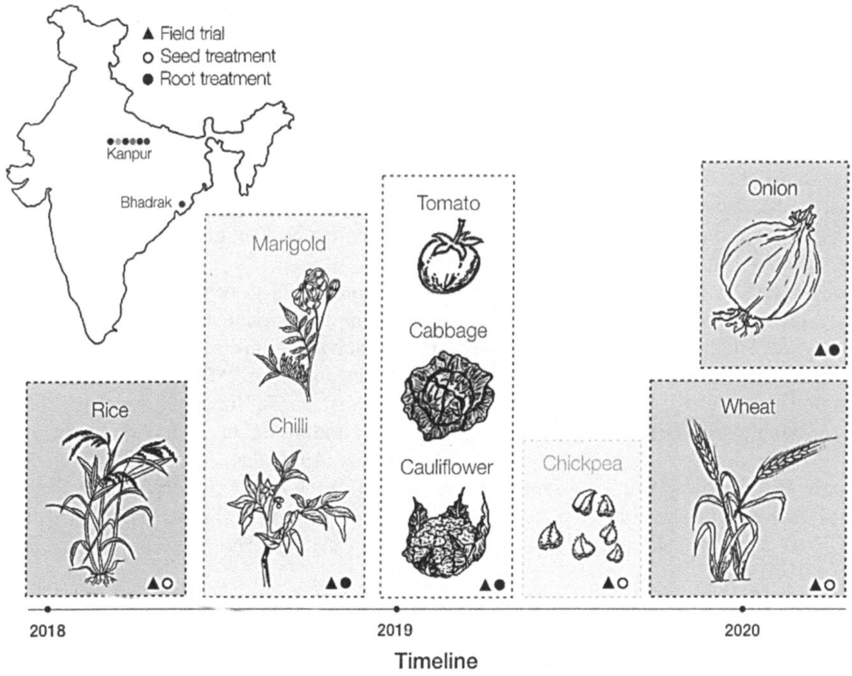

FIGURE 12.11 Graphical representation of the crop trials conducted in India as a part of the research between 2018–2020.

This was achieved by selectively fortifying the plant seeds or roots at the onset of their growth with nano iron disulphide (nano pyrite). We discovered that a seed or root treatment results in a more robust root architecture and better root foraging capacity. Improved root architecture offers these plants an ecological advantage to compete for nutrients. The nano pyrite seed treatment of the chickpea (pulse) improves root nodule formation – suggesting a better natural nitrogen-fixing ability (Jangir, 2019c). Thus, it is feasible to achieve sustainable growth while maintaining genetic diversity and a reduction in fertilizer use.

After establishing nano pyrite as a seed and root bio-stimulant through a series of experiments, the interaction of nanomaterials with seeds/roots can be explored. Particularly, the electrochemical changes that occur in the seed/root microenvironment during the treatment. The effect of nanomaterials on oxygen evolution from chloroplast can be evaluated. These questions can help establish a link between the aerobic (plant systems) and anaerobic (iron pyrite, a molecule from hydrothermal vents) world. Further experiments in salt-affected and nutrient-deficient soils and studies in controlled-defined hydroponic settings are to be conducted for a better understanding of this biro-stimulant effect of nano pyrite.

REFERENCES

Das, C. K., Srivastava, G., Dubey, A., Roy, M., Jain, S., Sethy, N. K., et al. (2016a). Nano iron pyrite seed dressing: A sustainable intervention to reduce fertilizer consumption in vegetable (beetroot, carrot), spice (fenugreek), fodder (alfalfa) and oilseed (mustard, sesamum) crops. *Nanotechnol. Environ. Eng.*, 1, 2.

Das, C. K., Srivastava, G., Dubey, A., Verma, S., Saxena, M., Roy, M., et al. (2016b). The seed stimulant effect of nano iron pyrite is compromised by nano cerium oxide: Regulation by the trace ionic species generated in the aqueous suspension of iron pyrite. *RSC Adv.*, 6, 67029–67038.

Das, C. K., Jangir, H., Kumar, J., Verma,, S., Mahapatra, S. P., Philip, D., et al. (2018). Nano pyrite seed dressing: A sustainable design for NPK-equivalent rice production. *Nanotechnol. Environ. Eng.*, 3, 14.

Das, M. (2015). Iron pyrite as seed treatment biostimulant: The new revolution? New AG International, June–July, pp. 41–42. www.newaginternational.com.

Dubey, A., Singh, S. K., Tulachan, B., Roy, M., Srivastava, G., Philip, D., et al. (2016). Nano iron pyrite (FeS_2) exhibits bi-functional electrode character. *RSC Adv.*, 6, 16859–16867.

Jangir, H., Das, C. K., Kumar, J., Mahapatra, S. S., Srivastava, F., Bhardwaj, A., et al. (2019a). Nano pyrite (FeS_2) root priming enhances chilli and marigold production in nutrients-deficient soil: A nano strategy for fertiliser tuning. *Appl. Nanosci.*, 9, 327.

Jangir, H., Bhardwaj, A., Srivastava, G., Das, M. (2019b). Nano pyrite driven root foraging increases production of the heavy feeders, viz., cauliflower, cabbage and tomato in nutrient deficient soil with no fertiliser application. *Adv. Nat. Sci: Nanosci. Nanotechnol.* 10, 035007.

Jangir, H., Bhardwaj, A., Das, M. (2019c). Larger root nodules, increased Fe, Mo, Mg, P, Ca, Mn, K in the roots and higher yield in chickpea grown from nano FeS_2 pre-treated seeds: Emulating nitrogenase. *Appl. Nanosci.*, 10(2), 445–454.

Jangir, H., Bhardwaj, A., Srivastava, G., Das, M. (2020a). Fertilizer free cultivation of wheat in nutrient deficient soil by treating the seeds with nano pyrite. *Nanotechnol. Environ. Eng.*, 5 (9).

Jangir, H. (2020b). Exploring sulfides to design strategies for sustainable agriculture and green energy. (Doctoral thesis), Indian Institute of Technology Kanpur, India. Electronic Thesis and Dissertations (ETD) link: http://172.28.64.70:8080/jspui/handle/123456789/18922.

Srivastava, G., Das, A., Kusurkar, T. S., Roy, M., Airan, S., Sharma, R. K., et al. (2014a). Iron pyrite, a potential photovoltaic material, increases plant biomass upon seed pretreatment. *Mater. Express*, 4 (1), 23–31.

Srivastava, G., Das, C, K., Das, A., Singh, S. K., Roy, M., Kim, H., et al. (2014b). Seed treatment with iron pyrite (FeS_2) nanoparticles increase the production of spinach. *RSC Adv.*, 4 (102), 58495–58504.

Srivastava, G. (2016). Iron pyrite: A seed biostimulant. Doctoral thesis, Indian Institute of Technology Kanpur, India. Electronic Thesis and Dissertations (ETD) link: http://172.28.64.70:8080/jspui/handle/123456789/15534

13 Plant Extract Guided Nanomaterials and Plant Disease Management

Zehra Karaagac, Seyma Dadi, Semiha Ekrikaya, and Ismail Ocsoy

CONTENTS

13.1 INTRODUCTION TO NANOMATERIALS

Nanomaterials (NMs) are a class of materials with at least one dimension between 1 to 100 nm. Recently, NMs have been increasingly utilized in a wide variety of scientific fields, such as medicine, electronics, and industry (Veiseh et al., 2010; Qiu et al., 2015; Prasad et al., 2017). For instance, in terms of nanomedicine, NMs have been designed as imaging tools (optical, fluorescent, and magnetic imaging), antimicrobial, anticancer, and antioxidant agents, and delivery vehicles (drug, gene, nucleic acid, photosensitizer, and dye, etc.) (Dreaden et al., 2012). The physico-chemical and biological properties of NMs rely on their dimensions, size, and shape, crystallinity, and solubility (Peng et al., 2014). Generally, NMs with the best properties are produced in organic solvents at high temperatures, but they are highly toxic for animals and humans (Hochella et al., 2015; Ocsoy et al., 2018a; Ankamwar et al., 2005). The toxicity of NMs is a vitally important issue and strictly limits their use in bioanalytical and biomedical applications due to the risk of exposure to living organisms.

13.1.1 Synthesis Approach and Methods

In general, NMs have been used in various fields due to their tunable and greatly enhanced physico-chemical characteristic properties, such as catalytic activity, electrical and thermal conductivity, light absorption, and scattering compared to their bulk counterparts. There are two essential approaches for the synthesis of metallic NMs: i) the physical approach or 'top-down' method is based on reducing a bulk material to the nanoscale and ii) the chemical approach relies on the assembly of atoms to create materials in nano size, which is also termed the 'bottom-up' approach. Although the top-down approach is suitable to produce NMs on large-scale for the need of industry, the lack of control of size, morphology, and size distribution of NMs are major drawbacks to this approach and result in remarkable reductions in their corresponding properties (Cheon et al., 2004; Lu et al., 2007). The shape, size, and size distribution control of NMs are achieved with a bottom-up approach, but the production yield is quite low compared to the top-down approach (Cedervall et al., 2007; Bruchez et al., 1998). Chemical methods are commonly used in the bottom-up approach, including micro-emulsion/colloidal, sonochemical, solvothermal, thermal decomposition, and pulse laser methods (Kumar and Kumbhat, 2016; Cheon et al., 2004; Lu et al., 2007).

The formation mechanism for the synthesis of NMs was first explained in the 1940s by La Mer and coworkers (Figure 13.1). The La Mer theorem explained that rapid nucleation was considered as a first step. Then after the formation of seeds in the nucleation process, the growth process started in the second step.

This theorem has used for producing colloidal materials with a narrow size distribution (LaMer and Dinegar, 1950). Although it was proposed for micro-sized

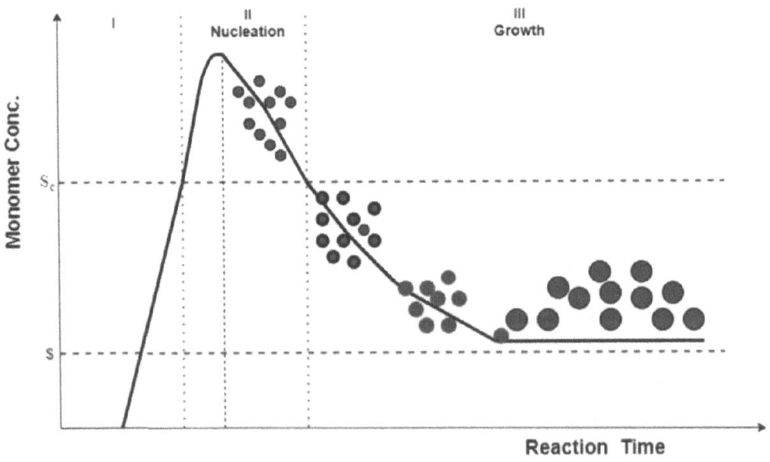

FIGURE 13.1 Illustration of La Mer mechanism for synthesis of NMs with nucleation and growth stages. Supersaturation (S) and critical supersaturation (Sc) are given on the plot.

materials, researchers applied the theory to elucidate NM formation. The formation mechanism of NMs is a thermodynamic procedure that includes reduction, decomposition, and condensation. The La Mer theory describes the three steps in the synthesis of NMs, such as the nucleation, growth, and completion (saturation) steps (Schladt et al., 2011). At the first stage of the reaction, monomer concentrations generated from the precursor material are quickly increased to reach the initiation point of nucleation. At the transition stage before the second stage, self-nucleation and excessive amounts of nuclei occur. Then, for the second stage, the concentration of monomers is decreased below the nucleation sub-limit, and the growth of the seeds starts. In the final stage, the growth of seeds is saturated, and the formation of the NMs is completed.

It is worth mentioning that for homogeneous nucleation to occur, monomer concentrations can be calculated with Fermi level energy or Gibbs free energy barrier, as shown in phase II. Although La Mer's theory was originally designed for the synthesis of micro-sized materials, it could be substantially adapted for the formation of NMs. However, while uncontrolled growth and particle agglomeration can occur during synthesis, due to the high free surface energy in NMs, it is not observed in the synthesis of micro materials (Alivisatos, 1996).

Other reaction conditions, including heating rate, reaction time, surfactant concentration in solution, also have a direct influence on the size, shape, and size distribution of the NMs, which play critical roles in the physical and chemical properties of the NMs. Generally, the most used methods of NM synthesis with the 'bottom-up' approach are the chemical and biological methods.

13.1.1.1 Chemical Methods

The chemical method mostly uses micro-emulsion, co-precipitation, thermal decomposition synthesis strategies to produce hydrophobic NMs in organic solvents with the surfactant molecules acting as capping and reducing agents (Qin et al., 2007; Shen et al., 1999; Kim et al., 2008). The monodisperse and highly crystalline NMs are obtained in these methods owing to high temperature and long-chain surfactant molecules. The other role of surfactants is to prevent the agglomeration of the NMs (De Palma et al., 2007). The surface of the NMs could be functionalized by various targeting ligand molecules (Weissleder et al., 2000), but hydrophobic NMs are not soluble in water and could not be used in bio-related applications owing to the rapid aggregation of the NMs in aqueous solutions. To use them in biomedical applications, they must be transferred to aqueous solutions with proper hydrophilic, amphiphilic ligands, and coating materials (Ocsoy et al., 2013a; Karaagac et al., 2020a; Peng et al., 2014). For instance, the surface ligand of iron (Fe) NMs synthesized in the organic phase was modified by changing the chimeric DNA molecule used as the hydrophilic ligand for its transition into the aqueous solution, as seen in Figure 13.2 (Chen et al., 2012).

The frequently used methods of surface modification are based on surfactant micelles, bi-functional (Amstad et al., 2009), or polymeric ligands and silica coatings (Shukoor et al., 2008). There are various types of capping ligands, including hydrophobic/hydrophilic and amphiphilic capping or reducing ligands with thiol (Sun et al., 2004), phosphines (Hyeon, 2003), amines (Cheon et al., 2004), and carboxylates (Park et al., 2004)

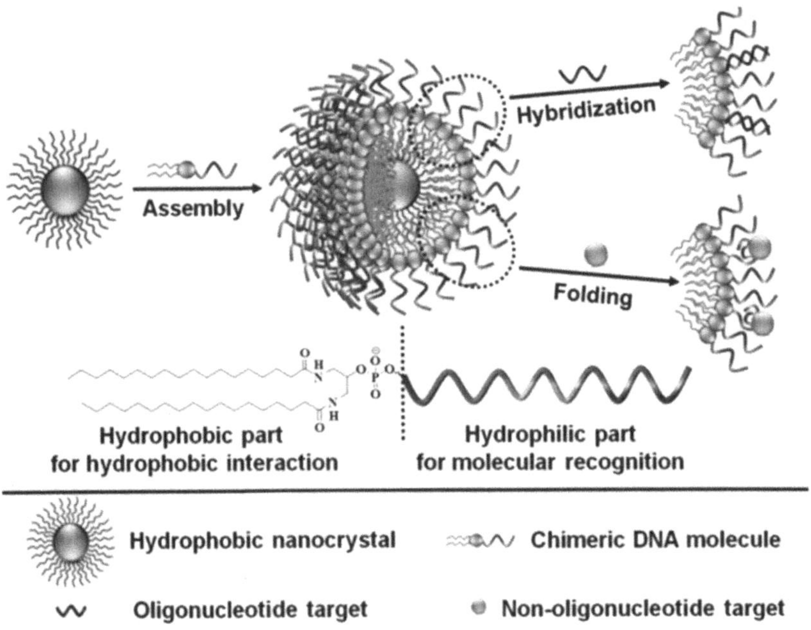

FIGURE 13.2 Surface transfer of hydrophobic nanoparticles with the chimeric DNA molecules, including both hydrophobic (diacyl lipid) and hydrophilic (oligonucleotide) sections. (Chen et al., 2012).

functional groups. In the amphiphilic micelle, the hydrophobic tails of amphiphilic ligands are integrated between the capping ligands on the surface of hydrophobic NMs, whereas the hydrophilic side tails through to the outside arranging into a hydrophilic surface (Figure 13.3). Another approach to ligand NMs is by exchanging a polymer ligand. The polymeric ligands contain several active groups, functional parts, and hydrophilic tails. These polymer ligand groups allow for the further binding of various ligand molecules onto the surface of NMs. The surface of NMs can be modified using bi-functional ligands. These ligands carry an anchor group conjugated to a hydrophilic binder that carries a functional end group. The exchange of hydrophobic ligands ensures the complete solubilization of NMs in aqueous solutions.

13.2 BIOLOGICAL METHODS OF NANOMATERIAL SYNTHESIS AND THEIR ANTIMICROBIAL ACTIVITIES

For the synthesis of NMs, there are various approaches, especially the chemical, physical, and biological methods (Wagner et al., 2014; Shukoor et al., 2012; Yasun et al., 2012; Li et al., 2013; Ocsoy et al., 2013b, 2013c, 2017a, 2016). The NMs with a good crystallinity are produced with chemical methods. But chemical methods mostly use toxic capping agents for the stabilization of the NMs. The toxicity is a major disadvantage of these NMs in biological applications (Wang et al., 2010).

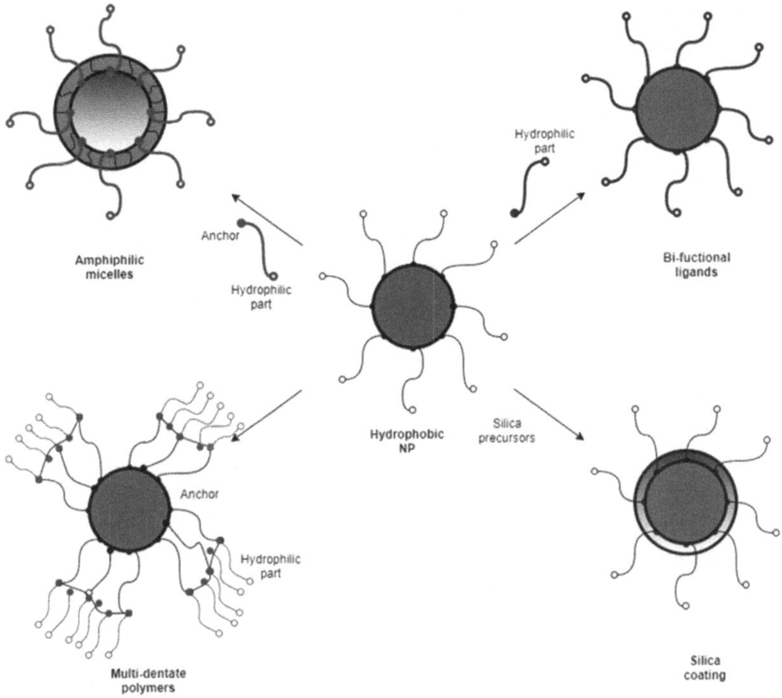

FIGURE 13.3 Most commonly used methods for surface exchange of hydrophobic NMs into the aqueous solutions (Thomas, 2010).

Therefore, non-toxic, ecofriendly synthesis methods are needed. In recent studies, lots of biomolecules have been used as reducing or stabilizing agents for the synthesis of NMs. Mainly these biomolecules are proteins, enzymes, nucleic acids (DNA and RNA), peptide, plant extracts, and standard molecules, and there can be some microorganisms, such as algae, fungi, actinomycetes, bacteria, and viruses (Zhang et al., 2010; Karaagac et al., 2020b; Ocsoy et al., 2013c, 2013d; Turek et al., 2013; Rayman et al., 2018; Altinsoy et al., 2019; Leng et al., 2016; Wei et al., 2011; Wirges et al., 2009). For instance, Watson-Crick base coupling and the design properties of DNA enable them to act as templates for the nucleation and growth of various NMs. (Ma et al., 2006; Rothemund, 2006; Pal et al., 2011; Ocsoy et al., 2013b; Strayer et al., 2016).

Studies indicated that metal ions favourably bind to the nucleotides or phosphates of nucleic acids (Gugliotti et al., 2004; Berti and Burley, 2008). Furthermore, DNA, proteins, and various plant molecules act as reducing agents for the synthesis of hydrophilic NMs. The metal ions may bind to various functional groups (amine, thiol, or carboxyl) to form primary nanocrystals or seeds.

Although the synthesis of biocompatible NMs is carried out with nucleic acids, proteins, and enzymes as biomolecules, the main disadvantages of biological ligands are their expense, low stability, and easy contamination. To address these disadvantages, plant extracts have been used in the synthesis of NMs (Demirbas et al., 2016, 2017, 2019a, 2019b; Karatoprak et al., 2017; Ma et al., 2006, 2009; Dogru et al., 2017; Ocsoy et al., 2017b, 2018b; Pazi et al., 2016; Some et al., 2019; Koca et al., 2020; Unal et al., 2020; Celik et al., 2019; Onbaslı et al., 2018). The main advantages of plant extracts are that they are widely reachable, easily prepared, cost-effective, at no risk of contamination, non-toxic, and ecofriendly (Njagi et al., 2011). Plant extracts can also be reducing and/or stabilizing agents during NMs synthesis due to the presence of polyphenols, polypeptides, enzymes, and proteins. To date, many types of NMs have been produced using plant extracts, such as red cabbage, blueberry, and blackberry (Demirbas et al., 2019a, 2019b). For instance, common plasmonic NMs, gold (Au), and silver (Ag) NMs were synthesized with plant extracts. Au and Ag NMs have been used in a large variety of scientific fields, especially in nanomedicine (Aromal and Philip, 2012), nano electronic systems (McLamore et al., 2016), bioimaging, and biosensing (Ocsoy et al., 2017c). In the synthesis of metallic NMs, experimental parameters including extracts and metal salt concentrations, reaction temperature and time, and stirring rate are generally optimized to produce uniform NMs.

Additionally, biomolecules incorporated into hybrid NMs have been recently developed using enzymes, proteins, and plant extracts as organic components, with some metal ions as inorganic components (Celik et al., 2018, 2020; Altinkaynak et al., 2016, 2017, 2019; Baldemir et al., 2018, 2017; Ildiz et al., 2017; Yilmaz et al., 2016; Somturk et al., 2016; Ocsoy et al., 2015; Ge et al., 2012; Zhu et al., 2013; Ackermann et al., 2011; Wang et al., 2013a, 2013b; Li et al., 2014; Somturk et al., 2016). the main mechanism in the formation of an organic–inorganic hybrid NMs is that organic molecules preferentially coordinate with metal ions through their functional groups. These metal–organic frameworks (MOFs) have porous flower-shaped morphologies and greatly enhanced stabilities, and catalytic and antimicrobial activities. The potential reasons for the superior biological activities of this organic–inorganic hybrid NMs are their i) high surface to volume ratio, ii) favourable conformation in their structure, iii) synergistic effects occurring between organic and inorganic molecules, iv) morphology dependent activity, and v) potentially effective Fenton reaction in the presence of hydrogen peroxide (H_2O_2) (when plant extracts and other bio-or-organic molecules are used instead of enzymes).

NMs have found various applications in dentistry because of their unique properties. The application of NMs in dentistry can be categorized in two ways: preventive dentistry and restorative dentistry (Bapat et al., 2019; Hanning and Hannig, 2012). NMs are actively used in all branches of dentistry due to their antibacterial activity (Bapat et al., 2018). Polymicrobial dental plaque in oral flora is the main source of dental caries and infections. In addition to cleaning, antimicrobial agents have a great role in the prevention of dental diseases (Baehni and Takeuchi, 2003). NMs are antimicrobials that play an active role in combating dental diseases with their biomedical properties (Allaker, 2010). NMs show bactericidal effects by disrupting the membrane structure of oral flora bacteria (Lu et al., 2013). Also, NMs promise to inhibit dental caries made by *S. Mutans* acid and enhance remineralization (Cheng et al., 2012; Zhang et al., 2012). Various other NMs, including metal and oxide NMs, such as ZnO, TiO_2, MgO, CaO, and CuO NMs and their antibacterial capabilities in dental resins, were investigated (Tavassoli et al., 2013; Welch et al.,

2010). These NMs are promising for incorporation into dental materials. Therefore, nanotechnology has the potential to significantly improve restorative and preventive dentistry (Kandaswamy et al., 2016; Stubbings et al., 2017; Cheng et al., 2015). NMs can adhere to and enter the cell walls of Gram-positive and Gram-negative bacteria and disrupt cell metabolism with their release of Ag ions. Ag NMs have been proven to show antimicrobial properties for many bacteria in oral flora (*S. Mutans, S. Oralis, S. Salivarius, S. Mitis, E. Faecalis, A. Actinomycetemcomitans, C. Albicans, P. Intermedia, P. gingivalis*, and *F. nucleatum*) (Noronha et al., 2017). For example, Wu et al., and Samiei et al., proved that Ag NPs are a quite effective antimicrobial and antibiofilm agents for *E. Feacalis.* Ag NMs are also used as effective irrigation agents for endodontic treatments with their antimicrobial activities. They prevent periapical lesions and infection formations with superior bactericidal effects on anaerobic bacteria (Wu et al., 2014; Samiei et al., 2016; Some et al., 2018).

13.2.1 Plant Extract Directed Metallic and Metal Oxide Nanomaterials

The green synthesis of inorganic NMs can be classified as metal oxide and metallic NMs. When it comes to obtaining NMs with the green synthesis approach, the synthesis mechanism occurs in the presence of plant molecules. Metal oxides are used in various biomedical applications, such as in drug-delivery, biosensing, and biodetection systems (Koca et al., 2018). Generally, they have been synthesized utilizing bio reduction, UV irradiation, and micro-emulsion methods. (Nazari et al., 2017). Also, these NMs can be synthesized chemically in aqueous solutions using chemical molecules as reducing agents like sodium borohydride ($NaBH_4$) (Duan et al., 2015). The chemical agents are often toxic, which threatens the environment, animal and human health. Thus, the green synthesis approach with plant extracts may decrease or eliminate the use of toxic reducing and capping agents (Zikalala et al., 2018). It is known that various organisms are also used for the synthesis of NMs in the biological method. Up to now, various metallic or metal oxide NMs (Ag, Au, Cu, CuO, Fe_3O_4) have been synthesized using plant extracts (Table 13.1).

Au NMs have been of great interest due to their optical, catalytic, and nontoxic properties, and they can be used in a wide spectrum of applications, including nanomedicine, nano-biosensing, and nanoelectronics (Aromal and Philip, 2012). Since a change in the concentration and content of the reducing agents in plant extracts causes different size and morphological nanoparticle synthesis, plant extracts have been used for the synthesis of Au NMs with various sizes and morphologies (Lee et al., 2016). For instance, while leaf extract from *Croton Caudatus Geisel* resulted in the formation of the spherical shape of Au NMs ranging from 20 to 50 nm, triangular and hexagonal shaped Au NMs with an average size of 11 nm were synthesized using leaf extract of *Rosarugosa* (Kumar et al., 2019; Dubey et al., 2010).

Ag NMs have also received considerable attention due to their antimicrobial and optical features that increase their preferability in a large variety of scientific fields (Aromal and Philip., 2012, Ocsoy et al., 2017d). Plant extracts provide biocompatible

TABLE 13.1
Plants Used in Green Synthesis of Plasmonic NMs (Demirbas et al., 2021)

Plants name(s)	NM(s)	Size (nm)	Shape(s)	Reference(s)
Achillea biebersteinii (Flower)	Ag	12±2	-	Some et al. (2019)
Andrographis paniculata leaf	ZnO	96-115	Spherical, hexagonal	Rajakumar et al. (2018)
Camellia sinensis (Leafs)	Cu^{2+}	-	Hybrid nanoflower	Baldemir et al. (2017)
Citrus limon (Fruits)	CuO	10	Spherical	Zikalala et al. (2018)
Croton caudatus Geisel	Gold	20	Spherical	Kumar et al. (2019)
Eucalyptus globulus bark	Cu	44-145	Nanowires	Pinto et al. (2019)
European black elderberry fruit	Ag	20-80	Spherical	David et al. (2019)
Heriteira fomes, *Sonneratiaa petala*	Zn, Ag	65-66	Spherical (Ag), hexagonal (Au)	Singh et al. (2017)
Leucasaspera extract	Ag	25-80	Clustere, irregular	Kumaran et al. (2017)
Litchi chinensis	Au, Ag	-	-	Murad et al. (2018)
Nephelium lappaceum L. (Peels)	ZnO	25–40	Spherical	Zikalala et al. (2018)
Perseaamericana bark	Pd-Ag	9-13-16	-	Meva et al. (2019)
Polygala tenuifolia (Roots)	ZnO	33-73.5	Spherical	Zikalala et al. (2018)
Prunus domestica gum	Au, Ag	7-30	Spherical	Islam et al. (2017)
Sansevieria raxburghiana leaf	Gold	-	-	Kumar et al. (2019)
Terminalia catappa, *Terminalia mellueri*, *Terminalia bentazoe*, *Terminalia bellerica* leaf	Ag	7-10-11	Spherical	El-Rafie et al. (2014)
Trianthema portulacastrum Linn. Leaf	ZnO	10-20	Spherical	Yadav et al. (2018)
Trifolium resupinatum (Seed)	Ag	17	Spherical	Some et al. (2019)
Viburnum opulus L. (Fruits)	Cu^{2+}	-	Hybrid nanoflower	Ildiz et al. (2017)
Zea mays	Ag	249.12	Spherical	Patra et al. (2016)
Zingiber officinale (Rhizome)	Ag	10–20	-	Some et al. (2019)

Ag NMs by reducing Ag^+ to Ag^0 compared to the chemical synthesis of Ag using chemical reducing agents (Figure 13.4). Quasi-spherical Ag NMs with an average particle size of ~54 nm were synthesized using the leaf extract of *Morusindica* L. V1 containing polyphenol constituents. According to the proposed mechanism for the

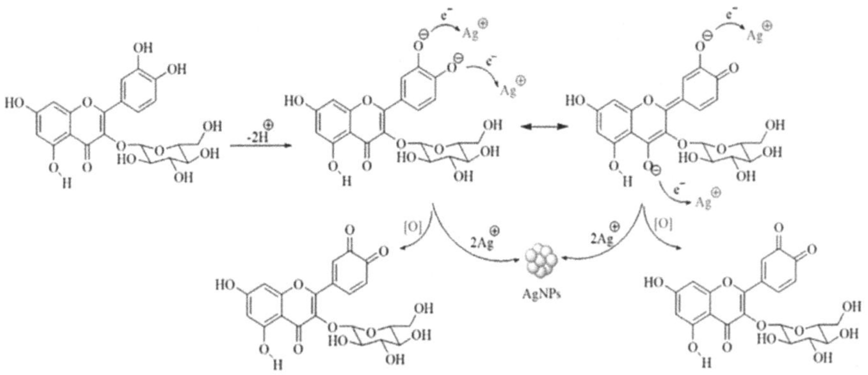

FIGURE 13.4 Proposed mechanism of plant-mediated synthesis of Ag NMs (Some et al., 2019).

synthesis of Ag NMs, flavonoids donate two electrons to Ag^+ ions to form a cluster together, which results in stable Ag NMs (Some et al., 2019).

Copper (Cu) has great properties such as electrical conductivity, catalytic activity, and surface-enhanced Raman scattering activity; therefore, Cu NMs have been an essential component for nanodevices. Cu NMs have applications in gas sensors, catalysis, and coatings on textiles (Chen et al., 2012). The synthesis of Cu NMs can be carried out with chemical and physical methods, but within the last 20 years biosynthesis of Cu NMs with plant extracts has been widely achieved (Kasana et al., 2017). Cu NMs with different sizes and morphologies can be obtained by using various plants. NMs of 40 nm and spherical in shape were synthesized using a flower broth of *Aloe vera* (Karimi and Mohsenzadeh, 2015) and Cu nanowires with average sizes ranging from 44 nm to 145 nm were obtained by using *Eucalyptus globulus* extract (Pinto et al., 2019).

In recent years, metal oxide NMs prepared with plant extracts have shown outstanding promise for a wide variety of applications. These metal oxide NMs have great bio-functionality and biocompatibility (Yuliarto et al., 2019). Zinc oxide (ZnO) NMs were synthesized using *Andrographis paniculata* leaf extract, and their antioxidant, anti-diabetic, and anti-inflammatory properties were investigated (Rajakumar et al., 2018). In another study, a anthocyanin-rich red cabbage extract was used for the synthesis of Fe_3O_4 NMs. The oxidation of the catecholamine groups in anthocyanin leads to a reduction of Fe^{2+} ions to Fe atoms, resulting in the synthesis of Fe_3O_4 NMs (Demirbas et al., 2019b).

13.3 PLANT DISEASE MANAGEMENT

Plant pests and pathogens lead to substantial reductions in crop production. Various techniques have been developed for the protection of crops from bacterial disease. The antibiotics streptomycin and kasugamycin were used to prevent the proliferation of bacteria on plants in the 1950s. However, antibiotics were ineffective due to a bacterial strain resistance. Plant diseases have been managed for many years using Cu bactericides incorporated with ethylene bisdithiocarbamate (EBDC) fungicides (e.g. mancozeb). However, the prevalent use of this combination caused Cu tolerance and a reduction in the effectiveness against diseases (Ocsoy et al., 2013c; Paret et al., 2013a).

Systematic acquired resistance (SAR) has been reported as an effective alternative tool for conventional Cu bactericides. Systemic acquired resistance is a stimulated defense mechanism that provides long-term protection against a wide spectrum of microorganisms (Durrant and Dong, 2004). Acibenzolar-S-methyl (ASM) as a potent SAR inducer activates defence-related genes and provides resistance to pathogens (Paret et al., 2013b). The use of ASM is effective against many plant diseases, including bacterial wilt (Pradhanang et al., 2007), powdery mildew (Görlach et al., 1996), and tomato spotted wilt (Csinos et al., 2007) The use of ASM in disease management has caused some problems since induced resistance is a host response and so it will be affected by many factors, such as the environment and host genotype; their use was highly limited due to these factors (Walters and Fountaine, 2009). Bacteria phages have been reported as an alternative to conventional Cu bactericides. However, when using bacteriophage, disease control efficacy may be influenced by limited phage viability and specific surrounding environment requirements for multiplication (Ocsoy et al., 2013c).

As a result, considerable efforts have been made to develop effective and biocompatible materials because it is difficult to manage bacterial diseases using conventional strategies.

Recently, nanotechnology enabled the development of NMs for plant disease management and creates new weapons against many factors that threaten crop health. In particular, the antimicrobial activity of metal and metal oxide NMs has attracted researchers' attention in plant disease management, especially those diseases caused by bacteria (Ocsoy et al., 2017d).

13.3.1 Antimicrobial Activity of Metal and Metal Oxide Nanomaterials

The protection of plants with metal-based NMs possessing unique antimicrobial activities represents one of the most recently implemented solutions that offers successful results for controlling bacterial and viral plant pathogens (Mohamed and Abd-Elsalam, 2018). Ag NMs have been one of the most effective biocides against a variety of pathogens, viruses, and fungi compared to other metallic NMs. According to the proposed mechanism, Ag NMs can attach to the surface of the cell membrane and lead to severe structural deformation by interacting with thiol, carboxyl, amino, hydroxyl, and phosphate groups on the bacterial membrane. Additionally, Ag NMs penetrate the cell through the cell membrane, which leads to the disruption function of the enzymes and finally results in cellular death (Ocsoy et al., 2013c).

The antimicrobial efficacy of Ag NMs has been confirmed as being related to shape, size, type of surface coating, and surface charge-dependent (Dal Lago et al., 2011; Kvitek et al., 2008; El Badawy et al., 2010). Additionally, aggregation of Ag NMs to Ag clusters have caused weakening antimicrobial activity because of decreasing particle–cell surface interaction. For instance, Ocsoy and coworkers showed that DNA directed Ag NMs grown on graphene oxide (GO) composites displayed high antibacterial activity toward *Xanthomonas perforans* using bacterial spot on tomato. The use of GO helps prevent the agglomeration of Ag NMs while maintaining their high surface reactivity. Additionally, the antibacterial activity of Ag@dsDNA–GO composites was enhanced by a synergistic effect between Ag NMs and GO (Ocsoy et al., 2013c).

Other NMs, including Cu, CuO, and ZnO, etc., are known to have efficient antimicrobial activity and can be widely used for the management of plant disease. For instance, Strayer-Scherer et al. reported the antibacterial effect of micro and nano Cu composites. Although micron-sized metallic Cu did not affect the Cu tolerant *Xanthomonas perforans*, all of the nano Cu composites dramatically reduced bacterial spot disease on tomato (Strayer-Scherer et al., 2018).

In another study, titanium dioxide (TiO_2) NMs containing Zn were evaluated for the management of *Xanthomonas* sp. infecting roses. TiO_2/Zn caused a suppression of the bacterial leaf spot on rose caused by *Xanthomonas* sp. when NMs were exposed to light for ten mins (Paret et al., 2013a).

13.4 CONCLUSION

The plant assisted synthesis of NMs is a developing area in nanotechnology that has important effects on the evolvement of nanoscience. The use of plant materials for the synthesis of NMs provides a more sustainable and environmentally friendly way compared to conventional chemical methods. NMs utilize antimicrobial activity using numerous mechanisms that can vary with individual properties such as size, shape, type of surface coating, and surface charge. Metal and metal oxide NMs with an antimicrobial activity can find many applications in the field of the medical industry in the future. The antimicrobial activity of NMs against a wide range of pathogens has attracted the attention of researchers for the synthesis and application of NMs for plant disease management. It is concluded that NMs especially synthesized from plants will play a major role in plant protection.

AUTHOR'S CONTRIBUTIONS

Z.K., and S.D., equally contributed to the manuscript with help from S.E. and I.O.

ADDITIONAL INFORMATION

Competing interests: The authors declare that there are no competing interests

REFERENCES

Ackermann, L., Dell'Acqua, M., Fenner, S., Vicente, R., Sandmann, R. 2011. Metal-free direct arylations of indoles and pyrroles with diaryliodonium salts. *Organic Letters* 13(9):2358–2360.

Alivisatos, A.P. 1996. Perspectives on the physical chemistry of semiconductor nanocrystals. *J. Phys. Chem.* 100(31):13226–13239.

Allaker, R.P. 2010. The use of nanoparticles to control oral biofilm formation. *Journal of Dental Research* 89:1175–1186.

Altinkaynak, C., Ildiz, N., Baldemir, A., Ozdemir, N., Yilmaz, V., Ocsoy, İ. 2019. Synthesis of organic-inorganic hybrid nanoflowers using *Trigonella foenum-graecum* seed extract and investigation of their antimicrobial activity. *Derim* 36(2):159–167.

Altinkaynak, C., Tavlasoglu, S., Kalin, R., Sadeghian, N., Ozdemir, H., Ocsoy, I., Özdemir, N. 2017. A hierarchical assembly of flower-like hybrid Turkish black radish peroxidase-Cu^{2+} nanobiocatalyst and its effective use in dye decolorization. *Chemosphere* 182:122–128.

Altinkaynak, C., Yilmaz, I., Koksal, Z., Özdemir, H., Ocsoy, I., Özdemir, N. 2016. Preparation of lactoperoxidase incorporated hybrid nanoflower and its excellent activity and stability. *International Journal of Biological Macromolecules*, 84:402–409.

Altinsoy, B.D., Karatoprak, G.Ş., Ocsoy, I. 2019. Extracellular directed Ag NPs formation and investigation of their antimicrobial and cytotoxic properties. *Saudi Pharmaceutical Journal* 27(1):9–16.

Amstad, E., Gillich, T., Bilecka, I., Textor, M., Reimhult, E. 2009. Transparent metal nanowire thin films prepared in mesostructured templates. *Nano Letters* 9:4042–4048.

Ankamwar, B., Damle, C., Ahmad, A., Sastry, M. 2005. Biosynthesis of gold and silver nanoparticles using Emblica officinalis fruit extract and their phase transfer and transmetallation in an organic solution. *Journal of Nanoscience and Nanotechnology* 5(10):1665–1671.

Aromal, S.A., and Philip, D. 2012. Green synthesis of gold nanoparticles using *Trigonella foenum-graecum* and its size-dependent catalytic activity. *Spectrochimica Acta Part A: Molecular and Biomolecular Spectroscopy* 97:1–5.

Asmathunisha, N., Kathiresan, K. 2013. A review on biosynthesis of nanoparticles by marine organisms. *Colloids and Surfaces B: Biointerfaces* 103:283–287.

Baehni, P.C., and Takeuchi, Y. 2003. Anti-plaque agents in the prevention of biofilm-associated oral diseases. *Oral Diseases* 9(1) Supplement 1:23–29.

Baharara, J., Ramezani, T., Mousavi, M., Asadi-Samani, M. 2017. Antioxidant and anti-inflammatory activity of green synthesized silver nanoparticles using *Salvia officinalis* extract. *Annals of TropicalMedicine and Public Health* 10:1265–1270.

Baldemir, A., Karaman, U., Yusufbeyoglu, S., Eken, A., Ildız, N., Ilgun, S., Cankaya, S. 2018. A new strategy for enhancing acanthamoebicidal activity with synthesis of nanoflower of *Lauroceraususus officinalis* Roemer (cherry laurel) fruit extracts. *Mikrobiyoloji Bulteni* 52(1):56–71.

Baldemir, A. Kose, N. B. Ildiz, N. Ilgun, S. Yusufbeyoglu, S. Yilmaz, V. Ocsoy, I. 2017. Synthesis and characterization of green tea (*Camellia sinensis* (L.) Kuntze) extract and its major components-based nanoflowers: A new strategy to enhance antimicrobial activity. *RSC Advances* 7(70):44303–44308.

Bapat, A.J., Chaubal, T.V., Joshi, C.P., Bapat, P.R., Choudhury, H., Pandey, M., Gorain, B., Kesharwanie, P. 2018. An overview of application of Ag nanoparticles for biomaterials in dentistry. *Materials Science and Engineering* 91:881–889.

Bapat, R.A., Joshi, C.P., Bapat, P., Chaudal, T.V., Pandurangappa, R., Jnanendrappa, N., Gorain, B., Khurana, S., Kesharwani, P. 2019. The use of nanoparticles as biomaterials in dentistry. *Drug Discovery Today* 24(1):85–98.

Baskaran, X., Vigila, A.V.G., Parimelazhagan, T., Muralidhara-Rao, D., Zhang, S. 2016. Biosynthesis, characterization, and evaluation of bioactivities of leaf extract-mediated biocompatible silver nanoparticles from an early tracheophyte, *Pteris tripartita* Sw. *International Journal of Nanomedicine* 11:5789–5805.

Berti, L., Burley, G.A. 2008. Nucleic acid and nucleotide-mediated synthesis of inorganic nanoparticles. *Nature Nanotechnology* 3:81–87.

Bruchez, M., Moronne, J.R., Gin, M., Weiss, P., Alivisatos, S. 1998. Semiconductor nanocrystals as fluorescent biological labels. *Science* 281:2013–2016.

Cedervall, T., Lynch, I., Lindman, S., Berggard, T., Thulin, E., Nilsson, H., Dawson, K. A., Linse, S. 2007. Understanding the nanoparticle–protein corona using methods to quantify exchange rates and affinities of proteins for nanoparticles. *Proceedings of the National Academy of Sciences* 104:2050–2055.

Celik, C., Ildiz, N., Ocsoy, I. 2020. Building block and rapid synthesis of catecholamines-inorganic nanoflowers with their peroxidase-mimicking and antimicrobial activities. *Scientific Reports* 10(1):1–11.

Celik, C., Tasdemir, D., Demirbas, A., Katı, A., Gul, O.T., Cimen, B., Ocsoy, I. 2018. Formation of functional nanobiocatalysts with a novel and encouraging immobilization approach and their versatile bioanalytical applications. *RSC Advances*, 8(45):25298–25303.

Celik, G.Y., Onbaslı, D., Ozbahar, Ö., Ocsoy, I. 2019. Pisolithus arrhizus Ekstraktı Kullanılarak Sentezlenen Titanyum Nanopartikülünün Antimikrobiyal Etkisi. *Erciyes Üniv Vet Fak Derg* 16:157–116.

Chen, T., Ocsoy, I., Yuan, Q., Wang, R., You, M., Zhao, Z., Song, E, Zhang, X., Tan, W. 2012. One-step facile surface engineering of hydrophobic nanocrystals with designer molecular recognition. *Journal of the American Chemical Society* 134(32):13164–13167.

Chen, Y., Wang, D., Zhu, X., et al. 2012. Long-term effects of copper nanoparticles on wastewater biological nutrient removal and N_2O generation in the activated sludge process. *Environmental Science & Technology* 46(22):12452–12458.

Cheng, L. Weir, M.D. Xu, H.H. Antonucci, J.M. Kraigsley, A.M. Lin, N.J. LinGibson, S., Zhou, X. 2012. Antibacterial amorphous calcium phosphate nanocomposites with a quaternary ammonium dimethacrylate and Ag nanoparticles. *Dent. Mater* 28(5):561–572.

Cheng, L., Zhang, K., Weir, M.D., Melo, M.A., Zhou, X., Xu, H.H. 2015. Nanotechnology strategies for antibacterial and remineralizing composites and adhesives to tackle dental caries. *Nanomedicine* 10:627–641.

Cheon, J., Kang, N.-J., Lee, S.-M., Lee, J.-H., Yoon, J.-H., Oh, S.J. 2004. Shape evolution of single-crystalline iron oxide nanocrystals. *Journal of the American Chemical Society* 126:1950–1951.

Csinos, A.S., Pappu, H.R., McPherson, R.M., et al. 2007. Management of tomato spotted wilt virus in flue-cured tobacco with acibenzolar-s-methyl and imidacloprid. *Plant Disease* 85:292–296.

Dal Lago, V., Franca de Oliveira, L., Almeida Goncalves, K., et al. 2011. Size-selective silver nanoparticles: future of biomedical devices with enhanced bactericidal properties. *Journal of Materials Chemistry* 21:12267–12273.

David, L., Moldovan, B., Vulcu, A., Olenic, L., Perde-Schrepler, M., Fischer-Fodor, E., Florea, A., Crisan, M., Chiorean, I., Clichici, S., Filip, G.A. 2014. Green synthesis, characterization and anti-inflammatory activity of silver nanoparticles using European black elderberry fruits extract. *Colloids and Surfaces B:Biointerfaces* 122:767–777.

De Palma, R., Peeters, S., van Bael, M.J., van den Rul, H., Bonroy, K., Laureyn, W., Mullens, J., Borghs, G., Maes, G. 2007. Silane ligand exchange to make hydrophobic superparamagnetic nanoparticles water-dispersible. *Chemistry of Materials* 19:1821–1831.

Demirbas, A., Welt, B.A., Ocsoy, I. 2016. Biosynthesis of red cabbage extract directed Ag NPs and their effect on the loss of antioxidant activity. *Materials Letters* 179:20–23.

Demirbas, A., Yilmaz, V., Ildiz, N., Baldemir, A., Ocsoy, I. 2017. Anthocyanins-rich berry extracts directed formation of Ag NPs with the investigation of their antioxidant and antimicrobial activities. *Journal of Molecular Liquids* 248, 1044–1049.

Demirbas, A., Buyukbezirci, K., Celik, C., Kislakci, E., Karaagac, Z., Gokturk, E. 2019a. Synthesis of long-term stable gold nanoparticles benefiting from red raspberry (*Rubus idaeus*), strawberry (*Fragaria ananassa*), and blackberry (*Rubus fruticosus*) extracts-gold ion complexation and investigation of reaction conditions. *ACS Omega* 4:18637–18644.

Demirbas, A., Kislakci, E., Karaagac, Z., Onal, I., Ildiz, N., Ocsoy, I. 2019b. Preparation of biocompatible and stable iron oxide nanoparticles using anthocyanin integrated hydrothermal method and their antimicrobial and antioxidant properties. *Materials Research Express* 6(12):125011.

Demirbas, A., Karaytug, T., Arabaci, N., Yilmaz, E.S., Ocsoy, I. 2021. Synthesis of metallic and metal oxide nanomaterials. In *Green Synthesis of Nanomaterials for Bioenergy Applications*. John Wiley & Sons.

Dogru, E., Demirbas, A., Altinsoy, B., Duman, F., Ocsoy, I. 2017. Formation of Matricaria chamomilla extract-incorporated Ag nanoparticles and size-dependent enhanced antimicrobial property. *Journal of Photochemistry and Photobiology B: Biology* 174:78–83.

Dreaden, E.C., Alkilany, A.M., Huang, X., Murphy, C.J., El-Sayed, M.A. 2012. The golden age: Gold nanoparticles for biomedicine. *Chemical Society Reviews* 41(7):2740–2779.

Duan, H., Wang, D., Yadong, L. 2015. Green chemistry for nanoparticle synthesis. *Chemical Society Reviews* **44**:5778–5792.

Dubey, S.P., Lahtinen, M., Sillanpää, M. 2010. Green synthesis and characterizations of silver and gold nanoparticles using leaf extract of *Rosa rugosa*. *Colloids and Surfaces A: Physicochemical and Engineering Aspects* 364:34–41

Durrant, W.E., and Dong, X. 2004. Systemic acquired resistance. *Annual Review of Phytopathology* 42:185–209.

El Badawy, A. M., Silva, R. G., Morris, B., et al. 2010. Surface charge-dependent toxicity of silver nanoparticles. *Environmental Science & Technology* 45:283–287.

Ge, J., Lei, J., Zare, R.N. 2012. Protein–inorganic hybrid nanoflowers. *Nature Nanotechnology*, 7(7):428–432.

Giridharan, T., Masi, C., Sindhu, S., Arumugam, P. 2014. Studies on green synthesis, characterization and anti-proliferative potential of silver nano particle using *Dodonaeaviscosa* and *Capparis decidua*. *Bioscience Biotechnology Research Asia* 11(2):665–673.

Görlach, J., Volrath, S., Knauf-Beiter, G. 1996. Benzothiadiazole, a novel class of inducers of systemic acquired resistance, activates gene expression and disease resistance in wheat. *Plant Cell* 8(4):629–643.

Govindappa, M., Hemashekhar, B., Arthikala, M.K., Ravishankar, Rai, V., Ramachanda, Y.L. 2018. Characterization, antibacterial, antioxidant, antidiabetic, anti-inflammatory and antityrosinase activity of green synthesized silver nanoparticles using *Calophyllumtomentosum* leaves extract. *Results in Physics* 9:400–408.

Gugliotti, L.A., Feldheim, D.L., Eaton, B.E. 2004. RNA-mediated metal-metal bond formation in the synthesis of hexagonal palladium nanoparticles. *Science* 304(5672):850–852.

Hannig, M. Hannig, C. 2012. Nanotechnology and its role in caries therapy. *Advances in Dental Research* 24(2):53–57.

Hochella, M.F., Jr, Spencer, M.G., Jones, K.L. 2015. Nanotechnology: nature's gift or scientists' brainchild. *Environmental Science: Nano* 2:114–119.

Hyeon, T. 2003. Chemical synthesis of magnetic nanoparticles. *Chemical Communications* 8:927–934.

Ildiz, N., Baldemir, A., Altinkaynak, C., Özdemir, N., Yilmaz, V., Ocsoy, I. 2017. Self assembled snowball-like hybrid nanostructures comprising *Viburnum opulus* L. extract and metal ions for antimicrobial and catalytic applications. *Enzyme and Microbial Technology*, 102:60–66.

Islam, N.U., Amin, R., Shahid, M., Amin, M., Zaib, S., Iqbal, J. 2017. A multi-target therapeutic potential of *Prunus domestica* gum stabilized nanoparticles exhibited prospective anticancer, antibacterial, urease-inhibition, anti-inflammatory and analgesic properties. *BMC Complement. Alternative Medicine* 17:1–17.

Joghee, S., Ganeshan, P., Vincent, A., Hong, S.I. 2019. Ecofriendly biosynthesis of zinc oxide and magnesium oxide particles from medicinal plant *Pisonia grandis* R. Br. leaf extract and their antimicrobial activity. *BioNanoScience* 9(1):141–154.

Kandaswamy, E., Nagendrababu, V., Deivanayagam, K. 2016. Antimicrobial effect of nanoparticles in endodontics. *Nanobiomaterials in Dentistry* 7:161–186.

Karaagac, Z., Gul, O.T., Ildız, N., Ocsoy, I. 2020a. Transfer of hydrophobic colloidal gold nanoparticles to aqueous phase using catecholamines. *Journal of Molecular Liquids* 315:113796.

Karaagac, Z., Onal, I., Ildiz, N., Ocsoy, I. 2020b. Simultaneous use of phenylboronic acid as a phase transfer agent and targeting ligand for gold nanoparticles. *Materials Letters* 280:128561.

Karatoprak, G.S., Aydin, G., Altinsoy, B., Altinkaynak, C., Koşar, M., Ocsoy, I. 2017. The effect of *Pelargonium endlicherianum* Fenzl. root extracts on formation of nanoparticles and their antimicrobial activities. *Enzyme and Microbial Technology* 97:21–26.

Karimi, J., Mohsenzadeh, S. 2015. Rapid, green, and eco-friendly biosynthesis of copper nanoparticles using flower extract of *Aloe Vera. Synthesis and Reactivity in Inorganic, Metal-Organic, and Nano-Metal Chemistry* 45:895–898.

Kasana, R.C., Panwar, N.R., Kaul, R.K. et al. 2017. Biosynthesis and effects of copper nanoparticles on plants. *Environmental Chemistry Letters* 15:233–240.

Kim, J., Kim, H.S., Lee, N., Kim, T., Kim, H., Yu, T., Song, I.C., Moon, W.K., Hyeon, T. 2008. Multifunctional uniform nanoparticles composed of a magnetite nanocrystal core and a mesoporous silica shell for magnetic resonance and fluorescence imaging and for drug delivery. *Angewandte Chemie International Edition* 47:8438–8441.

Koca, F.D., Demirezen Yilmaz, D., Ertas Onmaz, N., Yilmaz, E., Ocsoy, I. 2020. Green synthesis of allicin based hybrid nanoflowers with evaluation of their catalytic and antimicrobial activities. *Biotechnology Letters*. 1–8.

Koca, F.D., Demirezen Yilmaz, D., Duman, F., Ocsoy, I. 2018. Comparison of phytotoxic effects of biosynthesized copper oxide nanoparticle and ionic copper on *Elodea canadensis. Chemistry and Ecology* 34(9):839–853.

Kumar, N., and Kumbhat, S. 2016. Carbon-based nanomaterials essentials in nanoscience and nanotechnology. *Essentials in Nanoscience and Nanotechnology*, 189–236.

Kumar, P.V., Kala, M.J., Parakash, K.S. 2019. Green synthesis of gold nanoparticles using *Croton Caudatus Geisel* leaf extract and their biological studies. *Materials Letters* 236:19–22.

Kumaran, N.S., Vijayaraj, R. 2017. Biosynthesis of silver nano particles from *Leucas aspera* (Willd.) link and its anti-inflammatory potential against carrageen induced paw edema in rats. *International. Journal of Pharmaceutical Sciences and Research* 8:2588–2593.

Kvitek, L. Pana cek, A., Soukupova, J. 2008. Effect of surfactants and polymers on stability and antibacterial activity of silver nanoparticles. *The Journal of Physical Chemistry C* 112:5825–5834.

LaMer, V.K., Dinegar, R.H. 1950. Theory, production and mechanism of formation of monodispersed hydrosols. *Journal of the American Chemical Society* 72(11):4847–4854.

Lee, S. Y. Krishnamurthy, S. Cho, C. W. 2016. Biosynthesis of gold nanoparticles using *Ocimum sanctum* extracts by solvents with different polarity. *ACS Sustainable Chemistry and Engineering* 4, 5:2651–2659.

Leng, Y., Fu, L., Ye, L., Li, B., Xu, X., Xing, X., Ji, X. 2016. Protein-directed synthesis of highly monodispersed, spherical gold nanoparticles and their applications in multidimensional sensing. *Scientific Reports* 6:28900.

Li, C., Chen, T., Ocsoy, I., Zhu, G., Yasun, E., You, M., Wu, C., Zheng, J., Song, E., Huang, C.Z., Tan, W. 2013. Gold-coated Fe_3O_4 nanoroses with five unique functions for cancer cell targeting, imaging, and therapy. *Advanced Functional Materials* 24(12):1772–1780.

Lin, Z., Xiao, Y., Yin, Y., Hu, W., Liu, W., Yang, H. 2014. Facile synthesis of enzyme-inorganic hybrid nanoflowers and its application as a colorimetric platform for visual detection of hydrogen peroxide and phenol. *ACS Applied Materials & Interfaces*, 6(13):10775–10782.

Lu, A.-H. Salabas, E. L. Schüth, F. 2007. Magnetic nanoparticles: synthesis, protection, functionalization, and application. *Angewandte Chemie International Edition* 119:1242–1266.

Lu, Z. Rong, K. Li, J. Yang, H. Chen, R. 2013. Size dependent antibacterial activities of Ag nanoparticles against oral anaerobic pathogenic bacteria. *Journal of Materials Science: Materials in Medicine* 24(6):1465–1471.

Ma, N., Dooley, C.J., Kelley, S.O. 2006. RNA-templated semiconductor nanocrystals. *Journal of the American Chemical Society* 128(39):12598–12599.

Ma, N., Sargent, E.H., Kelley, S.O. 2009. One-step DNA-programmed growth of luminescent and biofunctionalized nanocrystals. *Nature Nanotechnology* 4(2):121.

McLamore, E.S., Convertino, M., Ocsoy, I., Vanegas, D.C., Taguchi, M., Rong, Y., Gomes, C., Chaturvedi, P., Claussen, J.C. 2016. Biomimetic fractal nanometals as a transducer layer in electrochemical biosensing. In *Semiconductor-based Sensors*, edited by F. Ren, 35–67. World Scientific Publishing.

Meva, F.E., Antoinette, A., Philippe, N., Ebanda, B., Edmond, K., Alexa, T., Laura, S., Maximilian, S., Moll, B.K., Kökcam-Demir, U., Albert, E., Leopold, M.M., Lehman, G., Janiak, C. 2019. Silver and palladium nanoparticles produced using a plant extract as reducing agent, stabilized with an ionic liquid: sizing by X-ray powder diffraction and dynamic light scattering. *Journal of Materials Research and Technology* 8(2):1991–2000.

Mohamed, M.A., and Abd-Elsalam, K. A. 2018. Nanoantimicrobials for plant pathogens control: Potential applications and mechanistic aspects. In *Nanobiotechnology Applications in Plant Protection*, edited by K. Abd-Elsalam and R. Prasad, 87–109. Springer, Cham.

Mohamed El-Rafie, H., and Hamed, M.A.A. 2014. Antioxidant and anti-inflammatory activities of silver nanoparticles biosynthesized from aqueous leaves extracts of four *Terminalia* species. *Advances in Natural Sciences: Nanoscience and Nanotechnology* 5:1–11.

Moldovan, B., David, L., Vulcu, A., Olenic, L., Perde-schrepler, M., Fischer-fodor, E., Baldea, I., Clichici, S., Adriana, G. 2017. In vitro and in vivo anti-inflammatory properties of green synthesized silver nanoparticles using *Viburnum opulus* L. fruits extract. *Materials Science and Engineering C* 79:720–727.

Murad, U., Ullah, B., Khan, S., Ibrar, M., Ullah, S., Khattak, U. 2018. Synthesis of silver and gold nanoparticles from leaf of *Litchi chinensis* and its biological activities. *Asian Pacific Journal of Tropical Biomedicine* 8(3):142–149.

Nazari, P., Gharibzadeh, S., Ansari, F., Nejand, B.A., Eskandari, M., Kohnehpoushi, S., Ahmadi, V., Salavati-Niasari, M. 2017. Facile green deposition of nanostructured porous NiO thin film by spray coating. *Materials Letters* 190:40–44.

Njagi, E.C., Huang, H., Stafford, L., Genuino, H., Galindo, H.M., Collins, J.B., Hoag, G.E., Suib, S.L. 2011. Biosynthesis of iron and silver nanoparticles at room temperature using aqueous sorghum bran extracts. *Langmuir* 27(1):264–271.

Noronha, V.T., Paula, A.J., Duran, G., Galembeck, A., Cogo, M.K., Franz, M.M., Duran, N. 2017. Ag nanoparticles in dentistry. *Dental Materials* 33:1110–1126.

Ocsoy, I., Gulbakan, B., Shukoor, M.I., Xiong, X., Chen, T., Powell, D.H., Tan, W. 2013a. Aptamer conjugated multifunctional nanoflowers as a platform for targeting, capture and detection in laser desorption ionization mass spectrometry. *ACS Nano* 7(1):417–427.

Ocsoy, I., Paret, M.L., Ocsoy, M., Kunwar, S., Chen, T., You, M., Tan, W. 2013b. Nanotechnology in plant disease management: DNA-directed silver nanoparticles on graphene oxide as an antibacterial against *Xanthomonas perforans*. *ACS Nano* 7(10):8972–8980.

Ocsoy, I. Gulbakan, B. Chen, T. Zhu, G. Chen, Z. Sari, M.M. Peng, L. Xiong, X, Tan W. 2013c. DNA-guided-metal nanoparticle formation on graphene oxide surface. *Advanced Materials* 25(16):2319–2325.

Ocsoy, I., Ocsoy, M.A., Yasun, E., Tan, W. 2013d. Nucleic acid-functionalized nanomaterials. *Nano Life* 3(1):1340004.

Ocsoy, I., Dogru, E., Usta, S. 2015. A new generation of flowerlike horseradish peroxides as a nanobiocatalyst for superior enzymatic activity. *Enzyme and Microbial Technology*, 75:25–29.

Ocsoy, I. Isiklan, N. Cansiz, S. Ozdemir, N. Tan, W. 2016. ICG Conjugated magnetic graphene oxide for dual photothermal and photodynamic therapy. *RSC Advances* 6:30285–30292.

Ocsoy, I., Yusufbeyoglu, S., Yılmaz, V., McLamore, E.S., Ildız, N., Ulgen, A. 2017a. DNA aptamer functionalized gold nanostructures for molecular recognition and photothermal inactivation of methicillin-Resistant *Staphylococcus aureus*. *Colloids and Surfaces B: Biointerfaces* 159:16–22.

Ocsoy, I., Temiz, M., Celik, C., Altinsoy, B., Yilmaz, V., Duman, F. 2017b. A green approach for formation of silver nanoparticles on magnetic graphene oxide and highly effective antimicrobial activity and reusability. *Journal of Molecular Liquids* 227:147–152.

Ocsoy, I., Demirbas, A., McLamore, E.S., Altinsoy, B., Ildiz, N., Baldemir, A. 2017c. Green synthesis with incorporated hydrothermal approaches for silver nanoparticles formation and enhanced antimicrobial activity against bacterial and fungal pathogens. *Journal of Molecular Liquids* 238:263–269.

Ocsoy, I., Tasdemir, D., Mazicioglu, S., Tan, W. 2017d. Nanotechnology in Plants. *Advances in Biochemical Engineering* 164:263–275.

Ocsoy, I., Tasdemir, D., Mazicioglu, S., Tan, W. 2018a. Nanotechnology in plants. In *Plant Genetics and Molecular Biology. Advances in Biochemical Engineering/Biotechnology*, edited by R. Varshney, M. Pandey, A. Chitikineni. Springer, Champ 164: 273–275.

Ocsoy, I., Tasdemir, D., Mazicioglu, S., Celik, C., Katı, A., Ulgen, F. 2018b. Biomolecules incorporated metallic nanoparticles synthesis and their biomedical applications. *Materials Letters*. 212:45–50.

Onbaslı, D. 2018. Kitosan-Gümüş Nanopartikülü ve Kitosan-Gümüş-Grafen Oksit Nanokompozitinin Sentezi, Karakterizasyonu ve Antimikrobiyal Aktivitelerinin Belirlenmesi. *Erciyes Univ Vet Fak Derg* 15:208–215.

Pal, S., Deng, Z., Wang, H., Zou, S., Liu, Y., Yan, H. 2011. DNA Directed self-assembly of anisotropic plasmonic nanostructures. *Journal of the American Chemical Society* 133(44):17606–17609.

Paret, M.L., Palmateer, A.J., Knox, G.W. 2013a. Evaluation of a light-activated nanoparticle formulation of titanium dioxide with zinc for management of bacterial leaf spot on Rosa 'Noare'. *Hortscience* 48(2):189–192.

Paret, M.L., Vallad, G.E., Averett, D.R. et al. 2013b. Photocatalysis: Effect of Light-activated nanoscale formulations of TiO_2 on *Xanthomonas perforans* and control of bacterial spot of tomato. *Phytopathology* 103:228–236.

Park, J., An, K., Hwang, Y., Park, J.G., Noh, H.J., Kim, J.Y., Park, J.H., Hwang, N.M., Hyeon, T. 2004. Ultra-large-scale syntheses of monodisperse nanocrystals. *Nature Materials* 3:891–895.

Park, J., Joo, J., Kwon, S., Jang, Y., Hyeon, T. 2007. Synthesis of monodisperse spherical nanocrystals. *Angewandte Chemie International Edition* 46:4630–4660.

Patra, J.K., Baek, K.H. 2016. Biosynthesis of silver nanoparticles using aqueous extract of silky hairs of corn and investigation of its antibacterial and anticandidal synergistic activity and antioxidant potential. *IET Nanobiotechnology* 10(5):326–333.

Pazi, I., Kucuksezgin, F., Kacar, A., Gonul, L.T. 2016. Assessment of the effect of environmental pollution in a marina on caged mussels using chemical and genotoxic analysis. *Chemistry and Ecology* 32(8):756–773.

Peng, L., You, M., Wu, C., Han, D., Ocsoy, I., Chen, T., Chen, Z., Tan, W. 2014. Reversible phase transfer of nanoparticles based on photoswitchable host-guest chemistry. *ACS Nano* 8(3):2555–2561.

Pinto, R.J.B., Lucas, J.M.F. et al. 2019. Bio-based synthesis of oxidation resistant copper nanowires using an aqueous plant extract. *Journal of Cleaner Production* 221:122–131.

Pradhanang, P.M., Ji, P., Momol, M.T. et al. 2007. Application of acibenzolar-s-methyl enhances host resistance in tomato against *Ralstonia solanacearum. Plant Disease* 89:989–993.

Prasad, R., Bhattacharyya, A., Nguyen, Q.D. 2017. Nanotechnology in sustainable agriculture: Recent developments, challenges, and perspectives. *Frontiers in Microbiology* 8:1014.

Qin, J., Laurent, S., Jo, Y., Roch, A., Mikhaylova, M., Bhujwalla, Z., Muller, R., Muhammed, M. 2007. A high-performance magnetic resonance imaging T_2 contrast agent. *Advanced Materials* 19:1874–1878.

Qiu, L., Chen, T., Ocsoy, I., Yasun, E., Wu, C., Zhu, G., You, M., Han, D., Jiang, J., Yu, R., Tan, W. 2015. A cell-targeted, size-photocontrollable, nuclear-uptake nanodrug delivery system for drug-resistant cancer therapy. *Nano Letters* 15(1):457–463.

Rajakumar, G., Thiruvengadam, M., Mydhili, G., et al. 2018. Green approach for synthesis of zinc oxide nanoparticles from *Andrographis paniculata* leaf extract and evaluation of their antioxidant, anti-diabetic, and anti-inflammatory activities. *Bioprocess and Biosystems Engineering* 41(1):21–30.

Rayaman, P., Ocsoy, I., Gurer, U.S. 2018. Green synthesis and characterization of silver nanoparticles using the fungus *A. niger* and bioactive potential against microorganisms and cancer cells. *Latin American Journal of Pharmacy* 37(5):979–986.

Rothemund, P.W.K. 2006. Folding DNA to create nanoscale shapes and patterns. *Nature* 440:297–302.

Samiei, M., Farjami, A., Maleki, D.S., Lotfipour, F. 2016. Nanoparticles for antimicrobial purposes in Endodontics: A systematic review of in vitro studies. *Materials Science and Engineering* 58:1269–78.

Schladt, T.D., Schneider, K., Schild, H., Tremel, W. 2011. Synthesis and bio-functionalization of magnetic nanoparticles for medical diagnosis and treatment. *Dalton Transactions* 40(24):6315–6343.

Seethalakshmi, A.M.K. 2015. Evaluation of in-vitro anti-inflammatory activity of silver nanoparticles synthesised using *Piper nigrum* extract. *Journal of Nanomedicine and Nanotechnology* 6(2):1–5.

Shen, L., Laibinis, P.E., Hatton, T.A. 1999. Bilayer surfactant stabilized magnetic fluids: Synthesis and interactions at interfaces *Langmuir* 15:447–453.

Shukoor, M.I., Natalio, F., Therese, H.A., Tahir, M.N., Ksenofontov, V., Panthöfer, M., Eberhardt, M., Theato, P., Schröder, H.C., Müller, W.E.G., Tremel, W. 2008. Fabrication of a silica coating on magnetic γ-Fe_2O_3 nanoparticles by an immobilized enzyme. *Chemistry of Materials* 20:3567–3573.

Shukoor, M.I., Altman, M.O., Han, D., Bayrac, A.T., Ocsoy, I., Zhu, Z., Tan, W. 2012. Aptamer-nanoparticle assembly for logic-based detection. *ACS Applied Materials & Interfaces* 4(6):3007–3011.

Singh, P., Kim, Y., Zhang, D., Yang, D. 2016. Biological synthesis of nanoparticles from plants and microorganisms. *Trends Biotechnology* 34(7):588–599.

Some, S., Bulut, O., Biswas, K., Kumar, A., Roy, A., Sen, I.K., Das, S. 2019. Effect of feed supplementation with biosynthesized silver nanoparticles using leaf extract of *Morus indica* L. V1 on *Bombyx mori* L. (Lepidoptera: Bombycidae). *Scientific Reports*, 9(1):1–13.

Some, S., Sen, I.K., Mandal, A., Aslan, T., Ustun, Y., Yilmaz, E.S., Katı, A., Demirbas, A., Mandal, A.K., Ocsoy, I. 2018. Biosynthesis of Ag nanoparticles and their versatile antimicrobial properties. *Journal Materials Research Express* 6(1):012001.

Somturk, B., Yilmaz, I., Altinkaynak, C., Karatepe, A., Özdemir, N., Ocsoy, I. 2016. Synthesis of urease hybrid nanoflowers and their enhanced catalytic properties. *Enzyme and Microbial Technology* 86:134–142.

Strayer, A., Ocsoy, I., Tan, W., Jones, J.B., Paret, M.L. 2016. Low concentrations of a silver-based nanocomposite to manage bacterial spot of tomato in the greenhouse. *Plant Disease* 100:1460–1465.

Strayer-Scherer, A., Liao, Y.Y., Young, M., et al. 2018. Advanced copper composites against copper-tolerant *Xanthomonas perforans* and tomato bacterial spot. *Phytopathology* 108(2):196–205.

Stubbings, J., Brown, J., Price, G.J. 2017. Sonochemical production of nanoparticle metal oxides for potential use in dentistry. *Ultrasonics Sonochemistry* 35:646–654.

Sun, S., Zeng, H., Robinson, D.B., Raoux, S., Rice, P.M., Wang, S.X., Li, G. 2004. Monodisperse MFe2O4 (M = Fe, Co, Mn) nanoparticles. *Journal of the American Chemical Society* 126:273–279.

Tavassoli, S., Alaghemand, H., Hamze, F., Ahmadian, B.F., RajabNia, R., Rezvani, M.B., Kaviani, M., Atai, M. 2013. Antibacterial, physical and mechanical properties of flowable resin composites containing zinc oxide nanoparticles. *Dental Materials* 29(5):495–505.

Turek, D., Van Simaeys, D., Johnson, J., Ocsoy, I., Tan, W. 2013. Molecular recognition of live methicillin-resistant *Staphylococcus aureus* cells using DNA aptamers. World Journal of Translational Medicine 2(3):67–74.

Unal, I.S., Demirbas, A., Onal, I., Ildiz, N., Ocsoy, I. 2020. One step preparation of stable gold nanoparticle using red cabbage extracts under UV light and its catalytic activity. *Journal of Photochemistry and Photobiology B: Biology* 204:111800.

Veiseh, O. Gunn, J.W. Zhang, M. 2010. Design and fabrication of magnetic nanoparticles for targeted drug delivery and imaging. *Advanced Drug Delivery Reviews* 62(3):284–304.

Wagner, S., Gondikas, A., Neubauer, E., Hofmann, T.V., Kammer, F., 2014. Spot the difference: engineered and natural nanoparticles in the environment--release, behavior, and fate. *Angewandte Chemie International Edition* 53(46):12398–12419.

Veiseh, O. Gunn, J.W., Zhang, M. 2010. Design and fabrication of magnetic nanoparticles for targeted drug delivery and imaging. *Advanced Drug Delivery Reviews* 62(3):284–304.

Valsalam, S., Agastian, P., Valan, M., Naif, A., Mohammed, A.D.A.K., Kaviyarasuc, G.K., Ravindran, B., Chang, S.W., Arokiyaraj, S. 2019. Rapid biosynthesis and characterization of silver nanoparticles from the leaf extract of *Tropaeolum majus* L. and its enhanced in-vitro antibacterial, antifungal, antioxidant and anticancer properties. *Journal of Photochemistry and Photobiology B: Biology*. 191:65–74.

Walters, D.R., and Fountaine, J.M. 2009. Practical application of induced resistance to plant diseases: an appraisal of effectiveness under field conditions. *The Journal of Agricultural Science* 147:523–535.

Wang, F., Han, Y., Lim, C., Lu, Y., Wang, J., Xu, J., Chen, H., Zhang, C., Hong, M., Liu, X. 2010. Ultra-large-scale syntheses of monodisperse nanocrystals. *Nature* 463: 1061–1065.

Wang, L.B., Wang, Y.C., He, R., Zhuang, A., Wang, X., Zeng, J., Hou, J.G. 2013a. A new nanobiocatalytic system based on allosteric effect with dramatically enhanced enzymatic performance. *Journal of the American Chemical Society*, 135(4):1272–1275.

Wang, R., Zhang, Y., Lu, D., Ge, J., Liu, Z., Zare, R.N. 2013b. Functional protein–organic/inorganic hybrid nanomaterials. *Wiley Interdisciplinary Reviews: Nanomedicine and Nanobiotechnology*, 5(4):320–328.

Wei, H., Wang, Z., Zhang, J., House, S., Gao, Y.G., Yang, L., Robertson, I.M. 2011. Time-dependent, protein-directed growth of gold nanoparticles within a single crystal of lysozyme. *Nature Nanotechnology* 6(2):93.

Weissleder, R., Moore, A., Mahmood, U., Bhorade, R., Benveniste, H., Chiocca, E., Basilion, J.P. 2000. In vivo magnetic resonance imaging of transgene expression. *Nature Medicine* 6:351–354.

Welch, K., Cai, Y., Engqvist, H., Stromme, M. 2010. Dental adhesives with bioactive and on-demand bactericidal properties. *Dental Materials* 26(5):491–499.

Wirges, C.T., Timper, J., Fischler, M., Sologubenko, A.S., Mayer, J., Simon, U., Carell, T. 2009. Controlled nucleation of DNA metallization. *Angewandte Chemie International Edition* 48(1):219–223.

Wu, D., Fan, W., Kishen, A., Gutmann, J.L., Fan, B. 2014. Evaluation of the antibacterial efficacy of Ag nanoparticles against. Enterococcus faecalis biofilm. *Journal of Endodontics* 40:285–290.

Yadav, E., Singh, D., Yadav, P., Verma, A. 2018. Ameliorative effect of biofabricated ZnO nanoparticles of *Trianthema portulacastrum* Linn. on dermal wounds via removal of oxidative stress and inflammation. *RSC Advances* 8(38):21621–21635.

Yasun, E., Gulbakan, B., Ocsoy, I., Yuan, Q., Shukoor, M.I., Li, C., Tan, W. 2012. Enrichment and detection of rare proteins with aptamer-conjugated gold nanorods. *Analytical Chemistry* 84(14):6008–6015.

Yilmaz, E., Ocsoy, I., Ozdemir, N., Soylak, M. 2016. Bovine serum albumin-Cu (II) hybrid nanoflowers: An effective adsorbent for solid phase extraction and slurry sampling flame atomic absorption spectrometric analysis of cadmium and lead in water, hair, food and cigarette samples. *Analytica Chimica Acta* 906:110–117.

Yuliarto, B., Septiani, N.L.W, et al. 2019. Green synthesis of metal oxide nanostructures using naturally occurring compounds for energy, environmental, and bio-related applications. *New Journal of Chemistry* 43:15846–15856.

Zhang, K., Melo, M.A., Cheng, L., Weir, M.D., Bai, Y., Xu, H.H. 2012. Effect of quaternary ammonium and Ag nanoparticle-containing adhesives on dentin bond strength and dental plaque microcosm biofilms. *Dental Materials* 28:842–52.

Zhang, Y., Chen, Y., Han, D., Ocsoy, I., Tan, W. 2010. Aptamers selected by cell-SELEX for application in cancer studies. *Bioanalysis* 2(5):907–918.

Zhu, L., Gong, L., Zhang, Y., Wang, R., Ge, J., Liu, Z., Zare, R.N. 2013. Rapid detection of phenol using a membrane containing laccase nanoflowers. *Chemistry–An Asian Journal* 8(10):2358–2360.

Zielonka, A., Ochab, M.K. 2017. Fungal synthesis of size-defined nanoparticles. *Advances in Natural Sciences: Nanoscience and Nanotechnology* 8(4):043001.

Zikalala, N., Matshetshe, K., Parani, S., Oluwafemi, O.S. 2018. Biosynthesis protocols for colloidal metal oxide nanoparticles. *Nano-Structures and Nano-Objects* 16:288–299.

14 Nanomaterials, an Eco-Friendly Replacement of Traditional Agricultural Plant Fungicides

Priyanka Debnath, Arghadip Mondal, and Naba Kumar Mondal

CONTENTS

14.1 INTRODUCTION

Pathogenic fungi can cause a range of diseases on many agronomic, horticultural, ornamental, and forest plant species (Pariona et al., 2019). Crop cultivators continuously battle with fungal diseases affecting their crops. Plant pathogenic fungi must be controlled if consumer demand in developed countries for premium quality and diverse foods is to be met; while high-quality cereals, fruits, and vegetables are an indicator of economic growth in developing countries. Commercial agriculture relies heavily upon high inputs of chemical pesticides to protect crops against pathogens and pests. Pests and plant pathogens' resistance against pesticides and fungicides is rapidly becoming a severe problem (Hahn, 2014). These practices are currently being re-evaluated due to concerns about possible health and environmental consequences. Growing consumer awareness towards using these chemical compounds is beginning to direct farmers towards the investigation of alternative chemical control measures. Nanotechnology can play a critical role in global food production, food security, and food safety. Nanotechnology applications in agriculture include fertilizers to increase plant growth and yield, pesticides for pest and disease management, and sensors for monitoring soil quality and plant health (Shang et al., 2019).

Nanotechnology refers to a nanoscale technology, which has promising applications in day-to-day life. Several nanotechnologies can improve existing fungi control protocols in the short to medium term (Alghuthaymi et al., 2015). Nanotechnology emphasizes the implications of individual atoms or molecules or submicron dimensions in their applications to physical, chemical, and biological systems and, eventually, their integration into larger complex systems(Pandit et al., 2016). There are different types of nanomaterials like copper, zinc, titanium, magnesium, gold, alginate, and silver, which have been developed. However, among them, silver nanoparticles (Nano-Ag) have proved to be most the effective as they exhibit potent antimicrobial efficacy against bacteria, viruses, and eukaryotic microorganisms (Wang et al., 2017). The antimicrobial properties of silver have been documented since 1000 BC when silver vessels were used to preserve water, but its applications in the field of agriculture have only gained momentum very recently (Abdelmalek and Salaheldin, 2016). Commendable efforts have been made to explore the antimicrobial properties of silver nanoparticles against human pathogens, but little research has been done to study its effects against overwhelming phytopathogens. Few studies have established the fungicidal effects of nanosilver against several phytopathogenic fungi, but this compound's fungicidal mechanism has not been elucidated (Huang et al., 2020). Considering the above facts, the present study focused mainly on silver nanoparticles and their potential antifungal activities on agricultural plants.

14.2 SYNTHESIS OF NANOPARTICLES

As mentioned earlier, nanoparticles are an agglomeration of atoms and molecules within the range of 1–100 nm. When a larger particle becomes a nanoparticle, the surface area to volume ratio for that particle increases, which increases its reactivity (Ameta et al., 2015). There are several methods, such as physical, chemical, and biological, by which nanoparticles can be synthesized. However, both physical and biological synthesis processes have drawbacks as they not only require a high amount of energy and chemicals as a precursor, but they also produces by-products that are harmful to the environment (Yousuf, 2020). In recent years, green and biological processes using mild reaction conditions and non-toxic precursors have been emphasized in the development of nanotechnology to promote environmental sustainability

(Shah et al., 2015; Mondal et al, 2021a). A brief description of all these three methods are as follows:

Physical methods: Physical methods apply mechanical pressure, high energy radiations, thermal energy, or electrical energy to cause material abrasion, melting, evaporation, or condensation to generate NPs. These methods mainly operate on a top-down strategy. They require a high amount of energy and are comparatively costly to carry out. High energy ball milling, laser ablation, electrospraying, inert gas condensation, physical vapour deposition, laser pyrolysis, flash spray pyrolysis, and melt mixing are some of the most commonly used physical methods used to generate NPs (Dhand et al., 2015).

Chemical methods: Chemical methods are generally based on bottom-up approaches. Chemical-based nanoparticle synthesis requires the use of chemicals as reducing agents, mainly as organic and inorganic solvents. The energy requirements for nanoparticle synthesis using these methods are comparatively lower than for the physical methods, but various toxic chemicals, such as capping and reducing agents, increase the pollution loaded into the environment. Additionally, every chemical method produces a by-product that ultimately goes into the environment. The sol-gel method, microemulsion technique, hydrothermal synthesis, polyol synthesis, chemical vapour synthesis, and plasma-enhanced chemical vapour deposition technique are some of the most commonly used chemical methods for NP synthesis (Singh et al., 2018).

Biological methods: The development of green chemistry methods for the synthesis of metal nanoparticles has become a major focus in recent times; hence, it has been discussed by many researchers (Singh et al., 2020). Green chemistry has been used to find an eco-friendly technique for the production of well-characterized nanoparticles. The nanoparticles which are being synthesized with these methods exhibit long-term stability. Moreover, there is a diversity in the shape and size of the nanoparticles. There are no toxic residues, and no environmental hazards as biological products act as both capping and reducing agents for nanoparticle synthesis. These methods employ biological materials like bacteria, fungi, viruses, yeasts, actinomycetes, plant extracts, etc., to synthesize metal and metal oxide NPs (Narayanan and Sakthivel, 2010). Bio-assisted methods can be broadly divided into two categories:

- *Biogenic synthesis using microorganisms*

Fungi, prokaryotic bacteria, actinomycetes, algae, and yeasts are extensively used as bioreactors for the synthesis of NPs. Enormous scientific efforts have been made to develop this strategy for producing various NPs (Ag, Au, ZnO, TiO_2, Cd, etc.). In this method, microorganisms grab target ions from their environment and then turn the metal ions into the elemental metal through enzymes generated by the cell activities for the nanoparticles' biosynthesis. According to the location of the nanoparticle formation, the synthesis procedure is divided into extracellular and intracellular synthesis. The intracellular method consists of transporting ions into the microbial cell to form nanoparticles in the presence of enzymes. Whereas the extracellular

TABLE 14.1
Synthesis of Nanoparticles from Various Microbiological Sources

Sl. No.	Microbiological agent	Synthesized NPs	Shape and size of the NPs	Reference
1.	*Aspergillusfumigatus*	Ag	32.8 nm; Spherical	Shahzad et al. (2019)
2.	*Fusariumoxyporum*	Ag	2–24 nm; Spherical	Hamedi et al. (2017)
3.	*Pseudomonas fluorescens*	Au	5–50 nm; Spherical	Syed et al. (2016)
4.	*Salmonella typhimurium*	Cu	10–49 nm; Spherical	Ghorbani et al. (2015)
5.	*Aspergillusniger*	ZnO	53–69 nm; Almost spherical	Kalpana et al. (2018)
6.	*Lactobacillus sp.*	Se	50–500 nm; Spherical	Sasidharan and Balakrishnaraja (2014)
7.	*Pochoniachlamydosporium*	Mg	20–60 nm; Spherical	Kaul et al. (2012)
8.	*Humicola sp.*	Ag	50–100 nm; Spherical	Syed et al. (2013)
9.	*Arthrobacterkerguelensis*	Ag	6–13 nm; Almost spherical	Shivaji et al. (2011)
10.	*Streptomyces hygroscopicus*	Ag	20–0 nm; Spherical	Sadhasivam et al. (2010)

synthesis of nanoparticles involves trapping the metal ions on the surface of the cells and reducing the ions in the presence of enzymes (Li et al., 2011). These biosynthesized nanoparticles have been used in various applications, including drug carriers for targeted delivery, cancer treatment, antibacterial agents, biosensors, separation science, magnetic resonance imaging (MRI), etc. (Li et al., 2011). Table 14.1 shows nanoparticle synthesis using different microbiological sources.

- *Biogenic synthesis using plant extracts*

The biosynthesis of NPs using plant extracts or plant biomass is a beneficial, rapid, clean, non-toxic, and eco-friendly method (Table 14.2). This method has been utilized extensively to synthesize the NPs of noble metals, metal oxides, bi-metallic alloys, etc. (Ali et al., 2020). Plant biometabolites significantly help in the preparation of NPs based on their valuable role as reducing agents and capping agents. Figure 14.1 displays the mechanisms for the synthesis of metal NPs using plant extracts.

14.3 CHARACTERIZATION TECHNIQUES OF NANOPARTICLES

Several techniques have been used to characterize the size, crystal structure, elemental composition, and various other physical properties of nanoparticles. There are microscopy-based techniques (e.g. transmission electron microscopy, FESEM,

TABLE 14.2
Synthesis of Nanoparticles from Various Biological Sources

Sl. No.	Biological agent	Synthesized NPs	Shape and size of NPs	Reference
1.	*Salvia spinosa* (plant extract)	Ag	19–125 nm; Spherical	Pirtarighat et al. (2019)
2.	*Punicagranatum* (seed extract)	Fe	25–55 nm; Semi-spherical	Bibi et al. (2019)
3.	*Garciniamangostana* (Fruit peel)	Au	32.96 nm; Mostly spherical and some triangular	Lee et al. (2016)
4.	*Botryococcusbraunii* (Green alga)	Cu	10–70 nm; Spherical and cubical	Arya et al. (2018)
5.	*Passifloracaerulea* (Fresh leaves)	ZnO	70 nm; spherical	Santhoshkumar et al. (2017)

and atomic force microscope, etc.), which provide information on the nanomaterials' size, morphology, and shape. Some other techniques include spectroscopy, scattering, and X-ray, which provide further information on the structure, elemental composition, optical properties, and other common and more specific physical properties of the nanoparticle samples. A brief description of some of these techniques, which provide preliminary evidence of nanoparticle synthesis, is given below.

14.3.1 Ultraviolet-Visible Spectroscopy (UV-Vis)

UV-vis spectroscopy is a facile and low-cost characterization method that gives a first testimony about a nanoparticle's synthesis. Basically, spectroscopy is related to

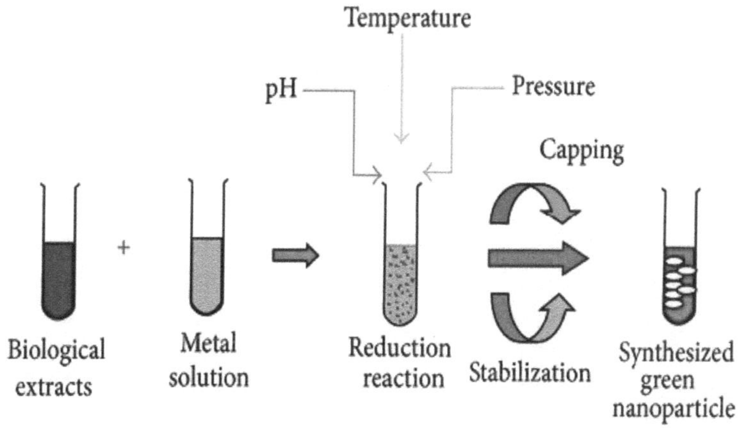

FIGURE 14.1 The basic principle of biologically synthesized nanoparticles.

the interaction of light with matter. UV spectroscopy obeys the Beer-Lambert law. It measures the intensity of light reflected from a sample and compares it to the intensity of light reflected from a reference material. Gold, silver, and copper are some nanoparticles that have optical properties due to the existence of an LSPR signal in the visible part of the spectrum. It has already been reported that silver nanoparticles have a particular wavelength range at which they give a distinct peak. It was recently reported that UV-visible absorption spectra indicated strong plasma resonance bands located at 408 and 400 nm for biosynthesized Ag NPs (Abdel-Raouf et al., 2019). Figure 14.2a shows a UV-vis image of biosynthesized silver nanoparticles.

14.3.2 Scanning Electron Microscopy (SEM)

To achieve high-resolution imaging from the upper-most surface of a nanoparticle, a SEM with advanced detectors is required. The high-resolution SEM is excellent for nanoparticle size and shape characterization because the necessary sample preparation and image acquisition are relatively quick and straightforward. The scanning electron microscope works on the same principle as an optical microscope, but it measures the electrons scattered from the sample rather than a photon. Because electric potential can accelerate electrons, the wavelength can be made shorter than one of the photons. This makes SEM capable of magnifying images up to 200,000 times (Lin et al., 2014). The following image shows the oval and cubical structure of biosynthesized Ag NPs (Figure 14.2b).

14.3.3 Transmission Electron Microscopy (TEM)

Transmission electron microscopy is a microscopy technique in which a beam of electrons is transmitted through an ultra-thin specimen, interacting with the specimen as it passes through. An image is formed from the interaction of the electrons transmitted through the specimen; the image is magnified and focused onto an imaging device, such as a fluorescent screen, on a layer of photographic film, or detected by a sensor such as a CCD camera. TEM is the preferred method to directly measure nanoparticle size, grain size, size distribution, and morphology (Williams and Carter, 2009; Mondal et al, 2019). Ag NPs with a spherical shape are represented in the following image (Figure 14.2c).

14.3.4 X-Ray Diffraction (XRD)

XRD is one of the most extensively used techniques for the characterization of NPs. Typically, XRD analysis provides information regarding the crystallinity, purity, and size of synthesized metal nanoparticles (Das et al., 2009). Apart from structure, the best thing for XRD analysis is to get the crystal size, and lattice distortion happens during the mechanical alloying. X-rays interact with the electronic clouds of the atoms: they are more absorbed and more scattered by atoms with a larger number of electrons (larger atomic number) (Waseda et al., 2011). The following image shows the XRD pattern of biosynthesized silver nanoparticles (Figure 14.2d).

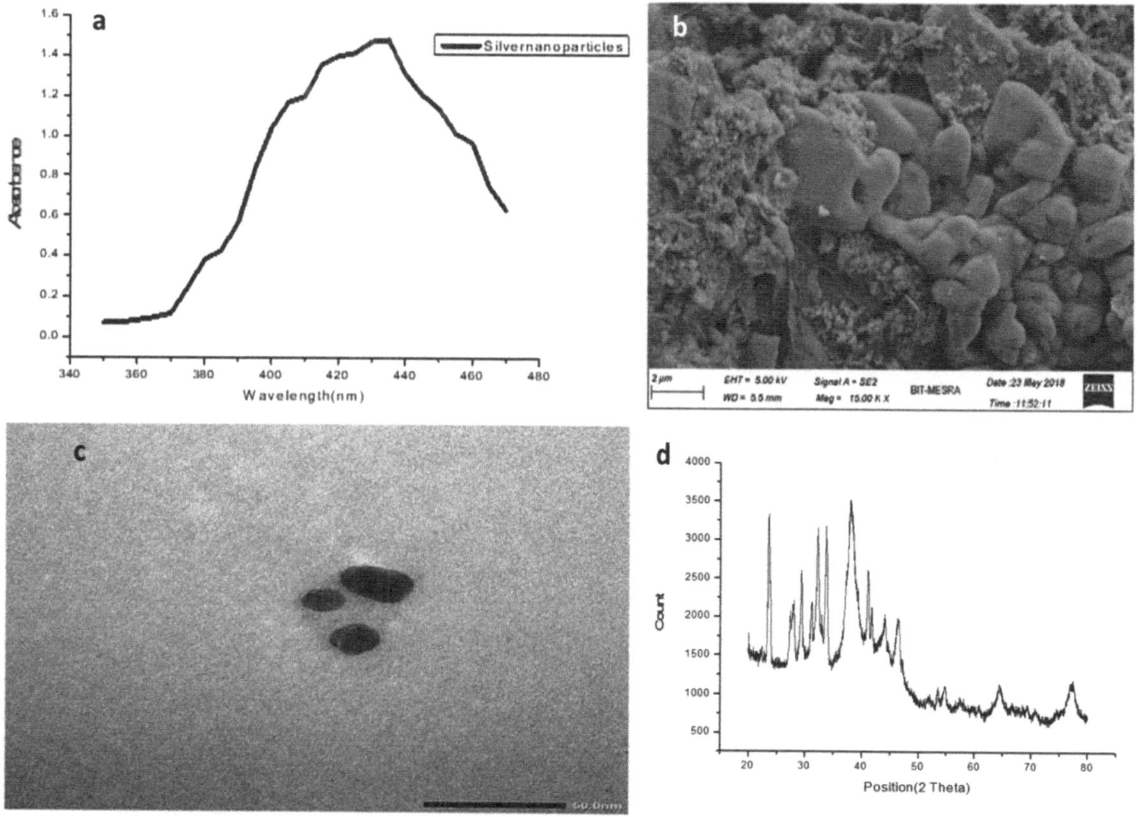

FIGURE 14.2 (a) UV-visible spectra of biosynthesized silver nanoparticle (Ag NP), (b) SEM image of biosynthesized silver nanoparticle (Ag NP), (c) TEM image of biosynthesized silver nanoparticle (Ag NP), and (d) XRD of biosynthesized silver nanoparticle (Ag NP) (Debnath et al., 2018).

14.3.5 Fourier-Transform Infrared Spectroscopy (FTIR)

FTIR enables the in-situ analysis of interfaces to examine functional groups' surface adsorption onto the nanoparticles (Thakkar et al., 2010). An advantage of FTIR is that it enables users to analyze a layer of nanoparticles coated on the ATR element, while also altering the overlying phase. An FTIR spectrometer simultaneously collects high-spectral-resolution data over a wide spectral range. As FTIR helps to study the functional groups present in the nanoparticles, this technique can be used for the qualitative analysis of materials. Each functional group possesses one or more characteristic peak at different wavenumbers (Hind et al., 2001; Mondal et al, 2021b).

14.3.6 Atomic Force Microscope (AFM)

AFM has become one of the most widely used tools for imaging, measuring, and manipulating matter at the nanoscale and has inspired various other scanning probe techniques. The AFM allows for 3D characterization of nanoparticles with a sub-nanometer resolution. Originally the AFM was used to image the topography of surfaces. However, by modifying the tip, it is possible to measure other quantities (for example, electric and magnetic properties, chemical potentials, friction, and so on) and perform various types of spectroscopy and analysis (Stephan et al., 2006).

14.4 APPLICATION OF NANOPARTICLES IN PLANT PATHOLOGY

Nanoparticles and nanosensors have wide applications in both the detection of microbial infections and the diagnosis of plant diseases.

14.4.1 Nanoparticles in the Detection of Plant Pathogens

Nanoparticles can be applied as a rapid diagnostic tool or as a biomarker for the detection of plant pathogenic bacteria (Boonham et al., 2008), viruses (Yao et al., 2009), and fungi (Chartuprayoon et al., 2010). Nanoparticles can either be directly modified for pathogen detection or as a diagnostic tool to detect specific chemicals synthesized under a diseased condition. The study reported successful and accurate salicylic acid measurement using the gold electrodes with a Cu NP sensor (Wang et al., 2011). Accordingly, there was a development of various types of nanostructured biosensors to recognize live cells, tissues, microbes, etc. (Abd-Elsalam and Prasad, 2019). The immobilization of the biorecognition element, such as DNA, antibodies, enzymes, etc., can be achieved using various approaches, including biomolecule adsorption, covalent attachment, encapsulation, or a sophisticated combination of these methods (Fang and Ramasamy, 2015). Moreover, there are other biological materials such as antibodies with which nanoparticles can be integrated to construct immunosensors to detect bacterial infections. Earlier, Yao et al. (2009) successfully prepared a silica nanoparticle biomarker using silica NPs along with antibodies and developed a silica NP probe to detect bacterial spot disease-causing bacteria, *Xanthomonasaxonopodis* pv. *vesicatoria*. Similarly, Gold nanoparticles are excellent markers when used in biosensors as several optical or electrochemical procedures can be modified to identify pathogens (Jiang et al., 2018). Many nanoparticle-based experiments have been performed to develop biomolecular detection with DNA or protein functionalized gold nanoparticles, which are used as the target-specific probes (Daniel and Astruc, 2003; Wang and Wang, 2014). Zhao et al. (2008) also reported that sensor-based approaches with transduction mechanisms, such as optical, electrochemical, and mass sensitive measurements, offered an accurate and reliable technology for detection of microorganisms. Whereas the study reported that DNA–gold nanoparticle probes could successfully detect an infection by *Pseudomonas syringae* pathogenic varieties (pathovars) (Vaseghi et al., 2013). Recently, further advancement has been made with nanosensors which could be placed directly into packaging material, where they would serve as 'electronic noses' to detect chemicals released during food spoilage (Panigrahi et al., 2017). Some nanosensors are based on microfluidic devices, and they can be used to identify pathogens proficiently in a short time with high sensitivity (Baeummer, 2004). Furthermore, copper oxide (CuO) nanoparticles and nanolayers were synthesized by sol-gel and spray pyrolysis methods, respectively. Both CuO nanoparticles and nanostructural layer biosensors were used for detecting the *A. niger* fungi (Etefagh et al., 2013).

14.4.2 Nanoparticles as a Management Tool for Plant Pathogens

Plants in nature are attacked by various pathogens that suppress their growth and productivity (Khan and Anwer, 2011). Innovative and advanced methods are needed to be integrated with conventional methods to enhance the efficiency of disease management modules. Nanoforms of metals and other materials have ample scope for exploitation in the management of plant diseases. Singh et al. (2013) reported that among nanoforms of 15 micronutrients, $CuSO_4$ and Na_2B4O_7 were the most effective at controlling field peas rust disease.

Another study report recently described that nanomaterials' simplest and most apparent treatment is direct application into the soil, seeds, or foliage to protect plants from pathogen invasions (Shang et al., 2019). Not long ago, Muñoz-Bonilla et al. (2019) reported that the combination of NPs with biocontrol agents, essential oils, and biopolymer-based cupric and sulphur NPs could show a synergistic effect to enhance antimicrobial activity and reduce their net quantity of usage. Whereas, in some studies, synergism was obtained when antimicrobial NPs were combined with bioorganic pesticides (Dar and Soytong, 2014; Xue et al., 2014). Moreover, there is an array of nano-chemicals or agents known that are proven pesticides. However, due to the associated problems of nanopesticide delivery and stability, efforts are underway to develop hybrid formulations like an encapsulation of nanoemulsions (Jerobin et al., 2012) or liposome coatings (Kang et al., 2012). Inorganic NPs are considered biologically more active and environmentally safer for nanopesticide development. The inorganic NPs include silica, titanium dioxide, silver, copper, and aluminium (Kah and Hofmann, 2014). The use of Al_2O_3 NPs (Shenashen et al., 2017), MgO NPs (Wani and Shah, 2012), TiO_2 NPs (Elmer and White, 2016), S NPs (Gogoi et al., 2013), and CeO_2 NPs (Adisa et al., 2018) have all shown efficacy in suppressing diseases in different systems. Keeping in mind the above discussion, it is clear that most studies have been able to correlate NP amendment with direct alteration of the host defence mechanisms. Although the report says that when the bulk or salt form of these elements/molecules was compared to the nanoform the later was consistently more effective in promoting health, often with significantly less of the active metal being applied (Wang et al., 2017).

14.5 SILVER NANOPARTICLES AS A POTENT FUNGICIDE

Silver NPs may inhibit plant pathogens in various ways (Clement and Jarret, 1994), and the disease management achieved may be more effective and safer than chemical fungicides (Park et al., 2006). The antifungal activity of different forms of silver nanoparticles has been recorded on ambrosia fungus, *Raffaelea* sp. (Kim et al., 2009), and sclerotium-forming fungi (Min et al., 2009). Mallmann et al. (2015) reported Ag NPs were easily prepared by green synthesis using ribose as a reducing agent and SDS as a stabilizer. Additionally, they showed high activity against *C. albicans* and *C. tropicalis*, a similar activity observed by the antifungal amphotericin B, and this may represent an alternative for treating fungal infections. Whereas Ragam and Mathew (2019) reported on the microwave-assisted production of silver nanoparticles using aqueous silver nitrate, and stem extracts of the plant *Cosciniumfenestratum* were exploited for the detection of fungicide thiram. Moreover, considering the particular optical properties of silver nanoparticles, the interaction between silver nanoparticles and sulphurazon-ethyl herbicide was investigated (Dubas and Pimpan, 2008). It was found that silver nanoparticles are sensitive to increased herbicide concentrations in a solution and this induced a variation in the nanoparticles' colour from yellow to orange-red and finally to purple. A quick description of silver nanoparticles' fungicidal activity is depicted in the following table (Table 14.3).

14.6 FUTURE PERSPECTIVES OF THE PRESENT STUDY

Nowadays, many researchers around the world are synthesizing different nanoparticles using different biological agents. However, the factors or secondary metabolites of a particular biological agent that is playing the pivotal role in the synthesis is still not adequately understood. Detection of these metabolites will make the synthesis technique easier and quicker. Additionally, numerous studies have established the fungicidal effects of nanosilver against several phytopathogenic fungi, but this compound's fungicidal mechanism has not been elucidated. To date, the biological synthesis of metallic nanoparticles has been mostly carried out on a laboratory scale. Industrial-scale optimization is required for large-scale production. With proper optimized conditions and suitable microorganisms, these 'bio-nanofactories' can produce stable nanoparticles with a well-defined size, composition, and morphology. The commercialized processes will result in a non-toxic biological system capable of producing metallic nanoparticles, which will be another milestone toward sustainable development.

TABLE 14.3
Fungicidal Activity of Biosynthesized Ag NPs

Biological agent	NP	Targeted fungal strain	Inhibition zone	Reference
Moringaoleifera	Ag	*Pseudomonas aeruginosa*	11 mm	Varghese and Gajera (2020)
Dionaeamuscipula	Ag	*Pectobacteriumatro-septicum,* *Pectobacteriumpar-mentieri* *Dickeyadadantii*	10.6 mm 10.6 mm 21.2 mm	Banasiuk et al. (2020)
Trichodermaviride	Ag	*T. mentagrophytes* *T. rubrum* *C. albicans*	Minimuminhibitoryconc.– 1–7μg ml^{-1}, 10μg $m1^{-1}$ and 0.4–3.3μg ml^{-1} respectively.	Reddy (2015)
Gelidiumcorneum	Ag	*C. albicans*	Minimum inhibitory conc. 0.51 μg ml^{-1}	Öztürk et al. (2020)
Rhizobium sp.	Ag	*Aspergillusniger*	15 mm	Sowmya et al. (2019)
Waste-grass	Ag	*Fusariumsolani* *Rhizoctoniasolani*	Minimuminhibitoryconc.– 20 μg ml^{-1} for both fungi	Khatami et al. (2018)
Amaranthus retroflexus	Ag	*Macrophominaphaseo-lina* *Alternariaalternata*	Minimuminhibitoryconc.– 159.80±14.49, 337.09±19.72 μg ml^{-1}	BahramiTeimooriet al. (2017)
Arthrodermaful-vum	Ag	*Candida* sp, *Aspergillus* sp. *Fusarium* sp.	Minimuminhibitoryconc.– 0.125, 1 and 2 μg ml^{-1} respectively	Xue et al. (2016)
Malvacrispa	Ag	*Penicilliumexpansum,* *Penicilliumcitrinum*	16 mm 12 mm	Krishnaraj et al. (2015)
Sargassumlongifo-lium	Ag	*Aspergillus fumigates* *Candida albicans*	13.10 mm 14.20 mm	Rajeshkumar et al. (2014)
Bacillus sp. GP-23	Ag	*Fusariumoxysporum*	Minimuminhibitoryconc.– 8 μg ml^{-1}	Gopinath and Velusamy (2013)
Ocimum sanctum	Ag	*C. albicans* *C. Kefyr* *C. krusei*	15.22 ± 0.04 13.00 ± 0.05 11.15 ± 0.11	Rout et al. (2012)
Gelidiellaacerosa	Ag	*Humicolainsolens* *Trichodermareesei*	12.2 mm 17.2 mm	Vivek et al. (2011)
Calotropisgigan-tea	Ag	*Candida albicans*	13 mm	Nasrollahi et al. (2011)
Argemonemexi-cana	Ag	*Aspergillusflavus*	10 ± 0.2 mm	Singh et al. (2010)

On the other hand, since nanomaterials are of recent origin and their side effects are unknown, the toxicity or biosafety concerns involved in their applications against plant pathogens is significant. The long-term effects of nanoformulations on human health and environmental quality are yet to be investigated.

14.7 CONCLUSION

This chapter clearly illustrated nanoparticle synthesis using different kinds of cheap, easily available biological products. Moreover, The use of different types of biological agents in the synthesis of NPs has encouraged the designing of simple, green, cost- and time-effective approaches, thereby minimizing the use of chemicals and solvents. The synthesized NPs can be characterized using various techniques like UV-vis, SEM, and TEM, etc. Furthermore, the present study so far has provided substantial evidence that warrants the potential application of nanomaterials in microbial suppression and plant disease management. Studies have shown that direct application of nanoparticles may suppress the plant pathogenic fungi tested so far without any detrimental effect on the ecosystem. However, biosafety aspects concerning NP accumulation and the phytotoxic and mammalian toxic effects of the nanomaterials remain as major concerns that have to be ruled out before commercial use of nanomaterials in plant disease management.

Despite apprehension about the various adverse effects of nanomaterials on plants, animals, and the environment, the development of nano fungicides is one potential future option for enhancing global agricultural production and reducing chemical inputs. Necessary research efforts are also needed to devise and standardize nanomaterial application methods to exploit the maximum potential of nanomaterials in plant disease management.

ACKNOWLEDGEMENTS

The authors extend their sincere gratitude to all the faculty members and staff at the Department of Environmental Science, The University of Burdwan, for their moral support and valuable suggestions for preparing this manuscript. The authors acknowledge their sincere thanks to the funding agency, SVMCMF, Govt. of India for providing necessary funds for conducting the present research.

REFERENCES

Abdelmalek GAM, Salaheldin TA. (2016). Silver nanoparticles as a potent fungicide for citrus phytopathogenic fungi. *Journal of Nanomedicine Research*, 3(5):1–8. DOI: 10.15406/jnmr.2016.03.00065.

Abdel-Raouf N, Al-Enazi NM, Ibraheem IBM, Alharbi RM, Alkhulaifi MM. (2019). Biosynthesis of silver nanoparticles by using of the marine brown alga Padinapavonia and their characterization. *Saudi Journal of Biological Sciences* 26:1207–1215.

Abd-Elsalam K, Prasad R. (2019). Nanobiotechnology Applications in Plant Protection: Volume 2. In *Nanotechnology in the Life Science*. Edited by Kamel A. Abd-Elsalam, Ram Prasad. Published by Springer, Cham. 2523–8027. doi: 10.1007/978-3-030-13296-5.

Adisa IOVLR, Pullagurala S, Rawat JA, Hernandez-Viezcas CO, Dimkpa WH, Elmer JC, White JR, Peralta-Videa JL. (2018). Role of cerium compounds in Fusarium suppression and growth enhancement in tomato (*Solanum lycopersicum*). Royal Chemical Society, Environmental Nano Sci.

Alghuthaymi MA, Almoammar H, Rai M, Said-Galiev E, Abd-Elsalam KA. (2015). Myconanoparticles: Synthesis and their role in phytopathogens management. *Biotechnology & Biotechnological Equipment* 29(2):221–236.

Ali S, Sharma AS, Ahmad W, Zareef M. (2020). Noble metals based bimetallic and trimetallic nanoparticles: Controlled synthesis, antimicrobial and anticancer applications. *Critical Reviews in Analytical Chemistry*. doi: 10.1080/10408347.2020.1743964.

Ameta SC, Punjabi PB, Ameta R, Ameta C. (2015). *Microwave-Assisted Organic Synthesis: A Green Chemical Approach*. CRC Press, Taylor & Francis Group.

Arya A, Gupta K, Chundawat TS, Vaya D. (2018). Biogenic synthesis of copper and silver nanoparticles using green alga *Botryococcusbraunii* and its antimicrobial activity. *Bioinorganic Chemistry and Applications*. 1–9.

Baeummer A. (2004). Nanosensors identify pathogens in food. *Food Technology* 58:51–55.

Bahrami-Teimoori B, Nikparast Y, Hojatianfar M, Akhlaghi M, Ghorbani R, Pourianfar HR. (2017). Characterisation and antifungal activity of silver nanoparticles biologically synthesised by Amaranthusretroflexus leaf extract. *Journal of Experimental Nanoscience* 12(1):129–139.

Banasiuk R, Krychowiak M, Swigon D, Tomaszewicz W, Michalak A, Chylewska A, Ziabka M, Lapinski M, Koscielsk B, Narajczyk M, Krolicka A. (2020). Carnivorous plants used for green synthesis of silver nanoparticles with broad-spectrum antimicrobial activity. *Arabian Journal of Chemistry* 13(1):1415–1428.

Bibi I, Nazar N, Ata S, Sultan M, Ali A, Abbas A, Jilani K, Kamal S, Sarim FM, Khan MI, Jalal F, Iqbal M. (2019). Green synthesis of iron oxide nanoparticles using pomegranate seeds extract and photocatalytic activity evaluation for the degradation of textile dye. *Journal of Materials Research and Technology* 8(6):6115–6124.

Boonham N, Glover R, Tomlinson J, Mumford R (2008) Exploiting generic platform technologies for the detection and identification of plant pathogens. In *Sustainable Disease Management in a European Context*. Springer, Berlin, pp. 355–363.

Chartuprayoon N, Rheem Y, Chen W, Myung N. (2010). Detection of plant pathogen using LPNE grown single conducting polymer Nanoribbon. In *Meeting Abstracts*. The Electrochemical Society, Pennington.

Clement JL, Jarrett PS. (1994).Antibacterial silver. *Metal-Based Drugs* 1:467–482.

Daniel M-C, Astruc D. (2003). Gold nanoparticles: Assembly, supramolecular chemistry, quantum-size-related properties, and applications toward biology, catalysis, and nanotechnology. *Chemical Reviews* 104:293–346.

Dar J, Soytong K. (2014). Construction and characterization of copolymer nanomaterials loaded with bioactive compounds from Chaetomium species. *Journal of Agriculture, Science and Technology* 10:823–831.

Das R, Nath SS, Chakdar D, Gope G, Bhattacharjee R. (2009).Preparation of silver nanoparticles and their characterization. *Journal of Nanotechnology* 5:1–6.

Debnath P, Mondal A, Hajra A, Das C, Mondal NK. (2018). Cytogenetic effects of silver and gold nanoparticles on *Allium cepa* roots. *Journal of Genetic Engineering and Biotechnology* 16(2):519–526.

Dhand C, Dwived N, Loh XJ, Ying ANJ, Verma NK, Beuerman RW, Lakshminarayanan R, Ramakrishna S. (2015). Methods and strategies for the synthesis of diverse nanoparticles and their applications: A comprehensive overview. *RSC Advances* 5:105003–105037.

Dubas S, Pimpan V. (2008). Green synthesis of silver nanoparticles for ammonia sensing. *Talanta* 76(1):29–33

Elmer WH, White J. (2016). Nanoparticles of CuO improves growth of eggplant and tomato in disease infested soils. *Environmental Science: Nano* 3:1072–1079.

Etefagh R, Azhir E, Shahtahmasebi N. (2013). Synthesis of CuO nanoparticles and fabrication of nanostructural layer biosensors for detecting *Aspergillusniger fungi. Science Iranian* 20(3):1055–1058.

Fang Y, Ramasamy R. (2015). Current and prospective methods for plant disease detection. *Biosensors* 5(3):537–561.

Ghorbani HR, Mehr FP, Poor AK. (2015). Extracellular synthesis of copper nanoparticles using culture supernatants of *Salmonella typhimurium. Oriental Journal of Chemistry* 31(1):1–7.

Gogoi R, Singh PK, Kumar R, Nair KK, Alam I, Srivastava, C. Yadav S, Gopal M, Roy SC, Goswamiet A. (2013). Suitability of nano-sulphur for biorational management of powdery mildewof okra (*Abelmoschus esculentus* Moench) caused by *Erysiphe cichoracearum. Journal of Plant Pathology & Microbiology* 4(4):71–175.

Gopinath V, Velusamy P. (2013). Extracellular biosynthesis of silver nanoparticles using Bacillus sp. GP-23 and evaluation of their antifungal activity towards *Fusarium oxysporum. SpectrochimicaActa Part A: Molecular and Biomolecular Spectroscopy* 106:170–174.

Hahn M. (2014). The rising threat of fungicide resistance in plant pathogenic fungi: Botrytis as a case study. *Journal of Chemical Biology* 7(4):133–141.

Hamedi S, Ghaseminezhad M, Shokrollahzadeh S, and Shojaosadati SA. (2017). Controlled biosynthesis of silver nanoparticles using nitrate reductase enzyme induction of filamentous fungus and their antibacterial evaluation. Artif.Cells Nanomed.Biotechnol. 45:1588–1596.

Hind AR, Bhargava SK, McKinnon A. (2001). At the solid/liquid interface: FTIR/ATR – The tool of choice. Artificial Cells, Nanomedicine, and Biotechnology 93:91–114.

Huang W, Yan M, Duan Ha, Bi Y, Cheng X, Yu H. (2020). Synergistic antifungal activity of green synthesized silver nanoparticles and epoxiconazole against *Setosphaeria turcica. Journal of Nanomaterials* 1–7.

Jerobin J, Sureshkumar R, Anjali C, Mukherjee A, Chandrasekaran N. (2012). Biodegradable polymer based encapsulation of neem oil nanoemulsion for controlled release of Aza-A. *Carbohydrate Polymers* 90(4):1750–1756.

Jiang P, Wang Y, Zhao L, Ji C, Chen D, Nie L. (2018). Applications of gold nanoparticles in non-optical biosensors. *Nanomaterials (Basel)* 8(12):977.

Kah M, Hofmann T. (2014). Nanopesticide research: Current trends and future priorities. Environment International 63:224–235.

KalpanaVN, Kataru BAS, Sravani N, Vigneshwari T, Panneerselvam A, Rajeswari VD. (2018). Biosynthesis of zinc oxide nanoparticles using culture filtrates of *Aspergillusniger*: Antimicrobial textiles and dye degradation studies. *OpenNano* 3:48–55.

Kang MA, Seo MJ, Hwang IC, Jang C, Park HJ, Yu YM, Youn YN. (2012). Insecticidal activity and feeding behavior of the green peach aphid, Myzuspersicae, after treatment with nano types of pyrifluquinazon. *Journal of Asia-Pacific Entomology* 15(4):533–541.

Kaul RK, Kumar P, Burman U, Joshi P, Agrawal A, Raliya R, Tarafdar JC. (2012). Magnesium and iron nanoparticles production using microorganisms and various salts. *Materials Science-Poland* 30(3):254–258.

Khan MR, Anwer A. (2011). Fungal bioinoculants for plant disease management. In Paul, M., Clinton, M., Ahmad, I. (Eds.), *Microbes and Microbial Technology*. Springer, pp. 447–488.

Khatami M, Sharifi I, Nobre MAL, Zafarnia N, Aflatoonian MR. (2018). Waste-grass-mediated green synthesis of silver nanoparticles and evaluation of their anticancer, antifungal and antibacterial activity. *Green Chemistry Letters and Reviews* 11(2):125–134.

Kim SW, Kim KS, Lamsal K, Kim YJ, Kim SB, Jung M, Sim SJ, Kim HS, Chang SJ, Kim JK. (2009). An in vitro study of the antifungal effect of silver nanoparticles on oak wilt pathogen Raffaelea sp. *Journal of Microbiology and Biotechnology* 19: 760–764.

Krishnaraj C, Harper SL, Choe HS, Kim K-P, Yun S-I. (2015). Mechanistic aspects of biologically synthesized silver nanoparticles against food- and water-borne microbes. *Bioprocess and Biosystems Engineering* 38:1943–1958.

Lee KX, Shameli K, Miyake M, Kuwano N, Khairudin BNBA,1 Mohamad SEB, Yew YP. (2016). Green synthesis of gold nanoparticles using aqueous extract of *Garciniamangostana* Fruit Peels. *Journal of Nanomaterials* 1–7.

Li X, Xu H, Chen Z-S, Chen G. (2011). Biosynthesis of nanoparticles by microorganisms and their applications. *Journal of Nanomaterials* 1–16.

Lin PC, Lin S, Wang PC, Sridhar R. (2014). Techniques for physicochemical characterization of nanomaterials. *Biotechnology Advances* 32:711–726.

Mallmann EJJ, Cunha FA, Castro BNMF, Maciel AM, Menezes EA, Fechine PBA. (2015). Antifungal activity of silver nanoparticles obtained by green synthesis. *Revista do Instituto de Medicina Tropical de São Paulo* 57(2):165–167.

Min JS, Kim KS, Kim SW, Jung JH, Lamsal K, Kim SB, Jung M, Lee YS.(2009). Effects of colloidal silver nanoparticles on sclerotium-forming phytopathogenic fungi. *The Plant Pathology Journal* 25:376–380.

Mondal A, Mondal NK, Hajra A, Shaikh W, Chakraborty S. (2019). Synthesis of silver nanoparticle with *Colocasia esculenta* (L.) stem and its larvicidal activity against *Culex quinquefasciatus* and *Chironomus* sp. *Asian Pacific Journal of Tropical Biomedicine* 9(12): 510. doi:10.4103/2221-1691.271724.

Mondal A, Debnath P, Mondal NK. (2021a). Nanoparticles: A new tool for control of mosquito larvae. *Intelligent Environmental Data Monitoring for Pollution Management* 49–70. doi:10.1016/b978-0-12-819671-7.00003-8.

Mondal A, Chowdhury S, Mondal NK, Shaikh WA, Debnath P, Chakraborty S. (2021b). Insecticidal and fungicidal performance of bio-fabricated silver and gold nanoparticles. *International Journal of Environmental Science and Technology.* 1–20. doi:10.1007/s13762-021-03181-w

Muñoz-Bonilla A, Echeverria C, Sonseca Á, Arrieta MP, Fernández-García M. (2019). Bio-based polymers with antimicrobial properties towards sustainable development. *Materials* 12:641.

Narayanan KB, Sakthivel N. (2010). Biological synthesis of metal nanoparticles by microbes. *Advances in Colloid and Interface Science* 156(1–2):1–13.

Nasrollahi A, Paurshamsian KH, Mansourkiaee P. (2011). Antifungal activity of silver nanoparticles on some of fungi. *International Journal of Nano Dimension* 1(3):233–239.

Öztürk BY, Gürsu BY, Dağ İ. (2020). Antibiofilm and antimicrobial activities of green synthesized silver nanoparticles using marine red algae *Gelidiumcorneum*. *Process Biochemistry* 89:208–219.

Pandit S, Dasgupta D, Dewan N, Ahmed P. (2016). Nanotechnology based biosensors and its application. *The Pharma Innovation Journal* 5(6):18–25.

Panigrahi SS, Sarkar P, Roy E, Banerjee P. (2017).Nanosensors in food safety. *Portable Biosensors and Point of Care Systems.* doi: 10.1049/PBHE003E_ch10.

Pariona N, Mtz-Enriquez AI, Sánchez-Rangela D, Carriónd G, Paraguay-Delgadoe F, Rosas-Saitoa G. (2019). Green-synthesized copper nanoparticles as a potential antifungal against plant pathogens. *RSC Advances* 9:18835–18843.

Park HJ, Kim SH, Kim HJ, Choi SH. (2006).A new composition of nanosized silica-silver for control of various plant diseases. *The Plant Pathology Journal* 22:295–302.

Pirtarighat S, Ghannadnia M, Baghshahi S. (2019).Green synthesis of silver nanoparticles using the plant extract of *Salvia spinosa* grown in vitro and their antibacterial activity assessment. *Journal of Nanostructure in Chemistry* 9:1–9.

Ragam PN, Mathew B. (2019). Unmodified silver nanoparticles for dual detection of dithiocarbamate fungicide and rapid degradation of water pollutants. *International Journal of Environmental Science and Technology*. doi: 10.1007/s13762-019-02454-9.

Rajeshkumar S, Malarkodi C, Paulkumar K, Vanaja M, Gnanajobitha G, Annadurai G. (2014). Algae mediated green fabrication of silver nanoparticles and examination of its antifungal activity against clinical pathogens. *International Journal of Metals* 2014:8.

Reddy SJ. Silver nanoparticles-synthesis, applications and toxic effects on humans: A review. *International Journal of Bioassays*. 2015;4:4563–73.

Rout Y, Behera S, Ojha AK, Nayak PL. (2012). Green synthesis of silver nanoparticles using Ocimum sanctum (Tulashi) and study of their antibacterial and antifungal activities. *Journal of Microbiology and Antimicrobials* 4(6):103–109.

Sadhasivam S, Shanmugam P, Yun KS. (2010). Biosynthesis of silver nanoparticles by Streptomyces hygroscopicus and antimicrobial activity against medically important pathogenic microorganisms. *Colloids and Surfaces B: Biointerfaces* 81(1):358–362.

Santhoshkumar J, VenkatKumar S, Rajeshkumar S. (2017). Synthesis of zinc oxide nanoparticles using plant leaf extract against urinary tract infection pathogen. *Resource-Efficient Technologies* 3(4):459–465.

Sasidharan S, Balakrishnaraja R. (2014). Comparison studies on the synthesis of selenium nanoparticles by various microorganisms. *International Journal of Pure & Applied Bioscience* 2 (1):112–117.

Shah M, Fawcett D, Sharma S, Tripathy SK, Eddy G, Poinern J. (2015). Green synthesis of metallic nanoparticles via biological entities. *Materials (Basel)* 8(11):7278–7308.

Shahzad, A., Saeed, H., Iqtedar, M., Hussain, S. Z., Kaleem, A., and Abdullah, R. (2019).Size-controlled production of silver nanoparticles by *Aspergillus fumigatus* BTCB10: Likely antibacterial and cytotoxic effects. *Journal of Nanomaterials* 2019:5168698.

Shang Y, Hasan MK, Ahammed GJ, Li M, Yin H, Zhou J. (2019). Applications of nanotechnology in plant growth and crop protection: A review. *Molecules* 24(14):2558.

Shenashen M, Derbalah A, Hamza A, Mohamed A, El-Safty S. (2017). Antifungal activity of fabricated mesoporous alumina nanoparticles against root rot disease of tomato caused by Fusariumoxysporum. *Pest Management Science* 73(6):1121–1126.

Shivaji S, Madhu S, Singh S. (2011). Extracellular synthesis of antibacterial silver nanoparticles using psychrophilic bacteria. *Process Biochemistry* 46(9):1800–1807.

Singh A, Gautam PK, Verma A, Singh V, Shivapriya PM, Shivalkar S, Sahoo AK, Samanta SK. (2020). Green synthesis of metallic nanoparticles as effective alternatives to treat antibiotics resistant bacterial infections: A review. *Biotechnology Reports (Amst)* 25:e00427.

Singh A, Jain D, Upadhyay MK, Khandelwal N, Verma HN. (2010). Green synthesis of silver nanoparticles using argemonemexicana leaf extract and evaluation of their antimicrobial activities. *Digest Journal of Nanomaterials and Biostructures* 5(2):483–489.

Singh D, Kumar A, Singh AK, Tripathi HS.(2013). Induction of resistance in field pea against rust disease through various chemicals/micronutrients and their impact on growth and yield. *The Plant Pathology Journal* 12:36–49.

Singh J, Dutta T, Kim K-H, Rawat M, Samddar P, Kumar P. (2018). 'Green' synthesis of metals and their oxide nanoparticles: Applications for environmental remediation. *Journal Nanobiotechnology* 16:84.

Sowmya B, Venkat Kumar S, Karpagambigai SP, Santhoshkumar J, Rajeshkumar, S. (2019). Antioxidant and antifungal activity of bacteria mediated silver nanoparticles using *Rhizobium* sp. *Indian Journal of Public Health Research & Development* 10(11):3622–3627.

Stephan TS, Scott EM, Anil KP, Marina AD.(2006). Preclinical characterization of engineered nanoparticles intended for cancer therapeutics. In *Nanotechnology for Cancer Therapy*. CRC Press; Boca Raton, FL, USA. 105–137.

Syed A, Saraswati S, Kundu GC, Ahmad A. (2013). Biological synthesis of silver nanoparticles using the fungus Humicola sp. and evaluation of their cytoxicity using normal and cancer cell lines. *Molecular and Biomolecular Spectroscopy* 114:144–147.

Syed B., Prasad N.M., Satish S. (2016) Endogenic mediated synthesis of gold nanoparticles bearing bactericidal activity. *Journal of Microscopy and Ultrastructure* 2016;4:162–166.

Thakkar KN, Mhatre SS, Parikh RY. (2010).Biological synthesis of metallic nanoparticles. *Nanomedicine* 6:257–262.

Varghese J, Gajera S. (2020). Green synthesis of Silver Nanoparticles from various leaves and its Antimicrobial activity. *Studies in Indian Place Names* 40(71):1667–1674.

Vaseghi A, Safaie N, Bakhshinejad B, Mohsenifar A, and Sadeghizadeh M. (2013). Detection of *Pseudomonas syringaepathovars* by thiol-linked DNA–Gold nanoparticle probes. *Sensors and Actuators B: Chemical* 181:644–651.

Vivek M, Kumar PS, Steffi S, Sudha S. (2011). Biogenic silver nanoparticles by *Gelidiella acerosa* extract and their antifungal effects. *Avicenna Journal of Medical Biotechnology* 3(3):143–148.

Wang EC, Wang AZ. (2014). Nanoparticles and their applications in cell and molecular biology. *Integrative Biology (Camb)* 6(1):9–26.

Wang L, Hu C, Shao L. (2017). The antimicrobial activity of nanoparticles: Present situation and prospects for the future. *International Journal of Nanomedicine* 12:1227–1249.

Wang S, Lawson R, Ray RC, Yu H. (2011). Toxic effects of gold nanoparticles on *Salmonella typhimurium* bacteria. *ToxicolInd Health* 27:547–554.

Wani AH, Shah MA. (2012). A unique and profound effect of MgO and ZnO nanoparticles on some plant pathogenic fungi. *Journal of Applied Pharmaceutical Science* 2(3):40–44.

Waseda Y, Matsubara E, Shinoda K. (2011). *X-ray Diffraction Crystallography: Introduction, Examples and Solved Problems*. Springer Verlag; Berlin, Germany.

Williams DB, Carter CB. (2009). *The Transmission Electron Microscope*. Springer Verlag; New York, NY.

Xue B, He D, Gao S, Wang D, Yokoyama K, Wang L. (2016). Biosynthesis of silver nanoparticles by the fungus Arthrodermafulvum and its antifungal activity against genera of Candida, Aspergillus and Fusarium. International Journal of Nanomedicine 11:1899–1906.

Xue J, Luo Z, Li P, Ding Y, Cui Y, Wu Q. (2014). A residue-free green synergistic antifungal nanotechnology for pesticide thiram by ZnO nanoparticles. *Scientific Reports* 4:5408.

Yao KS, Li SJ, Tzeng KC, Cheng TC, Chang CY, Chiu CY, Liao CY, Hsu JJ, Lin ZP. (2009). Fluorescence silica nanoprobe as a biomarker for rapid detection of plant pathogens. *Advanced Materials Research* 79:513–516.

Yousuf A. (2020). *Microalgae Cultivation for Biofuels Production*. Academic Press Elsevier.

Zhao W, Brook MA, Li Y. (2008). Design of gold nanoparticle-based colorimetric biosensing assays. *Chembiochem* 9:2363–2371.

15 Current Trends in Nanotechnological Approaches for Treatment of Plant Diseases

Shalja Verma, Anand Kumar Pandey, and Brijendra Kumar Kashyap

CONTENTS

15.1 INTRODUCTION

Though the term 'nanotechnology' was introduced in 1974 by Norio Taniguchi, the lecture 'There's Plenty of Room at the Bottom' by R.P. Feynman at a meeting of the American Physical Society at Caltech inaugurated the field of nanoscience in 1959, by displaying possibilities for the development of nanomachines which could be manipulated for different atomic arrangements according to the requirements (Hulla et al., 2015). Engineering of materials at one billionth of a metre and their enormous number of potential applications in a wide variety of fields constitute the branch of nanotechnology. The properties of a material at such a small scale varies from that of the bulk material due to the decrease in size and the increased significance of atom percentage on the material's surface (Khan et al., 2017). Also, interacting functional groups and molecules individually become important considerations, concerning the macroscopic bulk properties of materials at the nanoscale as it mediates control of the fundamental structure of the molecules and hence the physical and chemical macroscopic properties (Jeevanandam et al., 2018). The shape, size, and surface properties of different materials can be engineered to extract novel properties based on the requirements of the application (Sharma et al., 2019). Although nanotechnology has found many applications in different industries, such as the paint industries, in semiconductors, the auto industries, the fabric industry, cosmetics, packaging, electronics, energy etc., its contribution to biology is huge (Stark et al., 2015). With their exceptional properties and behaviour in biological systems, nanoparticles have been applied not only to reveal fundamental biological processes, but also to extensively deal with major problems like drug delivery or the development of specific delivery systems for different molecules into biological systems, imaging, sensing, diagnostics, and even as drugs themselves (Wong et al., 2013). These applications are not limited to human use but have also been implicated in plant systems (Sanzari et al., 2019). Certain fundamental factors like the composition of the core, surface properties, shape and size, stability, purity, and manufacturing method play essential roles in commanding the interactions between nanoparticles and biomolecules (Wang et al., 2016). Diverse groups of nanoparticles have been developed to date on the basis of their implementation in biological systems. Some major groups include carbon-based, metallic, lipid, semiconductor, polymer, and ceramic nanoparticles (Figure 15.1).

15.2 TYPES OF NANOPARTICLES

A variety of nanoparticles have been prepared depending on bulk material used. Nanoparticles of different materials show different properties and can be utilized in the varied type of applications. The major types of nanoparticles relevant in the field of biology are discussed as follows

15.2.1 Carbon-Based Nanoparticles

The most prevalent forms of carbon nanoparticles are fullerenes and nanotubes. Fullerenes are hollow globular cages similar to that of carbon allotropes, and

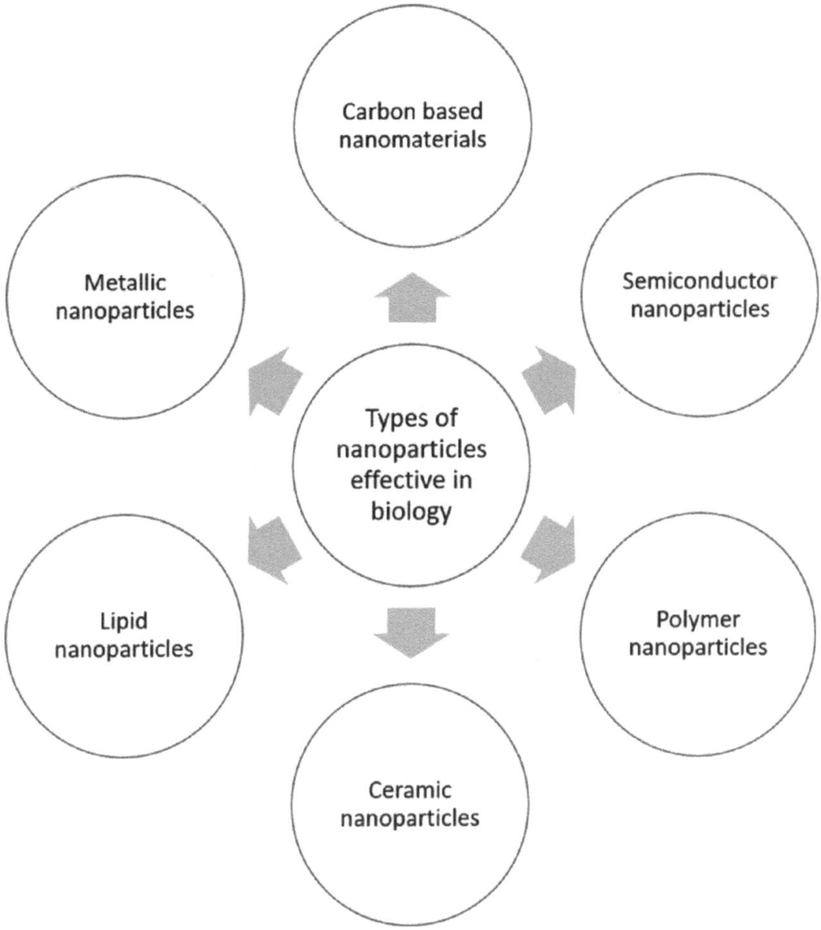

FIGURE 15.1 Different types of nanoparticles.

nanotubes are elongated tubelike structures with diameters in the range of 1 to 2 nm (Harris et al., 2004). These nanoparticles have found wide applications in biosensors, biomolecules, and drug delivery systems, imaging and diagnostic tools, cancer treatments, neuronal prosthesis, and tissue engineering (Maiti et al., 2019). Single-walled carbon nanotubes have been employed for the detection of adsorbed cytochrome c enzymes, which act as key enzymes in the mitochondrial respiratory chain (Ma et al., 2012). In another study, Carbon nanotubes have also been modified with biotin for the detection of streptavidin to biotin binding (Osojic and Hersam, 2012). A very high sensitivity resulted for electrical DNA detection by covalently coupled carbon nanotubes with alkaline phosphate. Modifications can also be made by applying the specific recognition of antigen by the antibodies for developing highly specific assays for antigen detection (Zhu and Lee, 2017). Biosensors for glucose detection in blood were developed by investigating the voltammetric changes that occurred in nanotubes adsorbed with glucose oxidase. The electrochemical properties of biomolecules, like dopamine, reduced NADH, and hydrogen peroxide, were also curated by employing the interaction of nanotubes with biomolecules (Sireesha et al., 2018). These nanotubes have also been used to develop drug, genetic material, and biomolecule delivery systems for targeted delivery. Carbon nanotubes possess the ability to cross cytoplasmic as well as nuclear membranes which offers them the capability to deliver a drug inside the cells with an insignificant reduction in the drug's concentration when compared to conventional drug therapy. They are widely investigated for cancer treatment. Their high surface-to-volume ratio is highly advantageous in this respect (He et al., 2013). DNA probes bound to these nanotubes have also been investigated as being more stable and protected from cleavage enzymes and DNA binding proteins, thus proving them to be highly advantageous for gene therapy (Wu et al., 2008). Pantarotto et al. (2004) developed single-walled nanotubes and DNA complexes and found an enhanced DNA expression than that of naked DNA. Carbon nanotubes have the ability to get biologically functionalized, resist biological degradation, and maintain a high biocompatibility, and this reveals their potential in tissue engineering and regeneration. They also convey a high mechanical strength and conductivity properties to the tissue scaffoldings (Pantarotto et al., 2004). MacDonald et al. in 2005 used single-walled carboxylated nanotubes combined with collagen to develop nanomaterial to be used for tissue engineering (MacDonald et al., 2005). Hence, all of the above findings convey highly efficient and effective applications of carbon nanotubes in biology.

15.2.2 Metallic Nanoparticles

Pure metals and their compounds at the nanoscale result in metallic nanoparticles. Metals like copper, gold, silver, titanium, platinum iron, thallium, and cerium are commonly used for the production of nanoparticles. Metallic hydroxides, oxides, phosphates, sulphates, chlorides, and fluorides have also been used for the preparation of nanoparticles to be used in a variety of fields. In the field of biology, the most prevalent application of metallic nanoparticles is their use as antibacterial agents. Copper and iron nanoparticles are well known for their antibacterial properties and have been investigated against many bacterial strains (Gold et al., 2018; Gahlawata and Chaudhary, 2019). Satyvaldiev et al. in 2018 prepared copper nanoparticles in the presence of gelatine, with a size distribution smaller than that synthesized in the absence of gelatine, and found an enhanced antibacterial effect on *Erwinia amylovora* (Satyvaldiev et al., 2018). Ansari et al. (2017) in their study reported antibacterial activities of iron oxide nanoparticles against *Klebsiella pneumonia* and *Bacillus cereus* using disc diffusion and agar dilution methods. Super-paramagnetic iron oxide nanoparticles have been extensively used in medicine after the functionalization of the surface with therapeutic agents. By applying the external magnetic field, these functionalized nanoparticles can be directed to specific sites of infection, which reduces side effects and enhances the efficiency of the therapeutics. This kind of therapeutic will also be beneficial for the reduction of resistant strains and biofilm formation and will increase the chances of diagnosis and treatment of parasitic diseases (Ansari et al., 2017). Another dominant application of iron nanoparticles is in water remediation, thus proving the effectiveness of iron nanoparticles in reducing water pollution and water treatment.

15.2.3 Lipid Nanoparticles

Solid lipid nanoparticles are monolayer lipid particles enclosing a lipid hydrophobic core at the nanoscale. They are prepared from lipids, emulsifiers, and water. Another type of lipid nanoparticles, called nanoliposomes, are bilayer nanoparticles with polar groups on both outer and inner core surfaces. This class of nanoparticles is highly favourable for drug delivery and gene therapy due to the nanoparticles' exceptional biodegradability and biocompatibility along with small size, high aspect ratio, enhanced drug loading, and phase interactions with interfaces. They improve the bioavailability, solubility, stability as well as specificity of bioactive agents encapsulated inside them (Puri et al., 2009). These nanoparticles have been extensively used for diagnosis, drug delivery, especially liposome-based chemotherapy, and gene therapy. They help to overcome the conventional challenges of gene delivery, like passing biological barriers for transfection across the cell, nuclease degradation, internalization by cells and intracellular trafficking, and low selective targeting. Therefore, they play an efficient role in gene transfection by acting as non-viral based vectors and provide effectiveness, safety, and high specificity in various gene-related or unrelated diseases (Battaglia and Ugazio, 2019; Garibay et al., 2015).

15.2.4 Semiconductor Nanoparticles

Semiconductor nanocrystals or quantum dots find wide applications in the preparation of semiconductor-based electronics systems. Along with this, their contribution to biology is also highly significant. Their exceptional electronic and optical properties of narrow and broad spectral ranges of emissions and excitations, respectively, long lifetime of fluorescence, tunability of emission peak, null photobleaching, and biological properties of high conjugation with proteins enhance their effective use in applications related to bioimaging (Hu and Zhu, 2015; Martynenko et al., 2017). They possess higher photostability and quantum yields compared to that of organic dyes. All these properties prove them to be a highly beneficial tool for SPT (single molecule targeting), HTS (high-throughput screening), FRET (fluorescence resonance energy transfer), and fluorescence multiplexing (Barroso, 2011). Quantum dots with a core made up of semiconductors like Se and Cd that are covered by a ZnS shell show enhanced optical properties. Thus, manipulation of core-shell quantum dots can be performed to produce dots of enhanced colloidal stability and solubility in aqueous solutions. These nanostructures can also be employed to study endocytic pathways mediated by receptors and their regulation in tissues and cells (Yan et al., 2019). Danglot et al. (2010) in their study of exocytosis pathway mediated by v-SNARE TI-VAMP used quantum dots and revealed the effective regulations which occurred due to the depletion of TI-VAMP (Danglot et al., 2010). Hanaki et al. (2003) reported long term endosome labelling with few negative effects using quantum dots

(Hanaki et al., 2003). Ranjan et al. (2008) at the nanoscale studied and visualized endocytic NGF receptor behaviour in real time employing NGF conjugated quantum dots (Ranjan et al., 2008). Quantum dots conjugated with peptides or toxins like shiga, Tfn, and ricin were found to be endocytosed in a clathrin-dependent manner, which shows their great potential for drug discovery (Tekle et al., 2008).

15.2.5 Polymer Nanoparticles

Macromolecules composed of repeating units in large numbers arranged in a chain-like fashion displaying a multiplicity of structures, compositions, and properties are known as polymers. Engineering of such substances at the nanoscale leads to the formation of polymer nanoparticles. A wide multiplicity and variety of properties encourage the use of these nanoparticles in various applications related to the biomedical field. The major applications of these nanoparticles are in the development of drug delivery systems, biosensors, and bioimaging techniques (Moreno-Vega et al., 2012). A huge amount of research has been conducted for the development of targeted drug delivery systems that are highly specific and efficient along with low or null toxic effects. These nanoparticles can have different structures, such as capsules, highly branched macromolecules, and core or shell amphiphilic micelles. Polymers present in nature, such as albumin, chitosan and heparin and synthetic polymers like polyethylene glycol, copolymer HPMA, polylactic acid, polylactic acid-glycolic acid and N-(2-hydroxypropyl)-methacrylamide, have been extensively used in delivery systems for delivering DNA, protein, oligonucleotide, and drugs (Nguyen and Alsberg, 2014). A nanoconjugate of albumin and paclitaxel showed significantly better results in a phase III clinical trial on metastatic cancer of the breast (Gradishar et al., 2005). The exceptional advantage with polymeric nanoparticles is their large capacities for drug and cell or tissue-specific protein loading, which enhances the specificity of the system. The ternary system made up of antibodies or ligands, polymers, and drugs have been developed by Cho et al. (2008) for the treatment of cancer (Cho et al., 2008). Many different therapies have focused on polymeric nanoparticles for the treatment of cancer due to their high specificity and low damage to healthy tissues compared to the conventional administration of the drug (Navya et al., 2019). Dendrimers, another type of polymeric nanometer range macromolecules made up of synthetic polymers, have monomers that are highly branched and emerge from the core radially. The major advantages associated with dendrimers are their high size control and surface modification, multivalency, controllable functionality, solubility in water, and availability of cavities in the internal core for the drug. Therefore, they are extensively used in drug delivery systems as well as bioimaging systems due to the advantage of a high drug or probe molecule holding capacity as compared to free administrations (Madaan et al., 2014; Trzepiński and Klajnert-Maculewicz, 2017).

15.2.6 Ceramic Nanoparticles

Ceramics like zirconia, hydroxyapatite, silica, alumina and titanium oxide at the nanoscale have enhanced physical and chemical properties leading to a reduction in cytotoxicity with *in vivo* systems. Drug release control, high loading capacity, the stability of different administration routes, and ease of incorporation towards both hydrophilic and hydrophobic systems makes them a good choice for drug and diagnostic agent delivery into biological systems (Moreno-Vega et al., 2012). These nanoparticles have also found applications in therapies employing photodynamics; for example, titanium dioxide has exceptional photooxidation properties, which promote its use as a therapeutic agent in new therapies for colon cancer. The cells of colon cancer, when exposed to nanoformulations of titanium dioxide, were destroyed when the nanoparticles were excited by specific light wavelengths (Behnam et al., 2018). Researchers reported the activation of titanium dioxide nanoparticles by ultrasonic waves in malignant gliomas, thus maintaining the foundation of sonodynamic therapy for gliomas (Yamaguchi et al., 2010). Their use is not limited to this extent, but hydroxyapatite nanocrystals have found great application in orthopaedic therapy, as these nanocrystals show good interactions with bone. Repair of bone defects and remineralization of teeth showed excellent outcomes when hydroxyapatite nanoparticles were used; this was due to their high biocompatibility, bioresorbability, bioactivity, osteoconductivity, morphology, size dimension, and functionalization of their surface (Kolmas et al., 2016).

Hence, such a diverse variety of nanoparticles and their varied applications proves the great potential of nanotechnology in the advancement of biology and biotechnology. Agriculture is an essential part of society, and fulfils the fundamental food requirements of the population on earth; thus, improvement in this branch of biology is a major focus of society as low crop production due to plant diseases, reduced fertility of the soil, development of mutants resistant against drought, growth of weeds, etc. leads to reduced crop productivity, and thus increasing the costs of production and needs of society (Preety, 2008). Although nanotechnology has contributed a lot in the fields of medicine, food, and cosmetics, its significant potential in agriculture is not low. Thus, the upcoming sections will emphasize the tools and techniques of nanotechnology developed for improvements in agriculture.

15.3 NANOTECHNOLOGY IN AGRICULTURE

The growing food insecurity due to population increases put forward an urgent need to improve agricultural practices. The major problems of reduced soil fertility and the nutrient uptake potential of crops, environmental stresses like droughts and floods, plant diseases, and uses of conventional monitoring systems are responsible for a high reduction in annual productivity. The application of synthetic fertilizers and pesticides to improve agricultural yield, in turn, reduces the soil quality and makes it barren and therefore useless for generations. Nanotechnology seems to be highly promising in this regard and has been extensively researched to treat these problems, but field trial studies are still lagging. Potential applications of nanotechnology in agriculture have been developed to improve soil quality, increase productivity, reduce the overgrowth of weeds, the development of drought and disease-resistant plant varieties, the stimulation of growth in plants, the development of smart monitoring systems, and the most significant applications are in the management of plant

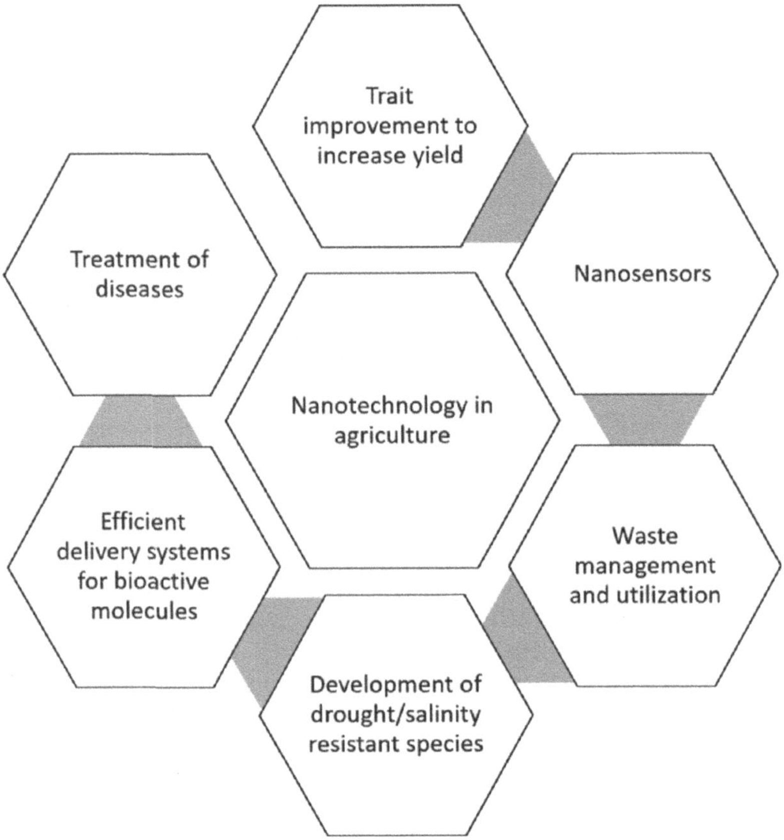

FIGURE 15.2 Applications of nanotechnology in agriculture.

diseases (Figure 15.2) (Singh et al., 2015). Nanoparticles can act as magical bullets to transfer their content, like chemicals, herbicides or genetic materials, into plant tissues in a controlled manner. Nanocapsules enhance effective herbicide penetration into cuticles and tissues, which allow for a constant and slow active compound release (Pérez-de-Luque et al., 2009). Torney et al. (2007) synthesized mesoporous silica nanoparticles with pores of 3 nm with honeycomb-like patterns on their surface to deliver inducing chemicals and DNA into the cells of plants and intact leaves. These nanoparticles were capped with gold nanoparticles to prevent leaching of the molecules. Uncapping these nanoparticles under controlled conditions resulted in chemical release, and the expression of genes was triggered (Torney et al., 2007). CH et al. (2012) prepared nanoemulsions of neem oil and Tween 20 with size means ranging from 31.03–251.43 nm and investigated their larvicidal effect on *Culex quinquefasciatus*; their results showed that the nanoemulsion of 31.03 nm was found to be the most effective in this respect (CH et al., 2012). Milani et al. (2012) in their dissolution kinetic study of Zn fertilizers revealed that the solubility of fertilizer monoammonium phosphate was increased when coated with ZnO nanoparticles, hence showing promising results for improvement of the effectiveness of fertilizers in agronomics (Milani et al., 2012). Pesticides released by farming are a major cause of environmental pollution in respect to agriculture. Treatment of these pesticides becomes important as they pose deadly threats to aquatic life.

McMurray et al. (2006) demonstrated atrazine degradation using a photocatalytic method by using a nanoparticulate film of titanium dioxide (McMurray et al., 2006). The application of nanotechnology to the development of biosensors is at its height in the present scenario. Vamvakaki and Chaniotakis. (2007) in their study used liposome for the development of nano-biosensors for pesticide detection. They detected paraoxon and dichlorvos. They internalized acetylcholinesterase enzymes in liposomes along with immobilization of the fluorescence producing dye pyranine and detected an increase in the pesticide amount through detecting low fluorescence due to a decrease in enzyme activity in the presence of the pesticide (Vamvakaki and Chaniotakis, 2007). Alemdar and Sain in 2008 isolated and characterized nanofibres from agricultural residues and investigated their potential to be used in biocomposite applications (Alemdar and Sain, 2008). Nanoformulations can be employed to immobilize nutrients and mediate the systemic controlled release of these nutrients in the soil. These nanosystems can reduce the problems of nutrient leaching and eutrophication and promote their uptake by plants. Nanohydrogels, nanozeolites, and nanoclays have been investigated for their improved capacities of holding water in the soil and thus can be potentially used for providing slow water release and therefore can reduce the periods of hydric shortage during cropping seasons (Adisa et al., 2019). Polymer, carbon-based, metal, and metal oxide nanoparticles are capable of absorbing contaminants present in the environment to a great extent, and hence they have the potential to be used for soil remediation, thereby reducing soil treatment costs and time (Guerra et al., 2018). Metal and metal oxide nanoparticles (like silicon dioxide, titanium dioxide, gold, and zinc oxide) and carbon-based nanoparticles, especially carbon nanotubes, have a great ability to enrich plant development as they enhance nutrient and elemental uptake in a manner dependent on their size, composition, charge, concentration, and chemical and physical properties (Rastogi et al., 2017). Carbon nanotubes have great potential to prevent plant diseases along with their growth-promoting and non-toxic nature as they are capable of injecting substances inside the seed during germination (Aslani et al., 2014). Some nanoformulations like nanofertilizers have been commercialized, such as Nano-Gro, Nano Green, Biozar-Nano Fertilizer, Nano-Ag answer, etc., that contain plant hormones, immunity regulators, antioxidants, electrolytes, and macro- and micronutrients to improve plant growth (Worrall et al., 2018). Nanopesticides composed of the nanoparticles of chitosan, alginate, silica, PGA, PAA, carboxymethyl chitosan, xyloglucan, and wheat gluten, etc., containing pesticides like atrazine, paraquat, azadirachtin, ethofumesate, tropicamide, etc. have been widely investigated (Prasad et al., 2017). Therefore, nanotechnology possess great potential and scope for the development of techniques to improve agriculture. Also, the major contributions of nanotechnology concerns to management of plant diseases to reduce agricultural losses, which will be enumerated in the following sections.

15.4 PLANT PATHOLOGIES AND CONVENTIONAL TREATMENTS

Microbes are well-known contributors to soil fertility by fixing nitrogen, breaking down complex nutrients in the soil into simpler ones, preparing humus, etc., but at the same time many species are highly infectious and can lead to highly adverse effects on plants, thus reducing yield. In agriculture, loss of crops due to bacterial, fungal, viral, and parasitic diseases is the most common and major problem. Knowledge of the pathologies behind these infections is of great importance. This section deals with major plant pathologies which lead to great losses on a global level.

Tomato, the most prevalent vegetable crop in the USA, covers around 99,000 acres of land. *Xanthomonas perforans* causes bacterial spot disease producing sun-scaled fruits and defoliation in tomatoes, and leads to a 10 to 50% reduction in yield under suitable disease occurrence conditions. Such a huge reduction results in great monetary losses along with increasing the condition of food insecurity. To tackle such diseased conditions, many cultural practices have been used along with the development of disease-resistant varieties, but their outcomes were found to be insignificant compared to the losses. Antibiotic treatment with streptomycin became successful in the 1950s, but soon streptomycin-resistant bacterial strains developed, thus making the treatment ineffective. After this, copper bactericides were introduced, which in combination with ethylene bisdithiocarbamate fungicides showed improved results due to an enhanced availability of free copper ions, but the widespread application of this treatment resulted in the development of copper resistance. Then, treatment with inducers to develop resistance against bacteria and bacteriophages came under investigation. ASM (acibenzolar-S-methyl), a systemic acquired resistance inducer which activates the defence system in plants by enhancing the rate of transcription of stress genes, gave better results. The use of bacteriophages also gave promising outcomes compared to copper bactericides, but the efficiency of the treatment posed a challenge due to the limited viability of phage and the specific conditions required for the multiplication of phage. Thus, the tomato industry is facing a great challenge due to bacterial spot disease (Ocsoy et al., 2013; Potnis et al., 2015; Sharma and Bhattarai, 2019).

Colletotrichum, a causative agent of the fungal disease anthracnose, poses a significant effect on the agricultural economy by invading a large variety of plants, like strawberry, perilla, bean, and some crop plants. Agrochemicals have been employed for a long time to treat this disease, but soon cases of resistant fungal strains came to light. Also, these agrochemicals are lethal to other microorganisms, which are beneficial for increasing soil fertility, and they can also invade the food chain resulting in accumulation inside the human body, causing various severe problems (Than et al., 2008).

Fungal disease powdery mildew, a foliar disease, is highly prevalent in cucurbits and is caused by two species of fungus *Sphaerotheca fusca* or *Golovinomyces cichoracearum*. These fungal species are obligatory ectoparasites which have a biotrophic nature. They result in inducing the same symptoms and can be identified under a microscope. These pathogens can grow and infect in a wide variety of environments, and hence huge vegetable crop varieties are victims of this disease, such as beans, carrot, beets, cucumbers, artichokes, lettuce, peas, melons, eggplant, pumpkins, radicchio, peppers, squash, turnips, radish, and tomatoes. However, conditions of temperature, light, and humidity are major players. The temperature of 75 to 85° Fahrenheit and relative humidity of 80 to 95% with no rainfall highly encourage disease development. Adverse infections can lead up to a 40% yield reduction in cucumber fruit, thereby leading to a great threat to food security and the economy. Leaf invasion by this disease pathogen reduces respiration and photosynthesis, resulting in reduced fruit yield, insufficient ripening, and bad flavour. This disease produces a powdery talcum-like growth of white colour on the leaves, stems, and petioles. This results in the early senescence of the leaves and thus plants. This disease is a major contributor to yield and quality reduction among different plant varieties. Development of genetic resistance in plants against the disease seems promising for controlling the disease, and hence resistant varieties are under development (Micali et al., 2008; Lamsal et al., 2011a; Hong et al., 2018; Liang et al., 2018).

Catastrophic bacterial wilt caused by *Ralstonia solanacearum* is a common plant disease in temperate regions like North America and Europe. *R. solanacearum* is a non-sporing soil-borne bacteria which conquers a variety of plant species, including tomatoes, potatoes, groundnuts, bananas, eggplant, olives, ginger and tobacco. This pathogen is stimulated by specific stimuli and proceeds in the direction of plant roots to attached to them and thus form a cluster on the vessels of xylem and block the vasculature by secreting polysaccharides and cell wall decomposing enzymes in the extracellular environment inside the tissue finally resulting in plant death. The various conventional techniques to treat this disease are by using chemicals, biocontrols, and breeding with resistant varieties. Although breeding is the most effective, it requires skilled labour. Inappropriate application of these control measures can lead to immense threat and risk to the environment. Thus, the development of new control techniques is highly demanded (Sarkar and Chaudhari, 2016; Phukan et al., 2019).

Downy mildew, a common disease of pearl millet, is caused by *Sclerospora graminicola*, a biotrophic oomycete, and it is highly devastating due to the resulting widespread economic and yield losses. *S. graminicola* has exceptional capabilities to adapt with new resistant hybrids, thus it can break the resistance of the host leading to the ineffectiveness of the treatment developed by incorporating resistance in the plant species. Chemical treatments pose a threat to humans and the environment; also use of chemicals for a poor crop is not economically feasible. Therefore, safe and effective treatments are in demand for innate immunity enhancements and the development of highly potent resistant varieties (Zoclanclounon et al., 2019; Jogaiah et al., 2007).

Citrus canker caused by the *Xanthomonas citri* bacterium in citrus plants leads to great yield losses. Grapefruits and oranges that are usually processed for fruit juice production are invaded by the disease and this results in the development of necrotic lesions on the stems, leaves, and fruits, thereby inducing symptoms like the premature dropping of fruit, defoliation, twig dieback, and hence decline of the tree. This disease becomes highly challenging if cropping is done in subtropical and tropical areas. Copper bactericides used for treatment becomes ineffective due to rains blown by the wind that directs bacteria to the stomata and bypasses the copper film made for protection on the surface of plants. Also, extended usage of copper causes toxicity in fruit and soil, posing severe health and environmental threats. Hence need to develop an effective treatment against the disease is increasing (Graham et al., 2004; Ference et al., 2018).

Tomato wilt virus invades thousands of plant species, especially tomatoes, bell peppers and potatoes. The common symptoms of this disease consist of ringspots, yellowing, line patterns, necrotic spots on the leaf, mosaics, dieback of the tip, mild mottling, and streaks on the stems and petioles of a black colour. To treat the disease-resistant strains, using the Tsw gene in pepper was developed, but soon the virus procured success in overcoming this treatment. Thus, an effective treatment is still needed in this scenario (Crescenzi et al., 2013).

Spot blotch disease is a common disease of the wheat plant, and its causative agent is *Bipolaris sorokiniana*. This occurs mainly in humid, warm regions favourable for wheat growth and in the Gangatic plains in South Asia. Numerous sources of resistance have been characterized and employed in breeding practices, but improvement in techniques is still required for field studies.

Hence, plant disease-causing pathogens and pests result in a vast reduction in production yield with 20 to 40% losses annually at a global scale. Current treatments for the management of pests rely on the use of fungicides, herbicides, and insecticides. Although these treatments are advantageous due to their fast action, reliability, and improved availability, the harmful outcomes of the pesticides on undesired organisms, the revival of pests, and resistance development in the pests against pesticides reduce the effectiveness and efficiency of the treatments. Also, the majority of the pesticide is lost after or during the treatment leading to economic loss. Thus, new techniques with high effectiveness and efficiency along with low side effects and economic losses are highly needed (Gupta et al., 2018; Kumar and Rai, 2019). Nanotechnological approaches for plant disease treatments have been extensively researched and seem to be highly promising for curing adverse conditions generated as by plant diseases, including those mentioned above.

15.5 NANOTECHNOLOGICAL APPROACHES IN PLANT PATHOLOGY

Nanoparticles, with their nanoscale size and ease of production, can be effectively used in the treatment of plant pathologies and their management (Figure 15.3). Such

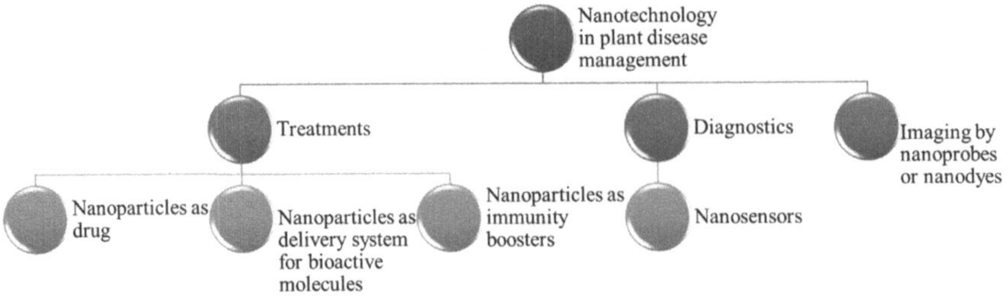

FIGURE 15.3 Role of nanotechnology in plant disease management.

an approach does not only focus on killing the causative agent, but in turn it increases plant immunity by providing nutrients, hormones, and genes of resistance, and thus prevents diseases. Many pieces of research have revealed that the nanoscale particles developed form the bulk, in themselves, possess properties of being antibacterial, antifungal, and antiviral. Thus, these properties divide the nanotechnology treatments into two categories, in which nanoparticles themselves act as either protectants or carriers of protectants.

15.5.1 Nanoparticles as Protectant against Plant Diseases

Many studies have been conducted which focus on the antibacterial, antifungal, and antiviral properties of nanoparticles. Such nanoparticles can play an important role in the treatment of various plant pathologies and can be highly effective at low cost. Panacek et al. (2006) produced colloid silver nanoparticles using a saccharide reduction of the silver complex. Out of the four saccharides used, maltose showed the narrowest distribution of size with a 25 nm average with high anti-bactericidal activity for Gram-negative and Gram-positive bacteria when employed even at very low concentrations of 1.69 μg/ml of silver. These were also found to be effective against methicillin-resistant strains. Lamsal et al. (2011a) in their study on powdery mildew disease of cucurbits, evaluated the antifungal properties of silver nanoparticles in their *in vivo* and *in vitro* experiments in different conditions of cultivation. Their results showed that a concentration of 100 ppm before and after the outbreak of the disease showed a comparatively high inhibition in pumpkins and cucumbers. This inhibition was expressed when fungal hyphae growth and germination of conidia was found to be retarded or stopped in *in vivo* trials. These results were also confirmed by SEM analysis.

Lamsal et al. (2011b) investigated the antifungal activity of silver nanoparticles against pepper anthracnose disease whose causative agent belongs to the *Colletotrichum* species. They applied silver nanoparticles at different concentrations at the lab and field scales and realised that a concentration of 100 ppm produced enhanced inhibition against both fungal hyphae growth and germination of conidia in comparison to the control. Their analysis through SEM showed that the silver nanoparticles had a lethal effect against *Colletotrichum* mycelial growth. Paret et al. (2013) studied the antibacterial effect on *Xanthomonas perforans* (causative agent of tomato spot disease of bacteria) of photocatalytic titanium dioxide, titanium dioxide doped with silver, and titanium dioxide doped with zinc. *In vitro* analysis of the correlation between the antibacterial properties and dose or photocatalytic activity was performed using nanoformulation coated coverslips of glass. The known strain of *X. perforans* was used as an inoculum for these glass slides, and the slides were illuminated by light ranging in the visible region of the spectrum. The results showed that titanium dioxide doped with silver and zinc were more photocatalytic for the organism concerned. Greenhouse experiments in turn displayed that infected plants, when treated with titanium dioxide doped zinc nanoparticles at a nearly 500 ppm to 800 ppm concentration, reduced the infection in a highly significant manner as compared to the control.

Ocsoy et al. (2013) also studied bacterial spot disease caused by *X. perforans*. They worked to enhance the specificity of the treatment and developed DNA directed nanocomposites of silver over graphene oxide. These composites showed high potential against *X. perforans* cells in *in vitro* cultures as well as on plants. They concluded that a low concentration of 16 ppm was enough to display a high antibacterial effect in culture experiments along with improved adsorption and stability. Greenhouse experiments also showed a significant decrease in infection compared to the standard current grower treatments.

Mishra et al. (2014) reported a silver nanoparticles synthesis in an extracellular medium collected from a culture of *Serratia* sp. BHU S4, an important microbe in respect to agriculture, and evaluated the effect of these nanoparticles on wheat infected with spot blotch disease. Characterization of these nanoparticles was done with many techniques such as XRD, microscopies (atomic and electron), spectroscopies (FTIR and energy dispersive X-ray) and thermogravimetric analyzer. Their evaluation proved that concentrations of 2, 4, and 10 μg/ml were effective to convey inhibition of the conidia germination completely. Greenhouse assays demonstrated a significant reduction of *B. sorokiniana* (the causative agent of spot blotch disease) infestation in wheat plants. Histochemical staining demonstrated the active role of these nanoparticles for the induction of the deposition of lignin in vascular bundles of wheat plant.

Elbeshehy et al. (2015) in their study produced silver nanoparticles from the supernatant of isolated strains of bacteria. Out of them, three *Bacillus* strain cultures (*B. pumilus*, *B. persicus*, and *B. licheniformis*) were found to be highly effective for nanoparticle production. The antimicrobial and antiviral effects of these biosynthesized nanoparticles were also studied, and results proved that nanoparticles produced from *B. licheniformis* were more stable and possessed considerable activity against the yellow mosaic virus infection of beans in fava beans in Africa.

Ying et al. (2016) synthesized chitosan Schiff based silver nanoparticles and chitosan silver nanoparticles and studied their effects against the tobacco mosaic virus. For evaluating inhibition, resistance index was determined, and experiments on the lesions were performed. They considered two species of plants *Nicotiana tabacum* and *Xanthi-NN*, and performed the experiments in which they found that 30 times diluted Shiff based silver nanoparticles provided better treatment against the disease due to a significantly high lesion reduction. Also, the activities of peroxidase, superoxide dismutase, catalase, and phenylalanine ammonia lyase were found to be promoted, resulting in an enhanced soluble protein content and chlorophyll along with a decrease in the content of malonaldehyde. Chen et al. (2016) evaluated the silver nanoparticle's antibacterial effect, after surfactant stabilization, on *Ralstonia solanacearum* compared to silver ions. A superior antibacterial effect was revealed for Tween 80 stabilized silver nanoparticles against the organism. Measurements with TEM analysis, SDS PAGE, and fluorescence microscopy proved that Tween 80 stabilized particles presented highly adverse effects when they directly came into contact with the cell by damaging the membrane of the cell along with a modification and destruction of proteins. Their results dictated 84.21, 90.11, and 96.71% of the efficiency of the control in 21, 14, 7 days of treatment, respectively.

Graham et al. (2016) studied the antimicrobial activity of formulations prepared from two different zinc oxide nanomaterials (Zinkicide SG6 particulate and Zinkicide SG4 plate-like) against *Xanthomonas citri*, which causes the highly devastating disease citrus canker in citrus plants. In *in vitro* trials, a 2 times lower concentration of Zinkicide SG4 was found to show similar inhibition effects against *X. alfalfae* and *E. coli* compared to that of copper hydroxide, copper sulphate, or zinc oxide/cuprous oxide. The minimum concentration for inhibition in the case of Zinkicide SG6 was found to be 7 times lower for *E. coli* and *X. alfalfae* in comparison to the above-mentioned compounds. A greenhouse assay was performed on leaves of *Citrus paradise* and *Citrus sinensis* by injecting them with *X. citri* and then spraying them with Zinkicide, and this reduced the lesions of citrus canker significantly. Field experiments in Florida in 2015 and 2014 proved a high potential for Zinkicide, rather than zinc oxide and cuprous oxide, for bactericidal effects by displaying a significantly high reduction of grapefruit canker. These formulations of Zinkicide also portrayed profound antifungal effects against melanose and citrus scab on grapefruit. Lastly, the most favourable quality of these Zinkicide formulations is that they show null phytotoxic effects.

El-shazly et al. (2017) prepared silver nanoparticles of 12 nm size from a solution of 1 mM $AgNO_3$ by using a chemical reduction method and characterized them with UV-visible spectroscopy, XRD, and TEM analysis. They then studied the antiviral effect of the above nanoparticle preparation on potato plants infected by potato virus Y. In their study, they found that the most significant trials of treatments were those in which nanoparticles of silver sprayed after 24 hours of infection, and nanoparticles alone and nanoparticles with salicylic acid before 3–7 days of infection, both at 0.1 µg/ml concentration. These treatments resulted in a significant reduction, leaving virus infection percentages of 10, 30, and 10%, respectively, which was much better than the infection percentages left after salicylic acid treatment only. A significant increase was also observed after these treatments in the height of the plant, diameter of the tuber, number of tubers harvested, length of the tuber, total sugars (soluble), and total yield of tuber as compared to healthy ones.

Cai et al. (2018) demonstrated the antibacterial activities of magnesium nanoparticles. Though magnesium in a bulk state is an essential mineral for plants and shows no toxicity against the organism, at the nanoscale particles of magnesium oxide showed significant activities against *Ralstonia solanacearum,* which causes bacterial wilt and leads to a global loss of a large variety of crops, especially tobacco. Their results proved that the activities of magnesium nanoparticles against the organism is concentration-dependent, and the minimum concentration required for inhibition is 200 µg/ml and for bactericidal effect it is 250 µg/ml. Further analysis by SEM and TEM revealed cell membrane injury, reduction in the formation of biofilm due to the attachment of nanoparticles on the cell surface, and reduced mobility were the major reasons for the inhibitory effect. Along with this, damage of the DNA due to excessive ROS accumulation can be another reason for such inhibition effects.

Siddaiah et al. (2018) in their work revealed high acetylation efficiency of chitosan nanoparticles to treat the most devastating disease of downy mildew caused by *Sclerospora graminicola* in pearl millet. Nanoparticles were prepared from chitosan, with small molecular weights. Experiments on seeds enhanced the vigour of the seedlings and the percentage of germination in comparison to the control. Treatment of seeds with chitosan nanoparticles resulted in a durable and systematic resistance and exposed a protective effect against mildew under greenhouse conditions when compared to the control. This protective effect was generated due to the upregulation of peroxidase, catalase, phenylalanine ammonia lyase, superoxide dismutase, and polyphenol oxidase. Also, the proteins related to pathogenesis, like PR1 and PR5, were found to be over-expressed. A nitric oxide presence in chitosan nanoparticle treatment modulated the protective effect against the disease, and it was displayed that nitric oxide inhibitors totally removed the effect by decreasing the defence enzymes and pathogenesis protein (PR) expression significantly. Along with this, the comparison of a chitosan nanoparticle treatment with chitosan resulted in a low dosage requirement of nanoparticles as compared to chitosan for the management of downy mildew.

Shafie et al. (2018) in their study on the tomato spotted wilt virus from infected potato plants, demonstrated the antiviral effects of silver nanoparticles. They found that application of different concentrations of nanoparticles on potato leaves for 24 hours resulted in improvement of the spongy tissue, vascular bundle length, and thickness of lamina compared to the control. Five concentrations of nanoparticles were analyzed for their antiviral effects, and all the concentrations were found to be inhibitory and reduced the number of local lesions. Highest inhibition of 90.4% was found on treatment 24 hours post-infection, but this inhibition was less in plants on which nanoparticle treatment was given before infection.

Numerous pieces of research have been done to investigate the potential of nanoparticles as drugs, and many are in progress. Thus, the role of nanotechnology is incomparable in plant disease treatment and can be a better option than the conventional treatments in respect to effectiveness, efficiency, cost of treatments, and safety issues related to the environment and the plants themselves. Further, many pieces of research have been conducted to explore the effects of nanoparticles as immunity boosters that can prevent the incidences of disease.

15.5.2 Nanoparticles as Immunity Boosters for Plants

The previous section demonstrated the potential of nanoparticles with different compositions to act as drugs, but their capabilities are not limited to treating diseases by killing microbes. They also possess a high efficiency to prevent or cure diseases by enhancing the activities of the plant's immune system. Although there is less research done in this respect, there is the opportunity for further investigation. Chandra et al. (2015) investigated the immunomodulatory effects of chitosan nanoparticles to enhance the immunity of plants against diseases. Treatment of these nanoparticles on leaves induced the enzymes responsible for various defence mechanisms and also upregulated the genes responsible for their synthesis. Many antioxidant enzymes, along with the increment of phenolic content, were observed after the treatment. They also noticed increased levels of nitric oxide, which plays a major role in signalling the pathways of defence. Hence, chitosan nanoparticles can be more effectively employed for disease control in plants compared to chitosan in its bulk form.

Chaudhary et al. (2017) evaluated the efficiency of copper–chitosan nanoparticles as immunity boosters against maize disease curvularia leaf spot and as a growth promoter. Plants treated with the respective nanoparticles were found to have high defence (phenylalanine ammonia lyase and polyphenol oxidase) and antioxidant enzymes (peroxidase and superoxide dismutase). The results showed that control of disease was significant in maize at 0.04 to 0.16% in pot treatments, and in the field the concentration of the nanoparticles was around 0.12 to 0.16%. Copper–chitosan nanoparticles also promoted the growth of plants in their height, the diameter of the stem, the length of roots, the number of roots and the content of chlorophyll. Such effects were also observed in field experiments hence suggesting their sustainable development in agriculture.

Though the research findings of using nanoparticles as immunity boosters are few, this approach still seems to be promising and will soon develop into a potent treatment to prevent and cure disease. Another, highly potent approach but not yet used in a commercial application is the use of nanoparticles in delivery systems for herbicides, pesticides, insecticides, and RNA/DNA molecules. The following section elaborates the details of the past and present research in this area.

15.5.3 Nanoparticles as Carriers of RNA/DNA Molecules, Pesticides, Herbicides, Insecticides, and Fertilizers

Conventional methods of spraying pesticides, insecticides, herbicides, and fertilizers lead to a non-uniform supply to the plants and also to accumulation of the chemicals resulting in toxicity issues. These methods also result in various problems related to environmental pollution, like soil and water pollution. But the incorporation of these compounds is necessary for plants to protect them from diseases cause by pests, insects, unwanted weeds and a deficiency of nutrients. Delivery of RNAi inside the cells is difficult using conventional spraying methods as the uptake of these molecules, and the specificity of the technique, will be extremely low. To deal with these situations, the application of nanoparticles in delivery systems by encapsulating, entrapping, absorbing, and attaching different molecules in/on the nanoparticles has been extensively researched and has gained much importance in agriculture and plant disease management, but these findings have still not reached the level of greenhouse or field trials, and therefore they have not yet achieved commercialization. To improve this scenario, scientists are continuously pushing forward. Many studies have been performed to develop nanoparticle-based delivery systems to deliver pesticides, insecticides, herbicides, or RNA/DNA into infected plants to treat a wide variety of diseases (Figures 15.4 and 15.5). The most crucial aspect of a potential delivery system is the slow, sustained, and specific delivery of the compound being delivered along with low toxicity due to the presence of the carrier. Different types of nanomaterials fulfil some or all these criteria and seem to be highly promising for use in plant disease treatments.

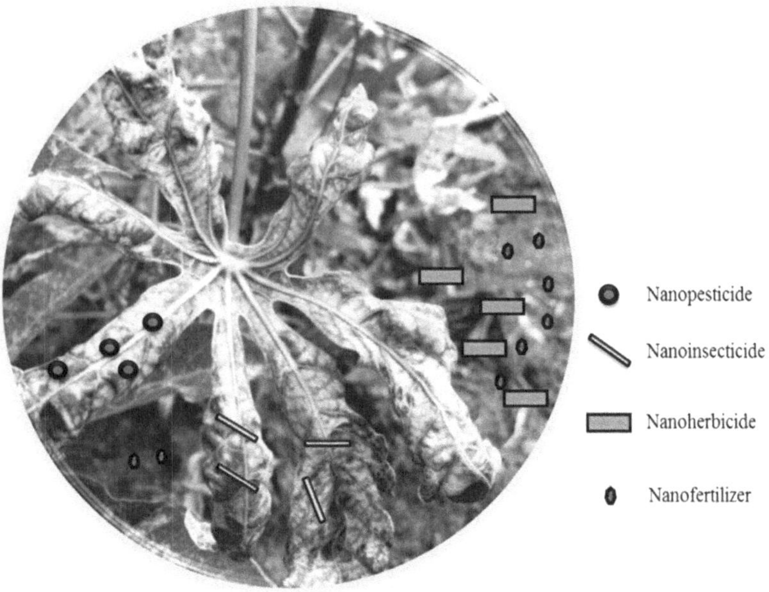

FIGURE 15.4 Application of nanoparticles to deliver different chemicals for plant disease treatment.

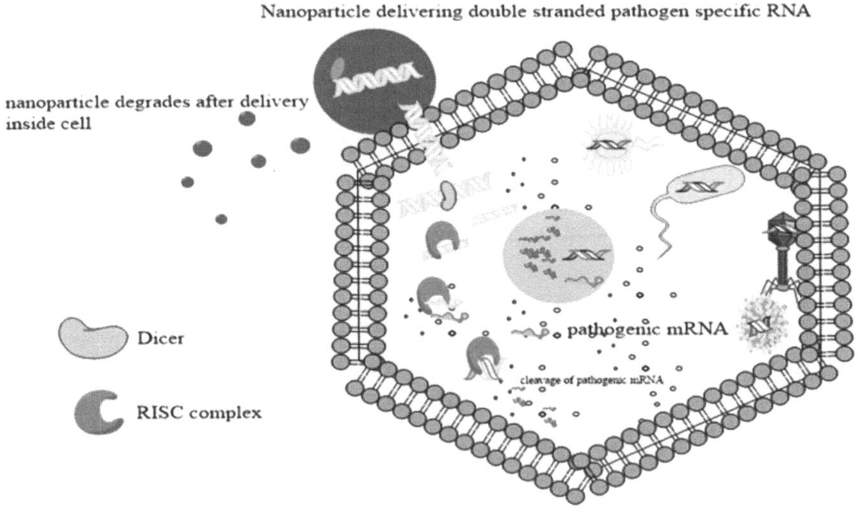

FIGURE 15.5 dsRNA delivery via nanoparticle inside the infected cell.

Li et al. (2007) prepared porous hollow silica nanoparticles to shield pesticides from degradation by ultraviolet light. They employed avermectin as a model for the pesticides. They concluded that nanoparticles with a 15 nm shell thickness and a pore diameter of 4–5 nm provided a capacity of 625 g/Kg for the used pesticide with exceptionally good UV-shielding. They also reported that UV-light intensity, temperature, and pH of medium were the affecting parameters for the release of pesticide. An increase in these parameters reduced the protection efficiency to a large extent. Avermectin, when enclosed in the inner core of these nanoformulations, showed slow release patterns ranging over a period of 30 days. Thus, they showed a high potential, such as silica nanoparticles.

Chitosan nanoparticles are the most promising among the present nanoformulations to be used as delivery vehicles. Their exceptionally high efficacy and efficiency lies in their ability for systemic activity induction, a large surface area, exceptional ability to form a film, low size, low toxicity, high mobility, and they are economically feasible. They have also enhanced biocompatibility and permeability and are totally biodegradable with low toxicity. Hasaneen et al. (2014) prepared chitosan nanoparticles using a methacrylic acid method and studied them using TEM and UV-visible spectroscopy. They achieved a size range of 17–25 nm with a homogeneous nature of morphology, high uniformity of nanosphere size, and pH sensitivity. NPK (nitrogen, phosphate, and potassium) fertilizers were loaded on these particles. They found that the stability of these nanoparticles was higher when potassium and nitrogen were loaded than when phosphorous was loaded, due to a higher calcium phosphate anion strength than urea and potassium chloride. The increase in the diameter means of these nanoparticles in the suspension after adding NPK compounds was different, and the order of increase was found to be $P > K > N$. This proved aggregation and loading of these compounds on the surface of nanoparticles.

Sarlak et al. (2014) conducted a study on pesticide encapsulation in a citric acid shell conjugated as a poly-citric acid graft on carbon nanotubes (multiwalled) and revealed high pesticide encapsulation and toxicity against fungi *Alternaria alternata*. They polymerized citric acid on the oxidized carbon nanotubes, and due to the presence of citric acid solubility of the material in water was enhanced along with their ability to entrap metal ions and chemicals, which are water soluble. Then pesticides like mancozeb and zineb in an aqueous solution of nanotube hybrid material were encapsulated into the hyper-branched shell of poly-citric acid. They also investigated the optimum conditions for the same to standardize the technique.

Mitter et al. (2017) developed a sustainable RNAi delivery system for crops to protect them from viral infections. They prepared double hydroxide layered nanosheets of clay and loaded double-stranded RNA on them, which resulted in a sustained release of the RNA with no wash off (which is also evident on the leaves after 30 days) or toxicity. These particles were degradable as in their experiments; they noticed uptake of double-stranded RNA and homologous RNA silencing when sprayed on the leaves. A one-time spray of this nanoformulation provided protection from a virus for 20 days when infected. Hence, this application of nanoparticles shows highly promising results.

Tong et al. (2017) in their study developed a pollution-free polymer nanoparticle formulation for delivery of a hydrophobic metolachlor pesticide, which enhanced the solubility of pesticides in water. A Polyethylene glycol and polylactic co glycolic acid nanoparticle formulation was prepared and characterized. This formulation improved the solubility of metolachlor in water, and the presence of an organic surfactant or solvent was absent, which usually act as pollutants. In the study of these nanoparticles (Cy5 labelled), absorption in the roots of rice provided the pathways responsible for transmittance of metolachlor and its enhanced utilization. Further tests of bioassays proved that this formulation was more effective at relatively low concentrations on *Digitaria sanguinalis* of *Oryza sativa*. Cytotoxicity effects on preosteoblast cells proved the low toxicity of the nanoformulation as compared to pesticide alone. Therefore, polymer nanoparticles can be the potential carriers of hydrophobic pesticides with low toxicity and pollution-related issues.

Bao et al. (2016) investigated a highly efficient method to facilitate the delivery of active molecules across the plant cell wall. Delaminated double hydroxide layered nanosheets of lactate were prepared to with a thickness of 0.5 to 2 nm and a diameter of 30 to 60 nm. These nanosheets showed a high capacity of absorption for biomolecules, which are negatively charged like fluorescent dyes and molecules of DNA, thus forming neutral conjugates. These conjugates can transport the dyes inside the plant cell cytosol with high efficiency. Also, they reported that internalization of these nanosheets was not challenged by the presence of endocytosis inhibitors and conditions of low temperature, depicting that they are internalized in an non-endocytic mode. Further, localization of these formulations inside the nucleus revealed their potential to be used as new gene carriers with no side effects.

Wang et al. (2007), using a two-step process, prepared a nanoemulsion and studied the ability of this formulation to dissolve pesticide betacypermethrin which is insoluble in water. They found better results as compared to the commercial microemulsion of this pesticide in terms of stability proving their formulation as an effective one. Guan et al. (2010) in their study to develop and investigate the dosage of a nanopestiside formulation of imidacloprid, which is widely used in China to control pests in vegetable plants, suggested that a dosage of 900 ml/hm is safe to us on fields.

Yang et al. (2009) studied the structure characterization and insecticidal activity of garlic oil loaded nanoparticles coated with PEG against *Tribolium castaneum*. Their study revealed many aspects. They used a melt-dispersion method for the preparation of the nanoparticles, which is simple and cost-effective and has a loading capacity of 80%. The shape and size of the particles produced were analyzed by TEM analysis and light scattering dynamics and they were found to be round and less than 240 nm, respectively. GCMS analysis showed that percentage and abundance content of the component of significance was not significantly different in nanoformulation and free form. The efficacy studies proved that this formulation's efficacy was above 80% even after a period of five months against *T. castaneum* in comparison to 11% of free garlic oil, mainly because of the slow and consistent release rate of the active compound form nanoparticles. Thus, the above-mentioned formulation shows a very high potential.

Shan et al. (2019) developed a nanosystem with the ability to release pesticide according to the stimuli present in the environment of the nanosystem. Their approach to developing such nanoparticles was based on electrostatic interactions instead of hydrophobic and hydrogen bond interactions. They worked to fabricate the mesoporous nanoparticle of silica which have a negative charge with sulfonate groups, by using a method of post grafting. Such fabrication was done by converting epoxy groups to sulfonate by applying bisulfite ion as a reagent for ring-opening. After fabrication, they loaded the well-known ammonium herbicide (quaternary) diquat dibromide into the sulfonated nanoparticles. Their results proved that such fabrication improved the loading capacity by 12.73% in comparison to non-fabricated nanoparticles. The herbicide release was also responsive to conditions of ionic strength and pH due to electrostatic interactions. They studied the activity of this herbicide nanoformulation against *Datura stramonium* and found exceptionally good results.

Liu et al. (2008) prepared polymeric stabilized nanoparticles for efficient and controlled release of a pesticide. They used pesticide bifenthrin to prepare nanoparticles with a process of flash nanoprecipitation. In this process, they employed micromixing rapidly for effective super-saturation and assemblage of the polymers for size control. Their nanoparticle size ranged from 100–200 nm as the loading of pesticide increased from 50–91%. Different polymers were tested as stabilizers, and the hydrophilic polymers hindered the aggregation of particles due to the resulting steric stability. These nanoparticles showed a significant efficiency coverage uniformity for the compound concerned, and the lowest worker exposure.

Olad et al. (2018) developed a usable hydrogel nanocomposite for the delivery of NPK fertilizers. They used the method of graft polymerization of sulfonated CMC in the presence of potassium, nitrogen, phosphorous, silica nanoparticles, and polyvinyl pyrrolidone, with acrylic acid. The incorporation status of NPK fertilizers was detected by XRD, FTIR, SEM analysis. The swelling abilities of these hydrogel nanoformulations in different saline and at different pH along with retention of water were estimated, and the results were in agreement with the Committee of European Normalization standards. This proves they have the properties of slow and consistent release and hence are capable of being used in fields.

Adak et al. (2012) synthesized nanosized micelles aggregated in an aqueous environment from copolymers that were amphiphilic, and encapsulated imidacloprid into them for development of a controlled release system. Higher power pf solubilization and a lower critical concentration of micelle formation of these polymers enhanced the formulation efficiency. They were characterized by light scattering dynamics, spectroscopy (infra-red), and microscopy (TEM). The release kinetics in water were highly efficient which was evident from the diffusion exponent of the pesticide, which ranges from 0.22–0.37 in water. The range of release time for 50% of the pesticide was 2.32–9.31 days. Hence, this formulation can be effectively and efficiently used for management of pests in crops.

de Oliveira et al. (2014) prepared nanoparticles of solid lipids encapsulating simazine and atrazine herbicides and performed their characterization and evaluation *in vitro* for testing herbicide release parameters, cytotoxicity, and activity. Stability testing was done for a duration of 120 days by analysing zeta potential, pH, size, poly dispersion index, and efficiency of encapsulation. They concluded in their results that the nanoparticles had high encapsulation and stability. The release rate was modified in water with an effective herbicidal effect on *Raphanus raphanistrum* (target species). Nonspecific activity was found to be null, and the toxicity of the herbicide was reduced along with dosage requirements compared to crude herbicide applications. Forim et al. (2013) developed nanoparticles for the delivery of biopesticide *Azadirachta indica*. They formulated dried powders and a suspension of colloids and characterized them for release rate, recovery, and efficiency of entrapment. Their efforts resulted in full recovery of an 100% efficiency of entrapment. Along with this, the bioassay revealed 100% mortality of the larva of *Plutella xylostella*. Also, the entrapment in the nanoparticles reduced sensitivity towards UV radiation and enhanced the aqueous phase dispersion.

Guo et al. (2014) in their work to develop a system of controlled release of the herbicide acetochlor, utilized a polybutyl methacrylate diacetone acrylamide formulation. These nanoformulations were in the form of nanocapsules, and their characterization of the distribution of size was performed with field-emission SEM and IR spectroscopy. The efficiency of loading and behaviour of swelling was also evaluated and was found to be 50%. The diffusion parameter (n) ranged from 0.28 to 0.61 indicating release of herbicide was controlled by diffusion. The release time of 50% herbicide was less when at high temperature and the diacetone acrylamide concentration hence showed a quick release at these conditions. Also, they depicted that slow release and improved incorporation of herbicides were connected to interactions of polymers with the herbicides. Application of nanotechnology for the development of carrier or delivery systems is not limited to the above mentioned; many efforts still needs to be studied. Although nanoparticles have been immensely explored for delivery, they have also been used in many other ways to tackle disease conditions. These other applications are discussed in the following section.

15.5.4 Miscellaneous Approaches

Nanodiagnosis is a term employed for the application of nanosized systems for detection of plant diseases and treatment status. The nanopore sequencing technique has advanced the application of nanotechnology for the detection of almost all plant diseases using DNA sequencing. On the spot detection of the presence of pathogenic DNA are also under research. But biosensors for quantification of pesticides, herbicides and insecticides in crops to detect the presence of the required quantity are also important and have been extensively researched, making the application of nanotechnology limitless. These nanosensors help to record the presence of the required chemicals to prevent diseases in plants. Following research in the upcoming section proves the broad range of applications of nanoparticles except for the use of nanomaterials as drug or carriers or immune booster.

Firrao et al. (2005) developed a nano-biotransducer for infection detection in samples from the field. These transducers were nanostructures linked to biomolecules with the capability to produce signals on the occurrence of a biologically relevant molecular event. They reported a probe formed of an oligonucleotide with a 5' end attached to fluorescein and a quencher with a 3' end with 2 nm gold particles. This biotransducer acts as a molecular beacon, thus emitting fluorescence strongly upon hybridisation with a DNA target. This approach is not limited by the constraints related to sequence requirement, as the proximity of the quencher and emitter is because of fluorophore adsorption on gold. This nanosystem was used to identify the amplification using PCR collected from extracted grapevine plant DNA affected by flavescence dorée.

Gong et al. (2012) developed a nanosensor for detection of an organophosphate pesticide, which they made by employing a modified carbon glassy electrode decorated with gold nanoparticles arranged on a nanosheet of graphene. This electrode worked for the extraction of the solid phase. This electrode film promoted the collection of nitroaromatic pesticides (with organophosphate) on its surface and sensed their oxidation using a stripping voltammetric method of detection by applying a model of methyl parathion. The performance of this method was evaluated by a square wave and cyclic voltammetry analysis. This combination of nanoassembly resulted in a simple, fast, and sensitive method based on electrochemistry for the detection of nitroaromatic pesticides with organophosphates. The linearity of detection ranged from 0.001 to 0.1 and 0.2 to 1 μg/ml concentration of methyl parathion, and had a limit of detection of 0.6 ng/ml.

Vishwanathan et al. (2009) developed electrochemistry-based sensors for insecticide detection. Organophosphorous-based chlorpyrifos and methyl parathion insecticides were considered, and a matrix for the immobilization of acetylcholinesterase was prepared. The matrix was of nanorange polyalanine made up of self-assembled carbon nanotubes with a single wall, which were wrapped by gold conjugated oligonucleotide terminated by thiol. This biosensor worked using an enzymatic reaction, which leads to a local pH change in close vicinity to the surface of an electrode. The presence of the pesticide inhibits the reaction and hence leads to detection. The limit of detection was $1 \times 10{-}12$ M, and the dynamic ranges for both considered insecticides showed good stability and reproducibility, thus proving the potential of these types of nano-biosensors.

15.6 DRAWBACKS

Nanotechnology has proved itself a great boon for plant disease treatment and management, but its utilization at the commercial level is still facing great challenges as most of the nanoformulations or nanosystems have one or the other negative effects which need through study for their reduction. Lipid solid nanoparticles are known for their ability to work as efficient carriers for lipophilic components, but the major shortcoming is that their efficiency of loading is low and there is the danger of active ingredient leakage from the nanoparticles (Tamjidi et al., 2013). Carbon-based nanoparticles like fullerene at increased concentrations reduce the biomass of soybean and corn (Mukherjee et al., 2016). Shen et al. (2010) reported that carbon nanotubes lead to oxidative stress, condensation of chromatin, and hence apoptosis in rice and *Arabidopsis* plants. Many studies have reported toxicity issues related to bare silver nanoparticles, which are found to have great potential in disease treatment and management. Silver nanoparticles showed toxic effects on the elongation of root growth of corn and a reduction of seedlings and oxidative stress in radish (Zuverza-Mena et al., 2016). The major shortcoming related to gold nanoparticles is of accumulation of these particles in the roots, resulting in hampered pathways and cell death (Sun et al., 2018). Copper nanoparticles are known for their antimicrobial properties, but researchers have proved that they get accumulated and dysregulate the expression of genes in the roots of cucumbers. They also decrease total biomass on their accumulation in some species and cause high toxicity in wheat, sorghum, and mung bean (Mosa et al., 2018). Other nanoparticles which are metal-based, like ZnS or CdSe or quantum dots, have shown increased oxidative stress with the production of ROS, transfer of nanoparticles in trophic chains, and reduction in the length of roots in some species (Navarro et al., 2011). Nanoparticles of CuO have been reported to cause phytotoxicity due to enhancement of ROS generating enzymes, and also they cause a reduction in root and shoot length in a dose-based manner (Rastogi et al., 2017). Fe_3O_4 nanoparticles, when supplied at high concentrations, are reported to generate spots of brown colour on corn plant leaves (Rui et al., 2016). Polymeric nanoparticles like dendrimers lag behind due to the formation of DNA liposome complexes, low stability, high toxicity, and high rate of clearance (Kretzmann et al., 2017). In turn there is a general threat related to all types of nanoparticles that they will be incorporated into the food chain of humans and animals and can result in high lethality (Exbrayat et al., 2015). Although various types of drawbacks are associated with different types of nanoparticles, research and treatments have shown that these drawbacks can be combated by performing different modifications to different types of nanoparticles.

15.7 PROSPECTS

Increasing food demand, along with an increasing population, indicates the urgent need for sustainable progress in agriculture. The occurrences of disease due to infections and pests leads to great losses and creates the highly adverse conditions at the global as well as local level. Nanotechnology has been extensively researched and applied to support animal and human support systems, but its contribution to plant science is still less than expectations have indicated. Specific applications of nanotechnology in the development of treatments and management of plant diseases has reached the greenhouse levels but it is still facing challenges to reach commercialization. So, advancement in this context is highly demanded. Improvements in effectiveness and reductions in toxic effects of nanomaterials is must, and efficient improvement in this context can lead to more enhanced applications of nanotechnology. Boosting of the immune system can be an exceptionally promising approach for the prevention and cure of plant diseases; hence, development strategies like preparation of new nanoformulations with immune system boosting abilities or optimizing the delivery systems to deliver immune boosting compounds or nucleic acid molecules that can contribute in this context, can lead to great advancement in the progress of treatments for plant diseases. Nano-biosensors are another highly promising approach for detection of infection and the presence of chemical compounds in plant systems, but to date not enough has been performed to commercialize these nanosystems; hence, exhaustive research is highly required. On-site, less time-consuming detection systems can be made by using nanotechnological approaches and can facilitate early and easy detection of disease in plants, resulting in a reduction of losses associated with crop production (Singh et al., 2015). Many other aspects like tendency to reduce antioxidant enzyme activity and production of ROS species require special attention in respect to metal nanoparticles. Further employment of genetic engineering approaches like CRISPER-CAS9 and high-efficiency RNAi, Recombinant DNA technology, and detection of disease-specific molecular biomarkers can be exceptionally supported by nanotechnology (https://innovativegenomics.org/projects/nanotechnologies-plant-genome-editing/) and can bring unexpectedly promising outcomes.

15.8 CONCLUSION

Plant diseases are a major contributor to agricultural losses at a global level, and thus create food insecurity and scarcity, which causes situations of famine. Highly efficient treatments to these diseases can reduce great economic and food losses and are demanded in the present scenario. Nanotechnology is emerging as a boon in a wide variety of scientific fields, especially in drug design and delivery, but its utilization in agricultural science is still not up to a commercial level, though the outcomes of nanotechnological applications in agriculture at a laboratory scale seem to be highly promising. To encourage nanoscience in agriculture with special focus on plant diseases, this chapter elaborated the current status of nanotechnology for the management and development of treatments against these adverse disease demons. Research has proven that the development of different varieties of nanomaterials shows potential to be employed in nanoformulations and nanosensors for treating diseases in plants. Nanoparticles in themselves show exceptional druglike properties, such as antibacterial, antifungal, and antiviral with high biocompatibility and bioavailability, thus they can be directly employed for treatments in plants. Many nanoformulations are capable of effectively transferring active agents inside the plants with very high specificity and hence can result in high-efficiency delivery systems for pesticides, insecticides, and herbicides with little or no wastage of chemicals or the emergence of pollution-related issues. The upcoming collaboration of genetic engineering with nanotechnology has not left behind the fields of agriculture and is being researched for development of highly effective treatments (by employing the concept of RNA interference) and of diagnostic tools for checking the presence of pest-related genes. Nanosensors have also been developed to detect the presence and quantification of different chemicals used for crop protection. In the upcoming years, on-site real-time detection and treatment of plant diseases seem to be becoming a reality. Hence, this chapter provides knowledge of the above mentioned and suggests the great potential of nanotechnological applications in plant disease management.

REFERENCES

Adak T, Kumar J, Shakil NA et al. (2012) Development of controlled release formulations of imidacloprid employing novel nano-ranged amphiphilic polymers. *Journal of Environmental Science and Health, Part B* 47:217–225.

Adisa IO, Pullagurala VLR, Peralta-Videa JR et al. (2019) Recent advances in nano-enabled fertilizers and pesticides: A critical review of mechanisms of action. *Environmental Science: Nano* 6:2002–2030.

Alemdar A, Sain M (2008) Isolation and characterization of nanofibers from agricultural residues: Wheat straw and soy hulls. *Bioresource Technology* 99:1664–1671.

Ansari SA, Oves M, Satar R et al. (2017) Antibacterial activity of iron oxide nanoparticles synthesized by co-precipitation technology against Bacillus cereus and *Klebsiella pneumoniae. Polish Journal of Chemical Technology* 19:110–115.

Aslani F, Bagheri S, Muhd Julkapli N et al. (2014) Effects of engineered nanomaterials on plants growth: An overview. *Scientific World Journal* 2014:1–28.

Bao W, Wang J, Wang Q et al. (2016) Layered double hydroxide nanotransporter for molecule delivery to intact plant cells. *Scientific Reports* 6:1–9.

Barroso MM (2011) Quantum dots in cell biology. *Journal of Histochemistry and Cytochemistry* 59:237–251.

Battaglia L, Ugazio E (2019) Lipid nano- and microparticles: An overview of patent-related research. *Journal of Nanomaterials* 2019:1–22.

Behnam MA, Emami F, Sobhani Z et al. (2018) The application of titanium dioxide (TiO_2) nanoparticles in the photo-thermal therapy of melanoma cancer model. *Iranian Journal of Basic Medical Sciences* 21:1133–1139.

Cai L, Chen J, Liu Z, Wang H et al. (2018) Magnesium oxide nanoparticles: Effective agricultural antibacterial agent against ralstonia solanacearum. *Frontiers in Microbiology* 9:1–19.

CH A, Sharma Y, Mukherjee A et al. (2012) Neem oil (*Azadirachta indica*) nanoemulsion – A potent larvicidal agent against Culex quinquefasciatus. *Pest Management Science* 68:158–163.

Chandra S, Chakraborty N, Dasgupta A et al. (2015) Chitosan nanoparticles: A positive modulator of innate immune responses in plants. *Scientific Reports* 5:1–14.

Chen J, Li S, Luo J et al. (2016) Enhancement of the antibacterial activity of silver nanoparticles against phytopathogenic bacterium *Ralstonia solanacearum* by stabilization. *Journal of Nanomaterials* 2016:1–15

Cho K, Wang X, Nie S (2008) Therapeutic nanoparticles for drug delivery in cancer. *Clinical Cancer Research* 14:1310–1316.

Choudhary RC, Kumaraswamy RV, Kumari S et al. (2017) Cu-chitosan nanoparticle boost defense responses and plant growth in maize (*Zea mays* L.). *Scientific Reports* 7: 1–11.

Crescenzi A, Viggiano A, Fanigliulo A (2013) Resistance breaking tomato spotted wilt virus isolates on resistant pepper varieties in Italy. *Communications in Agricultural and Applied Biological Sciences* 78:609–612.

Danglot L, Chaineau M, Dahan M (2010) Role of TI-VAMP and CD82 in EGFR cell-surface dynamics and signalling. *Journal of Cell Science* 123:723–735.

Elbeshehy EK, Elazzazy AM, Aggelis G (2015) Silver nanoparticles synthesis mediated by new isolates of *Bacillus* spp., nanoparticle characterization and their activity against bean yellow mosaic virus and human pathogens. *Frontiers in Microbiology* 13:1–13.

El-shazly M, Attia Y, Kabil F et al. (2017) Inhibitory effects of salicylic acid and silver nanoparticles on potato virus Y-infected potato plants in Egypt. *Middle East Journal of Agriculture Research* 6:835–848.

Exbrayat JM, Moudilou EN, Lapied E (2015) Harmful effects of nanoparticles on animals. *Journal of Nanotechnology* 2015:1–10.

Ference CM, Gochez AM, Behlau F et al. (2018) Pathogen profile recent advances in the understanding of *Xanthomonas citris* sp. citripathogenesis and citrus canker disease management. *Molecular Plant Pathology* 19:1302–1318.

Firrao G, Moretti M, Ruiz Rosquete M et al. (2005) Nanobiotransducer for detecting flavescence *Doree phytoplasma. Journal of Plant Pathology* 87:101–107.

Forim MR, Costa ES, da Silva MF et al. (2013) Development of a new method to prepare nano-/microparticles loaded with extracts of *Azadirachta indica*, their characterization and use in controlling *Plutella xylostella. Journal of Agricultural and Food Chemistry* 61:9131–9139.

Gahlawata G, Choudhury AR (2019) A review on the biosynthesis of metal and metal salt nanoparticles by microbes. *RSC Advances* 9:12944–12967.

Garibay PRDA, Ma S, Pozo-Rodríguez A et al. (2015) Solid lipid nanoparticles as non-viral vectors for gene transfection in a cell model of fabry disease. *Journal of Biomedical Nanotechnology* 11:500–511.

Gold K, Slay B, Knackstedt M et al. (2018) Antimicrobial activity of metal and metal-oxide based nanoparticles. *Advanced Therapeutics* 1:1–15.

Gong J, Miao X, Wan H et al. (2012) Facile synthesis of zirconia nanoparticles-decorated graphene hybrid nanosheets for an enzymeless methyl parathion sensor. *Sensors and Actuators B: Chemical* 162:341–347.

Gradishar WJ, Tjulandin S, Davidson N et al. (2005) Phase III trial of nanoparticle albumin-bound paclitaxel compared with polyethylated castor oil-based paclitaxel in women with breast cancer. *Journal of Clinical Oncology* 23:7794–7803.

Graham J, Gottwald T, Cubero J et al. (2004) *Xanthomonas axonopodis* pv. citri: Factors affecting successful eradication of citrus canker. *Molecular Plant Pathology* 5:1–15.

Graham J, Johnson E, Myers M et al. (2016) Potential of nano-formulated zinc oxide for control of citrus canker on grapefruit trees. *Plant Disease* 100:2442–2447.

Guan H, Chi D, Yu J et al. (2010) Dynamics of residues from a novel nano-imidacloprid formulation in soyabean fields. *Crop Protection* 29:942–946.

Guerra FD, Attia MF, Whitehead DC et al. (2018) Nanotechnology for environmental remediation: Materials and applications. *Molecules* 23:1–23.

Guo Y, Yang Q, Yan W et al. (2014) Controlled release of acetochlor from poly (butyl methacrylate-diacetone acrylamide) based formulation prepared by nanoemulsion polymerisation method and evaluation of the efficacy. *International Journal of Environmental Analytical Chemistry* 94:1001–1012.

Gupta PK, Chand R, Vasistha NK et al. (2018) Spot blotch disease of wheat: The current status of research on genetics and breeding. *Plant Pathology* 67:508–531.

Hanaki KI, Momo A, Oku T et al. (2003) Semiconductor quantum dot/albumin complex is a long-life and highly photostable endosome marker. *Biochemical and Biophysical Research Communications* 302:496–501.

Harris P, Hernández E, Yakobson B (2004) Carbon nanotubes and related structures: New materials for the twenty-first century. *American Journal of Physics* 72:414–415.

Hasaneen MNA, Abdel-Aziz HMM, El-Bialy DMA et al. (2014) Preparation of chitosan nanoparticles for loading with NPK fertilizer *African. Journal of Biotechnology* 13:3158–3164.

He H, Pham-Huy LA, Dramou P et al. (2013) Carbon nanotubes: Applications in pharmacy and medicine. *BioMed Research International* 2013:1–12.

Hong YJ, Hossain MR, Kim HT et al. (2018) Identification of two new races of *Podosphaera xanthii* causing powdery mildew in melon in south Korea. *The Plant Pathology Journal* 34:182–190.

Hu MZ, Zhu T (2015) Semiconductor nanocrystal quantum dot synthesis approaches towards large-scale industrial production for energy applications. *Nanoscale Research Letters* 10:1–15.

Hulla JE, Sahu SC, Hayes AW (2015) Nanotechnology: History and future. *Human and Experimental Toxicology* 34:1318–1321.

Jeevanandam J, Barhoum A, Chan YS et al. (2018) Review on nanoparticles and nanostructured materials: History, sources, toxicity and regulations. *Beilstein Journal of Nanotechnology* 9:1050–1074.

Jogaiah S, Mitani S, Nagaraj AK et al. (2007) Activity of cyazofamid against Sclerospora graminicola, a downy mildew disease of pearl millet. *Pest Management Science* 63:722–727.

Khan I, Saeed K, Khan I (2017) Nanoparticles: Properties, applications and toxicities. *Arabian Journal of Chemistry*. doi: 10.1016/j.arabjc.2017.05.011.

Kolmas J, Krukowski S, Laskus A et al. (2016) Synthetic hydroxyapatite in pharmaceutical applications. *Ceramics International* 42:2472–2487.

Kretzmann JA, Ho D, Evans CW et al. (2017) Synthetically controlling dendrimer flexibility improves delivery of large plasmid DNA. *Chemical Science* 8:2923–2930.

Kumar P, Rai RC (2019) Spot blotch: A threat to wheat in changing climate-an overview. *Journal of Pharmacognosy and Phytochemistry* 8:326–331.

Lamsal K, Kim SW, Jung JH et al. (2011a) Inhibition effects of silver nanoparticles against powdery mildews on cucumber and pumpkin. *Mycobiology* 39:26–32.

Lamsal K, Kim SW, Jung JH et al. (2011b) Application of silver nanoparticles for the control of colletotrichum species in vitro and pepper anthracnose disease in field. *Mycobiology* 39:194–199.

Li ZZ, Chen JF, Liu F et al. (2007) Study of UV-shielding properties of novel porous hollow silica nanoparticle carriers for avermectin. *Pest Management Science* 63: 241–246.

Liang P, Liu S, Xu F et al. (2018) Powdery mildews are characterized by contracted carbohydrate metabolism and diverse effectors to adapt to obligate biotrophic lifestyle. *Frontiers in Microbiology* 9:1–14.

Liu Y, Tong Z, Prud'homme RK (2008) Stabilized polymeric nanoparticles for controlled and efficient release of bifenthrin. *Pest Management Science* 64:808–812.

Ma X, Zhang LH, Wang LR et al. (2012) Single-walled carbon nanotubes alter cytochrome c electron transfer and modulate mitochondrial function. *ACS Nano* 6:10486–10496.

MacDonald R, Laurenzi, B, Viswanathan G et al. (2005) Collagen-carbon nanotube composite materials as scaffolds in tissue engineering. *Journal of Biomedical Materials Research. Part A* 74:489–496.

Madaan K, Kumar S, Poonia N et al. (2014) Dendrimers in drug delivery and targeting: Drug-dendrimer interactions and toxicity issues. *Journal of Pharmacy and Bioallied Sciences* 6:139–150.

Maiti D, Tong X, Mou X et al. (2019) Carbon-based nanomaterials for biomedical applications: A recent study. *Frontiers in Pharmacology* 9:1–16.

Martynenko IV, Litvin AP, Purcell-Milton F et al. (2017) Application of semiconductor quantum dots in bioimaging and biosensing. *Journal of Materials Chemistry B* 5:6701–6727.

McMurray TA, Dunlop P, Byrne J (2006) The photocatalytic degradation of atrazine on nanoparticulate TiO2 films. *Journal of Photochemistry and Photobiology A: Chemistry* 182:43–51.

Micali C, Göllner K, Humphry M et al. (2008) The powdery mildew disease of arabidopsis: A paradigm for the interaction between plants and biotrophic fungi. *Arabidopsis Book* 6:1–19.

Milani N, McLaughlin M, Stacey S et al. (2012) Dissolution kinetics of macronutrient fertilizers coated with manufactured zinc oxide nanoparticles. *Journal of Agricultural and Food Chemistry* 60:3991–3998.

Mishra S, Singh BR, Singh A et al. (2014) Biofabricated silver nanoparticles act as a strong fungicide against *Bipolaris sorokiniana* causing spot blotch disease in wheat. *PLoS One* 9:1–11.

Mitter N, Worrall E, Robinson K et al. (2017) Clay nanosheets for topical delivery of RNAi for sustained protection against plant viruses. *Nature Plants* 3:1–10.

Moreno-Vega AI, Gómez-Quintero T, Nuñez-Anita RE (2012) Polymeric and ceramic nanoparticles in biomedical applications. *Journal of Nanotechnology* 2012:1–10.

Mosa KA, El-Naggar M, Ramamoorthy K et al. (2018) Copper nanoparticles induced genotoxicty, oxidative stress, and changes in superoxide dismutase (SOD) gene expression in cucumber (*Cucumis sativus*) plants. *Frontiers in Plant Science* 9:1–13.

Mukherjee A, Majumdar S, Servin AD et al. (2016) Carbon nanomaterials in agriculture: A critical review. *Frontiers in Plant Science* 7:1–16.

Navarro D, Bisson M, Aga D (2011) Investigating uptake of water-dispersible CdSe/ZnS quantum dot nanoparticles by *Arabidopsis thaliana* plants. *Journal of Hazardous Materials* 212:427–435.

Navya PN, Kaphle A, Srinivas SP (2019) Current trends and challenges in cancer management and therapy using designer nanomaterials. *Nano Convergence* 6:1–30.

Nguyen MK, Alsberg E (2014) Bioactive factor delivery strategies from engineered polymer hydrogels for therapeutic medicine. *Progress in Polymer Science* 39:1236–1265.

Ocsoy I, Paret M, Ocsoy M (2013) Nanotechnology in plant disease management: DNA-directed silver nanoparticles on graphene oxide as an antibacterial against *Xanthomonas perforans*. *ACS Nano* 7:8972–8980.

Olad A, Zebhi H, Salari D et al. (2018) Slow-release NPK fertilizer encapsulated by carboxymethyl cellulose-based nanocomposite with the function of water retention in soil. *Materials Science and Engineering C: Materials for Biological Applications* 90:333–340.

Oliveira J, Campos E, Silva C et al. (2014) Solid lipid nanoparticles co-loaded with simazine and atrazine: Preparation, characterization and evaluation of herbicidal activity. *Journal of Agricultural and Food Chemistry* 63 422–432.

Osojic GN, Hersam MC (2012) Biomolecule-directed assembly of self-supported, nanoporous, conductive, and luminescent single-walled carbon nanotube scaffolds. *Small* 8:1840–1845.

Panacek A, Kvítek L, Prucek R et al. (2006) Silver colloid nanoparticles: Synthesis, characterization, and their antibacterial activity. *Journal of Physical Chemistry B* 110:16248–16253.

Pantarotto D, Singh R, McCarthy D et al. (2004) Functionalized carbon nanotubes for plasmid DNA gene delivery. *Angewandte Chemie International* 43:5242–5246.

Paret ML, Vallad GE, Averett DR et al. (2013) Photocatalysis: Effect of light-activated nanoscale formulations of TiO(2) on *Xanthomonas perforans* and control of bacterial spot of tomato. *Phytopathology* 103:228–236.

Pérez-de-Luque A, Rubiales D (2009) Nanotechnology for parasitic plant control. *Pest Management Science* 65:540–545.

Phukan T, Kabyashree K, Singh R (2019) Ralstonia solanacearum virulence in eggplant seedlings by the leaf-clip inoculation. *Ray Phytopathology Research* 1:1–11.

Potnis N, Timilsina S, Strayer A et al. (2015) Bacterial spot of tomato and pepper: Diverse Xanthomonas species with a wide variety of virulence factors posing a worldwide challenge. *Molecular Plant Pathology* 16:907–920.

Prasad R, Bhattacharyya A, Nguyen QD (2017) Nanotechnology in sustainable agriculture: Recent developments, challenges, and perspectives. *Frontiers in Microbiology* 8:1–13.

Pretty J (2008) Agricultural sustainability: Concepts, principles and evidence. *Philosophical Transactions of the Royal Society B: Biological Sciences* 363:447–465.

Puri A, Loomis K, Smith B et al. (2009) Lipid-based nanoparticles as pharmaceutical drug carriers: From concepts to clinic. *Critical Reviews™ in Therapeutic Drug Carrier Systems* 26:523–580.

Rajan S, Liu H, Vu T (2008) Ligand-bound quantum dot probes for studying the molecular scale dynamics of receptor endocytic trafficking in live cells. *ACS Nano* 2:1153–1166.

Rastogi A, Zivcak M, Sytar O et al. (2017) Impact of metal and metal oxide nanoparticles on plant: A critical review. *Frontiers in Chemistry* 5:1–16.

Rui M, Ma C, Hao Y et al. (2016) Iron oxide nanoparticles as a potential iron fertilizer for peanut (*Arachis hypogaea*). *Frontiers in Plant Science* 7:1–10.

Sanzari I, Leone A, Ambrosone A (2019) Nanotechnology in plant science: To make a long story short. *Frontiers in Bioengineering and Biotechnology* 7:1–12

Sarkar S, Chaudhuri S (2016) Bacterial wilt and its management. *Currents Science* 110:1439–1445.

Sarlak N, Taherifar A, Salehi F (2014) Synthesis of nanopesticides by encapsulating pesticide nanoparticles using functionalized carbon nanotubes and application of new nanocomposite for plant disease treatment. *Journal of Agricultural and Food Chemistry* 62:4833–4838.

Satyvaldiev AS, Zhasnakunov ZK, Omurzak E et al. (2018) Copper nanoparticles: Synthesis and biological activity. *IOP Conference Series: Materials Science and Engineering* 302:1–8.

Shafie RM, Salama AM, Farroh KY (2018) Silver nanoparticles activity against Tomato spotted wilt virus. *Middle East Journal of Agriculture Research* 7:1251–1267.

Shan Y, Cao L, Xu C et al. (2019) Sulfonate-functionalized mesoporous silica nanoparticles as carriers for controlled herbicide diquat dibromide release through electrostatic interaction. *International Journal of Molecular Sciences* 20:1–13.

Sharma G, Kumar A, Sharma S et al. (2019) Novel development of nanoparticles to bimetallic nanoparticles and their composites: A review. *Journal of King Saud University – Science* 31:257–269.

Sharma S, Bhattarai K (2019) Progress in developing bacterial spot resistance in tomato. *Agronomy* 26:1–11.

Shen CX, Zhang QF, Li J et al. (2010) Induction of programmed cell death in Arabidopsis and rice by single-wall carbon nanotubes. *American Journal of Botany* 97:1602–1609.

Siddaiah CN, Prasanth KVH, Satyanarayana NR et al. (2018) Chitosan nanoparticles having higher degree of acetylation induce resistance against pearl millet downy mildew through nitric oxide generation. *Scientific Reports* 8:1–14.

Singh S, Singh BK, Yadav SM et al. (2015) Applications of nanotechnology in agricultural and their role in disease management. *Research Journal of Nanoscience and Nanotechnology* 5:1–5.

Sireesha M, Babu VJ, Kiran ASK et al. (2018) A review on carbon nanotubes in biosensor devices and their applications in medicine. *Nanocomposites* 4:36–57.

Stark WJ, Stoessel PR, Wohllebenb W et al. (2015) Industrial applications of nanoparticles. *Chemical Society Reviews* 16:5793–5805.

Sun H, Jia J, Jiang C et al. (2018) Gold nanoparticle-induced cell death and potential applications in nanomedicine. *International Journal of Molecular Sciences* 19:1–20.

Tamjidi F, Shahedi M Varshosaz J et al. (2013) Nanostructured lipid carriers (NLC): A potential delivery system for bioactive food molecules. *Innovative Food Science and Emerging Technologies* 19:29–43.

Tekle C, Deurs B, Sandvig K et al. (2008) Cellular trafficking of quantum dot-ligand bioconjugates and their induction of changes in normal routing of unconjugated ligands *Nano Letters* 8:1858–1865.

Than PP, Prihastuti H, Phoulivong S et al. (2008) Chilli anthracnose disease caused by Colletotrichum species. *Journal of Zhejiang University Science B* 9:764–778.

Tong Y, Wu Y, Zhao C et al. (2017) Polymeric nanoparticles as a metolachlor carrier: Water-based formulation for hydrophobic pesticides and absorption by plants. *Journal of Agricultural and Food Chemistry* 65:7371–7378.

Torney F, Trewyn B, Lin V (2007) Mesoporous silica nanoparticles deliver DNA and chemicals into plants. *Nature Nanotechnology* 2:295–300.

Trzepiński P, Klajnert-Maculewicz B (2017) Dendrimers for fluorescence-based bioimaging. *Journal of Chemical Technology and Biotechnology* 92:1157–1166.

Vamvakaki V, Chaniotakis N (2007) Pesticide detection with a liposome-based nano-biosensor. *Biosensors & Bioelectronics* 22:2848–2853.

Viswanathan S, Radecka H, Radecki J (2009) Electro-chemical biosensor for pesticides based on acetylcholinesterase immobilized on polyaniline deposited on vertically assembled carbon nanotubes wrapped with ssDNA. *Biosensors and Bioelectronics* 24:2772–2777.

Wang L, Li X, Zhang G et al. (2007) Oil-in-water nanoemulsions for pesticide formulations. *Journal of Colloid and Interface Science* 314:230–235.

Wang Y, Sun C, Xu Cuilian et al. (2016) Preliminary experiments on nano-silver against tobacco mosaic virus and its mechanism. *Tobacco Science & Technology* 49:22–30.

Wang YC, Engelhard MH, Baer DR et al. (2016) Quantifying the impact of nanoparticle coatings and nonuniformities on XPS analysis: Gold/silver core-shell nanoparticles. *Analytical Chemistry* 88:3917–3925.

Wong IY, Bhatia SN, Toner M (2013) Nanotechnology: Emerging tools for biology and medicine. *Genes & Development* 27: 2397–2408.

Worrall EA, Hamid A, Mody KT et al. (2018) Nanotechnology for plant disease management. *Agronomy* 8:1–24.

Wu Y, Phillips JA, Liu H et al. (2008) Carbon nanotubes protect DNA strands during cellular delivery. *ACS Nano* 2:2023–2028.

Yamaguchi S, Kobayashi H, Narita T et al. (2011). Sonodynamic therapy using water-dispersed TiO2-polyethylene glycol compound on glioma cells: Comparison of cytotoxic mechanism with photodynamic therapy. *Ultrasonics Sonochemistry*, 18:1197–1204.

Yan T, Li Y, Song X et al. (2019) Highly luminescent NIR-emitting CuFeS2/ZnS core/shell quantum dots for optical imaging of inflamed tissue. *Journal of Materials Chemistry C* 7:7279–7287.

Yang FL, Li XG, Zhu F et al. (2009) Structural characterization of nanoparticles loaded with garlic essential oil and their insecticidal activity against *Tribolium castaneum* (Herbst) (Coleoptera: Tenebrionidae). *Journal of Agricultural and Food Chemistry* 57:10156–10162.

Zhu G, Lee HJ (2017) Electrochemical sandwich-type biosensors for α-1 antitrypsin with carbon nanotubes and alkaline phosphatase labeled antibody-silver nanoparticles. *Biosensors and Bioelectronics* 89:959–963.

Zoclanclounon YAB, Kanfany G, Kane A et al. (2019) Current status of pearl millet downy mildew prevalence across agroecological zones of Senegal. *The Scientific World Journal* 2019:1–8.

Zuverza-Mena N, Armendariz R, Peralta-videa J et al. (2016) Effects of silver nanoparticles on radish sprouts: Root growth reduction and modifications in the nutritional value. *Frontiers in Plant Science* 7:1–11.

16 Cellular Impacts of Plant Fabricated Silver Nanoparticles

Puja Kumari, Paritosh Patel, Suresh K. Verma, and M. Anwar Mallick

CONTENTS

16.1 INTRODUCTION

Due to the huge demand for silver nanoparticles in day-to-day life, their commercial products as well as their biomedical applications, synthesis, and production have increased in leaps and bounds (Tran, Nguyen, and Le, 2013a). Silver nanoparticles have been recognized as one of the best potential candidates in biomedical applications because of their antibacterial and medicinal properties (Abou El-Nour et al., 2010) (Figure 16.1).

The peculiar properties of silver nanoparticles have attracted a lot of biologists and physicists to explore their new applications based on their variable properties depending on the route of synthesis. Silver nanoparticles synthesized from different routes have been reported to possess significant variable properties (Tran, Nguyen, and Le, 2013b). The different methods for silver nanoparticles synthesis are the physical, chemical, and biological routes. Chemically and physically synthesized silver nanoparticles have shown promising applications in all fields; however, there is a limitation in their purity, biodegradability, as well as biocompatibility. Biological synthesis has proved to be a potential solution to these problems; moreover, they have been recognized as a green method because they are eco-friendly. Previous literature has reported the synthesis of silver nanoparticles from various plants and

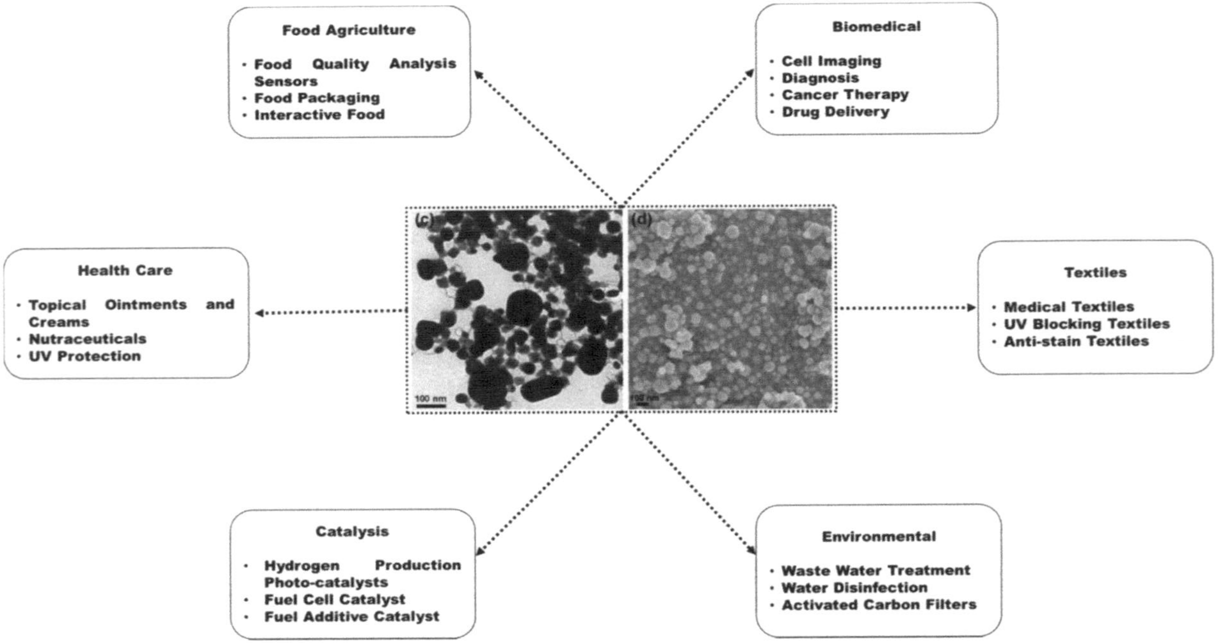

FIGURE 16.1 Applications of silver nanoparticles (Liao, Li, & Tjong, 2019). Transmission electron micrograph showing Ag NPs in spherical and stabilization of Ag NPs Scanning electron micrograph showing Ag NPs in a spherical form about 20-100 nm in diameter (Hossain et al., 2019).

their extracts (Thakkar, Mhatre, and Parikh, 2010). Though numerous plant extracts have been discovered to synthesize silver nanoparticles due to their capabilities of the reduction and stabilization of silver salts, medicinal plants have been given special emphasis. It is mandated to check the biocompatibility and cytotoxicity of a nanoparticle before considering it as a potential biomedical agent. To check the cytotoxicity of nanoparticles in the human system various live models have been used by researchers in the last few decades. One of the recent promising models is 'zebrafish' scientifically known as *Danio rerio*. Zebrafish and its embryonic stages have been considered as the most popular model for cytotoxicity testing because of their genetic similarities with the human model. Moreover, zebrafish are economic to rear and fast in development. Many nanotoxicologists have used the zebrafish model to determine the biocompatibility of nanomaterials in the last few years (Clemente et al., 2014). In this review, details of the biocompatibility of silver nanoparticles are described by evaluating their cytotoxicity with embryonic zebrafish along with the mechanism elucidated by the determination of induced reactive oxygen species and apoptosis as their consequences.

16.2 SYNTHESIS OF SILVER NANOPARTICLES

There are different types of approaches meant for the synthesis of nanoparticles. The two main approaches used are the top-down and bottom-up methods (Hasan, 2014).

16.2.1 Top-Down Method

As the name suggests, this method involves coming from high to low; hence, in this method, a bigger material is broken down into a smaller size (nano size). This method could be a green method if there isn't any involvement of harmful chemicals in the synthesis. A high energy ball milling method is an example of a top-down method where the synthesis of silver nanoparticles is achieved by mechanical grinding and has been successfully used by various scientists across the world and reported accordingly.

16.2.2 Bottom-Up Method

The top-down method does provide us with an easy method for the synthesis of nanoparticles without any usage of harmful chemicals, but concern for purity plays a major role in finding a new method for the synthesis of nanoparticles biologically. In this approach, nanoparticles are synthesized using a chemical reaction; the chemicals used are non-harmful and derived from different biological sources, such as plants, microbes, etc.

Various scientists have reported the synthesis of silver nanoparticles using various biological sources in which plants have gained a lot of attention for the synthesis of the desired nanoparticles as plant extracts work as reducing as well as capping agents because of the presence of biomolecules such as DNA, proteins, flavonoids, carotenoids, etc. Nanoparticles synthesized from plants are more stable than when synthesized using other methods.

16.2.3 Green Synthesis

As reported by various scientists, plant extracts are used for the synthesis of silver nanoparticles, and every new report conveys something new about these plants and green synthesized nanoparticles. And not only silver but many other metals or metal oxide nanoparticles have also been synthesized using plants. It doesn't have to only be a part of the plant, the whole plant can also be used to achieve the desired nanoparticles with fewer amounts of toxicity (Figure 16.2).

16.3 CHARACTERIZATION

After the reaction, the particle obtained needs to be analyzed for its confirmation and purity using various instruments. Commonly, UV-vis spectroscopy, DLS, FTIR, TEM, and SEM are used by the scientists. Figure 16.3 depicts an instrumental characterization of nanoparticles.

16.4 CYTOTOXICITY ANALYSIS

After instrumental analysis, biological experiments are conducted, including cytotoxicity analysis and other biological assays. Cytotoxic means being toxic or hazardous to the substance or surface it is exposed to, and hence cytotoxicity here refers to the induction of cell abnormalities leading to cell extermination because of the presence of nanoparticles. As studied, nanoparticles may lead to cell death; hence, cytotoxicity analysis proves to be one of the main experiments to dtermine its application in the clinical field. Cytotoxicity could be checked by both *in vitro* and *in vivo* models. The *in vitro* model refers to cytotoxicity analysis in different cell lines, such as the kidney cell line (HEK), whereas *in vivo* refers to the live animal models in which cytotoxicity is performed.

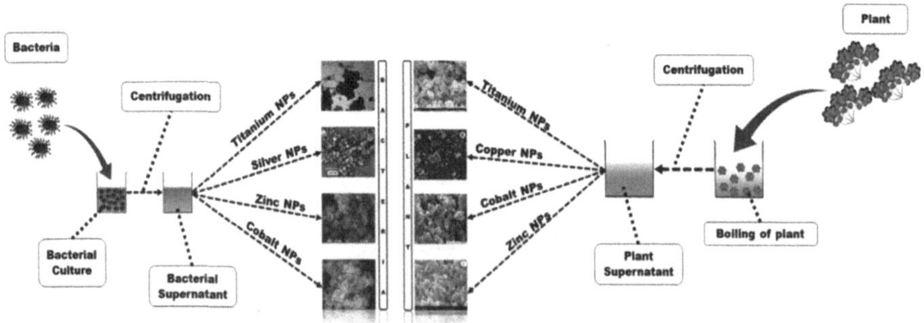

FIGURE 16.2 Mechanism of synthesis of silver nanoparticles (Hossain et al., 2019; Sangeetha, Rajeshwari, & Venckatesh, 2011; Rajamanickam, Viswanathan, & Muthusamy, 2012; Santhoshkumar et al., 2014; Khan & Fulekar, 2016; Jang et al., 2015; Bibi et al., 2017; Verma et al., 2018b; Verma et al., 2017b).

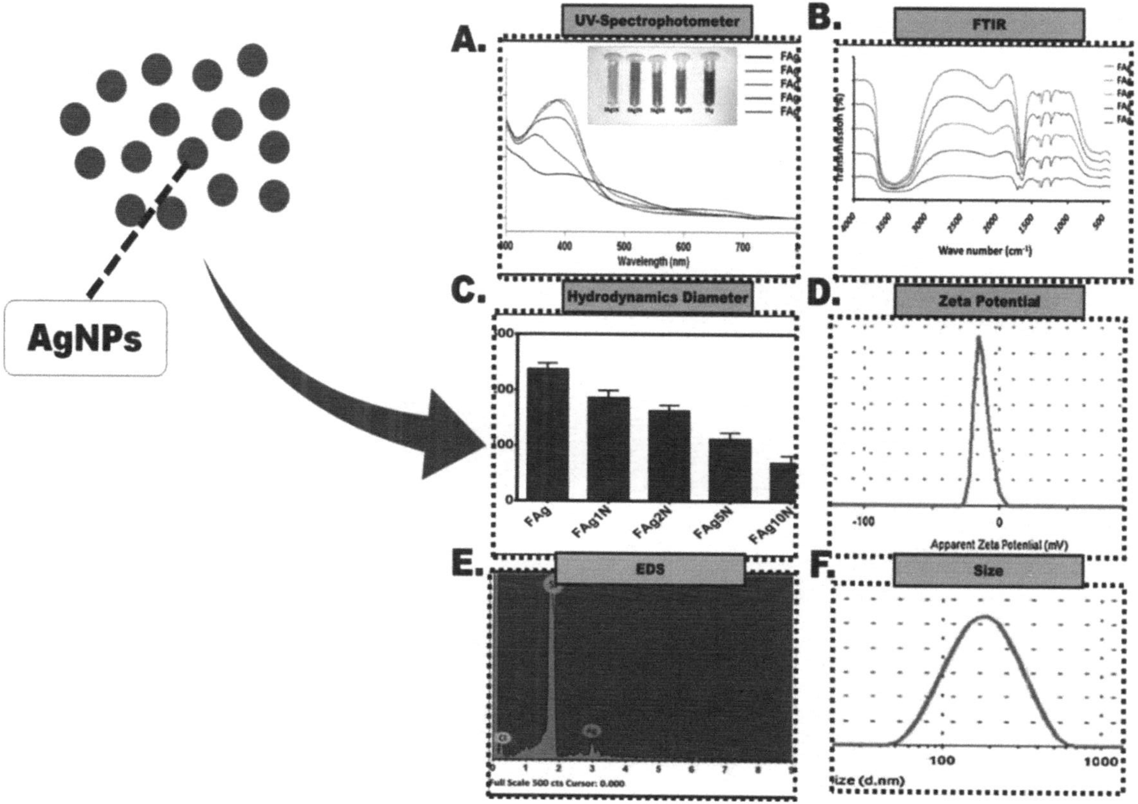

FIGURE 16.3 Physiochemical characterization of Silver nanoparticles: (A) UV-visible spectrum size determination by DLS; (B) FTIR analysis; (C) hydrodynamics diameter; (D) Zeta potential determination by DLS (Verma et al., 2018a, 2017a); (E) EDS analysis (Verma et al., 2018b; Journal et al., 2018); (F) size analysis (Sarkar et al., 2018).

When talking about live animal models, mice or rats were previously used in large amounts. But the era has changed: zebrafish (*Danio rerio*) have replaced rat and mouse models as they are more similar to humans, it is easier perform experiments, and it is comparatively cheaper than using mice.

Various studies on the cytotoxic effect of silver nanoparticles using various cell lines, such as rat liver cells, human lung epithelial cells, etc., have reported that silver nanoparticles enter into the body's surface through the endosomal pathway leading to cell damage which may be due to following reasons:

1. *Oxidative stress*: A higher level of ROS may lead to oxidative stress and a genotoxic effect.
2. *DNA Damage*: Various studies suggest that silver can damage the DNA either by increasing ROS production, which leads to the depletion of ATP production, or by directly breaking the DNA strands or damaging the DNA through overproduction of DNA repair proteins.
3. *Immunological dysfunction*: Silver nanoparticles are capable of producing stress, either stimulatory or suppressive depending upon the size of the nanoparticles and the amount and the cell type to which it is exposed. All these factors are responsible for immunological responses, such as inflammation, scars, etc. *In vivo* studies have also revealed that silver nanoparticles affect the circulatory system if injected or inhaled. In one of the studies, it has been shown that silver nanoparticles got accumulated inside the liver of mice. Not only this, but apoptotic cells have also been found in a stress condition with the exposure of silver nanoparticles.

   Irrespective of mice, cytotoxicity has also been reported in zebrafish embryos. With the exposure of silver nanoparticles to the embryos of zebrafish, their hatching rate, mortality rate, heart rate, etc. were affected depending upon the size, amount, or dose. Along with the hatching rate, mortality rate and changes in the morphology have also been observed such as slow or fast growth rate, organ disorders, oedema near the heart region, also known as pericardial oedema (Verma et al., 2018) (Figure 16.4).

   Figure 16.5 shows the survivability and hatching rate of embryonic zebrafish dtermined in the presence of different concentrations of silver nanoparticles. The survivability was found to be decreased with an increase in the concentration of silver nanoparticles in the HF medium used for the rearing of the embryos. Further, the hatching rate was determined in the presence of different concentrations, and it was found to be decreased with an increase in concentration.

When it comes to morphological analysis, Figure 16.6 depicts significant changes with the exposure of 100 μg/ml at 24 hpf whereas the embryos were found to be dead at 250 μg/ml concentration. At 96 hpf, the embryos were found hatched with some disorders such as bending of the body in 100 μg/ml concentration and pericardial oedema in 250 μg/ml concentration.

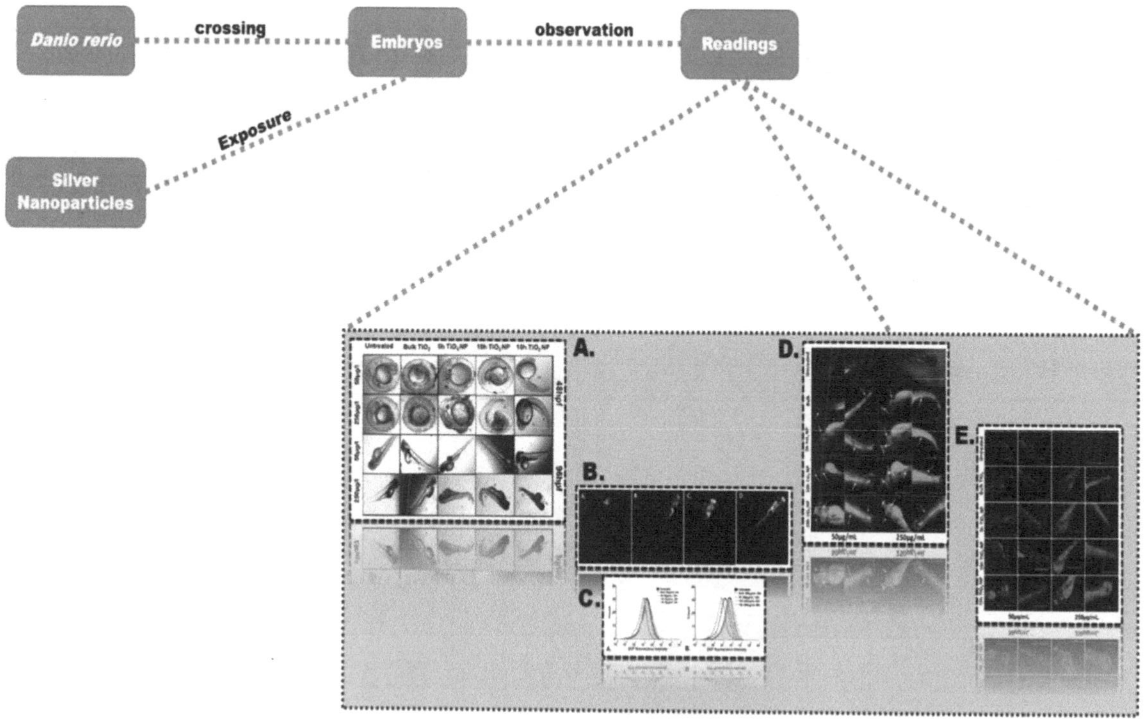

FIGURE 16.4 Effect of nanoparticles on zebrafish (Verma et al., 2018b; Kumari, Kumar, Pramanik, & Verma, 2019; Journal et al., 2018).

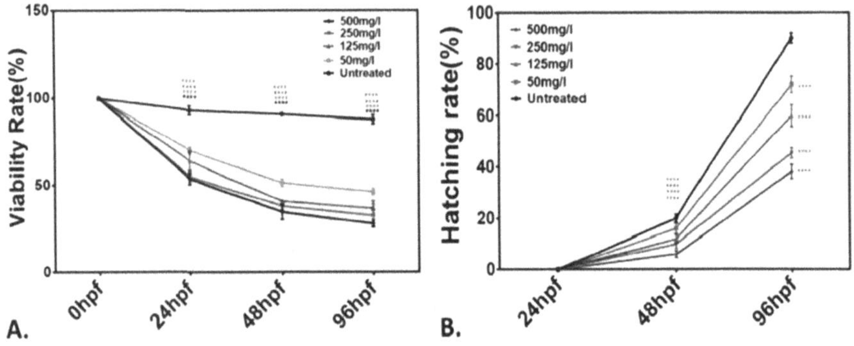

FIGURE 16.5 A. Survivability rate B: hatching rate of embryonic zebrafish (Kumari, Panda, Jha, Kumari, & Nisha, 2017; Verma et al., 2017c).

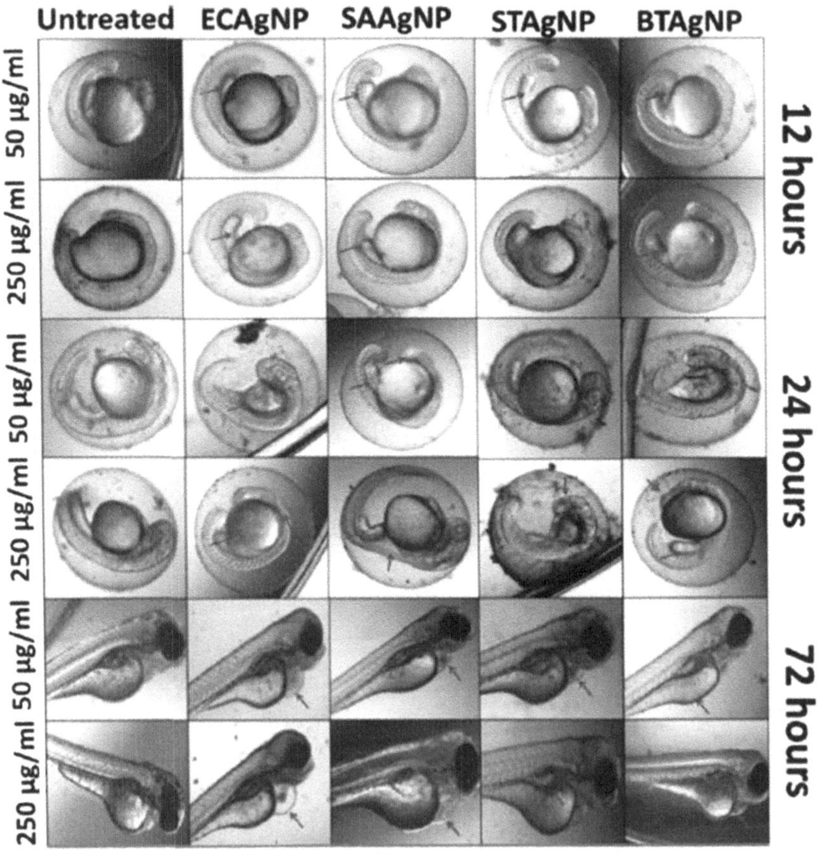

FIGURE 16.6 Morphological analysis of zebrafish embryos exposed to different concentration of Silver nanoparticles at different hour post-fertilization (HPF) (Verma et al., 2017a; Kumari et al., 2017).

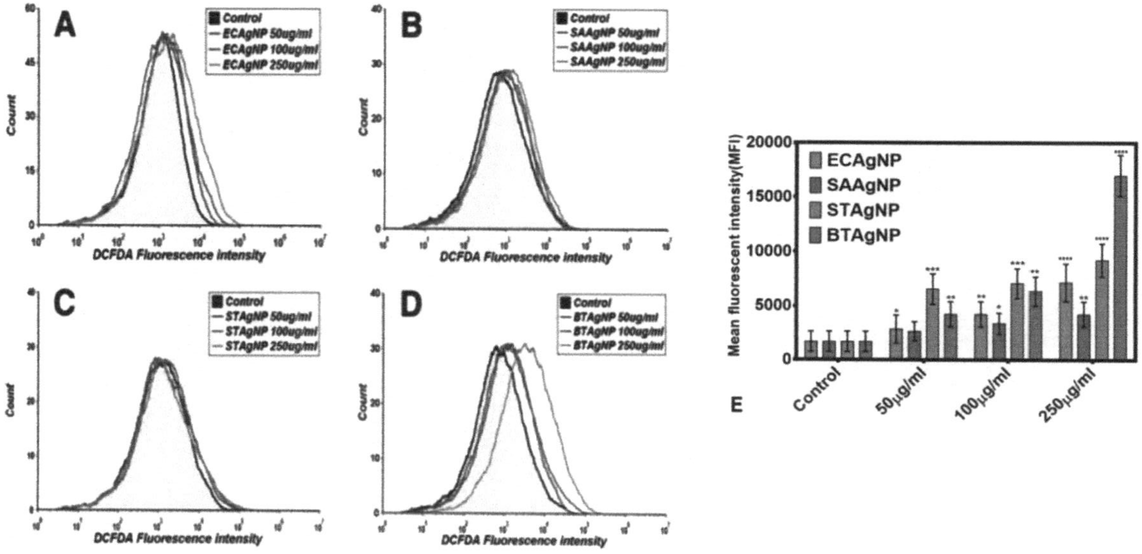

FIGURE 16.7 7 Reactive oxygen analysis of zebrafish exposed to different concentration of silver nanoparticles (Verma et al., 2017a, 2017c; Kumari et al., 2017, 2019).

To determine the mechanism of cytotoxicity of the silver nanoparticles, ROS analysis was performed as shown in Figure 16.7. ROS was found to be induced at a higher rate with an increase in the concentration of the exposed nanoparticles. ROS has been reported to induce various cytotoxic effects on living organisms (AshaRani et al., 2009). One of the most common effects reported is the enhanced induction of apoptosis in cells.

Apoptosis plays a vital role in embryogenesis, regulation of the immune system, and tissue homeostasis, and zebrafish constitute a powerful vertebrate model organism for the study of apoptosis, and it is widely used by the people all around the world. After various studies, a protocol was developed to visualize apoptosis in zebrafish *in vivo*. This protocol included the staining of zebrafish with AO (acridine orange), a nucleic acid, an intercalating dye that is capable of emitting green fluorescence when bound to dsDNA and stains the apoptotic cells. This apoptosis is the result of cytotoxicity due to the exposure of silver nanoparticles.

Figure 16.8 depicts apoptotic analysis of a zebrafish embryo visualized after AO staining at 250 μg/ml.

16.5 CONCLUSION

Since there have been threats to biotic and abiotic ecosystems with the advancement of nanotechnology, there is a need to make nanotechnology eco-friendly using environmental factors. Green nanotechnology has proven to be one of the potential

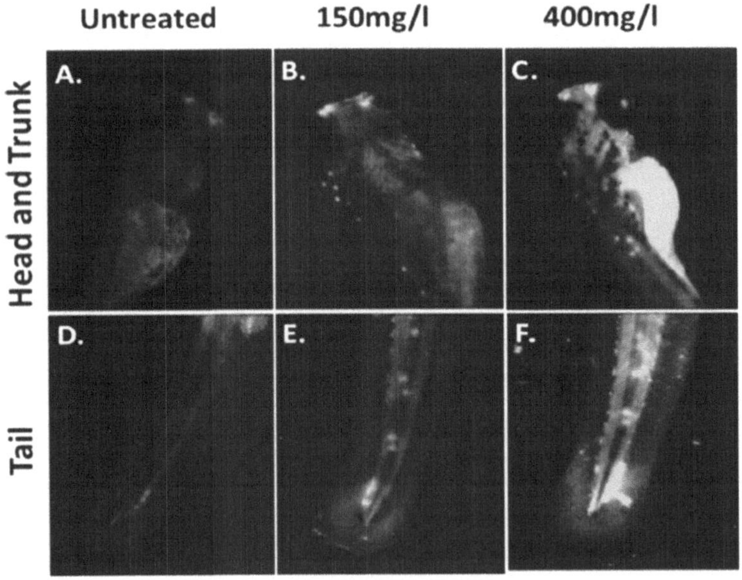

FIGURE 16.8 Apoptosis analysis of zebrafish embryo exposed to different concentrations of silver nanoparticles. (A), (B) Untreated, (C), (D) 250μg/ml (Kumari et al., 2017; Verma et al., 2018b, 2018c).

solutions for this using plant extracts, which act both as reducing and stabilizing agents. Metal nanoparticles and their composites have been bioengineered with the help of biological methods. Further, the qualities of synthesized nanoparticles in terms of their physicochemical properties have been found tuneable in terms of cytotoxicity.

REFERENCES

Abou El-Nour, K. M. M., Eftaiha, A., Al-Warthan, A., & Ammar, R. A. A. (2010). Synthesis and applications of silver nanoparticles. *Arabian Journal of Chemistry*, *3*(3), 135–140. doi: 10.1016/j.arabjc.2010.04.008

AshaRani, P. V., Mun, G. L. K., Hande, M. P., & Valiyaveettil, S. (2009). Cytotoxicity and genotoxicity of silver nanoparticles in human cells. *ACS Nano*, *3*(2), 279–290. doi: 10.1021/nn800596w

Bibi, I., Nazar, N., Iqbal, M., Kamal, S., Nawaz, H., Nouren, S., Abbas, M. (2017). Green and eco-friendly synthesis of cobalt-oxide nanoparticle: Characterization and photocatalytic activity. *Advanced Powder Technology*, *28*(9), 2035–2043. doi: 10.1016/j.apt.2017.05.008

Clemente, Z., Castro, V. L. S. S., Moura, M. A. M., Jonsson, C. M., & Fraceto, L. F. (2014). Toxicity assessment of TiO2 nanoparticles in zebrafish embryos under different exposure conditions. *Aquatic Toxicology*, *147*, 129–139. doi: 10.1016/j.aquatox.2013.12.024

Hasan, S. (2014). A review on nanoparticles: Their synthesis and types. *Research Journal of Recent Sciences*, *4*(4), 1–3.

Hossain, A., Hong, X., Ibrahim, E., Li, B., Sun, G., Meng, Y., … An, Q. (2019). Green synthesis of silver nanoparticles with culture supernatant of a bacterium pseudomonas rhodesiae and their antibacterial activity against soft rot pathogen dickeya dadantii. *Molecules*, *24*(12), 1–11. doi: 10.3390/molecules24122303

Jang, E., Shim, H. W., Ryu, B. H., An, D. R., Yoo, W. K., Kim, K. K., … Kim, T. D. (2015). Preparation of cobalt nanoparticles from polymorphic bacterial templates: A novel platform for biocatalysis. *International Journal of Biological Macromolecules*, *81*, 747–753. doi: 10.1016/j.ijbiomac.2015.09.009

Journal, A. I., Verma, S. K., Jha, E., Panda, P. K., Kumari, P., Kumari, S., … Kumari, S. (2018). Molecular investigation to RNA and protein based interaction induced in vivo biocompatibility of phytofabricated Au NP with embryonic zebrafish Molecular investigation to RNA and protein based interaction induced in vivo. *Artificial Cells, Nanomedicine, and Biotechnology*, 1–14. doi: 10.1080/21691401.2018.1505746.

Khan, R., & Fulekar, M. H. (2016). Biosynthesis of titanium dioxide nanoparticles using Bacillus amyloliquefaciens culture and enhancement of its photocatalytic activity for the degradation of a sulfonated textile dye Reactive Red 31. *Journal of Colloid and Interface Science*, *475*, 184–191. doi: 10.1016/j.jcis.2016.05.001.

Kumari, P., Panda, P. K., Jha, E., Kumari, K., & Nisha, K. (2017). Mechanistic insight to ROS and Apoptosis regulated cytotoxicity inferred by Green synthesized CuO nanoparticles from Calotropis gigantea to Embryonic Zebrafish. *Scientific Reports*, (September), 1–17. doi: 10.1038/s41598-017-16581-1.

Kumari, S., Kumar, P., Pramanik, N., & Verma, S. K. (2019). Environmental Nanotechnology, Monitoring & Management Molecular aspect of phytofabrication of gold nanoparticle from Andrographis peniculata photosystem II and their in vivo biological effect on embryonic zebrafish (*Danio rerio*). *Environmental Nanotechnology, Monitoring & Management*, *11*(December 2018), 100201. doi: 10.1016/j.enmm.2018.100201.

Liao, C., Li, Y., & Tjong, S. C. (2019). Bactericidal and cytotoxic properties of silver nanoparticles. *International Journal of Molecular Sciences*, *20*(2). doi: 10.3390/ijms20020449.

Rajamanickam, U., Viswanathan, S., & Muthusamy, P. (2012). Biosynthesis of Zinc Nanoparticles Using Actinomycetes for Antibacterial Food Packaging. International Conference on Nutrition and Food Sciences, *39*, 195–199.

Sangeetha, G., Rajeshwari, S., & Venckatesh, R. (2011). Green synthesis of zinc oxide nanoparticles by aloe barbadensis miller leaf extract: Structure and optical properties. *Materials Research Bulletin*, *46*(12), 2560–2566. doi: 10.1016/j.materresbull.2011.07.046

Santhoshkumar, T., Rahuman, A. A., Jayaseelan, C., Rajakumar, G., Marimuthu, S., Kirthi, A. V., Kim, S. K. (2014). Green synthesis of titanium dioxide nanoparticles using Psidium guajava extract and its antibacterial and antioxidant properties. *Asian Pacific Journal of Tropical Medicine*, *7*(12), 968–976. doi: 10.1016/S1995-7645(14)60171-1.

Sarkar, B., Verma, S. K., Akhtar, J., Prakash, S., Kr, S., Kumar, P., & Mukherjee, K. (2018). Chemosphere Molecular aspect of silver nanoparticles regulated embryonic development in Zebrafish (*Danio rerio*) by Oct-4 expression. *Chemosphere*, *206*, 560–567. doi: 10.1016/j.chemosphere.2018.05.018.

Thakkar, K. N., Mhatre, S. S., & Parikh, R. Y. (2010). Biological synthesis of metallic nanoparticles. *Nanomedicine: Nanotechnology, Biology, and Medicine*, *6*(2), 257–262. doi: 10.1016/j.nano.2009.07.002.

Tran, Q. H., Nguyen, V. Q., & Le, A.-T. (2013a). Silver nanoparticles: synthesis, properties, toxicology, applications and perspectives. *Advance in Natural Science, Nanosciences and Nanotechnology*, *4*, 20.

Tran, Q. H., Nguyen, V. Q., & Le, A.-T. (2013b). Silver nanoparticles: synthesis, properties, toxicology, applications and perspectives. *Advances in Natural Sciences: Nanoscience and Nanotechnology*, *4*(3), 033001. doi: 10.1088/2043-6262/4/3/033001

Verma, S. K., Jha, E., Kumar, P., Thirumurugan, A., & Patro, S. (2018). Materials Science & Engineering C Molecular insights to alkaline based bio-fabrication of silver nanoparticles for inverse cytotoxicity and enhanced antibacterial activity. *Materials Science & Engineering C*, *92*(July), 807–818. doi: 10.1016/j.msec.2018.07.037.

Verma, S. K., Jha, E., Panda, P. K., Mishra, A., Das, B., Parashar, S. K. S., & Suar, M. (2017c). Rapid novel facile biosynthesized silver nanoparticles from bacterial release induce biogenicity and concentration dependent in vivo cytotoxicity with embryonic zebrafish – A mechanistic insight, (December), 1–14. doi: 10.1093/toxsci/kfx204

Verma, S. K., Jha, E., Panda, P. K., Mishra, A., Thirumurugan, A., Das, B., … Suar, M. (2018a). Rapid novel facile biosynthesized silver nanoparticles from bacterial release induce biogenicity and concentration dependent in vivo cytotoxicity with embryonic Zebrafish-A mechanistic insight. *Toxicological Sciences*, *161*(1), 125–138. doi: 10.1093/toxsci/kfx204

Verma, S. K., Jha, E., Panda, P. K., Mukherjee, M., Thirumurugan, A., Makkar, H., … Suar, M. (2018b). Mechanistic insight into ROS and neutral lipid alteration induced toxicity in the human model with fins (*Danio rerio*) by industrially synthesized titanium dioxide nanoparticles. *Toxicology Research*, *7*(2), 244–257. doi: 10.1039/c7tx00300e

Verma, S. K., Mishra, A. K., Suar, M., & Parashar, S. K. S. (2017a). In vivo assessment of the impact of titanium oxide nanoparticle on zebrafish embryo In Vivo assessment of the impact of Titanium oxide nanoparticle on zebrafish embryo, *040030*. doi: 10.1063/1.4980232.

Verma, S. K., Panda, P. K., Jha, E., & Suar, M. (2017b). Altered physiochemical properties in industrially synthesized ZnO nanoparticles regulate oxidative stress; induce in vivo cytotoxicity in embryonic zebrafish by apoptosis. *Scientific Reports*, (May), 1–16. doi: 10.1038/s41598-017-14039-y.

17 Phytotoxic Study on Degraded Textile Dye (Methyl Red) Using Iron Oxide, Tungsten Oxide and Iron Tungstate Oxide Nanoparticles

A Case Report

Narendhran Sadasivam and Vinoth Velusamy

CONTENTS

17.1 INTRODUCTION

The environment is a sum of the entire living organism that occupies a particular habitat. The environment is affected by various types of pollution, which cause irreparable damage to all the living species in an ecosystem. Wastewater contamination with certain dyes from industrial effluents is one of the most significant problems that we are facing nowadays in our country. The contamination is caused mainly by the dye effluents released from the leather and textile industries. The major source of water pollution is caused by the release of hazardous dye-containing effluent into the environment without any processing. The art of adding color to textiles has improved their appearance, drawing the attention of all communities.

In ancient times, the dyes were made from natural fruits, vegetables, flowers, insects, and fish varieties. These naturally extracted dyes were used for colouring and gave only a dull range of colours to the fabrics. Because of its dull appearance, mordant was used, which helps in fixing the fabric and dye together by giving a wide range of colouring patterns with a bright appearance. The mordant quantities are very toxic and impact on the quality of the wastewater. This led to the discovery of synthetic dyes, which has attracted many consumers due to their brighter shades. This application of dyes on textile products has revolutionized nanotechnology in all industries (Whitaker and Willock, 1949).

With the growing consumer requirement for clothing, textile dyes are being manufactured and used more extensively than in the past. In the process of textile dying, only 85% of the dye fixes to the fabrics while the remaining is released along with the effluent (Aspland, 1997). The growing demand for textile products has led to an increase in the quality of wastewater discharged by the textile industries. The released effluent contains organic dyes, and their breakdown products have a toxic and mutagenic effect on the environment as well as human life (Weisburger, 2002).

The use of synthetic dyes in various industries, such as on paper, pulp, plastics, leather, printing, and fabric, has led to the release of effluent containing a residual amount of dye into the soil and marine ecosystem (Aksu, 2005). Most of the dyes employed in these industries are toxic and are very recalcitrant to the degradation of microbes (Pagga and Brown, 1986). The dyes in the effluents may undergo anaerobic degradation to form carcinogenic compounds, which affect the food chain (Banat et al., 1996). The high amount of dye colour in the wastewater blocks the penetration of sunlight and oxygen into the water leading to the death of marine organisms (Crini, 2006). The intermediate product resulting from the reduction of the dyes, such as aromatic amines, is genotoxic to living species.

When untreated effluents containing hazardous dyes are released into land or water bodies, they cause damage to the marine ecosystem leading to the death of phototrophs. They also decrease the dissolved oxygen in the water bodies, affecting the aquatic ecosystem and food chain. The organic reactive dyes regularly used in fabrics, cause allergic dermatitis, respiratory problems, and bladder cancer (Estlander, 1998; Hatch and Maibach, 1995; Gonzales et al., 1998). The dyes used in fabrication, when released along with effluent, can lead to bioaccumulation in plants and the marine ecosystem, causing biomagnification.

The treatment of dye effluent using traditional biological treatment is not sufficient because of its low biodegradation efficiency. A wide range of techniques are used for the withdrawal of dyes from the aquatic ecosystem, such as oxidation, irradiation, flocculation, membrane filtration, and electrokinetic coagulation. But these techniques are not effective in the removal of dyes, and a residual amount of dye still remains in the wastewater and exhibits toxic activity. These techniques are very cost-effective and cannot be performed all the time. So, adsorption and oxidation are considered to be a convenient technology for the removal of dyes which is not costlier than former techniques.

Photodegradation using semiconductors has an advantage over other conventional methods, such as the quick oxidation of dye effluents without the formation of polycyclic products. This method is considered as a rapid method for the withdrawal of pollutants from industrial effluents.

17.1.1 Azo Dyes in Textile Industries

Among the various types of dyes used in different industries, azo dyes are the most commonly used synthetic dyes in textile fabrication, food, and cosmetic industries. The merits of azo dyes are their attractive appearance, inexpensive cost, and the colour of the fabrics lasts for an extended period when compared to natural dyes. Azo dyes are a class of nitrogen-based synthetic dyes that produce bright red, orange, and yellow colours when coated on to fabrics, and this is the most commonly used type of dye in industries.

Azo dyes used on an industrial level cause severe problems to the environment and human life. The released effluent containing azo dyes into the water bodies changes the quality of the water, leading to the death of aquatic organisms and affecting the photosynthesis of aquatic plants. Many azo dyes and their derivatives are mutagenic and carcinogenic to various forms of life in our biological ecosystem. Some of the dyes released don't get degraded and if they do get degraded using certain techniques may lead to the production of harmful substances as they decompose. Azo dyes can cause symptoms in adults ranging from skin rashes, headaches, nausea, diarrhoea, muscle and joint pain, difficulty in breathing, irregular heartbeat, and seizures.

17.1.2 Nanoparticles in Decoloration of Industrial Effluents

The severe problem that are we are facing nowadays is the removal of colour from wastewater which is released by the dyeing industry. Azo dyes are the most commonly used coloured dyes in the textile industries, which is very toxic when exposed to the environment. So, the removal of colour from wastewater released from the industries is essential. Many techniques have already been employed to treat effluents, but the most successful and cost-effective method is adsorption using nanoparticles. Metal nanoparticles are most widely used for the decolourization of coloured effluents. The size and shape of the nanoparticles play a vital role in the decolourization of the effluents. This process can also be controlled with various physical and chemical mehods (Shen et al., 1999; Hergt et al., 2006; Naoki, 2008; Sakulchaicharoen et al., 2010).

17.1.3 Iron Oxide Nanoparticles

Nanotechnology is in rapid development that is superior to human activity in many areas (Salata, 2004). Iron chemically combines with oxygen and leads to the formation of iron oxide. There are approximately 16 different iron oxides present in the chemical table. They are inexpensive and play a vital role in various geological and biological processes. Iron oxides are used as catalysts and also as durable pigments in paints and coloured concretes (Laurent et al., 2008). Iron oxide nanoparticles are prepared using various methods such as wet chemical, dry process, or by microbiological techniques (Hasany et al., 2012). Each of these methods has its advantages and disadvantages, mainly in maintaining the particle size. In the wet chemical method of nanoparticle synthesis, particle size can be controlled by adjusting certain parameters. Some factors influence the size of the nanoparticles, such as mixing rate, temperature, agitation, and pH. Nanoparticles from a microbial origin were found to be low-cost with a high yielding potential (Xu et al., 2007; Narayanan et al., 2010). Iron oxide nanoparticles are preferred for the treatment of wastewater to a large extent due to their strong adsorption capacity, inertness, and property of super-paramagnetism. It is beneficial to coat the nanoparticles onto a matrix (effluent degrading bacteria) and let them degrade the effluent released by industries. Wastewater effluent treatment using nanotechnology is a potential technique, and it has benefits such as it is low-cost and environment-friendly (Huang et al., 2014; Oh et al., 2016).

17.1.4 Tungsten Oxide Nanoparticles

Tungsten oxide (WO_3) is a transition metal oxide n-type semiconductor with a widely tunable bandgap of about 2.6–2.8 eV at room temperature (Soultanidis et al., 2012). The organic dyes used in several industries are very carcinogenic to human and aquatic life, even at a low concentrations of about 0.001 mg/L (Slein and Sansone, 1980). The effluent has to be treated with a suitable technique to protect our ecosystem. So, many adsorption techniques have been developed as a conventional method to degrade dyes in the effluent. To overcome the limitation caused by the adsorption method, an advanced oxidation process using a non-toxic photocatalyst was developed (Daneshvar et al., 2004; Shirzad-Siboni et al., 2011). Tungsten trioxide nanoparticles are also referred to as a very important semiconductor with specific properties that can be used as a photocatalyst, and they can be used in the degradation of dye wastewater, so it can be assumed that more significant activity is visible under light irradiation on FeWO4 photocatalyst (Seddigi, 2010). The tungstate nanoparticles can be coated with other oxide metals to form a composite film that can be used to degrade various dyes. Tungstate nanoparticles in the form of nanowires are also employed for dye degradation. There are usually many active photocatalysts used in effluent treatment, among them tungsten oxide (WO_3) is considered as an ideal candidate because of its small bandgap (2.4–2.8 eV), nanotoxicity, and stability (Bamwenda et al., 1999; Morales et al., 2008; Xin et al., 2009).

17.1.5 Iron Tungstate Oxide Nanoparticles

A hydrothermal method was used for the synthesis of iron tungstate nanoparticles. This was prepared using sodium tungstate ($Na_2WO_4 \cdot 2H_2O$). In this preparation, sodium tungstate was used as a starting material. Iron tungstate oxide ($FeWO_4$) has a wide range of applications, such as in microwave ovens, humidity sensors, optical fibres, and also as photocatalysts. The degradation of dye present in the industrial effluent can be carried out using iron tungstate, in the process of photocatalysis. $FeWO_4$ nanoparticles can be synthesized in large amounts uisng various approaches, such as physical, chemical, and biological methods. A certain chemical can be mixed in a certain proportion to synthesize the nanoparticles with a particular particle size. They can also be manufactured using a plant source, which is called green synthesis. The nanoparticles produced by green synthesis help to overcome the toxicity which is caused by chemical synthesis. The chemical method is preferable for the synthesis of nanoparticles in large quantities in a short period as a capping agent can stabilize the size of the synthesized nanoparticles. The nanoparticles, such as iron oxide and tungstate oxide, are synthesized with an acid precipitation method while iron tungstate nanoparticles are synthesized using hydrothermal methods.

17.1.6 Impact of Nanoparticles on Plants

Nanoparticles which have been synthesized by various methods range in size between 1 to 100 nm and they have been used in the development of different promising technologies. This application is due to the peculiar physiochemical characteristics they possess (Handy et al., 2008). The massive release of nanoparticles into the environment has created a toxicological risk to living organisms by disturbing the ecological balance. This is because, the nanoparticles, when discharged into an environment, change their chemical properties by forming aggregates and complexes with the organic matters in the soil and water. Plants as an autotroph perform photosynthesis to maintain the oxygen content in the atmosphere as they supply the plant with organic compounds, which is the basis of the food chain. Plants have a more direct interaction with the soil when compared to other living organisms. Therefore, the impact of the nanoparticles on plants are faster. Plants have been used as a model system to analyze the possible toxicity of the nanoparticles.

Nanoparticles released into the atmosphere easily attach to the aerial part of plants and get integrated. Later, they interact with various cellular components such as the membrane, organelles, and nuclear proteins. This leads to damage at the chromosome level leading to an alteration in the genome. Nanoparticles affect the plasma membrane with the production of ROS and induce a change in antioxidant enzyme activities. This alteration in the plant cell leads to the accumulation of free radicals, which have a dangerous effect on the cell (Ray et al., 2012; Singh et al., 2017).

17.1.7 Phytotoxicity on Vigna Radiata

The plant *Vigna radiata* is well known as green gram, belonging to the family of *Fabaceae.* It has been consumed as a traditional food for more than 3500 years around the world. Mung bean was selected for the study of phytotoxicity due to its economic importance as a food legume. Mung bean was designated as a test species for the assessment of phytotoxicity by the Organization for Economic Corporation and Development.

The seeds of mung bean can be made to grow in the presence of different concentrations of nanoparticles under growth chamber conditions. The effect of the nanoparticles was studied on their plant growth and root elongation (Lequeux et al., 2010). In this study, the photocatalytic activity of iron oxide, tungstate oxide, and iron tungstate oxide nanoparticles synthesized using a chemical method, was used for the degradation of azo dye (Methyl red) released from textile and other dyeing industries. A phytotoxic study was carried out on the seeds of *Vigna radiata*, and their growth regulation was checked and recorded.

17.2 METHODS USED IN SYNTHESIS

17.2.1 Synthesis of Nanomaterials

Anhydrous iron (III) chloride ($FeCl_3$, 99%), sodium tungstate dihydrate (Na_2WO_4. H_2O, 99%), and methyl red (Table 17.1) were purchased from Sigma Aldrich. All other reagents used in this experiment were analytical grade. *Vigna radiate* was collected from the Tamil Nadu Agriculture University (TNAU), Coimbatore, India.

To synthesize the Fe_2O_3 NPs, a simple precipitation method was used. In brief, 0.1M of $FeCl_3$ was dissolved in 100 mL of deionized water by continuous stirring. Followed by the addition of 100 mL of 0.1M NaOH in a dropwise manner to the

TABLE 17.1
Chemical and Structural Properties of Methyl Red (MR) Dye

Dye	Molecular formula	Molecular weight	CAS register number	Structure
Methyl red	$C_{15}H_{15}N_3O_2$	269.304 g/mol	845- 10-3	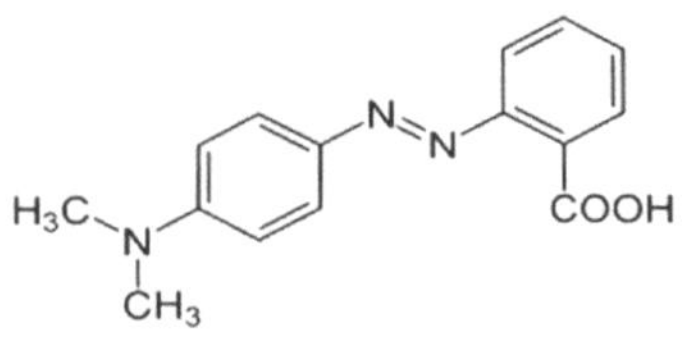

reaction mixture which was stirred for one hour. After stirring, the nanoparticle (Fe_2O_3) was separated by centrifugation at 5000 rpm for five minutes, and the pellet was collected. The purification of the precipitate was done to remove impurities by washing with deionized water and ethanol (70%) three times. After purification, annealing of the precipitate was done at 150°C for eight hours. The annealed precipitate was made into a fine powder using a pestle and mortar. The powdered product was stored in a screw cap bottle for future studies.

To synthesize WO_3 NPs, an acid precipitation method was used. In this precipitation technique, tungsten oxide (WO_3) nanoparticles were synthesized using sodium tungstate dihydrate (Na_2WO_4. $2H_2O$, 99%) and sulphuric acid (H_2SO_4); 0.1 M of Sodium tungstate was dissolved in 100 mL of deionized water by vigorous stirring. Followed by the addition of 6N concentrated sulphuric acid in a dropwise manner and incubation at 5°C for 60 minutes. The precipitation of the above mixture was done by incubating it at room temperature for about 24 hours. The obtained precipitate was washed using deionized water and centrifuged at 5000 rpm for five minutes. After washing, the precipitate was calcined at a temperature ranges from 180 to 200°C for 24 hours to get a powdered tungsten oxide nanoparticle. The obtained yellow coloured powder (tungsten oxide nanoparticle) was stored in a screw cap bottle for further studies.

To synthesize $FeWO_4$ NPs, a hydrothermal method was used. The synthesis of iron tungstate oxide nanoparticles was done by dissolving 0.1M of sodium tungstate dihydrate (Na_2WO_4. $2H_2O$, 99%) and 0.1M of $FeCl_3$, separately, using deionized water under vigorous stirring; followed by the addition of 0.1M sodium hydroxide solution to the mixture, which is resulted in a change in colour to dark brown. The above mixture was transferred to the crucible and dried in a hot air oven at 150°C for 24 hours to get the final product (iron tungstate oxide nanoparticles) in the form of powder. The obtained nanoparticle powder was washed thrice with deionized water and ethanol. As a result, the obtained brown colour precipitate was dried in a hot air oven at 100°C for two hours and transferred into an airtight container for further studies.

17.2.2 Characterization of Fe_2O_3, WO_3, and $FeWO_4$ NPs

Chemically synthesized nanoparticles (Fe_2O_3, WO_3, and $FeWO_4$) were characterized to determine their optical density and bandgap energy with UV-visible spectroscopy (V-650, JASCO). The crystalline nature of Fe_2O_3, WO_3, and $FeWO_4$ nanoparticles was observed by X-ray diffraction with Cu Kα radiation (Perkin-Elmer spectrum). Fourier-transfer infrared spectroscopy (FTIR, Perkin-Elmer 1725x) analysis was performed to study functional groups, which played a vital role in the reduction of Fe_2O_3, WO_3, and $FeWO_4$ nanoparticles. EDX (energy diffraction X-ray, model Quantax 200, Germany) and SEM (scanning electron microscope, model JSM 6390LV) analysis were done to determine the elemental composition and morphology of synthesized nanoparticles (Fe_2O_3, WO_3, and $FeWO_4$). Transmission electron microscopy (TEM) was used for methyl orange degraded nanoparticles.

17.2.3 Photocatalytic Activity of Fe_2O_3, WO_3, and $FeWO_4$ NPs

The photocatalytic activity of synthesized nanoparticles (Fe_2O_3, WO_3, and $FeWO_4$) was measured under UV light with the decomposition of organic dye methyl red (MR). A mercury lamp of about 500 W was used as a source of UV light, as it emits radiation over a wide wavelength. A catalyst powder of 100 mg (100 ppm) dissolved in 100 mL of dye solution (100 ppm) at room temperature, acts as a photocatalyst solution; it is magnetically stirred to get an absorption–desorption equilibrium between the photocatalyst and dye. A distance of about 35 cm was maintained between the UV light and the reaction mixture. For every 20 mins, 5 mL of suspension was taken and centrifuged at 1000 rpm to remove the nanoparticles. The solution was transferred to a cuvette to measure the absorption at $\lambda = 483$ nm. The dye degradation percentage was calculated as follows:

$$\text{Degradation rate}(\%) = \text{C0} - \text{Ct} / \text{C0} * 100$$

Where,

Ct– absorbance value of the dye solution at t.
C0– absorbance value of the dye solution at 0 min, by ultraviolet-visible (UV-vis) spectrometer, respectively.

The first-order kinetic formula has measured the rate constant:

$$\text{In}(\text{C/Co}) = -\text{kt}$$

Where,

C – initial concentration of dye.
Co– final concentration of dye.
k – rate constant.
T – time.

All parts of the experiment were performed in triplicate.

17.2.4 Assessing the Toxicity by *Vigna Radiata* Germination

After the photocatalytic degradation of methyl red dye, the nanoparticles were separated by centrifugation. The seeds of *Vigna radiata* was immersed in a 2.5% sodium hypochlorite solution and left undisturbed for 15 min to achieve a state of sterilization and increase the experimental consistency. The seeds were taken from the solution and washed thrice with deionized water. The washed seeds of *Vigna radiate* were soaked in nanoparticle treated methyl red for 24 hours. Deionized water and methyl red, used in the soaking process, act as a positive and negative control. Seed germination was recorded by counting the number of germinated seeds after seven days of treatment. Two parameters, such as relative root elongation and germination rate (%), were adopted to evaluate the condition of seed germination. All the data obtained were analyzed using the SPSS statistical tool at a significant level of ≤ 0.05 with the t-test to identify the significance of the experimental group from the control.

17.2.5 Cytotoxic Activity

The cytotoxic effect of the synthesized sample to the normal cells was determined with a MTT dye reduction assay. Treated cells of 100 µl were incubated with 50 µl of MTT at 37°C for three hours. After incubation, 200 µl of PBS was added to all the samples and aspirated carefully to remove excess MTT. Acid-propanol of 200 µl was added and left overnight in the dark for solubilization. The absorbance was read at 650 nm in a microtitre plate reader (Bio-Rad USA). The optical density of the control cells was fixed at 100% viable, and the per cent viability of the cells in the other treatment groups was calculated using the formula,

$$\text{Percent viability} = \frac{\text{Control OD} - \text{Sample OD}}{\text{Control OD}} \times 100$$

17.3 RESULTS AND DISCUSSION

17.3.1 XRD Analysis

The result of the XRD analysis pattern of the iron oxide nanoparticles is shown in Figures 17.1a and 17.1b. Fe_2O_3 nanoparticles showed an XRD pattern with 2θ values, such as 24.10, 33.16, 36.62, 40.81, 49.40, and 54.00. The peaks obtained through the analysis were labelled as (012), (104), (110), (113), (024), and (116). The 2θ value of the Fe_2O_3 nanoparticles was found to match with JCPDS-89-8104, confirming the crystalline nature of the iron oxide nanoparticles. Similarly, Xu et al. (2007) described a simple precipitation method for the synthesis of Fe_2O_3 nanoparticles using ferric chloride. Synthesis of the nanoparticles with this method aids in the formation of nanoparticles with a crystal structure.

The result of a XRD analysis pattern of tungsten oxide (WO_3) nanoparticles revealed the 2θ value of 16.20 = (112), 26.14 = (123), 32.16 = (120), 35.10 = (202), and 46.60 = (132) planes which was found to match JCPDS-83-0950, which confirms the crystalline structure of the tungsten oxide nanoparticles, as represented in Figure 17.1a. Morales et al. (2008) explained the synthesis of WO_3 nanoparticles using a unique HCl assisted precipitation method.

The XRD pattern of the iron tungstate oxide ($FeWO_4$) nanoparticles resulted in 2θ values, such as 15.80, 21.42, 26.10, 30.20, 37.42, 55.00, and 65.21 (Figure 17.1c). The $FeWO_4$ nanoparticles corresponded to a crystal plane of (010), (100), (011), (111), (021), (221), and (023), which were found to match with JCPDS-21-0439. Previously, Laurent et al. (2008) explained the crystal nature of $FeWO_4$ nanoparticles synthesized using a hydrothermal method. The resulting nanoparticles were found to be 150 nm in size with electrical properties, which most nanoparticles possess. Additionally, a simple solvothermal route can also be used for the synthesis of iron tungstate nanoparticles ($FeWO_4$), using ferric chloride and sodium tungstate

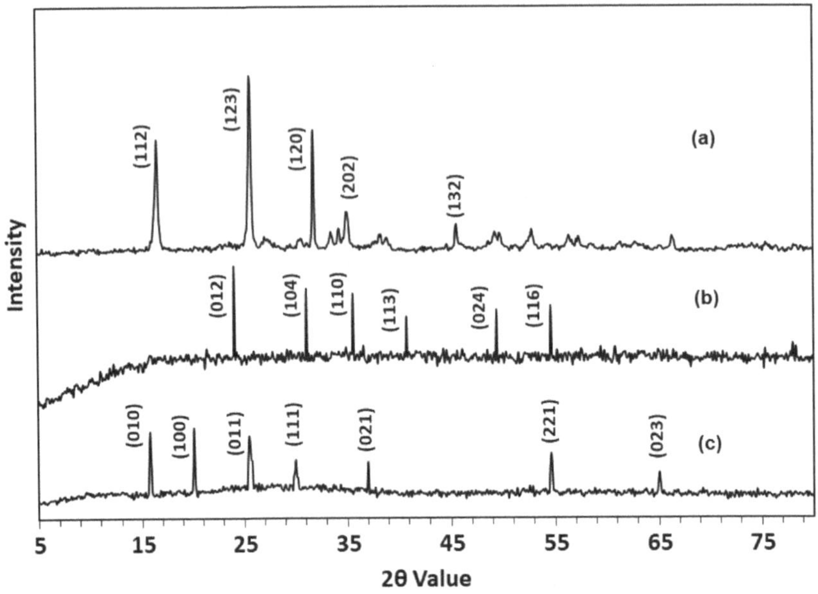

FIGURE 17.1 XRD pattern of (a) WO_3, (b) Fe_2O_3 and (c) $FeWO_4$ NPs.

solutions as precursors. The the Debye-Scherrer equation helps to determine the crystal nature of the nanoparticles.

Debye-Scherrer equation

$$D=0.94\ \lambda\ /\ \beta 1\ /\ 2\cos\theta$$

Where,

λ =1.54056 Å, as the wavelength of the incident X-ray source.

$\beta 1/2$=value of full width at half maxima of the diffracted peak.

θ =Bragg's angle.

Using the Debye-Scherrer equation, the nature of the synthesized nanoparticles can be determined. The results are tabulated in Table 17.2.

17.3.2 EDX and SEM Analysis

EDX along with SEM analysis helps to determine the elemental compositions and morphologies of nanoparticles. Figures 17.2–17.5 represent an EDX with SEM analysis of Fe_2O_3 (iron oxide), WO_3 (tungsten oxide), and $FeWO_4$ (iron tungstate oxide) nanoparticles. From the results, it was estimated that $FeWO_4$ nanoparticles were more irregularly agglomerated. However, Fe_2O_3 and WO_3 were slightly agglomerated without any collapse.

Further confirmation of Fe_2O_3, WO_3, $FeWO_4$ nanoparticles was performed by EDX analysis. The result of this analysis confirmed the presence of all three elements, Fe,

TABLE 17.2
The Average Size of Fe_2O_3, WO_3 and $FeWO_4$ NPs Estimated by Using the Scherrer Equation

Nanoparticles	Particle size (nm)
Fe_2O_3	20 ± 5
WO_3	10 ± 3
$FeWO_4$	25 ± 2

W, and O elements, as presented in Figure 17.2. The average size of the synthesized Fe_2O_3, WO_3 and $FeWO_4$ nanoparticles is represented in Table 17.3.

17.3.3 FTIR Analysis

FTIR studies are carried out for the identification of functional groups that the synthesized product possesses. The synthesized nanoparticles were determined using the FTIR spectrum in the range 400–4000 cm^{-1} (Perkin-Elmer 1725x) for the presence of functional groups. A FTIR spectrum analysis of Fe_2O_3, WO_3 and $FeWO_4$ nanoparticles is shown in Figure 17.6. The presence of alkynes, alkyl halide, amine, carboxylic acid, primary amine, aromatic, and alcoholic function groups was indicated by the spectrum results, represented in Table 17.4. The OH stretching of water molecules in Fe_2O_3, WO_3, and $FeWO_4$ nanoparticles was observed with distinguished peaks in the region of 3000–3600 cm^{-1}.

The sharp peaks at 1624 cm^{-1} and 1639 cm^{-1} may be attributed to the OH bending vibration of interstitial water. A W–O stretching vibration can be assigned to the bands in the region of 830–780 cm^{-1} and 900–825 cm^{-1}, respectively. The bands in the range of 700–600 cm^{-1} can be attributed to W–O stretching and bending in WO_3 and $FeWO_4$.

17.3.4 UV-Visible Spectroscopy and Bandgap

The UV-vis absorption spectrum of Fe_2O_3, WO_3 and $FeWO_4$ nanoparticles, as shown in Figure 17.6, lies between 200 and 800 nm. The bandgap energy is associated with the photocatalytic activity of the catalyst. By utilizing UV absorption spectra, band gaps of the synthesized nanoparticles were determined using Tauc's plot. The photocatalyst was employed to derive the bandgap of Fe_2O_3, WO_3, and $FeWO_4$ nanoparticles quantitatively using the formula

$$(Ahv/K)2 = hv - Eg$$

Where,

A – absorption.
K – absorption constant.
Eg – bandgap energy and the slope of the linear region to the x-axis interaction.

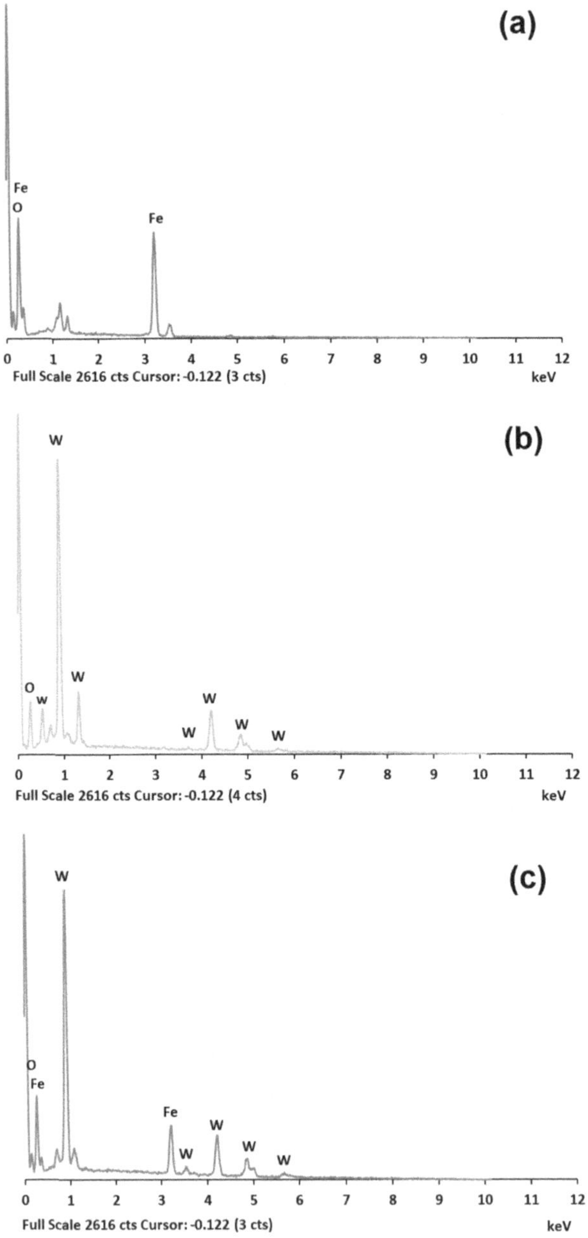

FIGURE 17.2 EDX pattern of (a) Fe_2O_3, (b) WO_3, and (c) $FeWO_4$ NPs.

FIGURE 17.3 SEM image of Fe_2O_3 NPs.

FIGURE 17.4 SEM image of WO_3 NPs.

FIGURE 17.5 SEM image of $FeWO_4$ NPs.

The bandgap energy of all the three synthesized nanoparticles (Fe_2O_3, WO_3, and $FeWO_4$) is presented in Figure 17.7. The results showed that the bandgaps of Fe_2O_3, WO_3, and $FeWO_4$ nanoparticles were 2.6 eV, 2.8 eV, and 2.96 eV, respectively. $FeWO_4$ nanoparticles showed a thriving redshift when compared with the Fe_2O_3 and WO_3 nanoparticles, which was attributed to narrowing or could indicate the formation of impurities in the band gaps. Therefore, it can be presumed that the FeWO4 photocatalyst can exhibit more significant activity when irradiated with visible light.

TABLE 17.3
EDX Analysis of Fe_2O_3, WO_3, and $FeWO_4$ NPs

	Atomic percentage (%)			Weight percentage (%)		
Elements	Fe	W	O	Fe	W	O
Fe_2O_3	68.76	–	31.27	39.74	–	60.26
WO_3	–	36.01	63.99	–	77.41	22.59
$FeWO_4$	11.01	13.74	75.26	14.15	58.13	27.72

TABLE 17.4
FTIR Analysis of Fe_2O_3, WO_3, and $FeWO_4$ NPs

Functional group and bond		Fe_2O_3	WO_3	$FeWO_4$
Alkynes	=C–H (Bending)	617.22	688.59	617.22
Alkyl halides	C–Cl stretch	–	–	831.32
carboxylic acids	O–H bend	–	941.26	–
Primary amines	N–H bend	1639.49	1624.06	1639.49
Amine	C–N (Stretch)	1128.36	1128.36	1130.29
Aromatic	C=C (Stretch)	1402.25	1402.25	1400.32
Alcohol	O–H (Stretch, H-bending)	3142.04	3379.29, 3140.11	3140.1

17.3.5 Photocatalytic Activity

A UV-visible irradiation method was used to determine the photocatalytic activity of methyl red dye in synthesized Fe_2O_3, WO_3, and $FeWO_4$ nanoparticles. The UV-visible spectrum of methyl red showed a strong absorption band at 510 nm. Figure 17.8a shows the time that depends on the optical density of the methyl red dye in the course of the photocatalytic reaction. It was found that the dye absorption band intensity decreased with time after the addition of nanoparticles as well as being exposed to visible light. In the presence of the nanoparticles, the degradation of methyl red (100 ppm) started at 1 hour 40 mins. About an 85% degradation of the dye was observed at 2 hours 40 mins using $FeWO_4$ nanoparticles and visible light when compared to WO_3 (79%) and Fe_2O_3 (82%) nanoparticles, as represented in Figure 17.8b. This study was compared with the previous report to find out the importance of the catalytic activity presented in Table 17.5. Most of the studies used visible light as a source for the photocatalytic degradation of azo dyes as the rate constant is slightly in this study.

Further, the Langmuir-Hinshelwood kinetic model examined the photocatalytic activity of MR dye induced by kinetic visible light. The photocatalytic degradation of MR dye using the kinetic reaction model is shown in Figure 17.8c.

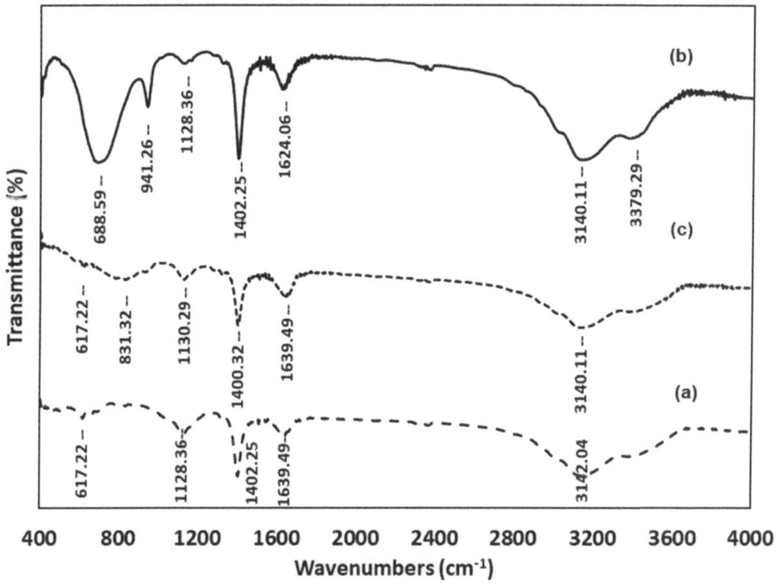

FIGURE 17.6 FTIR analysis of (a) Fe_2O_3, (c) WO_3, and (b) $FeWO_4$ NPs.

A pseudo-first-order linear fit was used for the calculation of regression coefficients (R2) and was found to be 0.95 for $FeWO_4$ nanoparticles, whereas WO_3 and Fe_2O_3 nanoparticles showed 0.89 and 0.86, respectively. $FeWO_4$ nanoparticles act as a potential catalyst for degrading methyl red dye based on the degradation rate (%) and rate constant (R2).

17.3.6 Toxicity and TEM Study

The photocatalytic process of the degradation of MR with $FeWO_4$ NPs was directly observed under TEM analysis. After absorption and degradation, there were many MR crystals as 10 nm molecules absorbed into the surface of $FeWO_4$ NPs, which resulted in a fuzzy image (Figure 17.9). The release of azo dye from industries on land may directly affect soil fertility, which may spoil the productivity of agricultural goods. Therefore, to protect our environment, it is vital to evaluate the toxic nature of the dye and degrade the dye products. A phytotoxicity study was carried out on the seeds of *Vigna radiata* against dyes and their degraded products. A germination of about 13% was observed in the seeds that were grown in the presence of MR dye. The mean root length of *Vigna radiata* was found to be 0.56 ± 0.12 cm, compared with seeds treated with distilled water which showed 100% germination and acted as a control. The mean root length of methyl red soaked *Vigna radiata* was found to be 5.92 ± 0.32 cm. The seeds treated with degraded products showed 72% germination, as represented in Figure 17.10. A similar result of phytotoxicity was observed with *Vigna radiata* and Reactive Red 141 and their degraded products (Akansha et al., 2019).

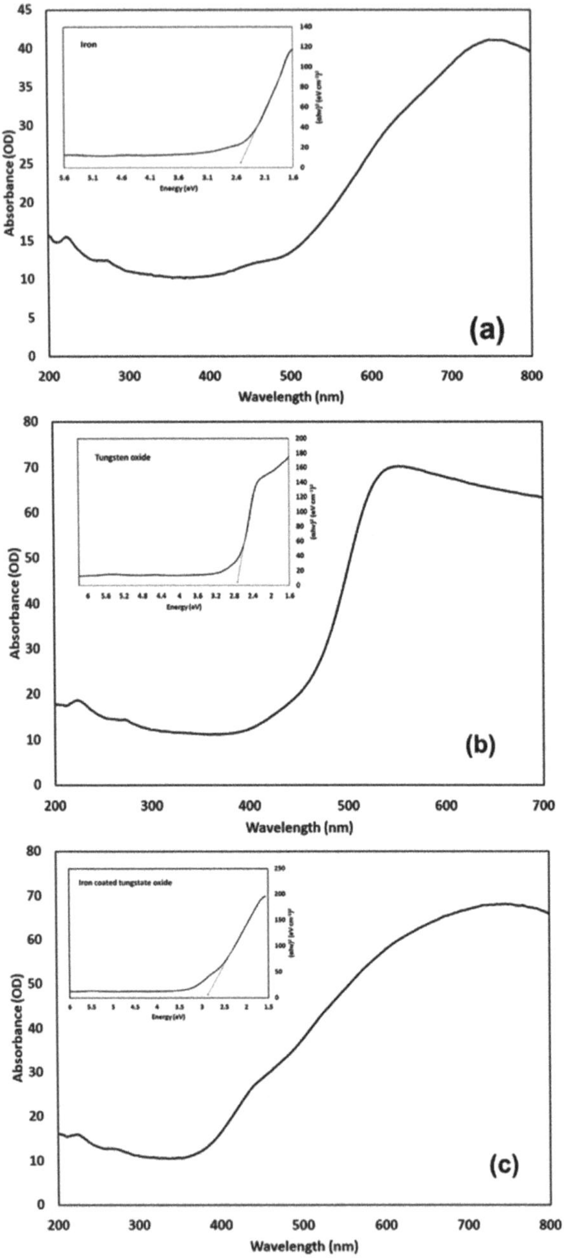

FIGURE 17.7 UV-Vis spectrum of (a) Fe_2O_3, (b) WO_3, and (c) $FeWO_4$ NPs absorption with the plot of (αhʋ)2 versus (hʋ).

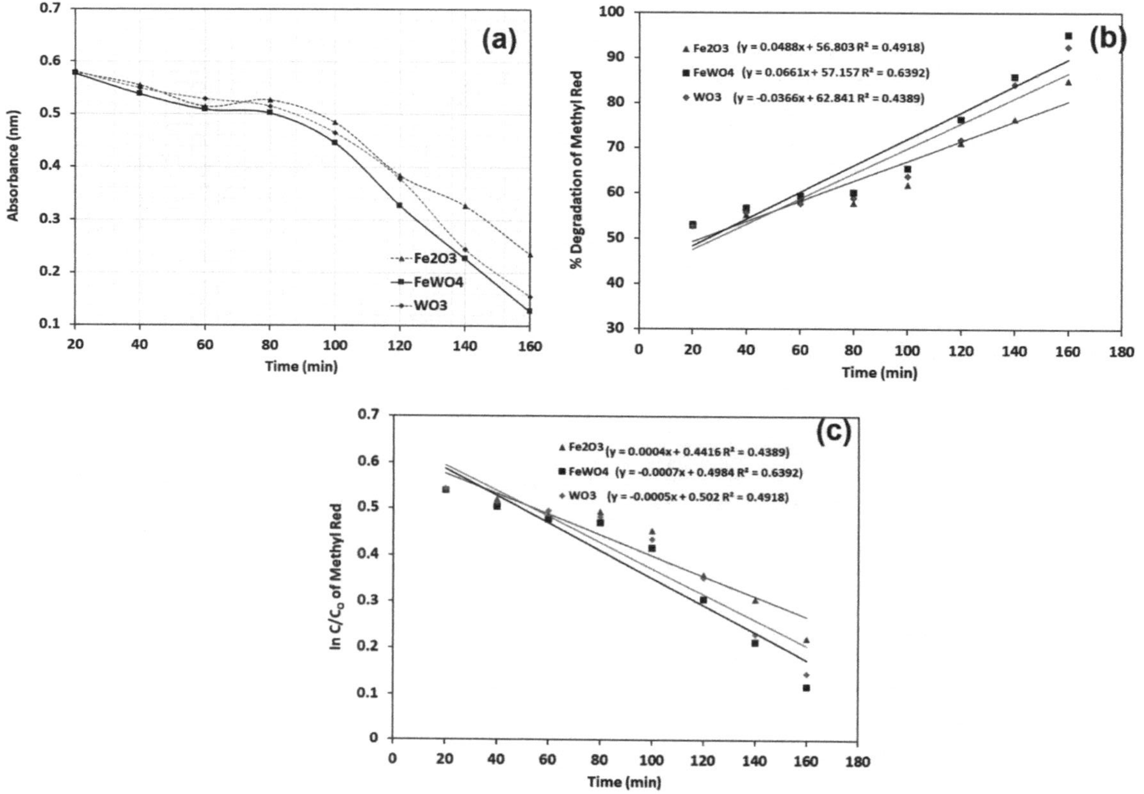

FIGURE 17.8 Degradation of methyl red (a) and kinetic study (b and c) of Fe_2O_3, WO_3, and $FeWO_4$ NPs under visible light irradiation.

TABLE 17.5
Degradation of Methyl Red Dye Compared with Previous Results

S.NO	Catalyst	Method of synthesis	Light source	Degradation efficiency (%)	Reference
1	Fe_2O_3	Co- precipitation	Visible light	95	Abbas Afkhami et al., 2010
2	Fe_2O_3	Hydrothermal	UV light	90	Mahapatra et al., 2012
3	Fe_2O_3	Precipitation	Visible light	79	Present study
4	WO_3	Acid precipitation	Visible light	82	Present study
5	WO_3	citric acid-assisted precipitation	UV light	82	Sanchez-Martinez et al., 2012
6	WO_3	functionalized (sulfated)	Visible light	90	Patil et al., 2018
7	$FeWO_4$	Hydrothermal	Visible light	85	Present study

The cytotoxic effects of methyl red dye and its degraded products against mice macrophage cell lines (RAW 264.7) were examined. A dose-dependent decrease in the viability of the RAW 264.7 cell line was observed with a methyl red degraded product, as shown in Figure 17.11 and Table 17.6. The result revealed that 25% of the macrophage cells were destroyed with a dose of 500 µg. These experiment results proved that degraded methyl orange products are less toxic.

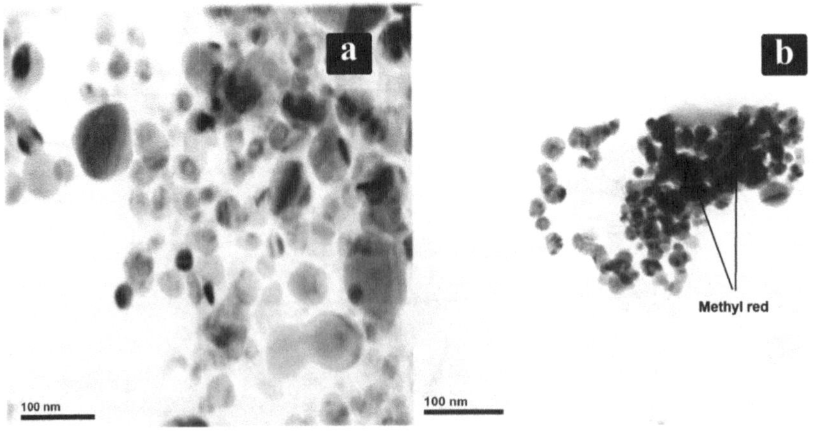

FIGURE 17.9 TEM image of (a) $FeWO_4$ NPs and (b) methyl orange degraded nanoparticle under visible light irradiation.

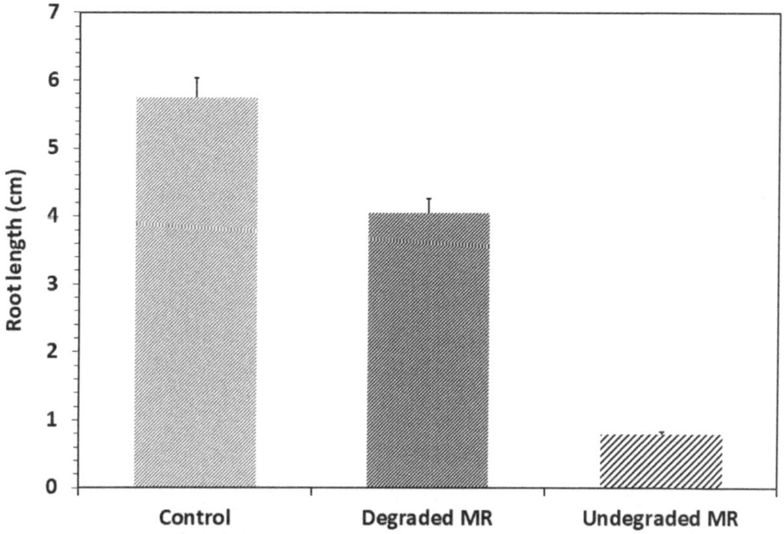

FIGURE 17.10 Phytotoxic effect of methyl red (MR) dye and its degraded product on *Vigna radiata* germination. The result is expressed in Mean ± SD and statistically significant ($P \leq 0.05$).

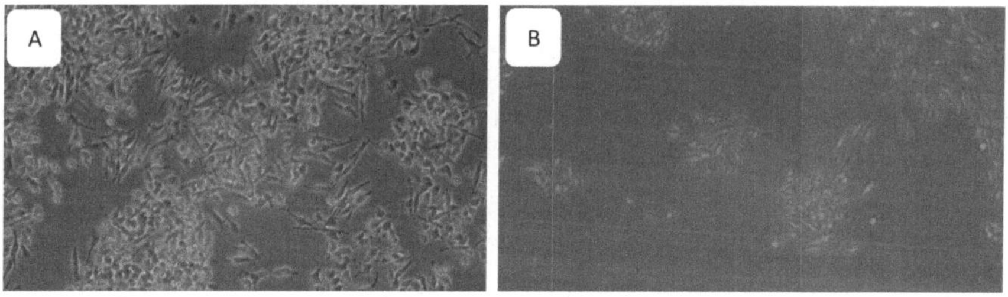

FIGURE 17.11 A) Control (without treatment); B) Cell treated with 500 μg/ml of degraded methyl red product for 24 hours.

17.4 CONCLUSION

Nanoparticles, such as iron oxide, tungstate oxide, and iron tungstate oxide, were successfully synthesized using acid precipitation and hydrothermal methods, respectively. They are usually synthesized without the use of any surfactant and capping agents. The purity and crystalline nature of the nanoparticles were confirmed by XRD and EDX analysis. Morphology plays an important role in the characterization of any nanoparticles, and this was done by analyzing the nanoparticles under SEM. The morphology of Fe_2O_3, WO_3, and $FeWO_4$ nanoparticles were found to be almost spherical with excellent optical properties and energy bandgaps. Photocatalytic degradation of methyl red under visible light irradiation was maximum when it was utilized with $FeWO_4$ nanoparticles. So, $FeWO_4$ can be considered as a significant

TABLE 17.6
Biocompatibility of Various Concentration of Methyl Red Dye Treated with $FeWO_3$ in Mice Macrophage Cell Line (RAW 264.7)

Treatment groups	OD values	% viability
Cells Alone (control)	0.996	100%
Cells + MO 100 µg	0.073	92.67
Cells + MO 200 µg	0.121	87.85
Cells +MO 300 µg	0.161	83.83
Cells + MO 400 µg	0.224	77.51
Cells + MO 500 µg	0.253	74.59

photocatalyst for degrading azo dye (methyl red). A phototoxic study was carried out on the seeds of *Vigna radiata* to check its germination and growth. A toxicity study on the germination of *Vigna radiata* and macrophage cell lines (RAW 264.7) confirmed the non-toxic nature of the degraded dye product. It was found that $FeWO_4$ nanoparticles were stable, efficient, and had a catalytic potential to treat effluent released from various dyeing industries.

17.5 ACKNOWLEDGEMENT

We thank the management of Sri Krishna Arts and Science College, Coimbatore, Tamil Nadu, India for providing necessary facilities to carry out this study.

17.6 REFERENCES

Akansha, K., Chakraborty, D. and Sachan, S.G. 2019. Decolorization and degradation of methyl orange by Bacillus stratosphericus SCA1007. *Biocatal. Agric. Biotechnol.* 18: 101044.

Aksu, Z. 2005. Application of biosorption for the removal of organic pollutants: A review. *Process Biochem* 40:997–1026.

Aspland, J.R. 1997. *Textile dyeing and coloration.* AATCC

Bamwenda, G.R., Sayama, K., Arakawa, H. 1999. The effect of selected reaction parameters on the photoproduction of oxygen and hydrogen from a WO3-Fe2+-Fe3+ aqueous suspension. *J. Photochem. Photobiol., A*, 122, 175–183.

Banat, I.M., Nigam, P., Singhm, D., Marchant, R. 1996. Microbial decolourization of textile-dye-containing effluents. *Bioresour. Technol.* 58:217–227.

Crini, G. 2006. Non-conventional low-cost adsorbents for dye removal. *Bioresour. Technol.* 97:1061–1085.

Daneshvar, N., Salari, D., Khataee, A.R. 2004. Photocatalytic degradation of azo dye acid red 14 in water on ZnO as an alternative catalyst to TiO_2. *J. Photochem. Photobiol. A: Chem.* 162:317–322.

Estlander, T. 1998. Allergic dermatoses and respiratory diseases from reactive dyes. *Contact Dermatitis* 18:290–297.

Gonzales, C.A., Riboli, E., Lopez-Abente, G.. 1998. Bladder cancer among workers in the textile industry: Results of a Spanish case-control study. *Am. J. Ind. Med.* 14:673–680.

Handy, R.D., von der Kammer, F., Lead, J.R., Hassello, M., Owen, R., Crane, M., 2008. The ecotoxicology and chemistry of manufactured nanoparticles. *Ecotoxicology* 17: 287e314.

Hasany, S., Ahmed, I., Rajan, J., Rehman, A. 2012. Systematic review of the preparation techniques of iron oxide magnetic nanoparticles. *Nanosci Nanotechnol.* 2(6):148–158.

Hatch, K.L. and Maibach, H.I. 1995. Textile dye dermatitis. *J. Am. Acad. Dermatol.* 32:631–639.

Hergt, R., Dutz, S., Mülle, R., Zeisberger, M. 2006. Magnetic particle hyperthermia: nanoparticle magnetism and materials development for cancer therapy. *J. Phy. Condensed Mat.* 18: S2919.

Huang, F., Ge, L., Zhang, B., Wang, Y., Tian, H., Zhao, L. P., et al. (2014). A fullerene colloidal suspension stimulates the growth and denitrification ability of wastewater treatment sludge-derived bacteria. *Chemosphere* 108:411–417. doi: 10.1016/j.chemosphere.2014.02.042.

Laurent, S., Forge, D., Port, M., et al. 2008. Magnetic iron oxide nanoparticles: synthesis, stabilization, vectorization, physicochemical characterizations, and biological applications. *Chem. Rev.* 108 (6):2064–2110.

Lequeux, H., Hermans, C., Lutts, S., Nathalie, V. 2010. Response to copper excess in Arabidopsis thaliana: Impact on the root system architecture, hormone distribution, lignin accumulation and mineral profile. *Plant Physiol. Biochem.* 48:673–682.

Morales, W., Cason, M., Aina, O., Tacconi, N. R. de, Rajeshwar, K. 2008. Combustion synthesis and characterization of nanocrystalline WO3. *J. Am. Chem. Soc.*, 130, 6318–6319.

Naoki, T. (2008). Color removal using nanoparticles. *Macromol. Symp.* 270, 27–39.

Narayanan, K.B., Sakthivel, N. 2010. Biological synthesis of metal nanoparticles by microbes. *Adv. Colloid Interface Sci.*, 156(1–2):1–13.

Narendhran, S., Rajiv, P., Rajeshwari, S. 2016. Biogenic ZnO nanoparticles synthesized using L. aculeate leaf extract and their antifungal activity against plant fungal pathogen. *Bull. Mater. Sci.* 39:415.

Oh, S.Y., Seo, Y.D., Kim, B., Kim, I.Y., Cha, D. K. 2016. Microbial reduction of nitrate in the presence of zero-valent iron and biochar. *Bioresour. Technol.* 200:891–896. doi: 10.1016/j.biortech.2015.11.021.

Pagga, U., Brown, D. 1986. The degradation of dyestuffs. *Chemosphere* 15:479–491.

Ray, P.D., Huang, B.-W., Tsuji, Y., 2012. Reactive oxygen species (ROS) homeostasis and redox regulation in cellular signaling. *Cell. Signal.* 24:981e990.

Sakulchaicharoen, N.A., O'Carroll, D.M.A., Herrera, J.E.B. 2010. Collagen polymer and magnetic collagen nanocomposite recycled from waste to reduce polluted water toxicity. *J. Cont. Hydrol.* 118(3–4), 117–127.

Salata, O.V. 2004. Applications of nanoparticles in biology and medicine. *J Nanobiotechnol.* 2(1):3.

Seddigi, Z.S. 2010. Removal of alizarin yellow dye from water using zinc doped WO_3 catalyst. *Bull Environ. Contam. Toxicol.* 84:564–567.

Shen, L, Laibinis, P.E., Hatton, T.A. 1999. Polymerization of olefin-terminated surfactant bilayers on magnetic fluid nanoparticles. *Langmui*:447.

Shirzad-Siboni, M., Samarghandi, M., Yang, J.-K., Lee, S.-M. 2011. Effect of organic matter on cyanide removal by illuminated titanium dioxide or zinc oxide nanoparticles. *Water Sci. Technol.* 64:1383–1387.

Singh, S., Vishwakarma, K., Singh, S., Sharma, S., Dubey, N.K., Singh, V.K., Liu, S., Tripathi, D.K., Chauhan, D.K., 2017. Understanding the plant and nanoparticle interface at transcriptomic and proteomic level: a concentric overview. *Plant Gene* 11, 265e272.

Slein, M.W., Sansone, E.B. 1980. *Degradation of Chemical Carcinogens*. New York, Van Nostrand Reinhold.

Soultanidis, N., Zhou, W., Kiely, Ch.J., Wong, M.S. 2012. Solvothermal synthesis of ultrasmall tungsten oxide nanoparticles. *Langmuir* 28:17771–17777.

Weisburger, J.H. 2002. Comments on the history and importance of aromatic and heterocyclic amines in public health. *Mutat. Res.* 506:9–20.

Whitaker, C.M., Willock, C.C. 1949. Dyeing with coal tar dyestuffs. *Tindall and Cox Bailliere* 5:1–7.

Xin, G., Guo, W., Ma, T. 2009. Effect of annealing temperature on the photocatalytic activity of WO3 for O2 evolution. *Appl. Surf. Sci.* 256:165–169.

Xu, J., Yang, H., Fu, W., et al. 2007. Preparation and magnetic properties of magnetite nanoparticles by sol-gel method. *J. Magn. Magn. Mater.* 309(2):307–311.

Index

H

I

L

M

N

O

P

Q

R

S

T

U

V

W

X

Y

Z